Diagnostics Development on the Route Towards Fusion Reactors

Didier Mazon
Editor

Diagnostics Development on the Route Towards Fusion Reactors

Originally published in the *Journal of Fusion Energy* Volume 38, Issue 3–4, as the Special issue "Diagnostics Development on the Route towards Fusion Reactors"

Springer

Editor
Dr. Didier Mazon
CEA Cadarache IRFM
13108 St Paul Lez Durance, France

ISBN 978-3-030-30588-8

This Springer imprint is published by the registered company Springer Nature Switzerland AG
The registered company address is: Gewerbestrasse 11, 6330 Cham, Switzerland

Contents

Journal of Fusion Energy (2019) 38:263
https://doi.org/10.1007/s10894-019-00217-5

GUEST EDITORIAL

Preface

Didier Mazon[1]

Published online: 23 March 2019

This special issue: "Diagnostics Development on the Route towards Fusion Reactors" contains selected invited papers that were presented at the occasion of two Summer Schools that I had the pleasure to organize and chair as Director:

- The 2nd PhDia Summer school focused on the 'Neutron and gamma diagnostics for Fusion plasma' held at Podlesice, Poland from 11 to 15 September 2017.
- The 16th Course International school of fusion reactors technology focusing on 'Diagnostics and Technology Developments in view of ITER and DEMO' held at Erice Italy from 28th April to 4th May 2017.

As it can be noted these two International Summer schools were not the first ones and were in the continuity of previous events more dedicated to particular Diagnostics topics. This time and due to the fact that I was involved in both of them it was a good time to try to make them complementary. The synergies between the two summer schools were obvious, both were discussing about diagnostics technology and treatments and in particular the diagnostics challenges of ITER (the International Thermonuclear Experimental Reactor, a joint venture between Europe, Japan, Russia, USA, China, India and South Korea) but also DEMO, the next step fusion reactor that will deliver electrical power from fusion. Indeed in both devices there will be the need to measure a wide range of plasma parameters in extreme conditions of temperature, neutron and gamma fluxes, these measurements being absolutely necessary to reach and sustain high levels of fusion power in a stationary manner and provide inputs to control systems with adequate reliability and long term stability. In order to characterize and understand fusion plasma we need to measure many plasma quantities or parameters in ITER different techniques are needed to measure the same parameters in different regions of plasma because of the very wide parameters ranges in fusion plasma, for example plasma temperature covers a range of 5 orders of magnitude. But in the case of DEMO, principally due to the breading blanket optimization, the number of diagnostics will be extremely limited.

These summer schools were a great opportunity to make a point on the present status in terms of Diagnostics both in ITER and DEMO, to check the capabilities of the present proposed systems to meet the requirements on the measurements. Critical issues related with the measurements of fast particles, fusion products, plasma facing components and radiation effects (including tungsten) were addressed as well as new concepts of detectors and developments. Finally real-time control of plasma parameters and the major role played by the diagnostics in this objective was a recurrent theme thanks to the fact that both ITER and DEMO are planning real-time measurements and control. The selected invited papers have gone through a rigorous reviewing process as required by JOFE.

To conclude I would like to have a thought and remember the great contribution of Pr Sergio Martellucci who helps me in the preparation of the event in Erice but unfortunately passed away just before the start of the 16th Course International school of fusion reactors technology Summer School. He was indeed a great scientist with lots of humility always ready to help the young generation which is the signature of great people. We will never forget him for all his contributions to plasma physics.

✉ Didier Mazon
Didier.mazon@cea.fr

[1] CEA, IRFM, 13108 St Paul lez Durance, France

Journal of Fusion Energy (2019) 38:264–282
https://doi.org/10.1007/s10894-018-0202-1

ORIGINAL RESEARCH

The Core-Plasma CXRS Diagnostic for ITER: An Introduction to the Current Design

Philippe Mertens[1]

Published online: 19 November 2018

Abstract
The development and design of the CXRS diagnostic for the core plasma of ITER is used as a pretext to elaborate on several of the main challenges of optical diagnostics on such a large fusion device. The idea was to confront the students of the 16th *Ettore Majorana School on Diagnostics and Technology Developments* with as many aspects of the design of optical diagnostics as possible. After an elementary review of the basics of charge-exchange, of typical spectra, of intended measurements and of the expected associated background, the DNB (diagnostic neutral beam) is briefly presented. The light collection and transport to the ex-vessel fibres and spectrometers constitutes the main part of this contribution: building on existing telescope and endoscope systems, the choice of a suitable optical layout for the core-plasma CXRS diagnostic at ITER is discussed. An account follows of the protective measures against the degradation of the first mirror, which will be exposed to high particle and heat fluxes: an appropriate duct, a shutter and, possibly, a cleaning discharge with "End-of-Cleaning Indicator".

Keywords Diagnostics · Tokamak · Charge-exchange · ITER

PACS 52.70.-m · 52.70.Kz · 52.40.Mj · 52.55.Pi

Introduction

The development and design of the CXRS diagnostic for the core plasma of ITER is used as a pretext to elaborate on several of the main challenges of optical diagnostics on such a large fusion device. The idea was to confront the students of the 16 *Ettore Majorana School on Diagnostics and Technology Developments* with as many aspects of the design of optical diagnostics as possible. Still, keeping the present contribution deliberately simple and abundantly illustrated, the elementary statements found here constitute an overview of the subject, evidently CXRS-biased, and provide literature references to give the opportunity to go deeper into specific fields. The main goal of the present article is to provide some insight in the design process of such complex, highly technological diagnostic systems.

The next section deals with the theoretical aspects and basics of the Charge eXchange Recombination Spectroscopy (CXRS), including the indispensable diagnostic beam (DNB). The main part of the present contribution, which follows in the "Light Collection and Transport" section, addresses the collection and transport of photons to the detector, from large devices—telescopes—on earlier tokamaks to the most recent considerations for ITER.

The realisation of the first mirror of optical diagnostics in ITER is one of the biggest challenges, discussed in "The First Mirror (M1)" section together with the main, essential protective measures, an appropriate duct, a shutter and an in situ cleaning system. Other challenges are just mentioned in the following chapter. A conclusion is given in the sixth section.

✉ Philippe Mertens
Ph.Mertens@fz-juelich.de

1 Institute of Climate and Energy Research (IEK-4, Plasma Physics), Forschungszentrum Jülich GmbH, 52425 Jülich, Germany

The CXR Process

Basics

In what follows, we shall deal mainly with the installation planned of a CXRS diagnostic for the core plasma of ITER. "CXRS" stands for Charge-Exchange Recombination Spectroscopy. It is based on the injection of a beam of neutral hydrogen atoms in the plasma, followed by the reaction

$H^0 + X^{z+1} \rightarrow H^+ + X^z(nlj)^*$, and the ion radiates as $X^z(nlj)^* \rightarrow X^z(n'l'j') + \hbar\omega$.

An impurity ion X captures an electron from an atom in the hydrogen beam and lowers its positive charge by one unit—hence the designation *charge exchange*. It ends up in an excited state, denoted by the asterisk. It decays with emission of a photon which constitutes the observed signal. To give a rough idea of the geometry of the arrangement foreseen on ITER, Fig. 1 shows the diagnostic beam, a neutral beam injected in the equatorial sector No. 4, and (red colour) the upper port plug No. 3 next to it, which contains the observation optics. The light is transported by fibres to spectrometers which are installed away from the torus in a zone less exposed to radiation.

Spectra

The following elementary figure (Fig. 2) helps in defining a typical spectral profile (photon flux versus frequency) where ν corresponds to the spectral axis (frequencies), ν_0 to the central unshifted wavelength and $\Delta\nu$ to the Full Width at Half Maximum (FWHM).

A Lorentzian profile for instance, like the natural line width of an optical transition, would write

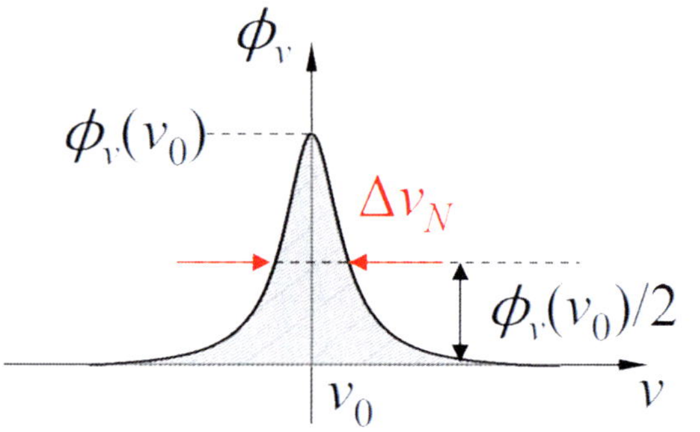

Fig. 2 The parameters which determine the spectral profiles defined in the text below. Here the case where the subscript 'N' denotes the natural line width

$$\phi(\nu)_N = \frac{1}{\pi}\frac{\Delta\nu_N/2}{(\nu-\nu_0)^2 + (\Delta\nu_N/2)^2} \tag{1}$$

whereas a Doppler-broadened profile would be given by

$$\phi(\nu)_D = \frac{2\sqrt{\ln 2}}{\sqrt{\pi}\Delta\nu_D} exp\left[-\left(\frac{2\sqrt{\ln 2}}{\Delta\nu_D}(\nu-\nu_0)\right)^2\right]. \tag{2}$$

Note that most observed spectral profiles are the result of a convolution according to

$$f(x) * g(x) = I(x) = \int_{-\infty}^{\infty} f(y)g(x-y)dy \tag{3}$$

so that the spectral profiles given above would merge in a so-called Voigt profile (after W. Voigt, Göttingen):

$$f(x)) = const. \int_{-\infty}^{\infty} \frac{e^{-y^2}}{(x-y)^2 + a^2} dy. \tag{4}$$

with

$$x = 2\sqrt{\ln 2}\frac{\nu-\nu_0}{\Delta\nu_D} \quad \text{and} \quad a = \sqrt{\ln 2}\frac{\Delta\nu_N}{\Delta\nu_D}. \tag{5}$$

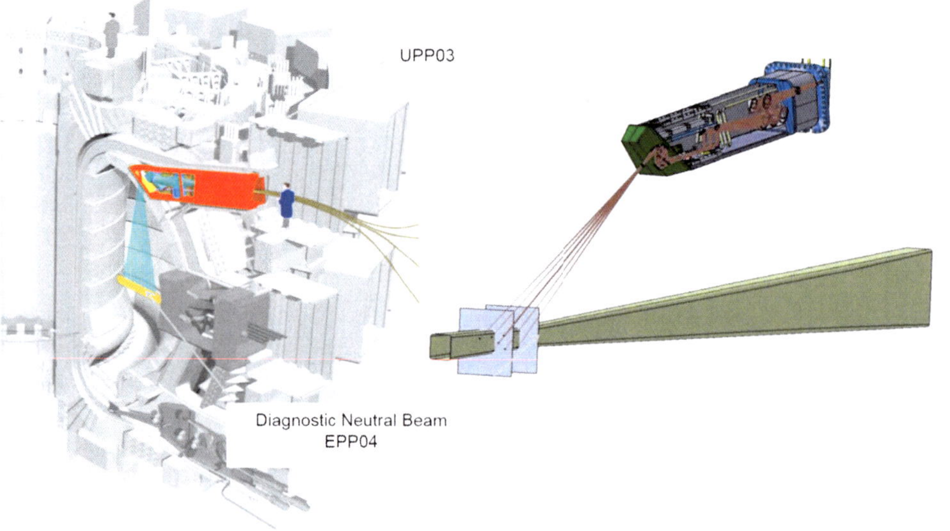

Fig. 1 Situation of the core-plasma CXRS diagnostic on ITER. The relative positions of the upper port plug and of the object (slices of the diagnostic neutral beam) are enlarged on the right

The consultation of [1], §IX.66 and §IX.65 is recommended for the details and use of this convolution, for instance when one of the profiles represents the finite spectral width of the measuring apparatus.

In general, one can deduce from the line shift a velocity and from the line intensity (integral over the relevant spectral range) a concentration or density of the species subject to an appropriate calibration of the signal amplitude. Accordingly, one expects in the first place from CXRS measurements in the core plasma the determination of several parameters like the concentration of the main fusion product (helium ash) and poloidal or toroidal rotation velocities of specific impurities.

The table shown in Fig. 3 lists a few of the physical parameters that can be measured by means of a core-plasma CXRS diagnostic installed on the diagnostic neutral beam (see "The Neutral Beam" section) together with expected ranges and presumably attainable accuracies.

A typical spectrum of the beam emission in JET, Fig. 4 around λ 656 nm [2], displays the main charge-exchange H_α line (the first of the Balmer series, ACX and PCX denoting the active—i.e. with the beam—and passive charge-exchange, respectively), and

- The components which correspond not only to the beam energy but also to one half or one third of the beam energy (all marked BES for Beam Emission Spectroscopy), the presence of which is due to the acceleration in the beam source of molecular ions H_2^+ and H_3^+;
- The lines of the $^2S - ^2P^o$ multiplet of singly ionised carbon at λ 658 nm, marked CII.

Moreover, all profiles are subject to a significant Stark splitting due to their velocity in the magnetic field of the tokamak (see insert b) and to a pronounced Doppler broadening at the high temperatures found in the core plasma which may wipe out the finer spectral structures

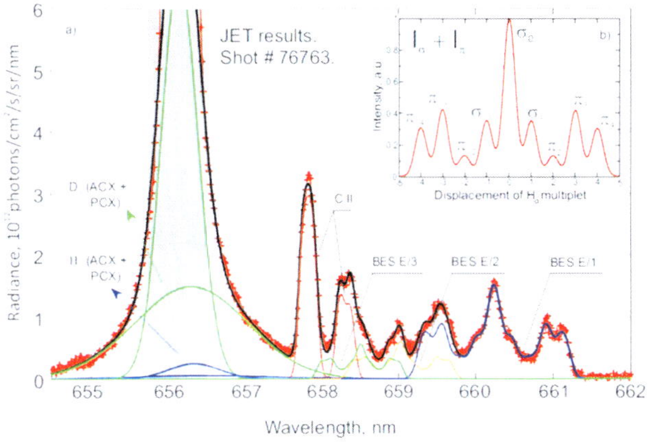

Fig. 4 Spectrum recorded on JET in the vicinity of H_α λ 656 nm, showing components from beam emission (BES) and charge-exchange (CX), after [2]

underneath as will be seen in some typical spectra presented below.

It will also be obvious in the following five figures (Figs. 4, 5, 6, 7, 8) that the CXRS signals may be affected by a high background. The main contributor to this background is the bremsstrahlung which comes from the acceleration of the charged particles in the plasma. The bremsstrahlung can advantageously be represented by formulas such as [3]

$$\Phi_{B\_LOS} = \int_{LOS} d\ell \int_0^\infty d\lambda\, C\, \bar{g}(\lambda, T_e, Z_{eff}) \frac{n_e^2 Z_{eff}}{\sqrt{k_B T_e}} \frac{exp\left(-\frac{hc}{\lambda k_B T_e}\right)}{\lambda^2} \tag{6}$$

for the photons delivered along a line of sight ℓ, or [4, 5]

$$\Phi_{B\_LOS} = 7.6 \cdot 10^{-21}\, Z_{eff}\, n_e^2 \left(2/\sqrt{T_e}\right) \Delta\lambda/\lambda \tag{7}$$

for the bremsstrahlung background caught by observation optics (in $m^{-2}s^{-1}sr^{-1}$), with $\bar{g} = 2$. In both equations, n_e represents the electron density in the plasma, T_e is the electron temperature, λ is the wavelength, h stands for

Measurement	No.	Phys. Parameter	Range to cover (IC3)	Time res.	Spatial res.	Accuracy	Conditions (IC3)	Diag. Contrib.
10. Plasma Rotation	018	V_{POL}	5 – 50 km.s^{-1}	10 ms	a/30 67 mm	max (30% ,10km.s^{-1})	r/a <0.70, Δt 200ms (DNB 5Hz) dir.: line-of-sight	Suppl. (1)
	019	V_{TOR}	5 – 300 km.s^{-1}	10 ms	a/30 67 mm	max (30% ,10km.s^{-1})	r/a <0.70, Δt 200ms (DNB 5Hz) dir.: line-of-sight	Primary (1)
28. Ion Temperature Profile	064	Core T_i	0.5 – 40 keV	100 ms	a/30 67 mm	max (10%,1.5 keV)	r/a <0.70	Primary
29. Core He Density Profile	066	nHe/ne	0.01-0.20	100 ms	a/10	max (0.01,0.1(n_{He}/n_e	r/a <0.70	Primary
	067	profile of 3He conc.	0.01-0.10	100 ms	a/10	max (0.01,0.1(n_{He}/n_e	r/a <0.70	Primary (2)
Notes:	(1)	Not lower than 5km/s; cannot discriminate between Vpol and Vtor over the full radial range						
	(2)	without discriminating between ^3He and ^4He						

Fig. 3 A short excerpt of the parameters that could be measured by the core-plasma CXRS diagnostic. *Source*: ITER core charge-exchange consortium IC^3

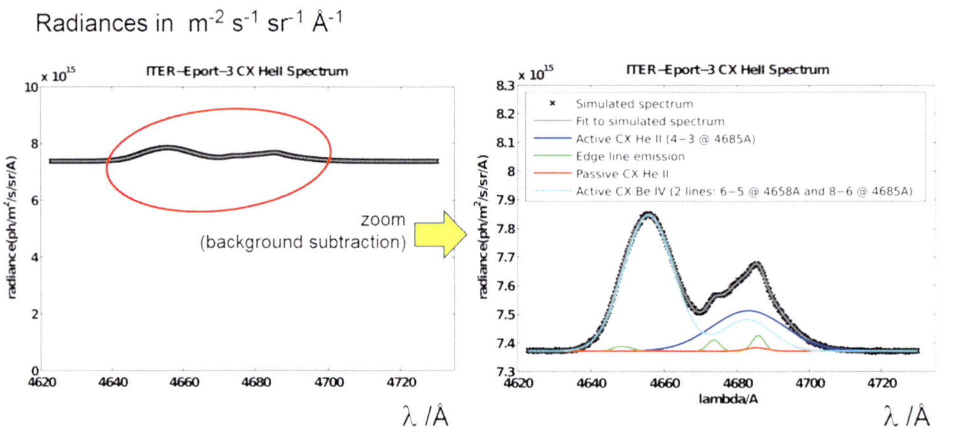

Fig. 5 Synthetic spectra calculated for ITER around the λ 468 nm (He II) transition. *Source*: [8]

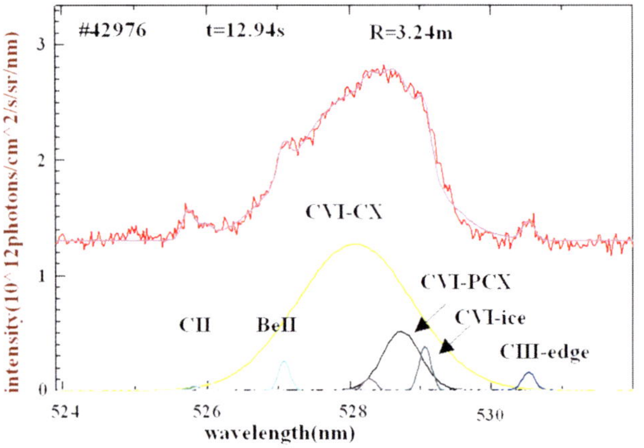

Fig. 6 The C VI charge-exchange line recorded on JET and the corresponding reconstruction (from [9])

Planck's constant and k_B stands for the Boltzmann constant. The effective charge is given by

$$Z_{eff} = \frac{\sum_i n_i Z_i^2}{\sum_i n_i Z_i} = \frac{\sum_i n_i Z_i^2}{n_e} \tag{8}$$

where n_i is the density of ion 'i' and Z_i the corresponding charge. $\bar{g}$ is the so-called Gaunt factor. For information on the latter and for the full formula which includes both free-free and free-bound transitions, please consult ([6], Chap. 8), or [7].

As mentioned above, one of the goals of the CXRS diagnostic in the plasma core is the measurement of helium, mainly of the radial profile of its concentration. Simulated spectra at λ 468 nm are shown in Fig. 5, where the vertical axes give radiances (in $m^{-2}s^{-1}sr^{-1}Å^{-1}$). As demonstrated from the left to the right figures, a background subtraction is indispensable. The right picture shows the different contributions to the profiles taken into account in the reconstruction [8].

Figure 6 presents a fit compared to a measured CX profile in the former JET-C (the JET tokamak with plasma-facing components made of carbon) which corresponds to

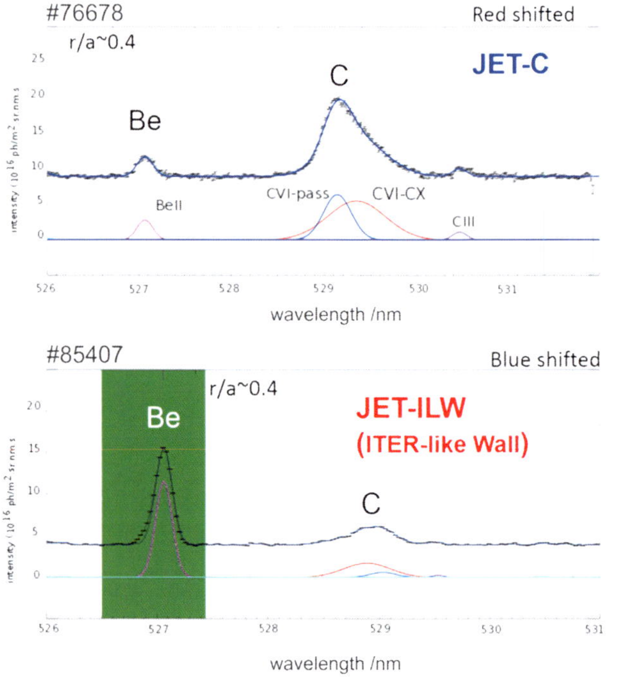

Fig. 7 Spectral profiles of Be II and C VI before and after passage to the JET ITER-like Wall (*cf.* [13])

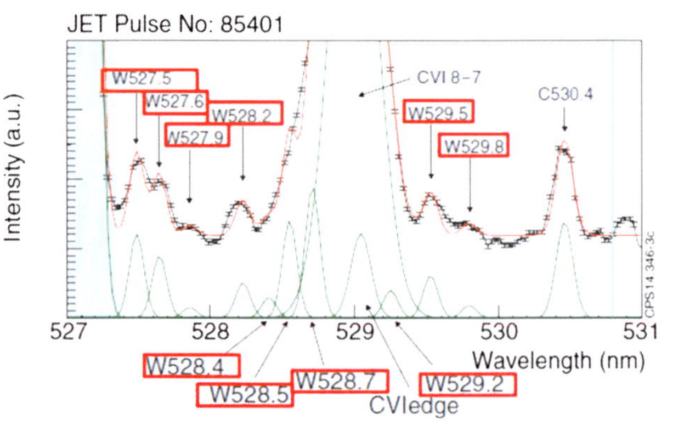

Fig. 8 Tungsten has a dense distribution of transitions in the visible spectral range, here around the carbon line at λ 529 nm. *Source*: [13]

the CVI ($n = 8 \rightarrow n' = 7$) transition at λ 529 nm, according to the spectroscopic notation where CI stands for the neutral atom and C VI represents the C^{5+} ion. A full explanation of the profiles can be found in [9].

Similar spectra are presented below (Fig. 7) which emphasize the increase of the beryllium-to-carbon ratio triggered by the move from a full carbon machine JET-C to a beryllium-tungsten device JET-ILW (for "ITER-like Wall"). As documented in [10–12], the concentration of carbon in the plasma has dropped significantly, *in fine* the residual flux of carbon at the edge was reduced on average by a factor of 20.

Finally, to close this short presentation of possible charge-exchange spectra, another simulation is given in Fig. 8 where the strong λ 529 nm CVI line is surrounded by a wealth of tungsten profiles—sometimes called a quasi-continuum—which may appear in the visible spectral range in a Be-W machine like ITER where the divertor may constitute a strong source of tungsten. For a recent account of CX spectroscopy on the CVI line in JET, especially in the frame of the ITER-like wall, one can advantageously consult [13].

The CXRS system on JET was enhanced before moving to the ITER-like Wall. For a full description of the improved system, please consult [14]. The system described therein is sketched on Fig. 9. Giroud and Meigs [15] and Bernardo et al. [16] give information on the impact of calibration techniques and on results—ion temperature and toroidal rotation, respectively.

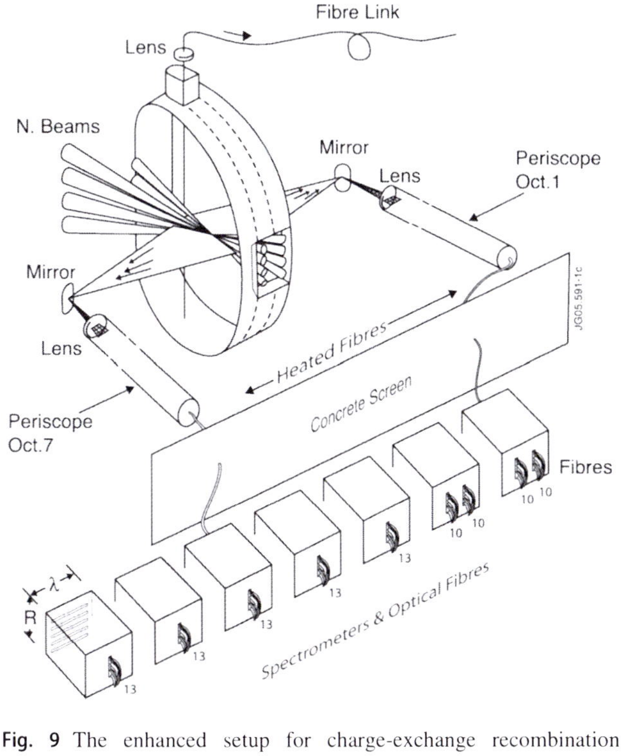

Fig. 9 The enhanced setup for charge-exchange recombination spectroscopy on JET. *Source*: [14]

Atomic Data and Evaluation Codes

The short glimpse in the previous pages presages a complex evaluation of the recorded spectra. Fortunately, the knowledge of the beam attenuation along the penetration radius, of the charge-exchange (CX) cross-sections, of the bremsstrahlung and of all other required quantities is well supported by different pieces of software that already exist. A main foundation is given by the Atomic Data and Analysis Structure (ADAS [17]) which provides a series of codes devoted to the processing of CX data: ADAS Series 3 covers Graph and Fit sections, the processing of effective coefficients with different resolutions (a state-selective cross-section database), the processing and tabulation of beam emission and beam stopping. Closely connected to this fundamental structure, several codes evaluate the data under different aspects, like the spectral analysis with CXSFIT [18] and CHEAP (Charge-Exchange Analysis Package, based on [19]), or the production of synthetic spectra, especially for performance analysis, with the "Simulation of Spectra" (SOS) software [9, 20]. One is thus basically equipped with software packages to write specific codes for the evaluation of recorded CXRS data, for instance for ITER. In parallel to those activities, analytical investigations of the CXRS signals deliver a wealth of information on the attainable accuracy or degree of plausibility of some of the input data [21, 22].

The Neutral Beam

Beams of neutral hydrogen are a workhorse for the heating of tokamak plasmas; Fig. 10 displays three of the main

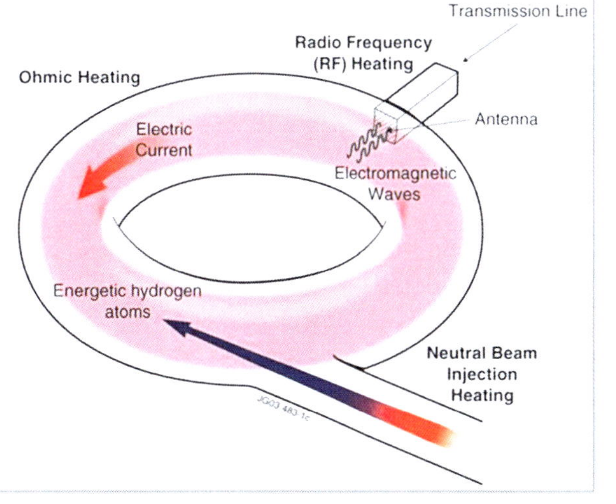

Fig. 10 Three of the main heating methods on a tokamak

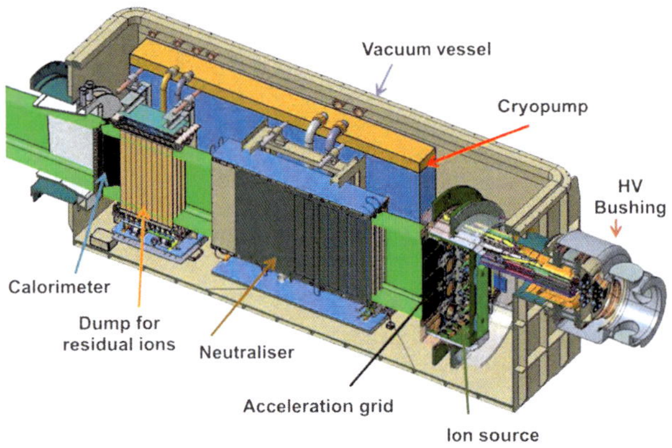

Fig. 11 Side view of the diagnostic neutral beam (DNB) shown in cross-section. It is located in the equatorial sector No. 4 whereas the CXRS observation optics is in the upper port No. 3 (Reprinted with permission Chakraboty et al. [23])

heating systems usually used: neutral beam injection (NBI), radio frequency heating (ICRH, e.g. tuned to the ion-cyclotron resonance $\omega_c = |q|B/m$ with q the electric charge, B the magnetic field strength and m the mass of the considered species), both of which come on top of the natural ohmic heating which decreases with the plasma temperature owing to the increasing electrical conductivity.

Although dealing with much lower currents than the heating beams, the neutral beams used for diagnostic purposes ("DNB" = diagnostic neutral beam) are a challenging endeavour especially with respect to the beam quality aimed at [23–25]: the section of the DNB which enters the plasma in a radial direction must be the sharply defined object which is observed by the CXRS optics while maintaining a high current density. The DNB planned for ITER aims at an energy of ≤ 100 keV/amu, $I \geq 36$ A at the entrance to the plasma chamber and a divergence limited to <7 mrad. The expected duty cycle is 3 s on/20 s off. A side cross-section can be seen in Fig. 11 and a top view in Fig. 12, next to the leftmost NB injector for plasma heating. A photo from inside the tokamak pit is shown in Fig. 13.

Light Collection and Transport

Obviously, the optical system used for the collection of the emitted photons needs a high throughput, which can be called the photon yield. More precisely, the efficiency of such systems may be described by the product of the overall transmission—from the entrance pupil to the detector—and the so-called *étendue*, which is a measure of the area times subtended solid angle on the observation side;

$$P = \mathscr{T} \times \mathscr{E} \quad in\ \mathrm{mm}^2 \cdot \mathrm{sr} \tag{9}$$

The main task is thus to maximize both the étendue and the transmission over the full visible range, keeping in mind that the lowest wavelength corresponds to the helium line in the blue spectral range (λ 468 nm as seen before) and the longest one is the Balmer-alpha line of the hydrogen isotopes at λ 656 nm in the red.

Telescopes and Endoscopes

To give a flavour of endoscopes and telescopes one can rely on for the development and design of an optical light collection system for the CXRS diagnostic in ITER, that is for integration in Upper Port Plug 3 (UPP3), here are a few examples on former tokamaks which have proved adequate.

Examples on the TEXTOR Tokamak and on JET

Figure 14 shows two telescopes which were installed on the TEXTOR tokamak for the radial and vertical observation of a small plasma volume, in the order of 1 cm^3 [26] where laser-induced fluorescence takes place. They were mainly used to study the profiles of atomic hydrogen near the first wall (liner), by excitation with a frequency-tripled laser at L_α (Lyman-alpha λ 121.6 nm).

Both are Cassegrainian systems with a primary concave mirror of diameters 400 mm and 300 mm, respectively. To limit the flow of impurities from the plasma to the mirrors during operation, apertures of ø50mm and ø40 mm, respectively, were installed at the very front of the telescopes. Note that the primary mirror of the vertical telescope can be tilted, as indicated by the curved arrows, to scan the profiles radially along the laser beam. This tilt is motorised and actuated by means of rotary feedthroughs. As all other systems presented below, these telescopes are purely achromatic since they are equipped exclusively with mirrors.

A similar arrangement which was installed to record fluorescence signals in front of a test limiter is shown in Fig. 15 [27]. The laser beam is shot from the top of the torus. The compact observation telescope, another Cassegrain, is driven by a motor for adjustable elevation in order again to scan radial profiles along the laser beam. In the case of optically reflecting limiter surfaces, the laser beam is reflected back, a useful property as will be demonstrated below in "The Optical Duct" section.

Turning now to JET, Figs. 16 and 17 exemplify a typical geometry for observation in a torus [28]. This device identified as the KL11 diagnostic is an achromatic endoscope for the spectroscopic observation of the divertor at

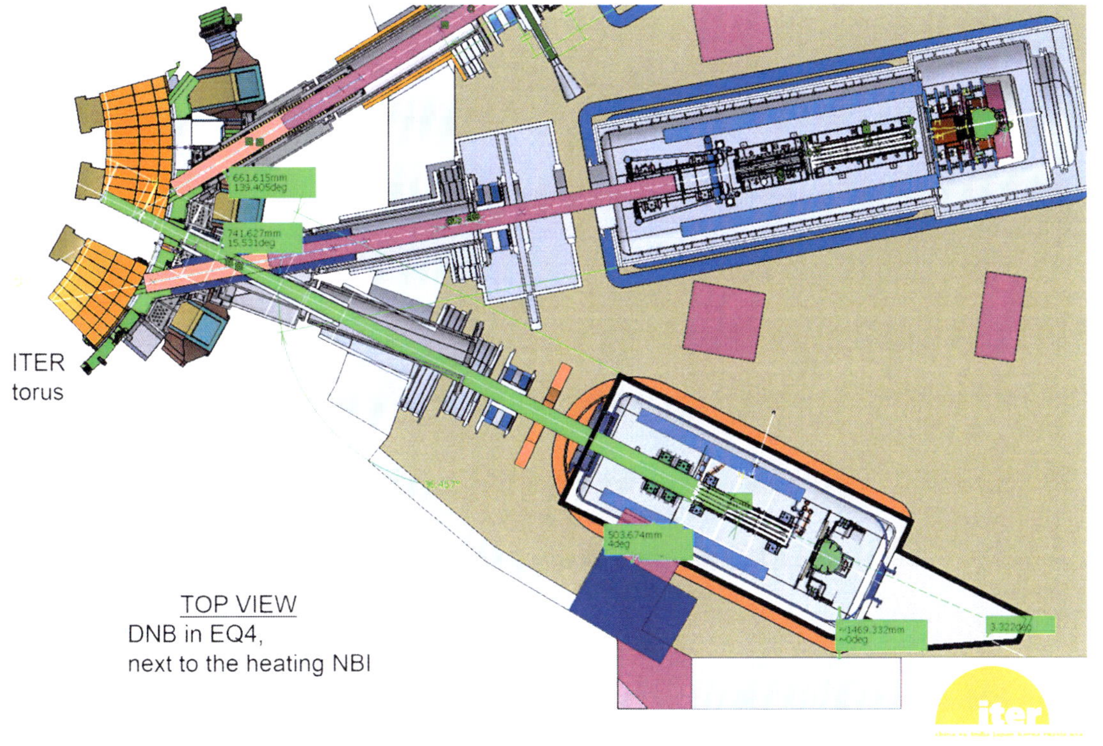

Fig. 12 Top view of the diagnostic neutral beam (DNB—green colour), next to two heating beams (NBI)

Fig. 13 The ducts of the three heating beams (left side of picture) and DNB duct (right, marked 1) viewed from the tokamak pit of ITER. '2' indicates the cell which corresponds to Upper Port No. 3

the bottom of the tokamak, about 4 m away from the elliptical entrance aperture.

Contrary to the previous telescopes, this optical device is equipped with a pneumatic shutter to protect the head optics from impurities which escape the plasma and

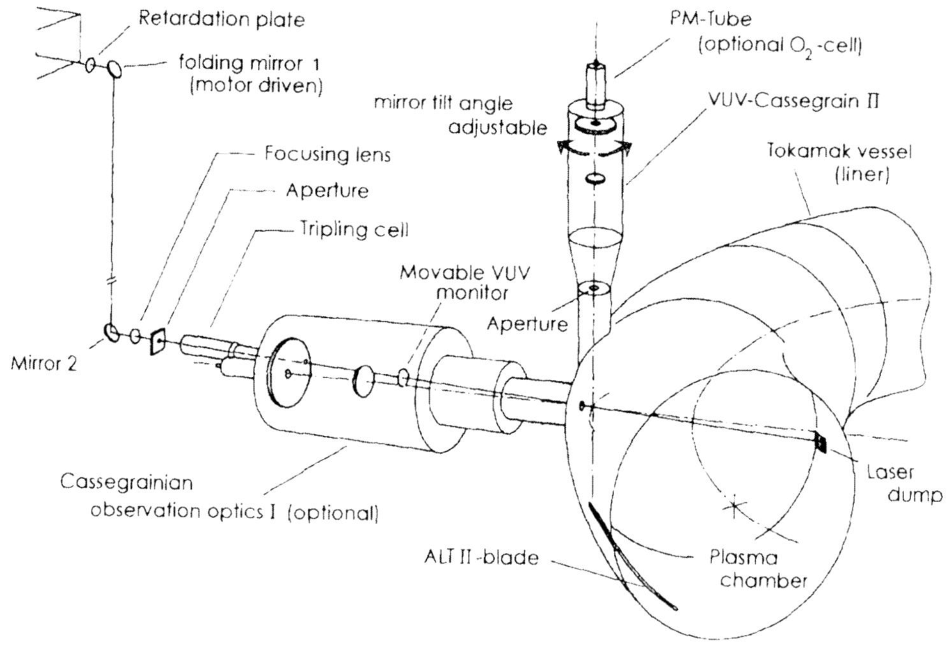

Fig. 14 Telescopes installed on TEXTOR for measurements with LIF (laser-induced fluorescence) in front of the first wall (Sector 11)

impinge on the first and second mirrors. A schematic model of the complete endoscope is given in Fig. 18 and the enlarged head construction in Fig. 19 [29].

The KL11 diagnostic provides a wide view of the divertor as demonstrated in Fig. 20 where the bulk tungsten row with >9000 lamellae distributed in four poloidal stacks is seen in the centre of the photograph [29]. The spatial resolution is high, in fact high enough to distinguish some of the single lamellae, which are only 5.5–6.0 mm thick in toroidal direction. For comparison, the CAD model of the JET-ILW divertor is shown in Fig. 21.

Schematics of a System for ITER

The optical layout of the observation system of the DNB for the core-plasma CXRS of ITER must fit in the Upper Port Plug No. 3 [31]. A possible set-up is sketched in Fig. 22. This simplified layout shows the main components which are necessary in principle to fulfil the expected functionality. Explicitly shown are:

1. The front aperture which is the entrance pupil of the optical system. It is installed inside the duct which is

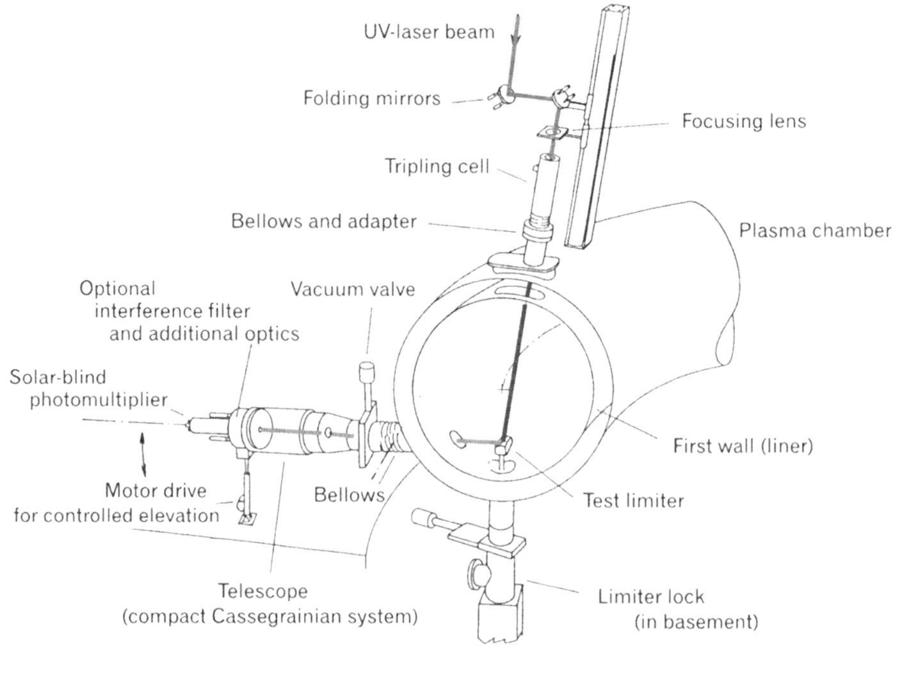

Fig. 15 A compact telescope on the TEXTOR tokamak for optical measurements in front of a test limiter (LIF)—Sector 10

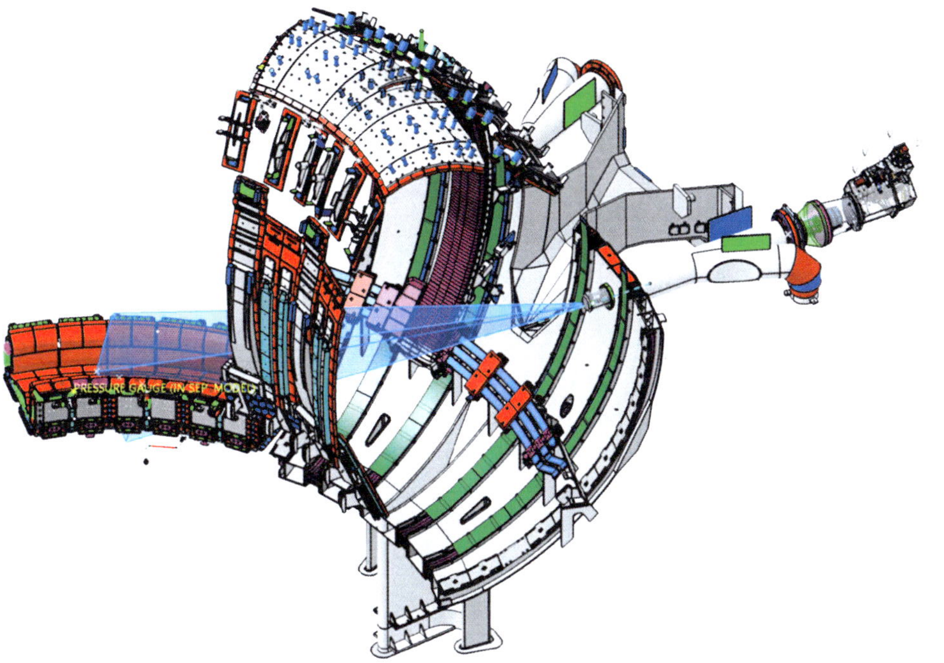

Fig. 16 The KL11 endoscope is an achromatic optical device for the spectroscopic observation of the JET divertor with a high spatial resolution. The solid viewing angle is shaded in light blue (Color figure online)

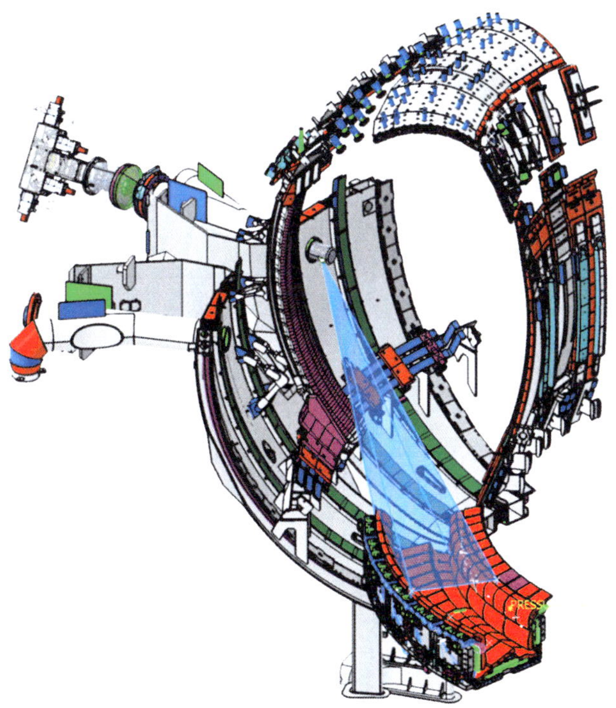

Fig. 17 The KL11 endoscope is an achromatic optical device for the spectroscopic observation of the JET divertor with a high spatial resolution. The solid viewing angle is shaded in light blue (Color figure online)

embedded in the first wall ("DFW": diagnostic first wall);

2. A shutter with blades located directly behind the entrance pupil. It protects the front mirrors from the particle fluxes which escape the plasma when the system is not measuring;
3. the first mirror;
4. A cleaning device which might be installed to clean the first mirror from deposits which would impede its proper operation;
5. Mirrors in the mirror chain to transport the light out of the torus;
6. The rear optics assembly, close to the output window(s).

Design Drivers and Performance Requirements

The main design drivers can be grouped in three classes (I–III) as in the following. Some of them indicate values which must be reached to fulfil the performance requirements.

Class I Collection, transmission and detection of enough light from the plasma (DNB):

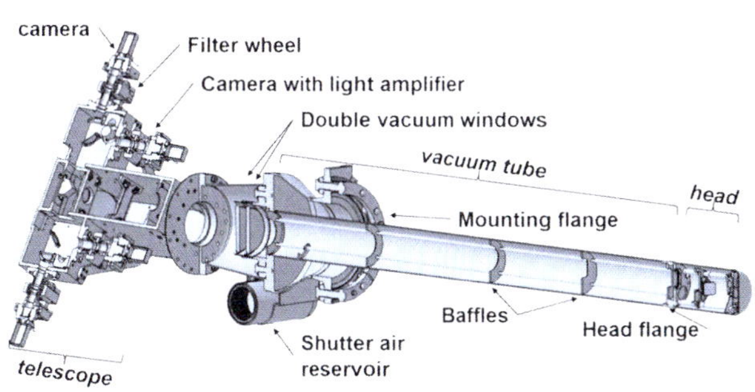

Fig. 18 A schematic view of the complete KL11 endoscope (JET)

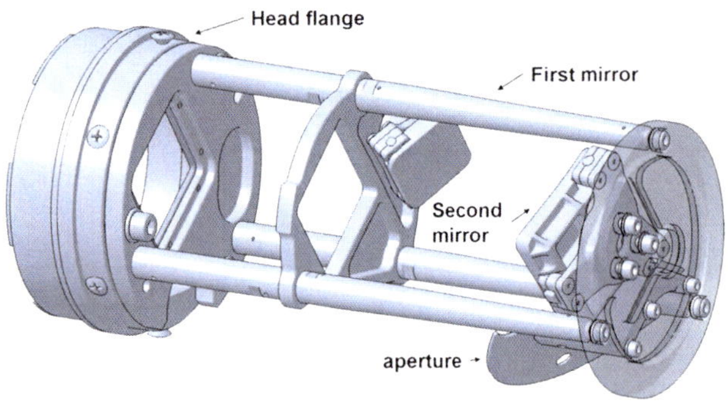

Fig. 19 3D model of the endoscope head (KL11 on JET)

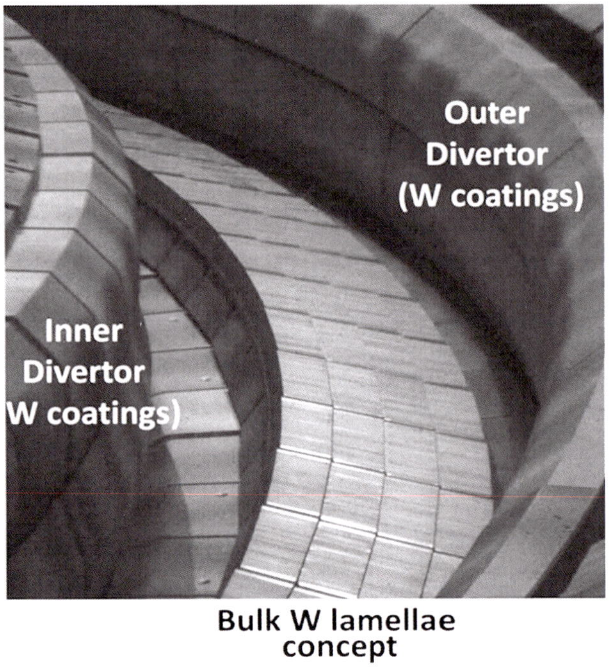

Fig. 20 View of the JET divertor through the KL11 endoscope. The solid tungsten row is in the centre [29]

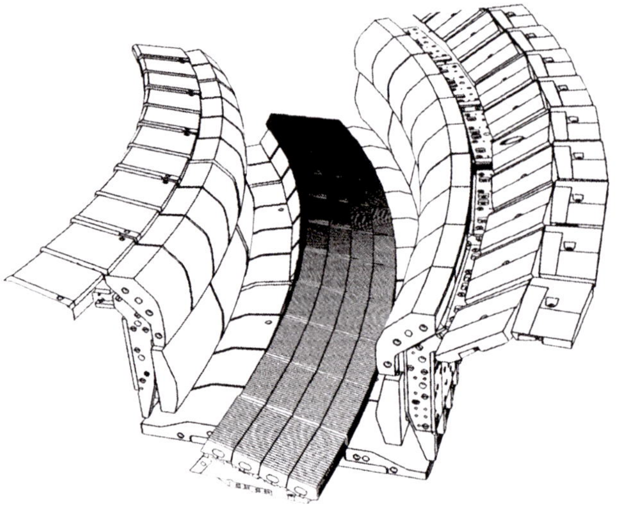

Fig. 21 A similar view as in the previous Fig. 20 given by the CAD model. See also [30]

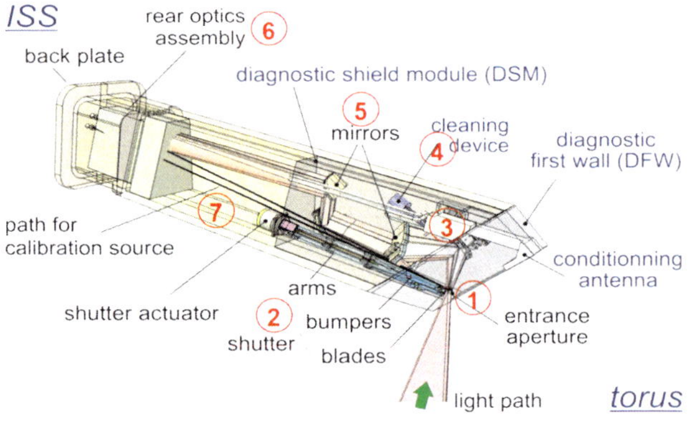

Fig. 22 Schematic view of a possible layout of all CXRS subsystems in UPP3

- An Etendue × Transmission of $0.3\text{mm}^2 \cdot \text{sr}$ is the performance target for an average channel through the plug optics. The design driver is obviously to maximise the transmission of the in-plug optics to the ultimate extent. How far this is quantitatively fulfilled will be checked with an optimised optical design;
- The image must be sharp enough to allow for the requirements on spatial resolution to be fulfilled and to be able to accommodate in a flexible way the ex-vessel components to be determined later. The design driver was given in form of a minimal MTF value (≥ 0.6) and a maximal spot size (in the order of 300μm).
- The field of view must be kept large enough both (i) in radial direction—along the DNB—to guarantee an overlap with the edge CXRS diagnostic and, at the same time, measure deeply enough in the core plasma (roughly r/a = [0.1–0.7], max. deviation 170 mm) and (ii) laterally with a margin of $+/-100$ mm to catch the maximum of the DNB intensity profile (cross-section) within an image region of sufficient quality.

Class II Supporting sub-systems and features for optics protection, calibration and monitoring:

- A shutter must be installed, which is normally closed (NC) and opens only during DNB 'on' time slices, that is, a duty rate of 3 s'open'/20 s 'closed' (see "The Neutral Beam" section). Derogating to this rule, the shutter may be opened whenever needed for alignment, monitoring or calibration of the system;
- A cleaning system is mandatory and its integration is one of the design drivers for the optics. The Front Mirror Unit (FMU) should withstand about 100 cleaning cycles of M1;
- The ducts, especially in front of mirrors in the FMU, must be designed to keep the mirror contamination as

low as possible. Accordingly, the ratio of the distance between entrance pupil and M1 to the diameter of M1 is to be kept high, $L/D \geq 5$, the larger the better;
- Sub-systems must be foreseen for re-alignment, calibration or at least monitoring of the optics degradation with time. A direct view from the closure plate to the back side of the shutter blade(s) must be provided if required by these sub-systems.

Class III Safety requirements and investment protection:

- The shielding efficiency of the system against neutrons must be maximised, this is a major design driver. The known criterion relates to the maximal SDDR (Shutdown Dose Rate) of 100μSv/h at 10^6s after shutdown and to the ALARA principle (As Low As Reasonably Achievable);
- The compatibility with remote handling (RH—mainly RH class 2) must be ensured for maintenance. Important in connection to this point is the ban on *scheduled* maintenance for UPP3;
- The structural integrity of the system must be guaranteed, which is evident. The structural integrity may be derived from conformance to the relevant codes and standards (SDC-IC / RCC-MRx);
- The overall cost is a major design driver too. Although the freedom to select between different realisations of a given concept is not always given for in plug components, the lowest cost/performance ratio is of course selected when the case arises. For the first mirror M1, materials which are considered expensive may be chosen if they bring much higher performances (reflectance) and the associated price stays in the order of 1–2% of the total estimated cost, within budget ceiling.

An Optical Design for ITER

A recent, but still conceptual design for the upper port plug #3 on ITER is presented below ("An Optical Design for ITER" and "Opto-Mechanical Components of a Recent Design"), followed by a discussion on selected subsystems.

Selection of a Layout

The design from scratch can often provide several possible options among which a ranking and eventually a selection takes place. The four optical layouts considered for the core CXRS diagnostic are sketched in Fig. 23. They rely on 3 to 6 mirrors to transmit the light from the object (DNB) to the output window at the plug closure plate. After unifying their field-of-views, the systems were optimised before ranking. Numerous criteria were used in order to rank and ultimately select a layout; they were grouped in six classes:

- optical performance, including among others the photon yield (see top of "Light Collection and Transport" section) and the image quality (Modulation Transfer Function MTF, absence of vignetting at the window, behaviour with respect to the polarisation properties…);
- radiation shielding, with minimisation of the neutron flux especially in the interspace, of the radiation impact on magnetic coils, etc.
- maintainability and logistics;
- ease of accommodation of external interfaces;
- design complexity and robustness;
- system lifetime.

On the basis of these criteria, the optical layout shown in Fig. 24, a modified design No. 3, was selected as a baseline.

Figure 25 is a representation of the footprint of the light beam at the exit windows assembly which shows that the "no-vignetting" criterion is fulfilled, leaving some margin for tolerances.

Situation in the Tokamak Building

The situation of the diagnostic, DNB and observation system in the upper plug, is illustrated in Figs. 26 and 27. A part of the ITER wall (blanket) is shown together with the portion of the DNB which penetrates the plasma.

Behind the upper port, the so-called interspace (IS) and, beyond the bio-shield, the port cell (PC) can be seen. The frames in yellow are the Interspace Support Structure (ISS) and the Port Cell Support Structure (PCSS). Note that the spectrometers are situated far from the port cell. The consequence is that the light must be transported by bundles of glass fibres about 60 m away, which brings an additional attenuation.

Opto-Mechanical Components of a Recent Design

Several opto-mechanical components were discussed during the school. Examples are given in Figs. 28 and 29.

The mirror mounts displayed must withstand very high reaction forces owing to the electromagnetic torques generated by eddy currents during disruptions or abrupt loss-of-plasma events (in the order of 5 kN·m with up to 100 T/s). The tilted and/or wedged window plates prevent the spurious multiple reflections that would appear with parallel surfaces.

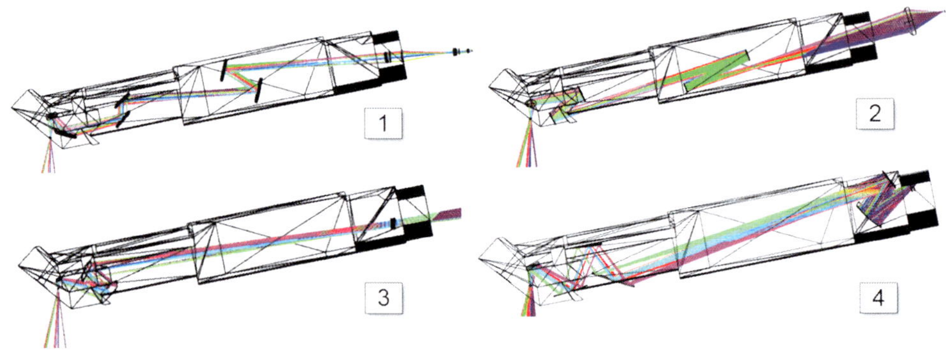

Fig. 23 Four optical layouts were considered as a first conceptual step for an ITER design (CXRS in UPP3)

The First Mirror (M1)

The first mirror is the most vulnerable part of the whole setup [32–34]. It is exposed to high particle fluxes, namely to all types of neutrals which escape the magnetic confinement and reach the mirror through the optical duct cut in the diagnostic first wall (DFW). Fluxes up to $10^{20}\mathrm{m}^{-2}\ \mathrm{s}^{-1}$ impinge on the mirror which is subject to erosion, especially by fast charge-exchange neutrals in the range 0.1–1 keV or more, and/or to deposition of impurities (corresponding flux rates up to 0.1 nm/s at the front). During cleaning, if a cleaning system is installed for recovery of the reflective properties, it may again be subject to physical sputtering by the discharge ions. From the point of view of the optical quality, a high specular reflectivity is demanded over the full spectral range—in the present case practically the complete visible region. If the mirror becomes hot, the reflectivity should be maintained and, most difficult of all requirements, the deformations of the front optical surface must be kept within the specified bounds of the order of 100 fringes (an interferometrically determined specification at a defined visible wavelength, for instance λ 632.8 nm of an He–Ne laser; 1 fringe corresponds to $\lambda/2$). Incidentally, the resistance to oxidation must also be high to avoid ruin in the occurrence of a water leak or steam ingress event.

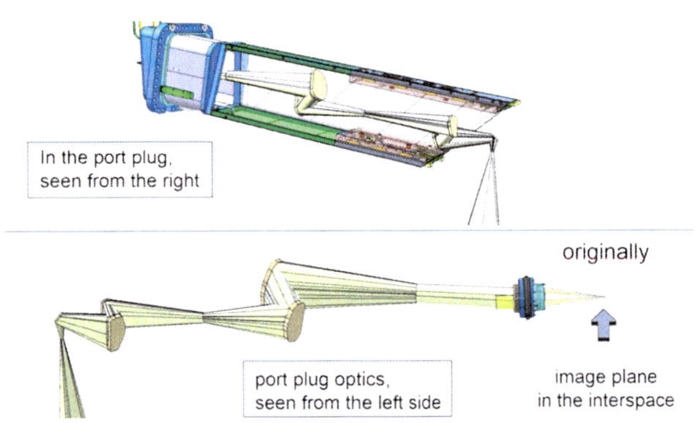

Fig. 24 Selected optical layout for the core plasma CXRS diagnostic

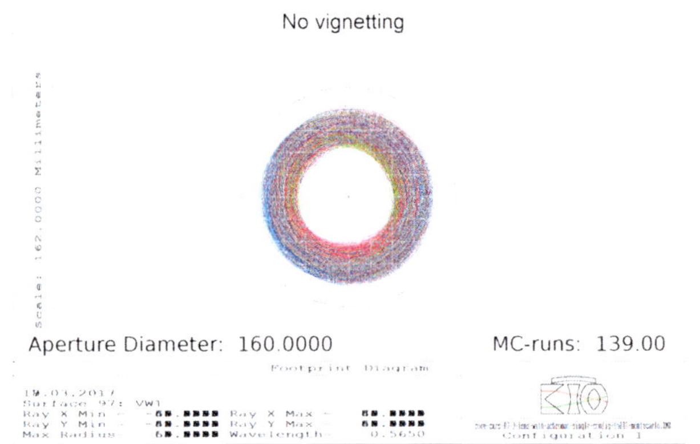

Fig. 25 Light footprint at a Ø160 mm window, showing some margin to the full aperture. *Source*: KTO GmbH

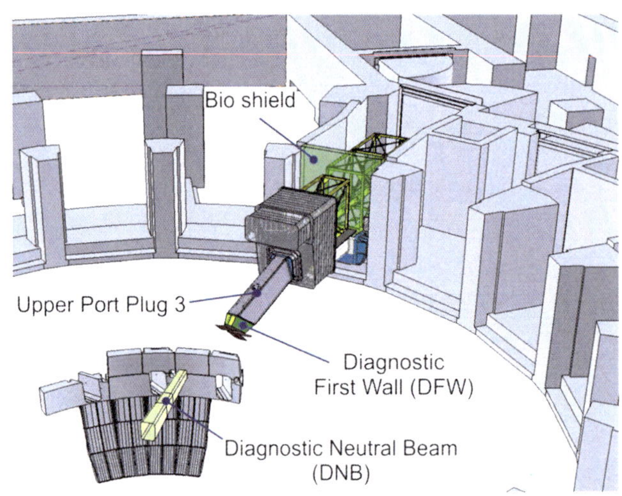

Fig. 26 Situation of the diagnostic neutral beam and of the upper port plug No. 3 in the tokamak building (view from the centre)

The current baseline for the core CXRS system is a flat mirror, easier to manufacture, made of rhodium (fallback solution: molybdenum) preferably in the form of a single crystal. The position and size of such a mirror can be estimated in Figs. 30 and 31, the latter showing the geometry of a possible enclosure as a front mirror unit. The mirror dimensions are about $85 \times 180\mathrm{mm}^2$. The expected reflectance should lie around $\mathscr{R}_{Rh} = 75\%$ over the full

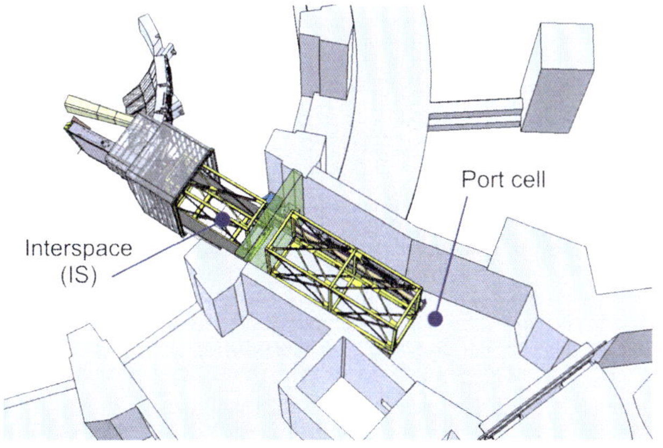

Fig. 27 Situation of the diagnostic neutral beam and of the upper port plug No. 3 in the tokamak building (view from the top). The bio-shield is shown in shaded green (Color figure online)

visible range [35] which is a reason for the choice of rhodium. More information on the first mirror for the core-plasma CXRS diagnostic can be found in [36, 37] and, on the specific use of rhodium, in [38]. Details on the production of coatings or on their resistance are given in [39–43]. A good example is provided by an all-reflective diagnostic in JET [44].

Protective Systems

As pointed out in the previous section, the first mirror must be protected for a reasonable lifetime (up to 30,000 ITER pulses, a large part of them at the highest performance) in the sense that the flux of incident particles must be reduced to a minimum. To put numbers on his assumption, the current view is that a flux of $10^{20} \mathrm{m}^{-2}\,\mathrm{s}^{-1}$ at the entrance ought to be reduced by a factor $\mathcal{O}(3 \times 10^4)$. This goal can only be achieved by combining all possible mitigation measures which can reasonably be taken to protect the diagnostic investment, namely

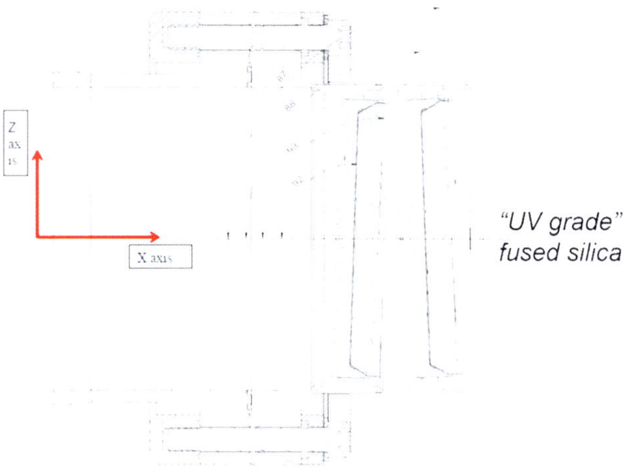

Fig. 29 Possible double-window assembly for the light path at the closure plate (tilted, wedged window plates). *Source*: ITER windows catalogue

- a properly designed optical duct in front of the mirror to mitigate the influence of the incident particle flux ("The Optical Duct" section);
- a shutter placed just behind the entrance pupil to close the entrance path to the largest achievable extent when the observation system is not in operation ("The Shutter" section);
- a cleaning system to recover the reflective properties of the mirror in case of excessive deposition ("A Possible Cleaning System" section). In order to spare the mirror surface—i.e. the reflective coating or the polished bulk material—from damage due to exaggerated cleaning, an "End-of-cleaning-indicator" is mandatory to stop the process timely ("End-of-Cleaning Indicator" section).

The Optical Duct

The optical duct is the first protective measure that can be taken against particle and heat fluxes to the first mirror. A schematic example with the introduction of fins (from [45])

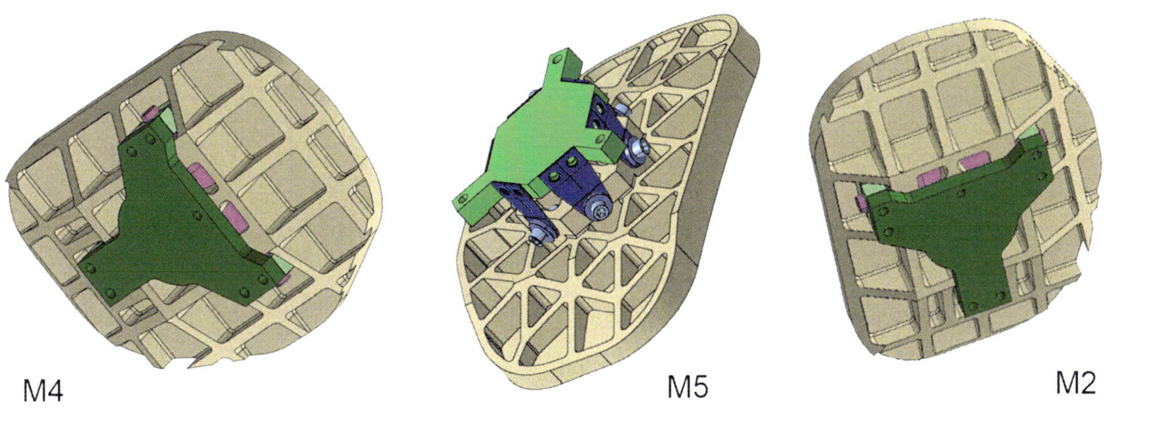

Fig. 28 Conceptual mirror mounts for M2 to M5

is shown in Fig. 32. As can be derived from [45], the attenuation of the particle flux can reach one or two orders of magnitude. The analysis in [46] considers the beryllium flux onto a modelled surface (surf. 37, see Fig. 33) which represents the front face of an upper port plug.

The advantage of implementing a cleverly designed duct, long enough (arbitrarily $L/D > 5$ in the present case, with L the duct length and D the free diameter) and consisting of a diverging frustum with appropriate fins [47], is supported and strengthened by the thermal analysis given in [48] which deals with the duct schematically presented in Fig. 34. In terms of attenuation of the particle flux, the duct could bring up to a factor of 10^2.

The Shutter

The shutter is the next protective measure which must be taken to limit the flux to the mirrors. In the case of core CXRS with a main modulation of the DNB specified as of 3 s 'on'/20 s 'off', the shutter opening and closure must be much faster than a system which would simply open ahead of the plasma pulse and close after it. To approach the targeted attenuation of the particle flux by about one order of magnitude, the opening and closing times must lie below 1 s. To avoid seizure of the actuating parts in vacuum, an incident which occurred on several other tokamaks, the goal is to use a frictionless mechanism. An example, based on the bending of long arms to keep the mechanical stresses within bounds, is shown in the photograph Fig. 35, Note that the blades in front are actively cooled with water flowing through the arms. Such a prototype has withstood 10^6 cycles (in air) without visible damage. This shutter was extensively documented in [49–51].

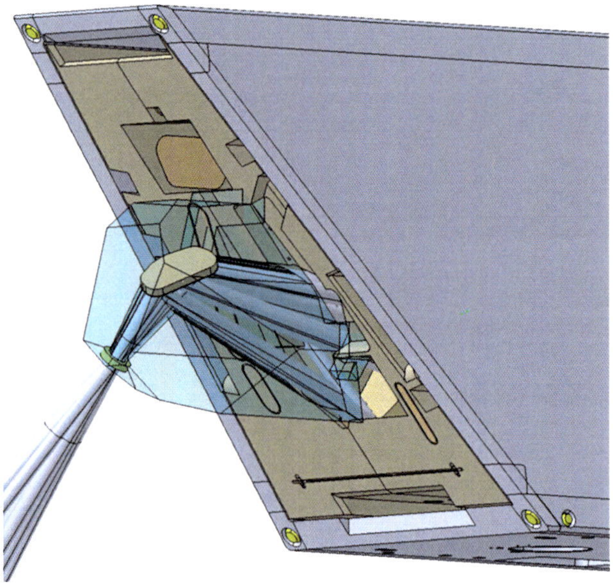

Fig. 31 Position and size of the first mirror in the '2017' design. The diagnostic first wall (DFW) is removed to show the optical path and the mirror. The shaded enclosure represents what could become a front mirror unit (FMU)

A Possible Cleaning System

The missing flux attenuation by another factor of 10 cannot be brought by additional components. A cleaning system is thus rather a remediation: it is supposed to restore the initial reflective properties of the mirror by removing the deposits without overly damaging the original mirror surface. The physical process which is resorted to is the physical sputtering of the deposited impurities. To this purpose, a DC discharge—preferably pulsed to remove

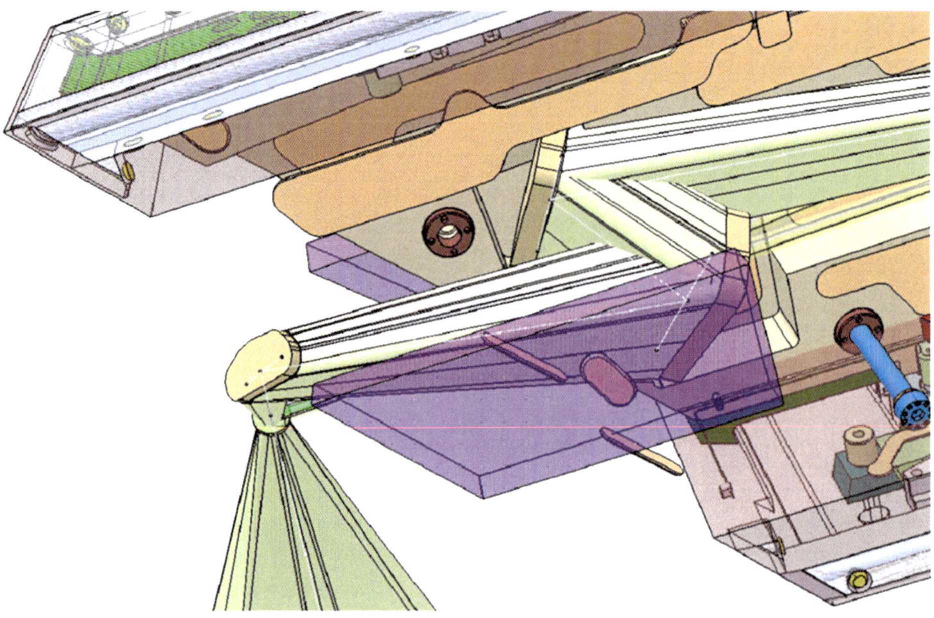

Fig. 30 Position and size of the first mirror in the '2017' design. The diagnostic first wall (DFW) is removed to show the optical path and the mirror. The purple blocks are reserved volumes for the attachment of the DFW (Color figure online)

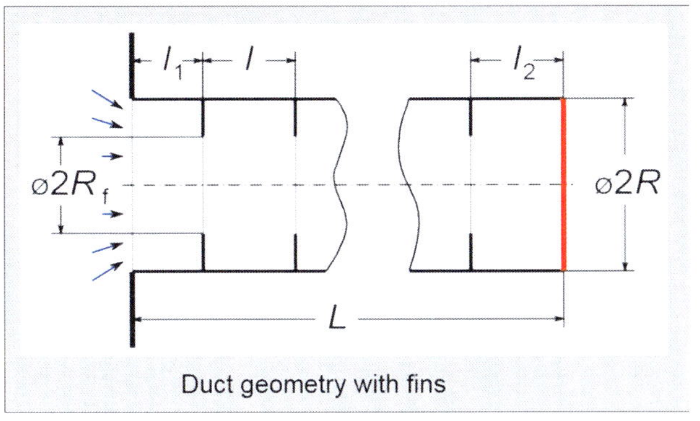

Fig. 32 Sketch of the duct analysed in [45]

hard ceramic-like impurity layers—or a capacitively coupled Radio-Frequency (RF) discharge in the VHF frequency range may be adequate. An artist view and an equivalent circuit are represented in Fig. 36. It shows the challenge of matching the impedance of the generator, electrical feedthrough, and even more of the feeding line and of the connection to the mirror which acts as a cathode. Accordingly, the mirror itself must be electrically insulated from the surroundings. In the case of RF, a workaround may be considered: to connect the mirror to the water feedthroughs by means of a λ-quarter line (length $\ell = \lambda/4$) which enables to realise the equivalent of an RF-insulator with conducting material, that is the water line might not necessarily be insulated at the penetration through the closure plate (see [52]).

There is a wealth of literature which documents different aspects of cleaning discharges, for instance [53–55]. These articles emphasize the subject of efficient cleaning, repetitive cleaning, choice of an RF frequency and power, and touch the selection of an adequate operative gas (for instance H_2, He, Ne, Ar).

End-of-Cleaning Indicator

As mentioned above, it is mandatory to stop the discharge as soon as the deposits have been removed to avoid damaging the reflecting surface. There is clearly a need for a kind of "End-of-Cleaning Indicator" (ECI). A direct option to this purpose is the use of spectroscopic detection; most of the optical diagnostics we refer to are or can be equipped with a simple spectrometer after all. If one takes the case of beryllium as an impurity, the main plasma-facing constituent of the ITER blanket, two challenges must be faced:

- the electron temperature T_e in the cleaning discharge must be high enough to allow excitation of the spectroscopic transition to be observed (e.g. Be II at λ527 nm)—compare to the ionisation step of about 9.3 eV. This condition may be only marginally fulfilled;
- other impurity species may be dominant so that the detection of beryllium in the discharge is not the relevant indicator.

Another scheme was proposed recently [56] which measures the reflectance of the mirror during cleaning and is thus independent of the type of impurities that are removed. The idea is to use a singular light source present in the discharge, namely fast atoms of hydrogen, and to derive the ratio of the light they emit directly in the direction of the observer to the light received after reflection on the mirror surface (Fig. 37).

Hydrogen atoms may be naturally present in the discharge or added as a trace in the operative noble gas. In the form of ions, they are accelerated in the potential drop just in front of the mirror which is the cathode of the discharge (see above). The wavelength λ of the observed light is Doppler shifted according to

Fig. 33 The incident flux of atomic beryllium on the upper port as a function of the toroidal angle (LD, MD and HD stand for low, medium and high density, respectively). See [46] for explanations

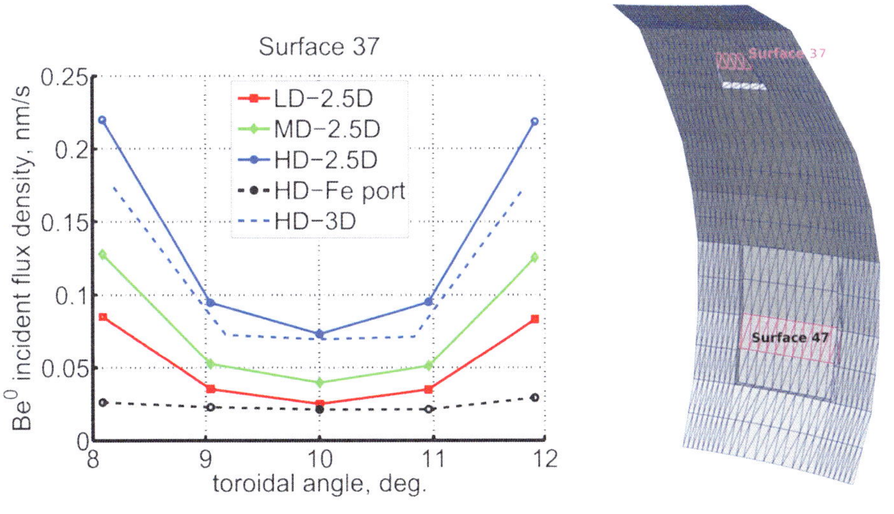

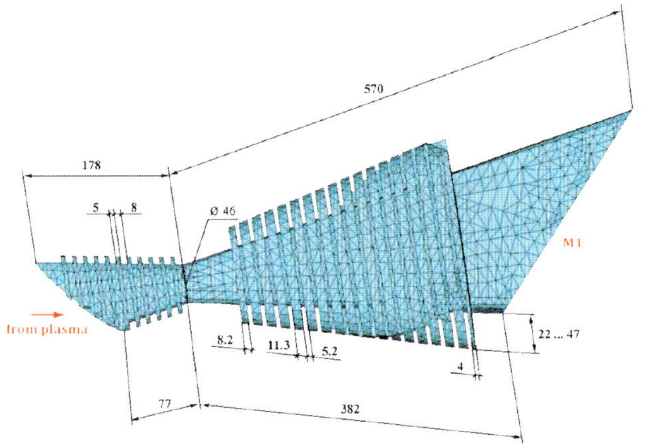

Fig. 34 The duct model, with mixed baffles or fins, analysed by Grigoriev in [48]

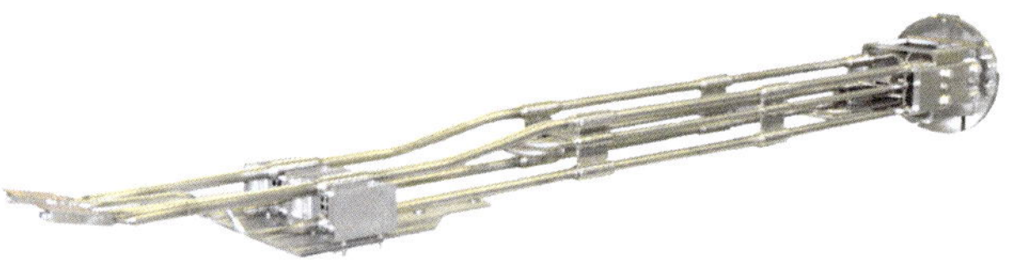

Fig. 35 The first successful prototype of a fast (0.7 s characteristic time) frictionless shutter

$$\frac{\Delta\lambda}{\lambda} = \frac{v}{c} \tag{10}$$

where v is the velocity of the atoms in the direction of observation and c the velocity of light.

The example of palladium (^{106}Pd close to ^{103}Rh) illustrates the profiles which are recorded at different bias voltages (Fig. 38). In all cases, the ratio of the red to blue spectral wings according to

$$\mathscr{R} = \frac{\int I_{red\_shift}\, d\lambda}{\int I_{blue\_shift}\, d\lambda} \tag{11}$$

with proper limits on the integrals gives the reflectance $\mathscr{R}$, measured in situ during the discharge [57]. With this method, it would be possible to follow the progress of the

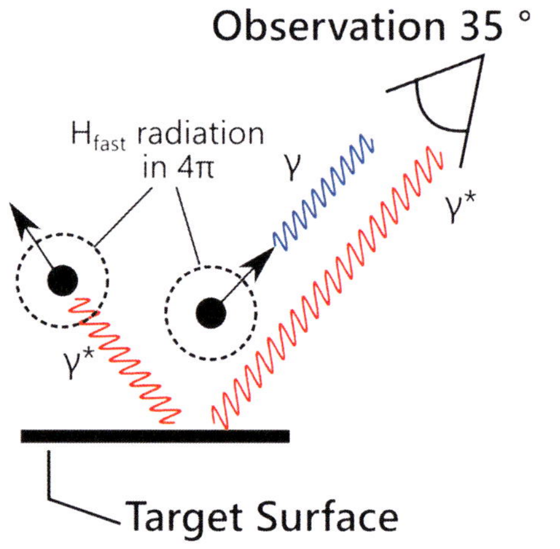

Fig. 37 Observation at an arbitrarily given angle (35°) of the light emitted by fast hydrogen atoms. The blue-shifted light is observed directly along the line-of-sight whereas the red-shifted contribution is observed after reflection on the mirror (Color figure online)

cleaning process to interrupt it as soon as the incremental improvement in reflectance becomes vanishingly small.

The method can be compared to the mirrored recording of the Thompson distributions [58] of atomic silicon released from a silicon carbide (SiC30) test limiter head in the TEXTOR tokamak:

$$L(\lambda-\lambda_0) = \frac{4}{\pi\,\Delta\lambda_0}\left(1-\frac{1}{\sqrt{1+(R/z)^2}}\right)^{-1} \times\left(\left(1+\frac{(\lambda-\lambda_0)^2}{(\Delta\lambda_0)^2}\right)^{-2}-\left(1+\left[1+(R/z)^2\right]\frac{(\lambda-\lambda_0)^2}{(\Delta\lambda_0)^2}\right)^{-2}\right) \tag{12}$$

The profiles were recorded by laser-induced fluorescence on the setup shown in Fig. 15, "Examples on the TEXTOR Tokamak and on JET" section: the laser beam is reflected on the shiny limiter head and records the distribution twice, i.e. forwards and backwards, as shown in Fig. 39,

Fig. 36 Envisaged radio-frequency discharge (left) and equivalent circuit (right)

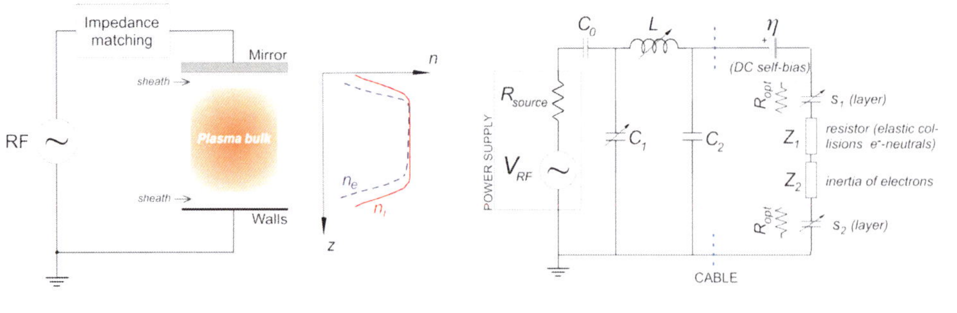

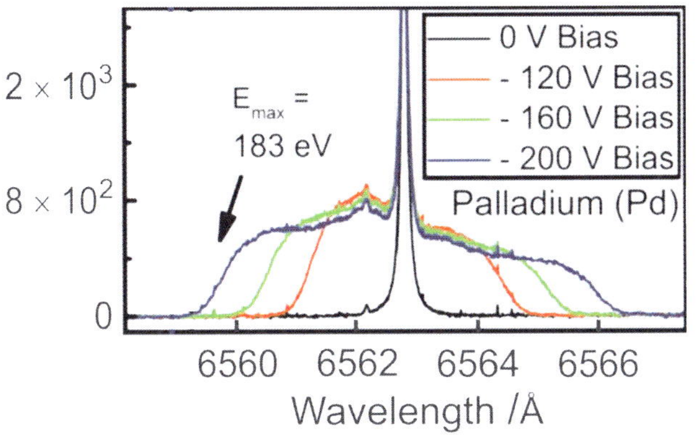

Fig. 38 Spectral profiles recorded upon reflection on a Pd mirror. Vertical scale: arbitrary counts

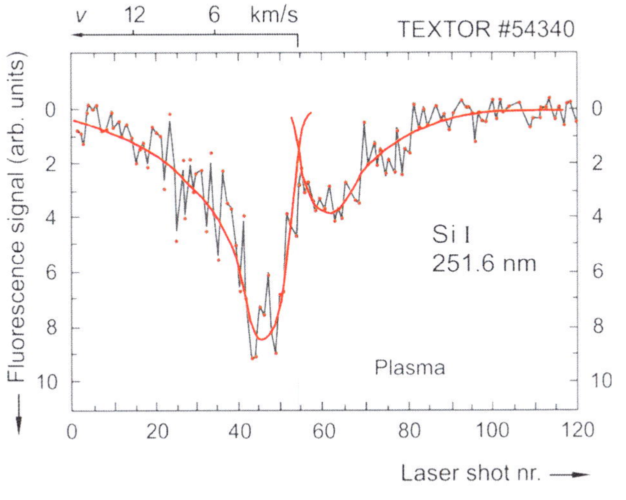

Fig. 39 Double record by LIF of the velocity distribution of Si-atoms released by physical sputtering from a limiter head in the TEXTOR tokamak: the mirrored image is due to the laser light which excites the atoms a second time after reflection on the limiter, hence the mirrored image with lower amplitude [59]

providing the reference wavelength at zero velocity as a valuable by-product [59].

Other Challenges

Neutronics

As reported in "Design Drivers and Performance Requirements" section under the 'Class III' heading, the shielding efficiency of the plug, particularly with respect to neutrons, is a main driver of the design. It was specially true during the selection of an optical layout ("Selection of a Layout" section). The corresponding calculations are well documented in [60]; compare also with [61]. A deeper discussion of this complex subject is out of the scope of the present article; please refer to the mentioned references.

Reflections from the Torus

Apart from bremsstrahlung ("Spectra" section, Fig. 5 as an example), the main contribution to the signal background is expected to come from spurious reflections of powerful light sources. Reflections of bright light sources, especially Balmer-α, from the divertor region—close to the targets or from the divertor throat in the case of detached plasmas [62]—might find their origin in protruding plasma-facing components or leading edges in the outer blanket region which is seen at the back of the CXRS Field of View (FoV), of course focussed on the DNB. Several numerical codes can be used to attempt a modelling of this effect. It is always a challenging task because the full detailed hardware components of the torus wall must be fed into the model to provide a representative view of the machine to be built and of the reflections that may occur. An example

Fig. 40 A comparison between an actual view of the JET torus with a wide angle camera (left) and a similar view reconstructed with *Raysect* (right). By courtesy of N. Hawkes

is given in Fig. 40 for the inner walls of JET. A similar simulation is intended for the FoV of the core CXRS observation system.

Conclusions

The development and design of the CXRS diagnostic for the core plasma of ITER was used as a pretext to elaborate on several of the main challenges of optical diagnostics on such a large fusion device. Following very basic considerations on CXRS and spectroscopy, a major chapter was devoted to light collection and transport to the detectors including several historical examples, as well as design drivers and performance requirements in the case of the core-plasma CXRS on ITER. A glimpse of the optical design for ITER was also embedded in that section. The next significant section dealt with the first mirror, probably the most challenging component in an optical diagnostic on a tokamak, and with the mandatory protective measures to reach the expected lifetime. The author hopes that some insight was provided to the students of the 16th *Ettore Majorana School on Diagnostics and Technology Developments* in the design process of such complex, highly technological diagnostic systems.

Acknowledgements This work was partly supported by the German Federal Ministry of Education and Research (BMBF Grant No. 03FUS0007) and by the ITER Organization in the framework of service contract ITER/CT/12/4300000546. A few of the later results were obtained under F4E-FPA-408 (DG), a partnership with *Fusion for Energy*, the European domestic agency to ITER. The views expressed here are the sole opinion of the author. He is indebted to his colleagues W. Biel, O. Marchuk and N. Hawkes for their constructive and helpful remarks. The support and contributions of all members of the IC^3 consortium (ITER Core Charge-exchange Consortium) are gratefully acknowledged—IC^3 is the ad hoc consortium to design and build the core-plasma CXRS diagnostic in the Upper Port Plug No. 3 of ITER.

References

1. A. Unsöld, *Physik der Sternatmosphären –mit besonderer Berücksichtigung der Sonne*, 2nd edn. (Springer, Berlin, 1955). **(in German)**
2. E. Delabie, M. Brix, C. Giroud et al., Consistency of atomic data for the interpretation of beam emission spectra. Plasma Phys. Control. Fusion **52**, 125008 (2010)
3. S.K. Rathgeber, R. Fischer, S. Fietz et al., Estimation of profiles of the effective ion charge at ASDEX upgrade with integrated data analysis. Plasma Phys. Control. Fusion **52**, 095008 (2010)
4. P. Bogen, E. Hintz, Plasma edge diagnostics using optical methods, in *Physics of Plasma–Wall Interactions in Controlled Fusion*, vol. 131, NATO ASI Series B: Physics, ed. by D.E. Post, R. Behrisch (Plenum, New York, 1986), pp. 211–280
5. M. De Bock, *VSRS Bremsstrahlung*, in Functional Breakdown and Summary of Requirements for 55.E6 VSRS, IO internal report, ITER_D_V5VMN6
6. H.R. Griem, *Plasma Spectroscopy* (McGraw-Hill, New York, 1964), p. 580
7. W.J. Karzas, R. Latter, *Electron Radiative Transitions in a Coulomb Field. Astrophysical Journal Supplementary Series*, vol. 6 (University of Chicago Press, Chicago, 1961), p. 167sqq. [(American Astronomical Society) provided by the NASA Astrophysics Data System]
8. M.F.M. De Bock, R. Barnsley, M. von Hellermann et al., Overview of active beam spectroscopy developments for ITER, in *ECPD : 2nd European Conference on Plasma Diagnostics (ECPD)* (2017)
9. M.G. von Hellermann, G. Bertschinger, W. Biel et al., Complex spectra in fusion plasmas. Phys. Scr. **T120**, 19–29 (2005)
10. S. Bresinzek, S. Jachmich, M.F. Stamp et al., Residual carbon content in the initial ITER-like wall experiments at JET. J. Nucl. Mater. **438**, S303–S308 (2013)
11. F. Romanelli, JET EFDA contributors. Overview of the JET results with the ITER-like wall. Nucl. Fusion **53**, 104002 (2013)
12. J.W. Coenen, M. Sertoli, S. Brezinsek et al., Long-term evolution of the impurity composition and impurity events with the ITER-like wall at JET. Nucl. Fusion **53**, 073043 (2013)
13. S. Menmuir, C. Giroud, T.M. Biewer et al., Carbon charge exchange analysis in the ITER-like wall environment. Rev. Sci. Instrum. **85**, 11E412 (2014)
14. C.R. Negus, C. Giroud, A.G. Meigs et al., Enhanced core charge exchange recombination spectroscopy system on joint European torus. Rev. Sci. Instrum. **77**, 10F102 (2006)
15. C. Giroud, A.G. Meigs, C.R. Negus et al., Impact of calibration technique on measurement accuracy for the JET core charge-exchange system. Rev. Sci. Instrum. **79**, 10F525 (2008)
16. J. Bernardo, M.F.F. Nave, C. Giroud et al., Ion temperature and toroidal rotation in JETs low torque plasmas. Rev. Sci. Instrum. **87**, 11E557 (2016)
17. Atomic data and analysis structure, see http://www.adas.ac.uk (c) 1995–2018 the ADAS project. Accessed 14 Nov 2018
18. A. Whiteford, M.G. von Hellermann, CXSFIT manual (2008)
19. M.G. von Hellermann, W. Mandl, H.P. Summers et al., Visible charge exchange spectroscopy at JET. Rev. Sci. Instrum. **61**, 3479 (1990)
20. M.G. von Hellermann, The SOS project (simulation of spectra code as modelling tool for active spectroscopy and its role for ITER diagnostics), ITER report ITER_D_U8M7S7 (2018)
21. O. Marchuk, The status of atomic models for beam emission spectroscopy in fusion plasmas. Phys. Scr. **89**, 114010 (2014)
22. O. Marchuk, D.R. Yu Ralchenko, Schultz, Non-statistical population distributions for hydrogen beams in fusion plasmas. Plasma Phys. Control. Fusion **54**, 095010 (2012)
23. M. Bandyopadhyay, M.J. Singh, C. Rotti, A. Chakraborty, R. Hemsworth, B. Schunke, Beamline optimization for 100keV diagnostic neutral beam (DNB) injector for ITER, in *23rd IEEE/NPSS Symposium on Fusion Engineering* (2009), pp. 1–4 (IEEE Transactions on Plasma Science **38**, 242 (2010))
24. C. Rotti, A. Chakraborty, M. Bandyopadhyay et al., Exploring the engineering performance limits of DNB, in *24th IAEA Fusion Energy Conference-IAEA CN-197, ITR/P5-43* (2012)
25. B. Schunke, I. Ahmed, M. Bandyopadhyay et al., Status of the negative ion based diagnostic neutral beam for ITER, in *24th IAEA Fusion Energy Conference CN-197, ITR/P1-03* (2012)
26. Ph Mertens, M. Silz, Radial profiles of atomic deuterium measured in the boundary of TEXTOR 94 with laser-induced fluorescence. J. Nucl. Mater. **241–243**, 842–847 (1997)
27. Ph Mertens, A. Pospieszczyk, Radial and spectral profiles of atomic deuterium in front of a limiter in TEXTOR 94: Results of

laser-induced fluorescence at Lyman-α. J. Nucl. Mater. **266–269**, 884–889 (1999)
28. A. Huber, S. Brezinsek, Ph Mertens et al., A new radiation-hard endoscope for divertor spectroscopy on JET. Fusion Eng. Des. **88**, 1361–1365 (2013)
29. A. Huber, S. Brezinsek, Ph Mertens et al., Development of a mirror-based endoscope for divertor spectroscopy on JET with the new ITER-like wall. Rev. Sci. Instrum. **83**, 10D511 (2012)
30. Ph Mertens, H. Altmann, T. Hirai et al., A bulk tungsten divertor row for the outer strike point in JET. Fusion Eng. Des. **84**, 1289–1293 (2009)
31. Ph Mertens, D.A. Castaño Bardawil, T. Baross et al., Status of the R&D activities to the design of an ITER core CXRS diagnostic system. Fusion Eng. Des. **96–97**, 129–135 (2015)
32. M. Walsh, P. Andrew, R. Barnsley, et al., ITER diagnostic challenges, in *IEEE/NPSS 24th Symposium on Fusion Engineering, SO3A-1* (2011)
33. V.S. Voitsenya, A.F. Bardamid, A.I. Belyaeva et al., Diagnostic first mirrors for burning plasma experiments, in *Advanced Diagnostics for Magnetic and Inertial Fusion*, ed. by P.E. Stott, A. Wootton, G. Gorini, E. Sindoni, D. Batani (Springer, Berlin, 2002), pp. 285–294
34. K. Vukolov, A. Bardamid, A. Gorshov et al., Experimental study of first mirror degradation under influence of ITER relevant conditions, in *Advanced Diagnostics for Magnetic and Inertial Fusion*, ed. by P.E. Stott, A. Wootton, G. Gorini, E. Sindoni, D. Batani (Springer, Berlin, 2002), pp. 299–302
35. J.H. Weaver, H.P.R. Frederikse, Optical properties of metals and semiconductors, in *Handbook of Chemistry and Physics*, 79th edn, ed. by D.R. Lide (CRC Press, Boca Raton) (1998–1999), pp. 12–141
36. A. Litnovsky, Yu. Krasikov, M. Rasinski et al., First direct comparative test of single crystal rhodium and molybdenum mirrors for ITER diagnostics. Fusion Eng. Des. **123**, 674–677 (2017)
37. Yu. Krasikov, A. Panin, A. Litnovsky, Ph Mertens, M. Schrader, Specific design and structural issues of single crystalline first mirrors for diagnostics. Fusion Eng. Des. **124**, 548–552 (2017)
38. Ph. Mertens, S. Dickheuer, Yu. Krasikov, A. Krimmer, A. Litnovsky, Ch. Linsmeier, M. De Bock, F.Le Guern, On the use of rhodium mirrors for optical diagnostics in ITER, submitted to Fusion Eng. Des. (Proceedings of 30th SOFT conference, 2018)
39. L. Marot, G. De Temmerman, P. Oelhafen et al., Rhodium coated mirrors deposited by magnetron sputtering for fusion applications. Rev. Sci. Instrum. **78**, 103507 (2007)
40. M. Joanny, J.M. Travère, S. Salasca, L. Marot et al., Achievements on engineering and manufacturing of ITER first-mirror mock-ups. IEEE Trans. Plasma Sci. **40**, 692–696 (2012)
41. L. Marot, G. Arnoux, A. Huber et al., Optical coatings as mirrors for optical diagnostics. J. Coat. Sci. Technol. **2**, 72–78 (2015)
42. I. Orlovskiy, A. Alekseev, E. Andreenko et al., Thermal testing of the first mirror unit mock-up for H-alpha and visible spectroscopy in ITER. Fusion Eng. Des. **96–97**, 899–902 (2015)
43. V.S. Voitsenya, A.F. Bardamid, A.J.H. Donné, Experimental simulation of the behaviour of diagnostic first mirrors fabricated of different metals for ITER conditions. Open Phys. J. **3**, 23–54 (2016)
44. M. Clever, G. Arnoux, N. Balshaw et al., A wide angle view imaging diagnostic with all reflective, in-vessel optics at JET. Fusion Eng. Des. **88**, 1342–1346 (2013)
45. V. Kotov, Report ITER C4T/09/37/OLT CHD/Diagnostic, Fusion Eng. Des. **86**, pp. 1583–1586 and ITER_D_AHMDZX (2011)
46. V. Kotov, Engineering estimates of impurity fluxes on the ITER port plugs. Nucl. Fusion **56**, 106027 (2016)
47. B. Kim, C. Seon, S.-G. Oh et al., A passive mitigation strategy of impurity deposition on the first mirrors using duct with baffles: a case study at a port of KSTAR with in-situ deposition monitoring. Fusion Eng. Des. **129**, 269–276 (2018)
48. S. Grigoriev, V. Tanchuk, K. Senik et al., Thermal analysis for optimization of the optical duct of the ITER core CXRS diagnostics. Fusion Eng. Des. **96–97**, 790–794 (2015)
49. D.A. Castaño Bardawil, Yu. Krasikov, A. Panin, O. Neubauer, W. Biel, Fast shutter concepts for the new ITER core CXRS upper port plug baseline considering the actuator located inside and outside the port plug. Fusion Eng. Des. **88**, 2073–2076 (2013)
50. D.A. Castaño Bardawil, Ph Mertens, G. Offermanns et al., Design overview of the ITER core CXRS fast shutter and manufacturing implications during the detailed design work. Fusion Eng. Des. **96–97**, 746–750 (2015)
51. S. Friese, A. Panin, Yu. Krasikov, Ph Mertens, J. Aßmann, Experimental and numerical studies of the shutter dynamics for the ITER core CXRS diagnostic. Fusion Eng. Des. **123**, 722–726 (2017)
52. F. Leipold, R. Reichle, C. Vorpahl et al., Cleaning of first mirrors in ITER by means of radio frequency discharges. Rev. Sci. Instrum. **87**, 11D439 (2016)
53. E.E. Mukhin et al., First mirrors in ITER: material choice and deposition prevention/cleaning techniques. Nucl. Fusion. **52**, 013017 (2012)
54. A.V. Rogov, YuV Kapustin, A.G. Alekseev, Application of the penning discharge for cleaning mirrors in optical diagnostics of the ITER. Instr. Exp. Tech. **58**, 161–166 (2015)
55. L. Moser, L. Marot, R. Steiner et al., Plasma cleaning of ITER first mirrors. Phys. Scr. **T170**, 014047 (2017)
56. O. Marchuk, C. Brandt, A. Pospieszczyk et al., Emission of fast hydrogen atoms at a plasma-solid interface in a low density plasma containing noble gases. J. Phys. B: At. Mol. Opt. Phys. **51**, 025702 (2018)
57. S. Dickheuer, O. Marchuk, C. Brandt et al., In-situ measurements of specular reflectance of metallic mirrors at the H_α line in a low density Ar-H plasma. Rev. Sci. Instr. **89**, 063112 (2018)
58. M.W. Thompson, The energy spectrum of ejected atoms during the high energy sputtering of gold. Philos. Mag. **18**, 377–414 (1968)
59. Ph Mertens, P. Bogen, B. Schweer, Study of sputtered Si-atoms in front of a TEXTOR-limiter by laser-induced fluorescence, in Proceedings 20th EPS Conference on Controlled Fusion and Plasma Physics, vol. 17C/III (1993), pp. 1123–1126
60. B. Weinhorst, U. Fischer, G. Grossetti, Ph Mertens, D.A. Castaño Bardawil, ITER core CXRS diagnostic: Assessment of different optical designs with respect to neutronics criteria. Fus. Eng. Des. **123**, 927–931 (2017)
61. B. Weinhorst et al., Shut-down dose rate analyses for the ITER electron cyclotron-heating upper launcher. Fus. Eng. Des. **89**, 1899–1904 (2014)
62. C. Guillemaut, R.A. Pitts, A.S. Kukushkin, M. O'Mullane, Radiative power loading in the ITER divertor. Fus. Eng. Des. **86**, 2954–2964 (2011)

Journal of Fusion Energy (2019) 38:283–290
https://doi.org/10.1007/s10894-019-00220-w

ORIGINAL RESEARCH

Present Status of ITER Neutron Diagnostics Development

L. Bertalot[1] · V. Krasilnikov[2] · L. Core[3] · A. Saxena[4] · N. Yukhnov[2] · R. Barnsley[1] · M. Walsh[1]

Published online: 11 July 2019

Abstract

Neutron diagnostic systems are needed to monitor some important ITER parameters, such as fusion power, power density or ion temperature. The aim of this paper is to describe various systems, such as neutron cameras, internal and external flux monitors, activation system and spectrometers. Also, the current status of in situ neutron calibration strategy is reported. ITER is currently under construction in Cadarache (France) and neutron diagnostic systems are progressing from design stage to manufacturing.

Keywords ITER · Neutron diagnostic · Fusion reactor · Fusion power measurement

Introduction

ITER is a tokamak-based nuclear reactor aimed to demonstrate the feasibility of nuclear fusion power, designed to produce 500 MW for extended periods of time [1]. The success of ITER will come through its ability to produce large amount of energetic neutrons, i.e. fusion power. These measurements are carried out by means of the neutron diagnostic systems. Measurements of neutron emission and fusion power are essential for achieving ITER goals, in particular the power gain factor, Q, related to the reactor performance as well for plasma control, machine protection and for plasma optimization. The work described below is related to the project, integration, engineering support needed for the neutron diagnostics. In order to meet ITER programme goals, a wide range of measurement will be performed, with the associated requirements shown in Table 1.

ITER will have a harsh environment due to very high neutron fluxes, which are expected to count up to 10^{14} n/cm^2 s with energy of 14 MeV. Plasma radiation and nuclear heating will induce elevated temperatures (up to 1000°) on the plasma facing components and magnetic fields up to 6 T will be present, along with significant electromagnetic noise due to the auxiliary radiofrequency heating system. In these severe environment conditions, a set of neutron diagnostic systems is planned to be installed to fulfil measurement requirements [2]. They will play a key role to measure the neutron emission and thus the fusion power, to provide both machine protection and plasma optimization, along with physics-dedicated measurements.

Disclaimer: The views and opinions expressed herein do not necessarily reflect those of the ITER Organization.

✉ L. Bertalot
Luciano.Bertalot@iter.org

1 ITER Organization, Route de Vinon-sur-Verdon, CS 90 046, 13067 St. Paul Lez Durance Cedex, France

2 Tokamak Energy Ltd, 120A Olympic Avenue, Milton Park, Abingdon OX14 4SA, UK

3 Arkadia Technology, 255, rue Paul Langevin, 13 100 Aix-en-provence, France

4 FIRCROFT, Lingley House, 120 Birchwood Point, Birchwood Blvd., Warrington, Cheshire WA3 7QH, UK

ITER Neutron Diagnostic

Neutron diagnostic in ITER are charged to measure key parameters like fusion power, fuel ratio n_T/n_D, profile for neutron or alpha sources, neutron fluence on the plasma facing components and energy distribution of reacting particles, like thermal and fast ions. These parameters play a key role not only in physics measurement, but also in machine protection and plasma optimization [3]. Figure 1 summarizes the relationships between the primary measurement parameters and the various systems. Diagnostics can have a primary or a supplementary role in performing measurements: "primary" means that the diagnostic is

Table 1 ITER measurement requirements

Measurement	Parameter	Range	Resolution		Accuracy (%)
			Time (ms)	Spatial	
Neutron flux and emissivity	Fusion power density	1 kW/m^3–15 MW/m^3	1	a/10	10
	Bulk ion profile	10^{14}–6 × 10^{18} n/(m^{3}s)	1	a/10	10
	Fusion power	100 kW–1.5 GW	1	Integral	10
	Total neutron flux	10^{14}–7.5 × 10^{20} n/s	1	Integral	10
Confined alphas and fast ions	Alpha density profile	(0.1–2) × 10^{18} α/m^3	100	a/10	20
Fuel ratio in plasma core	nT/nD, core	0.01–10	100	r/a < 0.85 a/10	20
Ion temperature profile	Ti, core	0.5–40 keV	100	r/a < 0.85 a/30	10

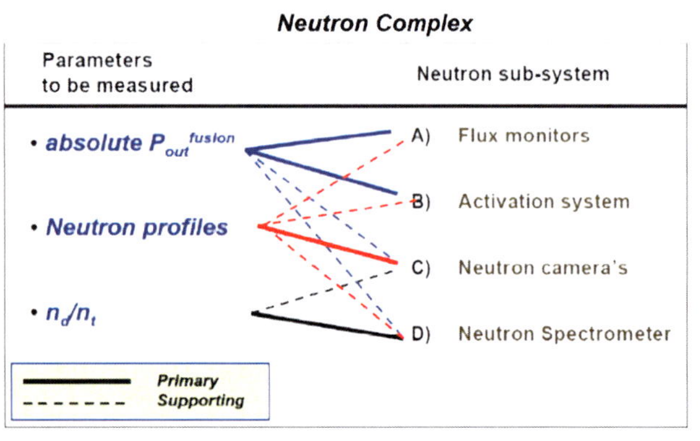

Fig. 1 Parameters to be measured associated to the corresponding neutron diagnostic sub-systems

dedicated to perform that measurement, while "supplementary" signifies that the system has some limitations in measurement, and thus it cannot assume the complete task, but just validate or calibrate primary systems.

In ITER a wide neutron emission rate is expected covering 7 orders of magnitude from 10^{14} n/s during DD plasma to 10^{21} n/s with DT plasma. The required accuracy for neutron emission is within 10% with a temporal resolution of 10^{-3} s [4].

Each set of neutron diagnostic can be identified with its dedicated measurement:

- Fusion power: it is associated with neutron flux monitors, plus a foil-based activation system;
- Bulk ion profiles: cameras installed in vertical and equatorial ports;
- Fuel ratio: dedicated spectrometers.

These systems will be described in details in the following paragraphs.

Neutron Flux Monitors

Neutron flux monitors must carry on the measurement of total neutron emission during all phase of ITER operation. In order to fulfil the task, fission chambers with different uranium content (both ^{235}U and ^{238}U) are employed at multiple locations, operating in counting, Campbell and current modes which allow covering the whole ITER dynamic range for neutron emission. Combinations of these detectors provide the measurement of the global neutron strength, used to estimate the relative fusion power [5]. Because this measurement is strictly bound to fusion power measurement and total neutron yield emission, redundancy is considered a key factor to assure the measurement in ITER project.

Neutron Flux Monitors (NFM) is the main system used for all phases of ITER operation. Its fission chambers containing different ^{235}U content are placed in different toroidal locations, both in-vessel and ex-vessel. In order adjust the sensitivity to DD or DT plasma neutrons, different moderating materials have been employed. NFM modules in equatorial ports #1 and #7 are dedicated to DD reactions for low fusion power phase [6]. NFMs installed in ports #8 and #17 will operate during DT phase for full fusion power neutron emission. The most sensitive fission chambers are the ones installed in NFM EQ#1 Port Plug inside the Vacuum Vessel. Moderating materials chosen are beryllium, graphite and B_4C, with different thickness optimized to be sensitive in the wide range of the neutron emission. To achieve a correct measurement, various techniques will be used for signal acquisition and data processing during the different operational modes: pulse counting is preferred for low flux levels, while during intermediated flux levels Campbell mode is chosen, and high flux operations require current mode. Figure 2 shows

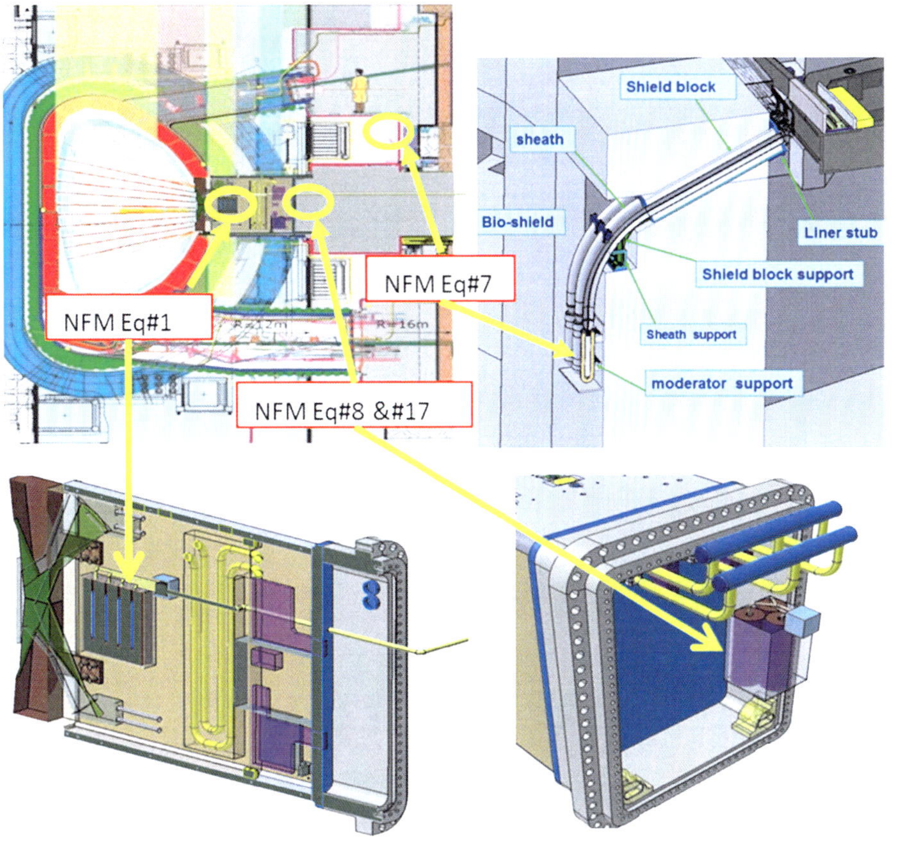

Fig. 2 Model of the different NFM modules and their location inside ITER machine

the distribution of the different NFM modules in the tokamak.

Micro-fission Chambers (MFC) are special pencil-size detector, with very low ^{235}U content (~ 10 mg for each detector). They are designed to measure high levels of neutron flux corresponding to high fusion power levels during DT operation phase. MFC will be installed in Port (#12) and Port (#3) outboard, between the blankets modules and the inner shell of the Vacuum Vessel [7]. At each port position, two sets made of MFC detectors are installed. One set is located at the upper part of the vacuum vessel (VV) whereas the other set in the VV lower part. Locations for MFCs are shown in Fig. 3 and have been chosen after investigation through neutron transport calculation, demonstrating that the average measurement coming from two MFCs modules is not sensitive to variations of plasma shape and position. Studies have been necessary to solve several issues like the rate of gas leakage inside the vacuum vessel (gas leakage could cause a plasma disruption and cannot be pumped out promptly), standing the elevated operational temperature, tolerances between MFC components and blanket modules, or assuring the proper functioning of MFCs during disruption events. Currently this diagnostic system is stepping in manufacturing phase for some of its components and test on prototypes are being performed [8]. The issue of gas leakage could affect also NFM in Equatorial Port Plug #1 and Diverton Neutron Flux Monitor, which are installed inside the vacuum vessel.

Another diagnostic system in charge of total emission rate is the Divertor Neutron Flux Monitor (DNFM). It is made up by three modules, each one composed by two detectors units (Fig. 4 on the left): the first detector units have three fission chambers containing different amounts of ^{238}U (500, 50 and 5 mg), while the second detector unit is equipped with three fission chamber with ^{235}U (50, 10 and 2 mg), all separated by ceramic insulators. It should be noted that the sensitivity and the energy response of ^{238}U FC will change during the ITER operation time because of uranium transmutation into ^{239}Pu, which has a large fission cross section under thermal neutron flux. Another important effect to be taken into account is the burn-up or consumption of ^{235}U due to the high neutron flux under the divertor dome. All six fission chambers have similar design and material and are wrapped in two cylinders to divide the ^{235}U chambers from the ^{238}U ones [9]. Combinations of these two kinds of fission chambers with various sensitivities will provide the measurement of neutron flux in the wide dynamic range of seven orders of magnitude, with the required time resolution. DNFM modules are located at three different toroidal positions in the lower part of ITER Vacuum Vessel, under the domes of Divertor cassettes. Because of its location, DNFM must withstand a

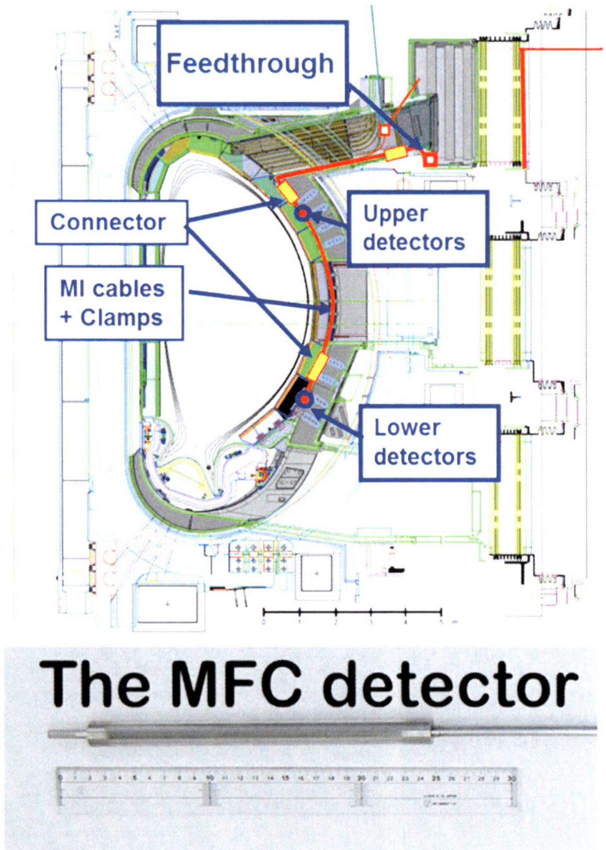

Fig. 3 Above: locations of MFC modules. Below: MFC detector unit

Neutron Activation System

Neutron Activation System (NAS) is based on pneumatic post system that transfers foil samples of different materials to a number of dedicated irradiation ends in different positions. It aims to provide time integrated measurement of the neutron fluence and the total fusion power through neutronic calculation. Encapsulated foil samples are irradiated near ITER First Wall and then are transferred via pneumatic apparatus, to counting station located in the Tritium Building, where the neutron induced gamma emission of the foil is measured. Irradiation ends are positioned between various blanket modules, on the vacuum vessel inner wall as well in the upper and equatorial ports. Figure 5 shows the distribution of NAS irradiation ending and the transfer lines leading to counting stations. Due to their location, irradiation ends must endure temperatures up to 800° because of plasma radiation, along with nuclear heating and strong electromagnetic forces [10].

NAS foils will be made of different materials such as ^{28}Si, ^{63}Cu or ^{48}Ti, in order to maximize the detection of neutron flux. On the other hand, detailed design and selection of gamma spectrometry system is still object of discussion. For the moment the choice is on NaI or HPGe detectors, but other candidates such as scintillator, gas chambers or semiconductor detectors are not excluded.

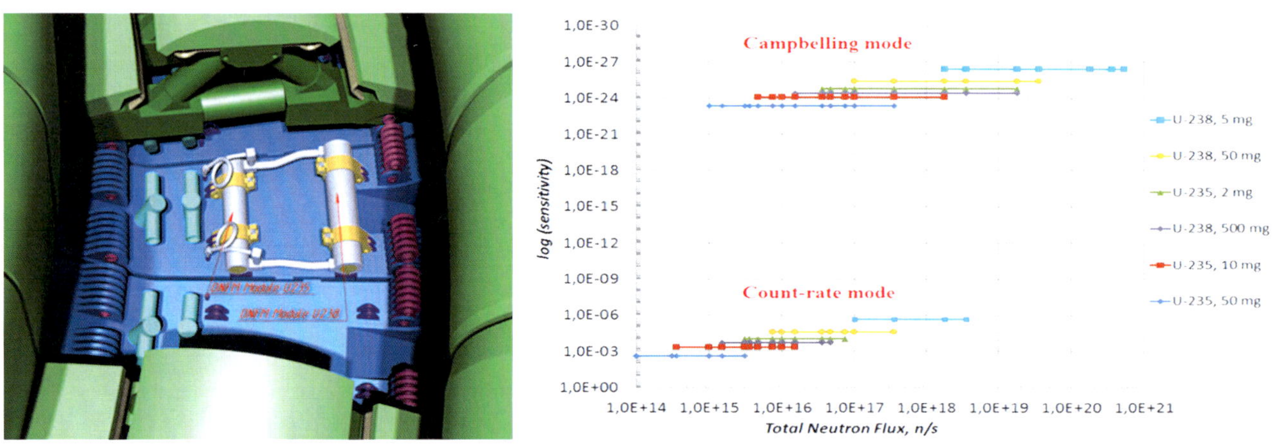

Fig. 4 Left: model of DNFM module. Right: performance of DNFM in measuring neutron flux, in both count-rate and Campbelling mode

challenging environment, with high nuclear and radiation power loads, high transient magnetic field with associated high forces and moments acting on the DNFM structures. DNFM modules must be cooled actively through the steel cylinder housing which allow the temperature control. Fission chambers and cables have to withstand temperatures up to 350° and the structure must avoid water cooling leakages.

Diagnostic Dedicated to Neutron Emission Profile

Two systems are mainly dedicated to neutron emission profile: Vertical Neutron Camera (VNC) and Radial Neutron Camera (RNC). Both of these systems represent multichannel neutron collimators aiming to characterise plasma neutron source. They are designed to measure the time-resolved neutron/alpha emission profile for both DD

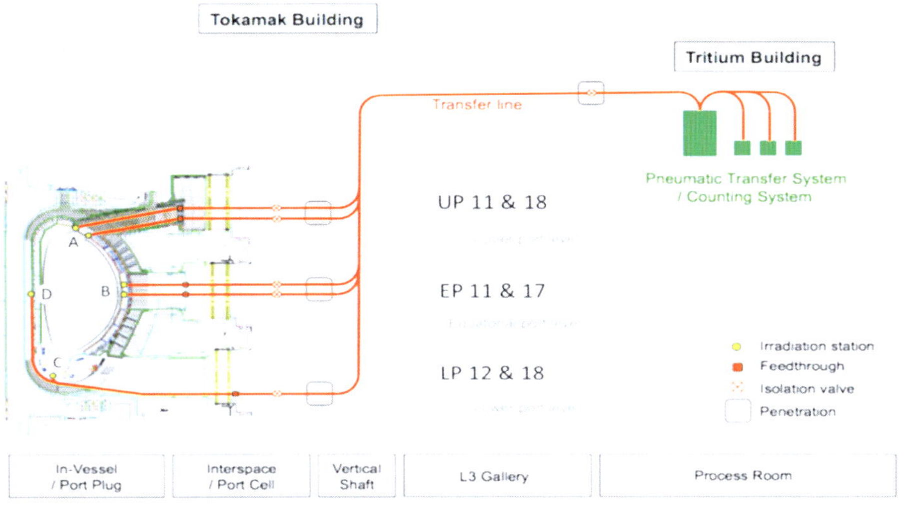

Fig. 5 Location of NAS irradiation ends, with transfer lines until Tritium Building where counting stations are located

and DT plasma, along with the evaluation of fusion power density, alpha source density, neutron and alpha emissivity profile, ion temperature profile. Total neutron flux and fusion power are provided by these systems as supplementary measurements.

Vertical Neutron Camera is made up by two sets each one including six detectors modules with vertical collimators alignment, one in lower port 14 and the other in upper port 18 (Fig. 6). Detector modules for VNC are composed by two CVD diamonds and two fission chambers with high purity ^{238}U [11]. Fission chambers are sensitive only to fast neutrons with a threshold of 1 MeV, while diamond detectors measure neutron energy distribution, particularly the 12–16 MeV range. VNC will have collimator to observe directly the plasma and the expected neutron flux is estimated to be from 10^5 up to 10^{10} n/cm^2 s. Nevertheless it is mandatory to develop CVD detector to improve their lifetime under neutron irradiation. In order to avoid the risk of gas leakage in the torus vacuum, detectors will be enclosed in a secondary vacuum container.

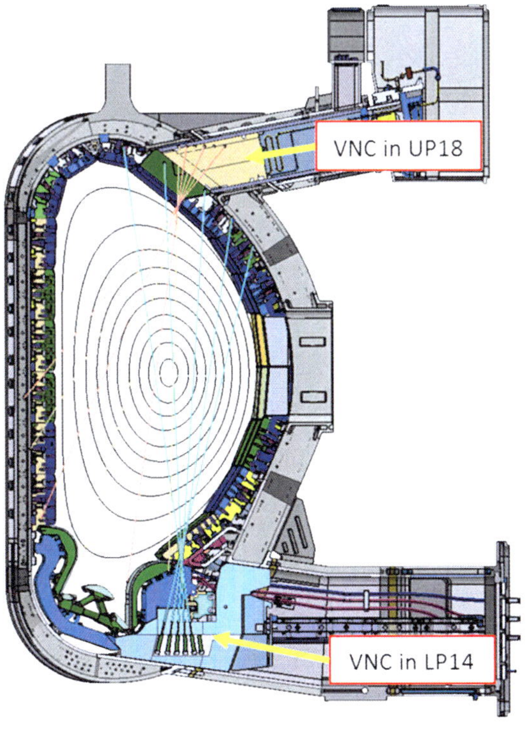

Fig. 6 Location of VNC in the upper and lower ports. Lines of sight are also shown with the effective cross section of the tokamak

The radial profile of neutron emissivity is measured by the RNC, which is composed by a fan-shaped array of collimators, with detectors arranged at two locations: in the port plug for plasma edge measurement, and in the port inter-space to observe the plasma core [12]. Both systems are located in Equatorial Port #1. Each collimator corresponds to a detector unit composed by a set of detectors designed to be active in a wide range of neutron emissivity. There will be in total 20 collimators for the ex-port systems and 6 for the in-port one, all with radial lines of sight on the poloidal plane. Also for RNC detector modules the environmental conditions to withstand are harsh, with dose up to 10^5 Gy, local temperature up to 250° and strong magnetic fields from few mT to few T (from in-port to ex-port region). Detector modules should be sensitive both to DD and DT neutrons, with a dynamic range of four orders of magnitude and a threshold able to reject neutrons from scattering events. It is foreseen to use plastic or glass scintillators with neutron/gamma discrimination capability as ex-port-plug detectors, and a combination of matrix diamond detectors and ^{238}U fission chambers for the in-port-plug ones. With these detectors, RNC will be able to provide some neutron energy spectrum as secondary measurement (Fig. 7).

Fig. 7 Arrangement of Radial Neutron Camera, with the two detector locations. Lines of sight are also shown with effective plasma cross section

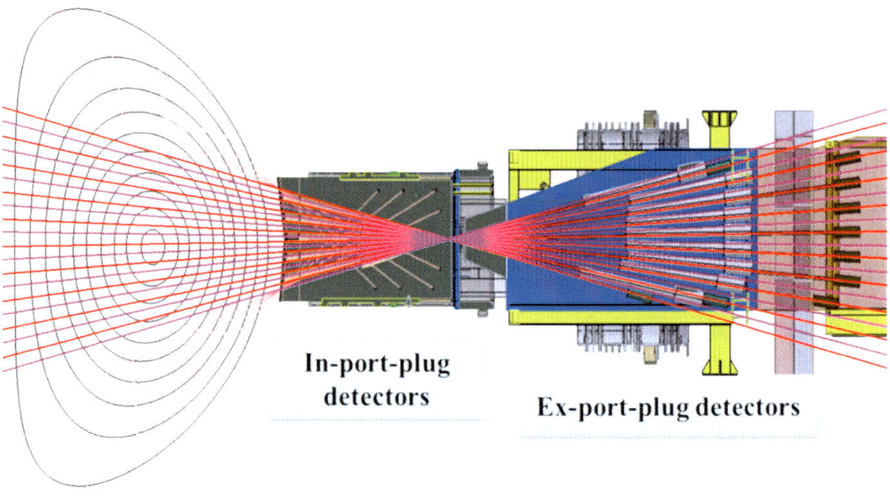

Spectrometers

The main goal of neutron spectroscopy in ITER is the measurement of the fuel ratio n_T/n_D from 2.5 and 14 MeV neutron emissions. The main diagnostic dedicated to this parameter is the High Resolution Neutron Spectrometer (HRNS), which is currently under development like all others. It will be composed by a single collimating channel for 2.5 and 14 MeV neutrons combined, and installed in the port cell of equatorial port #1. The actual design is composed by three sub-systems, based on thin-foil proton recoil (TPR) and time of flight (ToF) techniques [13]. The TPR detects the elastic neutron-proton scattering, generating a signal proportional to proton recoil energy. ToF system will include both conventional ToF and back scattering ToF, arranged as shown in Fig. 8. The first ToF is composed by fast organic scintillators placed in coincidence with plastic scintillators and coupled with photomultipliers. Back-scattering ToF relies on backward elastic scattering of neutrons on D; the measurement is achieved using a plastic scintillator for the direct neutron and an ordinary scintillator, placed in the backward direction, for the scattered neutron [14]. This technique allows the simultaneous measurement of 2.5 and 14 MeV neutrons. This diagnostic system is able to measure fuel ion ration spanning over an order of magnitude in fusion power $40 < P_{fus} < 500$ MW. The most common issues to be faced, along with the intense magnetic field, are high electromagnetic stress, high temperatures and nuclear loads and radiation hardness to assure front-end electronics lifetime (20 years) and stability.

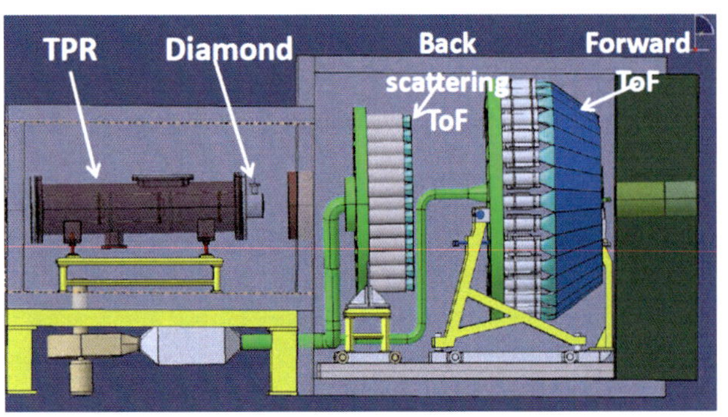

Fig. 8 General block-diagram and configuration of the HRNS

The Tangential Neutron Spectrometer (TNS) is a diamond detector based spectrometer with tangential lines of sight dedicated to studies of fast ion behaviour [15]. It will measure with fast charge-exchange atoms directly in the non-nuclear phase and spectra of DT neutrons in the nuclear phase [16]. Its measurements, dedicated to physics studies, will allow determining fast ions' distribution function in the presence of MHD events like sawteeth, fishbones, NTMs and TAE modes. The TNS diagnostic will have three lines of sight corresponding to the three detector module. These three lines of sight allow observing the plasma core, 1/4 of minor radius and 1/2 of minor radius.

Gamma Spectroscopy

ITER is also equipped with gamma-ray diagnostics, whose primary application is the time-resolved spatial measurement of gamma rays. From this measurement it is possible to derive alpha particles birth profile by means of reaction $D(t,\gamma)^5He$, profile of confined alpha particles, fast ion confinement effects during ICRF or NBI heating, bremsstrahlung hard X-ray profile for runaway electrons.

The diagnostic system dedicated to gamma measurement is the Radial Gamma Ray Spectrometer, which is composed by an array of detectors facing the plasma through 6 lines of sight filled with LiH to attenuate neutron background [17]. The detectors of RGRS are located behind the RNC ones in the Port Cell EQ#1. The detectors employed are $LaBr_3$ scintillator crystals and HPGe. The choice of these scintillators has been driven by need of operation at high and low counting rate, radiation hardness, high energy resolution, fast time resolution (10 ms) and an accuracy of 20% for runaway electrons detection. Detector size has been optimized to maximize gamma ray efficiency and minimize the collection of neutrons. The RGRS can provide information on both runaway electrons (in the energy range 1–100 MeV and current range 1–15 MA) and confined alpha. Fusion confined alphas will interact with fuel ions (D, T) producing gamma rays with specific line emission. Detection of these lines provides information on the reacting alpha-ions population [17]. It can also allow some supplementary measurement on impurities concentration, alpha energy spectrum and neutron/alpha source profile. In order to allow functioning of PMTs coupled with scintillators, inside the detector box the magnetic field must be reduced to 0.1 mT in the same direction of the PMT axis and to 0.02 mT in the orthogonal direction. To achieve this condition, a magnetic shielding is mandatory: the shielding has been designed to represent a series of layers of different materials, in order to reduce gravity load.

Neutron Calibration Project

ITER requires accuracy on fusion power evaluation inferior to 10% and thus the relation between neutron diagnostic output and the total neutron emission rate must have a higher accuracy level. For this reason, a neutron in situ calibration must be performed in order to evaluate the absolute calibration factor for each neutron flux monitor. Each measurement can be schematized as:

$$\text{Measurement} = Y_n \times \text{Det. Response Function} \times \text{Calibration Factor}$$

where the detector response function should be determined by the diagnostic developer, while the calibration faction can only be measured in situ [18].

For in situ calibration it is necessary to have a neutron source, either a neutron generator or a radioactive source, able to travel around the vessel or to be placed at different toroidal and poloidal positions. The source should calibrate the most sensitive detectors of the neutron diagnostics, like Neutron Flux Monitors (in equatorial ports #1 and #7), Divertor Neutron Flux Monitor, Micro-fission Chambers and Neutron Activation System. Three calibration campaigns are foreseen for ITER:

1. "Partial campaign", with a ^{252}Cf source or DD Neutron Generator with strength of 10^9 n/s, at the end Machine Assembly Phase II, just before hydrogen plasma. Only a partial scan in toroidal and poloidal positions will be carried out by moving the source between Eq. Ports 18, 1 and 2. The aim of this short calibration is the cross-check between experimental data and neutron transport simulations, in order to optimize the latter ones before the following stages.
2. 2.5 MeV neutron campaign, full toroidal/poloidal scans (360° and 180°, respectively) with either ^{252}Cf source or DD neutron generator, scheduled before helium operation phase.
3. 14 MeV neutron campaign (full toroidal/poloidal scans), with DT neutron generator, just before the Full Power Operation Phase

All other less sensitive detectors will be calibrated during ITER DD and DT phases, using plasma reference shots and previously calibrated systems, in particular NAS as a reference for cross-calibration [19].

In order to provide an uncertainty of 1% in the counting statistics, the radiation time should be adequate. Also, other accuracies must be fulfilled like a instability for neutron generator less than 5%, reduced neutron emission anisotropy, low hardware scattering. Moreover, concerning the 2.5 MeV neutron generators, these requirements are not met by any commercial device. Its feasibility is currently under study.

Another major issue for in situ neutron calibration is the transportation system to move the source inside the Vessel. Several aspects must be taken into account when choosing a transportation system, such as the total gravity load that the system can withstand, the neutron scattering, the Vessel portion that the system could cover and availability of the system according to the ITER general schedule. In the ideal case, in situ calibration would have a dedicated transporter to be deployed from upper ports and with an arm able to reach the chosen calibration points. Currently this system is not available. The actual remote handling available system is the so-called In-Vessel Transporter (IVT), mainly dedicated to blanket operations, relying on a rail at equatorial level and with a 360° configuration. Its scattering contribution and the absorbed dose during calibration campaign are still being investigated, as well as its availability. A possible alternative is the exploitation of the In-Vessel staging, with an appropriate tool to handle and move the source/neutron generator. The last option foresees to employ if available, the Multi-Purpose Deployer. This RH system can easily manage the weight of the neutron source plus a mandatory interface device and can access all the Vacuum Vessel surfaces by switching the equatorial ports. The main drawback of this option is the huge impact on

ITER schedule. All the listed option are currently being evaluated in terms of scattering contribution, vessel coverage, post access, rescue possibility and schedule effect.

A major role in calibration is played by neutron transport calculations. Codes are an important tool to evaluate parameters like irradiation time, scattering from RH mean and surrounding environment, effects of operational condition on detection efficiency [20]. These evaluations are currently ongoing to establish the final calibration strategy.

Conclusions

ITER is now stepping into the construction phase. All the listed neutron diagnostic systems have been conceptually designed and part of them is undergoing manufacturing, with the beginning of operations expected in 2025. This paper provides an overview of the neutron diagnostic systems and the associated engineering issues. The current in situ calibration strategy is under development.

Acknowledgements The authors wish to thank ITER Diagnostics Team and the colleagues of ITER Members of Japan, Korea, China, Europe, United State of America, India and Russian Federation for useful discussions and collaboration. The work described was supported and carried out by ITER Organization together with ITER Members and Domestic Agencies of Japan, Korea, China, Europe, United State of America, India and Russian Federation. Part of this activity was performed within the frame work of the European Fusion Development Agreement.

References

1. ITER website https://www.iter.org/
2. G. Vayakis, E.R. Hodgson, V. Voitsenya, Chapter 12: generic diagnostic issues for a burning plasma experiment. Fusion Sci. Technol. **53**, 699–750 (2008)
3. M. Sasao, T. Nishitani, A. Krasilnikov, S. Popovichev, V. Kiptily et al., Fusion product diagnostics. Fusion Sci. Technol. **53**, 604–639 (2008)
4. L. Bertalot, et al., Fusion neutron diagnostics on ITER tokamak, in *Proceedings of the FNDA-2011* (2011)
5. M. Sasao et al., Issues on the absolute neutron emission measurement at ITER. Plasma Fusion Res. **8**, 2402127 (2013)
6. J. Yang et al., Fusion neutron flux monitor for ITER. Plasma Sci. Technol. **10**, 141 (2008)
7. M. Ishikawa et al., Design of microfission chambers for ITER operations. Rev. Sci. Instrum. **79**, 10E507 (2008)
8. A. Encheva, et al., Structural integrity report for ITER microfission chambers, ITER Doc. ITER_D_3TDURL (2011)
9. Y. Kashchuk, A. Krasilnikov, D. Prosvirin, A conceptual project for a divertor monitor of the neutron yield in the ITER. Instrum. Exp. Technol. **49**, 179–186 (2006)
10. M.S. Cheon et al., In-Vessel design of ITER diagnostic neutron activation system. Rev. Sci. Instrum. **79**, 10E505 (2008)
11. A. Krasilnikov, C.I. Walker, Y. Kashchuk, D. Prosvirin, A multichannel neutron collimator for the ITER tokamak. Instrum. Exp. Technol. **47**, 139 (2004)
12. L. Petrizzi et al., Neutronic design of the ITER radial neutron camera. Fusion Sci. Eng. **82**, 1308–1314 (2007)
13. G. Ericsson, et al., Conceptual design of a high resolution neutron spectrometer system for ITER, in *26th IAEA Fusion Energy Conference, 2016* (2016)
14. C. Hellesen et al., Conceptual design of a BackTOF neutron spectrometer for fuel ion ratio measurements at ITER. Nulc. Fusion **57**, 066021 (2017)
15. V. Krasilnikov, et al., Conceptual design of the ITER tangential neutral spectrometer, in *High Temperature Plasma Diagnostic Conference, 2018* (2018)
16. T. Kormilitsyn, et al., Assessment of the fast particle spectra for tangential spectrometer for H/He and DT ITER operation, in *44th EPS Conference on Plasma Physics, 2017* (2017)
17. M. Nocente, M. Tardocchi et al., Conceptual design of the Radial Gamma Ray Spectrometer system for α particle and runaway electron measurements at ITER. Nucl. Fusion **57**, 076016 (2017)
18. L. Bertalot, et al., A strategy for calibrating the neutron system at ITER, in *Proceedings of the 35th Plasma Physics Conference, 2008* (2008)
19. M. Sasao, L. Bertalot, M. Ishikawa, S. Popovichev, Strategy for the absolute neutron emission measurement on ITER. Rev. Sci. Instrum. **81**, 10D329 (2010)
20. M. Ishikawa, T. Kondoh, K. Takeda, K. Itami, Neutron Transport Analysis of the process affecting the in situ calibration of ITER In-Vessel neutron flux monitor equipped with a micro-fission chamber system. Plasma Fusion Res. **11**, 1402118 (2016)

Journal of Fusion Energy (2019) 38:291–298
https://doi.org/10.1007/s10894-018-0206-x

ORIGINAL RESEARCH

Energetic Particle Diagnostics in Present Tokamaks and Challenges Towards a Burning Plasma

M. Nocente[1]

Published online: 30 November 2018

Abstract
The paper presents a brief overview of energetic particle diagnostics in fusion plasmas. Differences between techniques used in nuclear and non-nuclear machines are in particular underlined and the challenges faced by fast ion diagnostics in a burning plasma are discussed.

Keywords Energetic particles · Diagnostics · Fusion plasmas

Introduction

Fast ions are an ubiquitous minority in tokamak plasmas [1]. They are generated by the auxiliary heating systems, for example neutral beam injection (NBI) or radio-frequency (RF) waves, or they can be born as products of the main fusion reactions, such as ^{3}He ions from the $d(d,n)^3$He reaction in deuterium (D) plasmas or α particles from $t(d,n)^4$He in deuterium–tritium (DT). In both cases, depending on the size of the device, their typical energies are in the range from some tens of keV to a few MeV, and therefore much larger than the bulk plasma temperature values found in present day tokamaks, say between 1 and 10 keV. In order to sustain the fusion burn, fast ions must be well confined to ensure that they transfer their energy to the bulk plasma and can keep it at the required temperature. A number of confined and lost fast ion diagnostics has therefore been developed to study the confinement of energetic ions in fusion plasmas.

Depending on the energy range, diagnostics based on different physics principles are preferred, where a major distinction is between systems for non-nuclear and nuclear devices. In the former case, the machine is not designed to confine fusion products, which are predominantly lost. Hence, the fast ion energies of interest are only those from the auxiliary heating systems, which are typically of few hundreds of keV or less. A very powerful method to study ions in this range comes from the application of charge exchange based reactions, for example by means of neutral particle analyzers or the fast ion D_α diagnostics (see "Charge Exchange Based Systems" section). Nuclear devices can instead confine also fusion products and, in general, ions up to a few MeV. In this case, diagnostics that exploit spontaneous nuclear reactions that involve the MeV range fast ions are of most importance, also because charge exchange based methods present some limitations (see below). Often, nuclear devices have a larger size, as this helps at ensuring the confinement of MeV range ions. This is the design choice behind the ITER tokamak, the next generation experimental fusion reactor presently under construction in the south of France. Good fast ion confinement could also be achieved by keeping the device more compact and exploiting higher magnetic fields, but this line of development is less pursued nowadays. For this reason, in the following we will often use the terms large and medium size tokamaks in lieu of nuclear and non nuclear devices.

The paper is organized as follows. Charge exchange based diagnostics are presented in "Charge Exchange Based Systems" section , where a description of the two most widespread systems is provided. "Elements of Fast Ion Physics" section presents a very concise introduction to the physics of fast ions in tokamaks, which is used to further motivate the need for energetic ion diagnostics, in

✉ M. Nocente
massimo.nocente@mib.infn.it

1 Dipartimento di Fisica "G. Occhialini", Università di Milano Bicocca, Milan, Italy

particular lost fast ion measurements. Nuclear based diagnostics are finally presented. For conciseness reasons, the principles and limitations of each diagnostic system are discussed only, and little is said on the required hardware. The interested reader is referred to the comprehensive review paper [2] for more details.

Charge Exchange Based Systems

In a charge exchange (CX) reaction between a fast ion (F) and a thermal neutral (T) an electron is transferred from T to F. The end result is the birth of a fast neutral particle and a thermal ion. From the diagnostic point of view, since the electron mass is much lighter than that of the ion, the fast neutral preserves all of the kinematic properties of the fast ion it comes from. This means that a measurement of the fast neutral population born from charge exchange reactions between F and T is a probe of the fast ion distribution function. The main advantage of neutral particles is that they are unconfined by the tokamak magnetic fields and, provided they do not get ionized as they travel from the plasma core to the edge, they can be measured by diagnostic probes at the edge.

The probability for a charge exchange reaction to occur depends on the reaction cross section σ. This is a monotonically decreasing function of the F–T relative energy E_{rel}, which has a maximum of $\sigma \approx 10^{-18}$ m^2 when $E_{rel} = 0$ keV [3] and gently falls down to $\sigma \approx 10^{-19}$ m^2 when E_{rel} is of a few tens of keV, followed by a sharp drop at higher energies. For this reason, CX reactions can be used to probe ions with energies up to about 100 keV, unless specific techniques are devised to overcome this limitation (see the paragraph below).

When CX reactions occur, they most of the times lead to a ground state neutral, which can be analyzed by a so called Neutral Particle Analyzer (NPA). In some cases, however, CX reactions can lead to a neutral in an excited state and, whenever this occurs, one can measure instead the spectroscopic properties of the light emitted by the de-excitation of the neutral as a way to probe its kinematic properties. This is the principle behind Fast-ion D_α (FIDA) measurements.

Neutral Particle Analyzer

NPAs measure the kinematic properties of neutrals born from CX reactions between fast ions in the core and a neutral. In one of the most conventional setups, a neutral that reaches the first wall enters a stripping cell, where it gets ionized (see Fig. 1). Parallel electric and magnetic fields ($E\|B$)are used to bend the trajectories of the ions

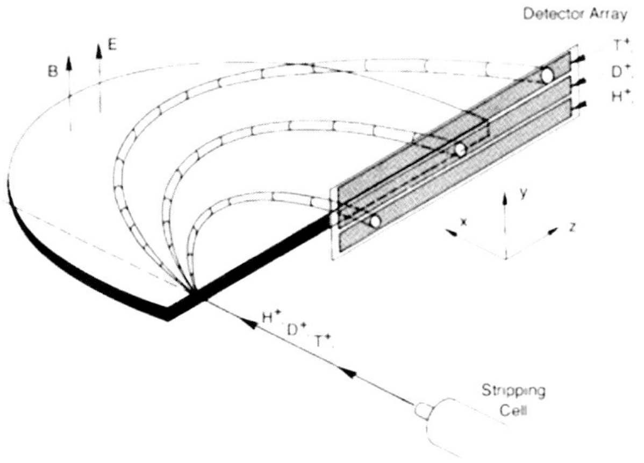

Fig. 1 Schematics of the $E\|B$ neutral particle analyzer. Neutrals enter a stripping cell, where they get ionized. An electric and a magnetic field are used to separate ions with different charge to mass ratios and energies on a detector array. Figure taken from [4]

after the stripping cell on a charged particle detector array. Ions with different charge to mass ratios are displaced on horizontal stripes at a different vertical position so that, for example, different hydrogen isotopes (protons, deuterons, tritons) can be separated. The horizontal position of the ions along each stripe is instead indicative of the ion energy, as the magnetic field bends the ion trajectory with a curvature radius that is proportional to its linear momentum [4]. The distribution function of the fast ions is finally measured by the intensity pattern on each stripe. In a more recent application, the $E\|B$ setup has been in some cases replaced by solid state detectors [5] which, albeit offering a poorer energy resolution, have the advantage of being compact and more practical for mid-size devices.

NPA measurements have been fundamental for studies of fast ions generated by the auxiliary heating systems in mid size devices, both neutral beam injection and radio frequency heating [4]. However, they also present some limitations when projected to large, high density devices. One comes from the location where the neutralisation takes place. Fusion products, for instance, are predominantly born in the core and the neutrals of interest are those born in charge exchange reactions in the core. In large size devices, however, spontaneous charge exchange reactions predominantly occur at the edge, so that the signal from the fast ions is weakened compared to that from thermal ions. On one hand this happens because there is a much larger population of neutrals at the edge compared to the core. On the other hand, as the size of the machine becomes larger, a fast neutral born from a F–T charge exchange has an increased probability to be re-ionised along its path from the core to the edge.

Another limitation comes from the MeV energy range of the fast ions in large size tokamaks. As mentioned above, charge exchange cross sections have a sharp drop as the relative F–T energy exceeds few hundred keVs, which implies that an F–T charge exchange is less likely to occur for ions in the MeV range. This limitation can be partly compensated by active charge exchange, i.e. by inducing on purpose additional charge exchange reactions with the help of a diagnostic beam [3].

Finally, the charged particle detectors must be screened from the radiation background induced by the fusion neutrons, which are more abundant by some 2–3 orders of magnitude in a large size compared to medium size tokamak. Albeit these limitations, an NPA system is foreseen at ITER [6], where however it is expected to mostly contribute to the determination of the D–T fuel io ratio, rather than to measurements of the fast ion distribution function.

Fast Ion D_α

The FIDA diagnostic technique [7] is an application of charge exchange recombination spectroscopy and is based on measurements of the visible light emitted from a Balmer-α ($n = 3$ to $n = 2$) transition that occurs when an F–T charge exchange reaction leads to a fast neutral in the $n = 3$ excited state in a deuterium plasma. The FIDA emission is distinguished from the significantly more abundant Balmer-α emission arising from thermal CX reactions by its low and high wavelength Doppler shift. Figure 2 shows a diagram of the different components that make up the spectrum of the Balmer-α emission. The centre of the spectrum is at a wavelength of 656.1 nm, where the thermal (halo) component, that comes from CX reactions between the neutral beam and the surrounding thermal deuterons [3], is predominant. There is then a beam emission spectral feature, which is born instead from CX reactions between the full, half and one third energy components of the beam ions and thermal neutrals. The weaker FIDA component emerges from the relatively flat visible bremsstrahlung background on a logarithmic scale and consists of high and low energy tails that are born from the Doppler shift of the D_α light along the viewing chord and due to the fast ion motion. The shape of the tail is thus representative of the projection of the fast ion velocity distribution on the line of sight. As FIDA measures visible light, it is relatively easy to devise a relatively large number of lines of sight in a practical installation. This means that the diagnostic can be set up to measure the fast ion profile by means of space resolved measurements of the D_α light intensity. Recent measurements of this type have clearly shown that the fast ion profile is often non classical [8] and have contributed to understanding many aspects of the underlying mechanisms that affect fast ion transport in tokamaks (see also "Elements of Fast Ion Physics" section). Another possibility offered by the simultaneous use of many viewing chords is to observe the same plasma volume but with different angles with respect to the magnetic field. In this case, a time dependent high definition tomographic image of the fast ion distribution function, which can be resolved both in the energy and in the pitch angle dimensions, can be obtained [9]. This is one of the most recent and promising methods to study the distribution function of the fast ions and its evolution during the discharge.

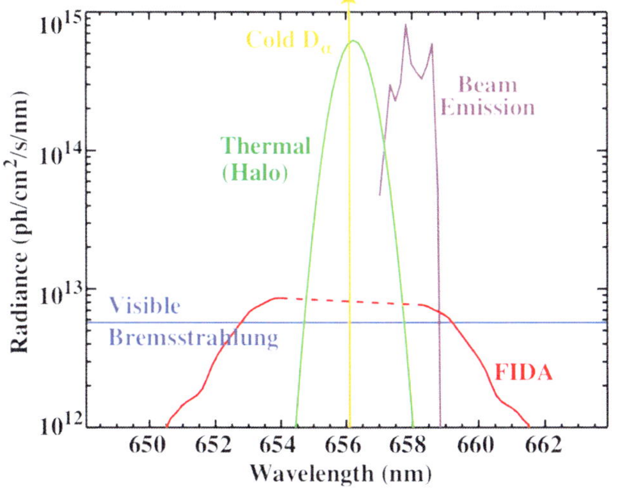

Fig. 2 Components of the spectrum from the D_α visible light born from CX reactions. The FIDA contribution is the low and high energy tails that barely emerge from the background due to the visible bremsstrahlung. Reprinted from [7], with the permission of AIP publishing

While the power of the FIDA diagnostic as one of the leading methods to study fast ions in mid-size devices is undoubted, its application to large size tokamaks is more questionable. Being a CX based method, the technique suffers from the same limitation of the NPA, namely the fact that CX cross sections experience a sharp drop for MeV range ions. Another more specific limitation comes from the extent of the visible bremsstrahlung, which increases as n_e^2 (n_e is here the plasma electron density) and must be subtracted from the data to highlight the weak FIDA tails. In large size devices this background might be simply too high compared to the signal, leading to large uncertainties in the subtraction that might make it very difficult to exploit the fast ion information from the FIDA tail.

Elements of Fast Ion Physics

The interest in fast ion measurements is not only motivated by the need to assess the effectiveness of the auxiliary heating systems, but is also (and, nowadays, primarily) driven by the need to investigate the physics effects that energetic ions can induce in a thermonuclear plasma. The field of fast ion physics in tokamak plasmas is very broad and of great relevance in view of ITER and DEMO, especially with respect to the largely unknown effects that α particles may have in a reactor grade plasma. A basic introduction to the field is presented in [10], while a recent, more comprehensive review can be found in [1] and references therein. Here we only limit ourselves to introducing the key ideas.

Alfvèn waves are naturally found in a magnetized plasma. Of particular relevance is the shear Alfvèn wave, which is a transverse low frequency wave that propagates along the magnetic field. A typical analogy is with the wave that is generated when a string of a guitar is plucked. The wave propagates along the string, which has a transverse motion with respect to the direction of propagation. In a similar way, shear Alfvèn waves propagate along the magnetic field direction as the field line itself has a transverse oscillation. The propagation velocity is given by the Alfvèn speed v_A, which depends on the magnetic field and the i-th plasma ion species mass m_i and density n_i as

$$v_A = \frac{B}{\sqrt{\mu_0 \sum n_i m_i}} \tag{1}$$

Here μ_0 is the vacuum permeability and the summation is taken over all the plasma ions. The corresponding wave frequency ω is given by

$$\omega = k_{\|} v_A \tag{2}$$

where $k_{\|}$ is the component of the wave vector parallel to the magnetic field.

In a uniform plasma, the group velocity $v_g = \frac{\partial \omega}{\partial k}$ derived from 2 equals the phase velocity $v_f = \frac{\omega}{k}$ and the wave packet is dispersionless. In a tokamak, instead, due the periodic nature of the system, $k_{\|}$ has a discrete spectrum and is given by

$$k_{\|} = \frac{(n - mq)}{R} \tag{3}$$

where n, m are integer numbers, q is the safety profile and R is the major radius. As both q and R are position dependent, shear Alfvèn waves in a tokamak are dispersive. The actual Alfvèn spectrum in a realistic geometry is rather complicated and, like in many periodic systems as solid lattices, is characterized by bands, where a large number of frequencies can reside (continuum), and gaps where no frequency can occur. Waves with frequencies in the continuum, even if excited, are easily damped due to their large dispersive nature and are typically of no concern.

A fundamental theoretical and later experimental observation has however been that relatively dispersionless, discrete modes can reside naturally within the gaps. In general terms, these discrete Alfvèn modes (also called Alfvèn Eigenmodes, AEs) are born from "defects" of the magnetic system such as, for instance, the coupling of nearby poloidal harmonics, a non monotonic q profile etc. and can be excited by resonances between the orbital frequencies of the fast ions and that of the modes. As the discrete modes are relatively dispersionless, they are also long lived and can modify the fast ion transport significantly, which in turn changes the excitation dynamics of the discrete modes in a non linear fashion. A significant fraction of the present research on energetic ions is devoted to studies of the non linear interplay between the excitation of AEs and the fast ion population and the effect that this has on the confinement of the suprathermal ions. When projected to a reactor, in the most benign case, AEs may simply result in a change of the energetic ion profile, with little effect on the performance of the fusion burn. In the worst case scenario, instead, the interactions with the AEs might be so severe to lead to a substantial loss of the energetic ions, with detrimental effects on the fusion performance.

There are then other aspects of the broad field of fast ion physics which are currently a subject of active research and which we just limit to mention without further discussion. These are the interaction between the energetic ions and magnetic perturbations at the edge [11, 12]; the role that the fast ions might have on suppressing the plasma turbulence [13]; the additional instabilities that the fast ions can drive when their pressure is a non negligible fraction of the thermal plasma pressure [14] and many others [1]. From the diagnostic point of view, all of these phenomena provide additional motivations to study the properties of the confined fast ions, but also point to the need to measure the properties of those fast ions that are unconfined by the instabilities. This can be accomplished by dedicated fast ion loss detectors (FILD).

Lost fast ion detectors

In every fusion plasma there is always a fraction of fast ions which are unconfined, even in the absence of MHD perturbations. This so called "prompt" lost ion population is due to energetic ions that are born on unconfined orbits. A typical example comes from the application of NBI. In this case, a fraction of neutrals gets ionized in the scrape off layer on unconfined orbits and their measurement can be used, for example, to verify models of neutral beam heating. In addition to prompt losses, there can then be additional losses caused by the excitation of a broad range on fast ion driven instabilities, some of which have been briefly mentioned above, or by intrinsic and extrinsic magnetic perturbations. In both cases, it is of great interest to get access to the lost ion velocity space, as this is often of help for identifying and characterizing the cause of the losses.

Experimentally, the detection of lost fast ions is done by a thin (a few microns) scintillator plate, which is protected by a graphite collimator and is installed at the edge of the plasma, often on a manipulator [15, 16]. The collimator lets a fraction of the lost ions impinge on the scintillator plate. Depending on their gyroradius and pitch angle (see Fig. 3), the escaped ions hit the plate in a different position, so that the spatial map of the light from the scintillator plate is an image of the lost fast ion velocity space. In practical implementations, the light is measured by a CCD camera installed in vicinity of the probe. Most often, a beam splitter is also used to transport a fraction of the light a few meters away from the diagnostics by means a bundle of optical fibers. Here the light is read by a set of photomultiplier tubes and a digitizer at a sampling rate of the order of 1 MHz. In this way, the frequency spectrogram of the light from the lost ions in a specific region of the plate can be built and losses at characteristic frequencies are identified [17, 18]. This very powerful diagnostic method, which has been made possible by the recent use of scintillators with some ns decay time for the FILD, is of paramount importance for fast ion research as it allows obtaining a direct identification of the modes responsible for the observed losses. This is done by comparing the frequencies measured with the FILD spectrogram with those derived by the spectrogram of the signal from conventional magnetic diagnostics, for example Mirnov coils.

While the role of FILD in advancing fast ion physics in mid-size tokamaks is undoubted, the application of this diagnostics to a large power tokamak is particularly challenging. A first difficulty comes from the severe thermal loads to which the detector head is exposed. In present tokamaks, where the discharge duration is of few seconds and the thermal loads at the edge are acceptable, time resolved FILD measurements can be taken over the full pulse duration. In high performance devices, instead, the FILD system can be exposed to the plasma only for few seconds and must be later cooled down. This means that the system must be installed on a reciprocating probe and a cycle must be envisaged where the detector is inserted periodically in and out of the plasma. Hence, continuous measurements for the full duration of the discharge are likely not to be feasible. An additional challenge is represented by the neutron and gamma-ray background. At ITER, for instance, the neutron flux at the first wall is 10^{14} n/s/cm^2 and an equivalent gamma-ray flux can be expected. Neutron and gamma-rays provide a background signal on the scintillator plate, which must be taken into account. Another aspect is the nuclear load that the radiation induces on the whole FILD system so that, even when the detector is not inserted in the plasma, active cooling might be necessary. In general, the implementation of a FILD in a high performance tokamak is certainly a very complex engineering task but, at least at ITER, it is presently attempted given the highly valuable information that this system can provide to understand fast ion physics.

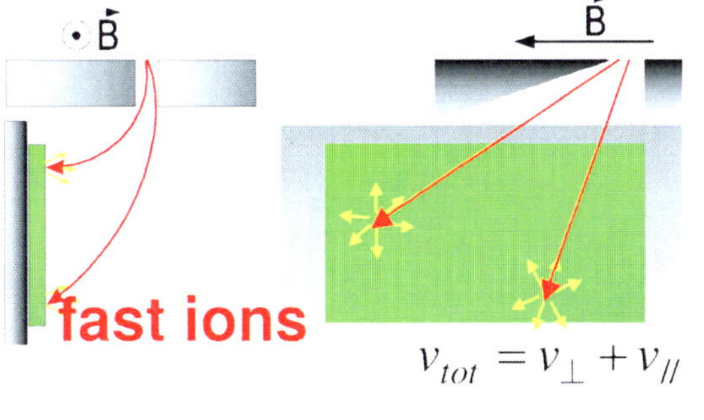

Fig. 3 Schematics of the detection principle of the fast ion loss detector. A collimator lets a fraction of the lost ions impinge on a rectangular scintillator. The impact position of the lost ions on the scintillator is a measurement of their pitch angle and gyroradius. Figure taken from [19]

Fast Ion Diagnostics in Large Tokamak Devices

As discussed in the previous paragraphs, most of the fast ion diagnostics adopted in mid-size tokamaks face experimental challenges when applied to large size devices. On the other hand, a large size tokamak is a strong source of nuclear radiation, as it produces neutrons by the main fusion reactions and gamma-rays by nuclear reactions between ions in the MeV range and impurities. This implies that diagnostics based on nuclear radiation measurements are naturally suited for large tokamak devices. Besides, these diagnostic methods do not suffer from most of the limitations presented by other systems such as, for example, the high thermal load at the edge for FILD detectors or the order of magnitude decrease of the cross section for charge exchange based systems. On the other hand, nuclear radiation measurements are indirect and careful modeling is often required to gain quantitative information, particularly on the fast ions. In this case, information is often contained in the energy spectrum of the radiation, which has motivated the development of different neutron and gamma-ray spectroscopy techniques, particularly with the advent of large tokamaks in the last 20 years.

In this paragraph we present a short summary of gamma-ray spectroscopy measurements. Concerning neutron measurements, we leave the reader to the dedicated lecture of reference [20] for an introduction to this field in fusion plasmas. Here, we only explicitly mention about the possibility to measure α particles by neutron spectroscopy in a nuclear grade DT plasma.

Gamma Ray Spectroscopy Measurements

In fusion plasmas gamma-rays are produced by two different processes. One is the main fusion reactions, which often have a weak gamma-ray emission branch, with a probability in the range between 10^{-4} and 10^{-5} to occur compared to their corresponding neutron emission process. An example is $t(d,\gamma)^5$He in a plasma of deuterium–tritium [21], which emits gamma-rays with an average energy of about 17 MeV and is the weaker branch of the more intense $t(d,n)^4$He reaction. For fast ion applications, however, the so called two step processes are of most interest. In these reactions a fast ion interacts with an impurity (for example, ^{9}Be from the first wall at JET and ITER), which leads to the production of a heavy nucleus in an excited state (first step). This in turn decays on a time scale faster than nanoseconds, leading to the emission of a gamma-ray (second step). A prominent example is the $^9\mathrm{Be}(\alpha,n\gamma)^{12}C$ reaction between α particles and impurities and which will be one of the main diagnostic methods to study confined αs in deuterium–tritium plasmas at JET and ITER. Besides $^9\mathrm{Be}(\alpha,n\gamma)^{12}\mathrm{C}$ for α particle studies, a large number of reactions is available and has been exploited to study fast ions in deuterium plasmas [22]. This often leads to a rather complex spectrum, an example of which is shown in Fig. 4 for an experiment at JET on ^{3}He ion acceleration by radiofrequency waves [23].

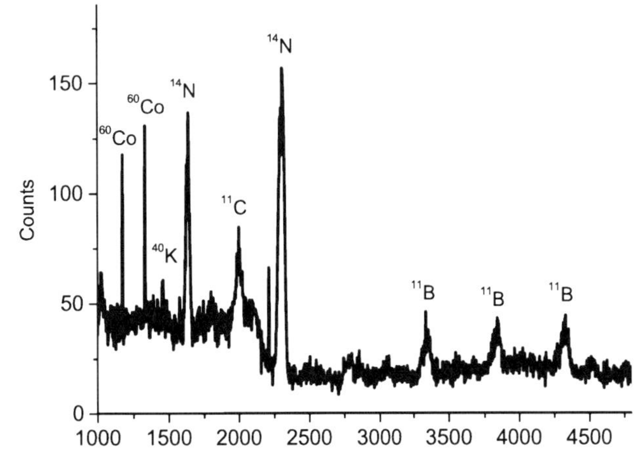

Fig. 4 Gamma-ray emission spectrum measured at JET with a high purity Germanium detector in a experiment aimed at accelerating ^{3}He ions with radio-frequency heating [23]. Peaks produced by the $^{12}C(^3\mathrm{He},p\gamma)^{14}$N, $^{12}\mathrm{C}(^3\mathrm{He},n\gamma)^{11}$C and $^{12}\mathrm{C}(^3\mathrm{He},p\gamma)^{11}$B reactions between ^{3}He ions and ^{12}C impurities are marked in the figure by the corresponding excited state that leads to the emission. Additional peaks coming from a ^{60}Co and ^{40}K radioactive sources put in vicinity of the detector (and used for energy calibration) are shown. Lines from nuclear reactions induced by the ^{3}He ions are broadened by the kinematics

Depending on the accuracy of the measurements, information with a different level of detail can be extracted from the gamma-ray spectrum. At the most basic level, one can identify the energies of the visible peaks as they indicate the excited nuclear state responsible for the emission and, hence, the reaction that has produced it. As many reactions require a minimum fast ion energy to occur, the identification of lines at a characteristic energy testifies the presence of fast ions with energies exceeding the production thresholds of the corresponding reactions in the plasma. This is useful when additional heating is used, for example to help understanding the propagation of radiofrequency waves in the plasma and their coupling to the different possible resonant species [24]. More detailed information can be extracted by detailed modeling of the gamma-ray emission spectrum and depending on the complexity (and knowledge) of the reaction cross sections. For example, in some cases peaks at different energies can be emitted by the same excited nucleus and their ratio yields information on the fast ion distribution function [25]. If instruments with a sufficiently high energy resolution are used, measurements can resolve the detailed shape of the emission peaks and use those to constrain the fast ion distribution function [26]. In the most advanced application, high resolution gamma-ray data can be combined with neutron spectroscopy measurements to infer the full fast ion distribution function in the MeV range in some scenarios [27].

As for neutron measurements, gamma-ray spectrometers can be placed along multiple sightlines in a so called camera arrangement. A practical limitation in this case comes from the limited space often available, which has however been recently overcome by the development of compacts detectors that combine good energy resolution, high counting rate capability and insensitivity to magnetic fields [28, 29]. After tomographic inversion, this type of measurements allows obtaining the spatial profile of the gamma-ray emission in the poloidal plane, from which, in most cases, the profile of different fast ions in the plasma can be extracted. This is done by integrating the spectrum in the energy bands that correspond to peaks from reactions induced by different ions, for example $^{12}\mathrm{C}(d,p\gamma)^{13}$C and $^9\mathrm{Be}(\alpha,n\gamma)^{12}$C when deuterons and ^{4}He ions are simultaneously heated by radio-frequency waves to mock up α particles in a ^{4}He plasma [30].

Given the rich blend of information that, albeit indirect, can be extracted from the gamma-ray emission spectrum of a fusion plasma, and the increased gamma-ray fluxes expected in a fusion reactor compared to present day devices, gamma-ray based diagnostics are considered of primary importance for fast ion studies in next step devices. At ITER, for instance, a radial camera of gamma-ray spectrometers [31] is presently under design, primarily to study confined α particles in deuterium–tritium plasmas as well as runaway electrons by measurements of their bremsstrahlung emission spectrum [32].

α Particle Measurements by Neutron Spectroscopy in DT Plasmas

An interesting, albeit less widespread, application of neutron spectroscopy is the potential this technique has to measure α particles in a DT plasma by the observation of a distinctive low amplitude, high energy tail in the 14 MeV neutron spectrum. This so-called alpha knock-on component of the neutron emission (AKN) is a signature of the way α particles slow down in a DT plasma and is predominantly born from rare, head-on nuclear elastic collisions between α particles and the bulk plasma. When a fusion reaction occurs, an α particle with a mean energy of 3.5 MeV is born, which subsequently slows down in the bulk plasma. The dominant slowing down mechanism is the cumulative effect of multiple small angle Coulomb collisions, but there is also some small chance that α particles will experience nuclear elastic collisions (NECs) with deuterium and tritium ions as they thermalize. As NECs are particularly effective at transferring energy to the bulk species, they induce a deviation to the Maxwellian distribution of the bulk plasma at high energies. In other words, at a detailed level, the distribution function of the bulk species is made of two components: a Maxwellian at the bulk plasma temperature T_i, say $T_i < 20$ keV depending on the detailed plasma conditions, and an additional high energy component, whose features reflect NECs with α particles [33, 34]. This high energy deviation of the reactants distribution function manifests itself in the fusion neutron spectrum which, besides featuring a main peak at 14 MeV born from thermal reactions, additionally has a low amplitude (about 10^{-4} and 10^{-5} compared to the main thermal peak), high energy component, so called AKN [35]. The number of events in the tail is proportional to the α particle pressure, while the shape of the AKN component depends on the details of the corresponding energetic tail of the fuel ion distribution function, which in turn reflects the way α particles slow down in the plasma [36]. The AKN component is thus potentially a diagnostic of confined α particles.

Albeit very small, this component has been experimentally observed by summing neutron spectra from several deuterium–tritium discharges in high power experiments at the Joint European Torus. The measurement was made possible by the use of a dedicated spectrometer based on the magnetic proton recoil principle (MPR), which allows obtaining a dynamic range of about 5 orders of magnitude with a signal to noise ratio of 10^4:1 [37]. In principle, the same technique could be used to diagnose confined α particles at the ITER tokamak. However, this would require the development of an MPR type spectrometer which, due to its massive radiation shielding and relatively bulky dimensions, is presently not considered for installation.

Conclusion

A concise overview of the most widespread diagnostics for energetic ions in nuclear and non-nuclear grade tokamak plasmas has been presented, including a brief introduction to some aspects of the physics of fast ions in tokamak plasmas. Charge exchange based systems play a dominant role in non-nuclear devices, but they experience some limitations when applied to high performance tokamaks. Here, nuclear diagnostics based on neutron and gamma-ray measurements play a major role, as they are naturally targeted for studies of confined energetic ions in the MeV range. The importance of lost fast ion measurements for advancing fast ion physics has been also highlighted and the special challenges that these measurements face when applied to a reactor relevant plasma have been discussed.

References

1. N.N. Gorelenkov, S. Pinches, K. Toi, Energetic particle physics in fusion research in preparation for burning plasma experiments. Nuclear Fusion **54**(12), 125001 (2014)
2. D. Moseev, M. Salewski, M. Garcia-Munoz, B. Geiger, M. Nocente, Fast-ion diagnostics for magnetically confined plasmas. Rev. Mod. Plasma Phys. **2**, 7 (2018)
3. I.H. Hutchinson, *Principles of Plasma Diagnostics* (Cambridge University Press, Cambridge, 2002)
4. A. Kislyakov, A. Donne, L. Krupnik, S. Medley, M. Petrov, Chapter 8: Particle diagnostics. Fusion Sci. Technol. **53**(2), 577–603 (2008)
5. K. Shinohara, D. Darrow, A. Roquemore, S. Medley, F. Cecil, Solid state neutral particle analyzer array on National Spherical Torus Experiment. Rev. Sci. Instrum. **75**(10), 3640–3642 (2004)
6. V. Afanasyev, F. Chernyshev, A. Kislyakov, S. Kozlovski, B. Lyublin, M. Mironov, A. Melnik, V. Nesenevich, M. Petrov, S.Y. Petrov, Neutral particle analysis on ITER: present status and prospects. Nuclear Instrum. Methods Phys. Res. Sect. A Accel. Spectrom. Detect. Assoc. Equip. **621**(1–3), 456–467 (2010)
7. W. Heidbrink, Fast-ion d_α measurements of the fast-ion distribution. Rev. Sci. Instrum. **81**(10), 10D727 (2010)
8. W. Heidbrink, C. Collins, M. Podestà, G. Kramer, D. Pace, C. Petty, L. Stagner, M. Van Zeeland, R. White, Y. Zhu, Fast-ion

transport by Alfvén eigenmodes above a critical gradient threshold. Phys. Plasmas **24**(5), 056109 (2017)
9. M. Salewski, B. Geiger, A. Jacobsen, P.C. Hansen, W. Heidbrink, S.B. Korsholm, F. Leipold, J. Madsen, D. Moseev, S.K. Nielsen et al., High-definition velocity-space tomography of fast-ion dynamics. Nuclear Fusion **56**(10), 106024 (2016)
10. W. Heidbrink, Basic physics of Alfvén instabilities driven by energetic particles in toroidally confined plasmas. Phys. Plasmas **15**(5), 055501 (2008)
11. M. Garcia-Munoz, S. Äkäslompolo, O. Asunta, J. Boom, X. Chen, I. Classen, R. Dux, T. Evans, S. Fietz, R. Fisher et al., Fast-ion redistribution and loss due to edge perturbations in the ASDEX Upgrade, DIII-D and KSTAR tokamaks. Nuclear Fusion **53**(12), 123008 (2013)
12. M. Garcia-Munoz, S. Äkäslompolo, P. de Marne, M. Dunne, R. Dux, T. Evans, N. Ferraro, S. Fietz, C. Fuchs, B. Geiger et al., Fast-ion losses induced by ELMs and externally applied magnetic perturbations in the ASDEX Upgrade tokamak. Plasma Phys. Controll. Fusion **55**(12), 124014 (2013)
13. J. Garcia, T. Görler, F. Jenko, Isotope and fast ions turbulence suppression effects: consequences for high-β ITER plasmas. Phys. Plasmas **25**(5), 055902 (2018)
14. W. Heidbrink, N. Gorelenkov, M. Murakami, Beam-driven energetic particle modes in advanced tokamak plasmas. Nuclear Fusion **42**(8), 972 (2002)
15. S. Zweben, Pitch angle resolved measurements of escaping charged fusion products in TFTR. Nuclear Fusion **29**(5), 825 (1989)
16. M. Garcia-Munoz, H.-U. Fahrbach, H. Zohm, A.U. Team, Scintillator based detector for fast-ion losses induced by magnetohydrodynamic instabilities in the ASDEX upgrade tokamak. Rev. Sci. Instrum. **80**(5), 053503 (2009)
17. M. García-Muñoz, H.-U. Fahrbach, S. Günter, V. Igochine, M.J. Mantsinen, M. Maraschek, P. Martin, P. Piovesan, K. Sassenberg, H. Zohm, Fast-ion losses due to high-frequency MHD perturbations in the ASDEX upgrade tokamak. Phys. Rev. Lett. **100**(5), 055005 (2008)
18. A.U. Team, M. García-Muñoz, N. Hicks, R. van Voornveld, I. Classen, R. Bilato, V. Bobkov, M. Bruedgam, H.-U. Fahrbach, V. Igochine et al., Convective and diffusive energetic particle losses induced by Shear Alfvén waves in the ASDEX Upgrade tokamak. Phys. Rev. Lett. **104**(18), 185002 (2010)
19. Garcia-Munoz, M.: Fast response scintillator based detector for MHD induced energetic ion losses in ASDEX Upgrade, Ph.D. thesis, Max-Planck-Institut für Plasmaphysik (2006)
20. M. Cecconello et al., The 2.5 MeV neutron flux monitor for MAST. Nuclear Instrum. Methods Phys. Res. Sect. A Accel. Spectrom. Detect. Assoc. Equip. **753**, 72–83 (2014)
21. M. Nocente, J. Källne, M. Salewski, M. Tardocchi, G. Gorini, Gamma-ray emission spectrum from thermonuclear fusion reactions without intrinsic broadening. Nuclear Fusion **55**(12), 123009 (2015)
22. V. Kiptily, F. Cecil, O. Jarvis, M. Mantsinen, S. Sharapov, L. Bertalot, S. Conroy, L. Ingesson, T. Johnson, K. Lawson et al., γ-ray diagnostics of energetic ions in JET. Nuclear Fusion **42**(8), 999 (2002)
23. M. Tardocchi, M. Nocente, I. Proverbio, V. Kiptily, P. Blanchard, S. Conroy, M. Fontanesi, G. Grosso, K. Kneupner, E. Lerche et al., Spectral broadening of characteristic γ-ray emission peaks from $^{12}C(^{3}He,p\gamma)^{14}N$ reactions in fusion plasmas. Phys. Rev. Lett. **107**(20), 205002 (2011)
24. Y.O. Kazakov, J. Ongena, J. Wright, S. Wukitch, E. Lerche, M. Mantsinen, D. Van Eester, T. Craciunescu, V. Kiptily, Y. Lin et al., Efficient generation of energetic ions in multi-ion plasmas by radio-frequency heating. Nature Phys. **13**(10), 973 (2017)
25. M. Tardocchi, M. Nocente, G. Gorini, Diagnosis of physical parameters of fast particles in high power fusion plasmas with high resolution neutron and gamma-ray spectroscopy. Plasma Phys. Controll. Fusion **55**(7), 074014 (2013)
26. M. Nocente, M. Tardocchi, V. Kiptily, P. Blanchard, I. Chugunov, S. Conroy, T. Edlington, A. Fernandes, G. Ericsson, M.G. Johnson et al., High-resolution gamma ray spectroscopy measurements of the fast ion energy distribution in JET ^{4}He plasmas. Nuclear Fusion **52**(6), 063009 (2012)
27. M. Salewski, M. Nocente, A.S. Jacobsen, F. Binda, C. Cazzaniga, G. Ericsson, J. Eriksson, G. Gorini, C. Hellesen, A. Hjalmarsson et al., MeV-range velocity-space tomography from gamma-ray and neutron emission spectrometry measurements at JET. Nuclear Fusion **57**(5), 056001 (2017)
28. D. Rigamonti, A. Muraro, M. Nocente, V. Perseo, G. Boltruczyk, A. Fernandes, J. Figueiredo, L. Giacomelli, G. Gorini, M. Gosk et al., Performance of the prototype $LaBr_3$ spectrometer developed for the JET gamma-ray camera upgrade. Rev. Sci. Instrum. **87**(11), 11E717 (2016)
29. M. Nocente, D. Rigamonti, V. Perseo, M. Tardocchi, G. Boltruczyk, A. Broslawski, A. Cremona, G. Croci, M. Gosk, V. Kiptily et al., Gamma-ray spectroscopy at MHz counting rates with a compact $LaBr_3$ detector and silicon photomultipliers for fusion plasma applications. Rev. Sci. Instrum. **87**(11), 11E714 (2016)
30. V. Kiptily, J. Adams, L. Bertalot, A. Murari, S. Sharapov, V. Yavorskij, B. Alper, R. Barnsley, P. de Vries, C. Gowers et al., Gamma-ray imaging of D and ^{4}He ions accelerated by ion-cyclotron-resonance heating in JET plasmas. Nuclear Fusion **45**(5), L21 (2005)
31. M. Nocente, M. Tardocchi, R. Barnsley, L. Bertalot, B. Brichard, G. Croci, G. Brolatti, L. Di Pace, A. Fernandes, L. Giacomelli et al., Conceptual design of the radial gamma ray spectrometers system for α particle and runaway electron measurements at ITER. Nuclear Fusion **57**(7), 076016 (2017)
32. M. Gobbin, L. Li, Y. Liu, L. Marrelli, M. Nocente, G. Papp, G. Pautasso, P. Piovesan, M. Valisa, D. Carnevale et al., Runaway electron mitigation by 3D fields in the ASDEX-upgrade experiment. Plasma Phys. Controll. Fusion **60**(1), 014036 (2017)
33. D. Ryutov, Energetic ion population formed in close collision with fusion alpha-particles. Phys. Scr. **45**(2), 153 (1992)
34. P. Helander, M. Lisak, D. Ryutov, Formation of hot ion populations in fusion plasmas by close collisions with fast particles. Plasma Phys. Controll. Fusion **35**(3), 363 (1993)
35. G. Gorini, L. Ballabio, J. Källne, Alpha-particle kinetic effects in the neutron emission of burning dt plasmas. Rev. Sci. Instrum. **66**(1), 936–938 (1995)
36. L. Ballabio, G. Gorini, J. Källne, α-particle knock-on signature in the neutron emission of dt plasmas. Phys. Rev. E **55**(3), 3358 (1997)
37. J. Källne, L. Ballabio, J. Frenje, S. Conroy, G. Ericsson, M. Tardocchi, E. Traneus, G. Gorini, Observation of the alpha particle knock-on neutron emission from magnetically confined dt fusion plasmas. Phys. Rev. Lett. **85**(6), 1246 (2000)

Journal of Fusion Energy (2019) 38:299–314
https://doi.org/10.1007/s10894-019-00218-4

ORIGINAL RESEARCH

Optical Time-Resolved Diagnostics of Laser-Produced Plasmas

D. Batani[1,7] · J. Santos[1] · P. Forestier-Colleoni[1] · D. Mancelli[1,8] · M. Ehret[1] · J. Trela[1] · A. Morace[2] · K. Jakubowska[3] · L. Antonelli[4] · D. del Sorbo[5] · M. Manclossi[6] · M. Veltcheva[6]

Published online: 13 May 2019

Abstract
The paper presents the general challenges of diagnostics for laser-produced plasmas related to their small dimensions and short duration but also to the presence of very strong temperature and density gradient in the plasma. In particular, we focus on more detailed examples related to optical diagnostics for the undercritical region of the plasma: shadowgraphy, schlieren, interferometry, polarimetry.

Keywords Laser-produced plasmas · Inertial confinement fusion · Optical diagnostics

Introduction

Laser produced plasmas are relevant for basic physics and for applications, as well as inertial confinement fusion. Because of this large interest, many diagnostics have been developed and are used to study the physical processes taking place during laser–plasma interaction.

Apart from the general challenges related to plasma diagnostic, the diagnostics of laser-produced plasmas pose some specific challenges. First of all, laser produced plasmas are small and have a short duration. Typically, the size is comparable to the laser spot dimension (μm to mm) and the time life is comparable to the laser pulse duration (after the end of the laser pulse the plasma rapidly cools down and vanishes). Hence diagnostics need to have high spatial and temporal resolution.

But the main challenge arises from the fact that the temperature and density of laser-produced plasmas span over several orders of magnitude and are characterized by the presence of very distinct plasma "regions" all simultaneously present.

1. First the "undercritical region" (or plasma corona) is defined by having a density smaller than the laser critical density, in practical units:

$$n_{cr}\left(\mathrm{cm}^{-3}\right) = 1.1 \times 10^{21}/\lambda^2\,(\mu\mathrm{m}) \tag{1}$$

   In this region, the density decreases (due to plasma expansion) going away from the target surface. Also, this region is directly heated by the laser, which can propagate up to n_{cr}, and therefore it reaches high temperatures: a few 100 eV up to a few keV.
2. Then we have the "conduction region", or overcritical region, where the laser cannot propagate. Here heat is transported inwards by electrons or by XUV—soft X-ray radiation generated in the corona. From the value reached in the corona, the temperature drops down to a few eV. The electron density increases from the n_{cr} value up to the electronic density of solid matter.
3. Finally we have the "shocked material". Plasma expansion produces, due to momentum conservation, a shock travelling inward and compressing the target.

✉ D. Batani
dimitri.batani@u-bordeaux.fr

1 CEA, CNRS, CELIA (Centre Laser Intense at Applications), UMR 5107, University Bordeaux, 33405 Talence, France

2 ILE, Institute of Laser Engineering, University of Osaka, Osaka, Japan

3 IPPLM, Warsaw, Poland

4 University of York, York, UK

5 High Energy Density Science Division, SLAC National Accelerator Laboratory, Menlo Park, CA 94025, USA

6 LOA, ENSTA, Palaiseau, France

7 Department of Plasma Physics, National Research Nuclear University MEPhI, Moscow, Russia

8 Donostia International Physics Center (DIPC), Donostia-San Sebastian, Basque, Spain

The generated pressure can be as high as a few MBars and compression heats the material up to a few eV. Ahead of the shocked region we have the still unperturbed solid.

Due to the differences of several orders of magnitude in density and temperature among the different regions, different diagnostics are needed.

For instance, the plasma corona, where a probe laser beam can propagate, can be characterized by optical diagnostics: interferometry [1], shadowgraphy [2, 3], Faraday rotation [4], etc. Also, this region is very hot and hence emits X-rays. Therefore, one can use X-ray spectroscopy [5–9]. Optical spectroscopy can be useful to characterize parametric instabilities taking place in laser–plasma interaction [10–12]. And finally, one can use particle detection (charge collectors, Faraday's cups, etc.) [13].

The overcritical region is the main source of soft X-rays and XUV radiation because, even if the temperature is on average lower than in the corona, the density is much larger and then there are more collisions and photon emission. Hence one can sue emission X-ray spectroscopy and absorption X-ray spectroscopy [14, 15]. However, unlike the plasma corona, this region is optically thick which implies that there is a strong re-absorption of emitted photons and hence opacity effects must be taken into account. In principle one can use also XUV interferometry (for instance with XUV laser sources) but in reality, this is quite challenging [16].

The shocked region is particularly difficult to be probed. Due to its low temperature, there is no strong emission of radiation and this is concentrated at long wavelength, also this region is very opaque due to its large density. One can use X-ray radiography with hard X-rays emitted from a secondary laser-produced source [17], or proton radiography [18]. With sufficiently thin targets, it is also possible to use optical diagnostics: when the shock breaks out on the target "rear side", matter is heated and intense thermal radiation is emitted which can be recorder and time resolved, for instance, suing a streak camera [19–21].

To complete the panorama, one should cite all diagnostics which are typical of laser-driven inertial fusion experiments: neutron measurements [22], neutron spectroscopy [23], neutron imaging [24].

Of course, it is not possible to deal with all such different diagnostics in a single paper. Therefore, we will just concentrate on a few examples and, in particular, we will focus on optical diagnostics for the undercritical region of the plasma: shadowgraphy, schlieren, interferometry, polarimetry, where the presence of the plasma impacts the behavior of an electromagnetic (e.m.) wave of angular frequency ω.

Optical Shadowgraphy and Schlieren

Plasma of electronic density n_e (in the limit of low collisionality and when no external magnetic fields are present) has a refraction index n given by

$$n = \sqrt{1 - \frac{\omega_p^2}{\omega^2}} = \sqrt{1 - \frac{n_e}{n_{cr}}}$$

where ω is the frequency of the e.m wave (i.e. the laser in our case). For an undercritical plasma ($n_e < n_{cr}$) the refraction index is real and < 1, and for the case of much underdense plasmas ($n_e \ll n_{cr}$) it becomes

$$n \approx 1 - \frac{n_e}{2n_{cr}}$$

The gradients of density, which are always present and strong in laser-produced plasmas, then induce a gradient of refraction index $\nabla n \approx -\nabla n_e / 2n_{cr}$. According to Snell–Descartes law of refraction, such gradient induces a deviation of the light ray by an angle

$$\theta = \frac{1}{n_{cr}} \frac{\partial}{\partial y} \int n \, dl \tag{2}$$

This is at the base of the diagnostics techniques of shadowgraphy and schlieren (a complete analysis, see for instance [25] shows that shadowgraphy accounts for the second derivative of the density gradients in the plasma whereas schlieren photography records the first derivative of the plasma density gradients).

One usual approach is to use a laser probe beam (at intensity low enough so to not perturb the plasma by inducing additional ionization or heating), which "backlights" the plasma to be probed. In this case one can even use a slow detector to make the image of the plasma: in any case the detector will "see" the plasma only as long as the backlighting source is on. Hence the system will provide a "snapshot", i.e. a 2D plasma image obtained with a time resolution equal to the duration of the pulse of the probe laser beam. In the case of shadowgraphy for instance, we will see a "shadow" of the plasma because the refracted rays will be deviated away. This "normal" shadowgraphy (without a lens) differs from the so called "focused shadowgraphy" which we will describe in the following. Indeed in the cases of large density gradients, such as in laser generated plasmas, normal shadowgraphy may fail providing useful result for several reasons.

First, the image provided by the backlighter will superimpose to the image created on the detector by the plasma self-emission which of course is not time-resolved but lasts as long as the plasma is sufficiently hot to emits significantly in the range determined by the spectral sensitivity of the detector. We therefore need to eliminate plasma self-emission.

Second, the plasma will very likely be almost completely transparent to the incoming laser light, therefore the detector will be illuminated everywhere with little contrast. We therefore need to use a diagnostic approach, which evidences the presence of density gradients in the plasma.

Third, even if the rays which go through the plasma are deviated according to Eq. 2, then these are "brought back" to the detector when we use an imaging system (which is usually the case). For instance, if we use a lens which makes an image of the plasma on the detector, it brings back the refracted light to the image point corresponding to the object point where the ray originates, i.e. it has been deviated. So, the imaging system "kills" the shadow.

Therefore, the complete experimental set-up of shadowgraphy includes a spatial filter in the focal plane of the imaging lens (see Fig. 1). This diaphragm will cut all the rays, which have been strongly refracted and will let the un-deviated rays to go through. These are typically the rays, which go "around" the plasma (as shown again in Fig. 1) and which therefore are normally focused in the lens focal point. The result will be a "shadow" of the plasma: the plasma will look darker because most of the rays going though it are cut by the diaphragm. The recorded image is therefore called a "negative image" in opposition to the case of normal shadowgraphy where no lenses and no pin-hole are used. An alternative (up to a point depending on the optical system used and the density gradients of the plasma) is to "de-focus" the image by placing the lens not exactly at its focal position. This will enable a combination of the non-focused shadowgraphy with the focused shadowgraphy giving images suitable for analysis. Furthermore, by knowing the defocusing length, quantitative analysis can be performed giving absolute results for the plasma density. This is called quantitative shadowgraphy [26].

It is also possible to use the refraction fringes surrounding the accessible region of the plasma in order to calculate the density at the boundary of the corona plasma. This pattern is due to the refraction of the laser rays at the steep density gradient boundary of the plasma-vacuum interface [27, 28].

Normally strong gradients exist mainly at the edge of the plasma and thus shadowgraphy allows obtaining an image of the edges of the plasma. By changing the delay Δt between the beam creating the plasma and the probe beam, one can obtain a series of "frozen images" (i.e. 2D images integrated in time over the, short, duration of the probing laser beam) which allows reconstructing the plasma dynamics i.e. for instance measuring the plasma size at different times, thus allowing determining the expansion velocity of the plasma.

In this sense, optical shadowgraphy and schlieren are part of the techniques known as "pump and probe" diagnostics.

Figure 2 shows an example. Here a laser beam (a Nd:YAG with $E \approx 100$ mJ, $\lambda = 1.064$ µm, $\tau \approx 10$ ns) is focused on a solid target (an iron needle) in air. The probe beam is an unconverted fraction of the main beam (i.e. the probe is also infrared in this case) and is sent across the plasma with a variable delay. A hot and dense plasma is created at the surface of the target. Such plasma begins a rapid expansion compressing the air around it and generating a shock wave, which travels in air. The figure shows the evolution of the generated shock wave at three different times following irradiation [The use of a cylindrical needle as a target helps avoiding the problems related to possible small tilts of the target. In the case of a planar target, this must be perfectly parallel to the probe rays, otherwise it will mask some of them shadowing the part of the plasma closer to the surface, which is the denser one. Of course, this problem becomes bigger and bigger as the lateral dimension of the planar foil increases].

Figure 3 shows instead an image of plasma created by focusing a laser beam in air. In this case the plasma is elongated along the direction of propagation of the laser beam (the same laser of Fig. 2 is used here).

Fig. 1 In-principle scheme of optical shadowgraphy with the pin-hole placed in the focal point of the imaging lens

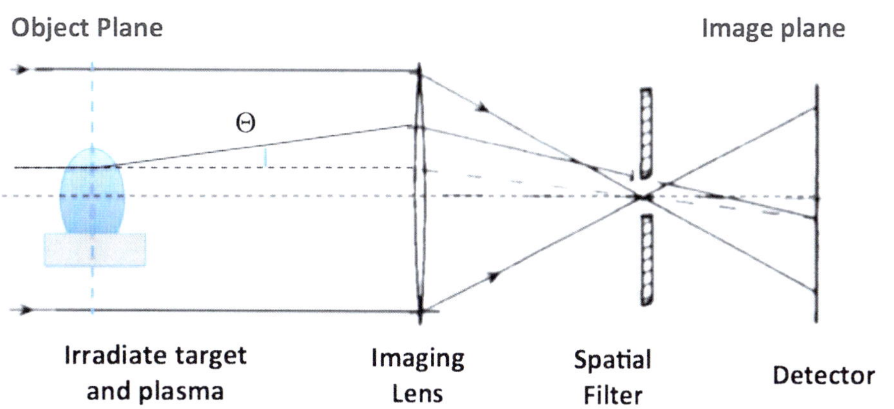

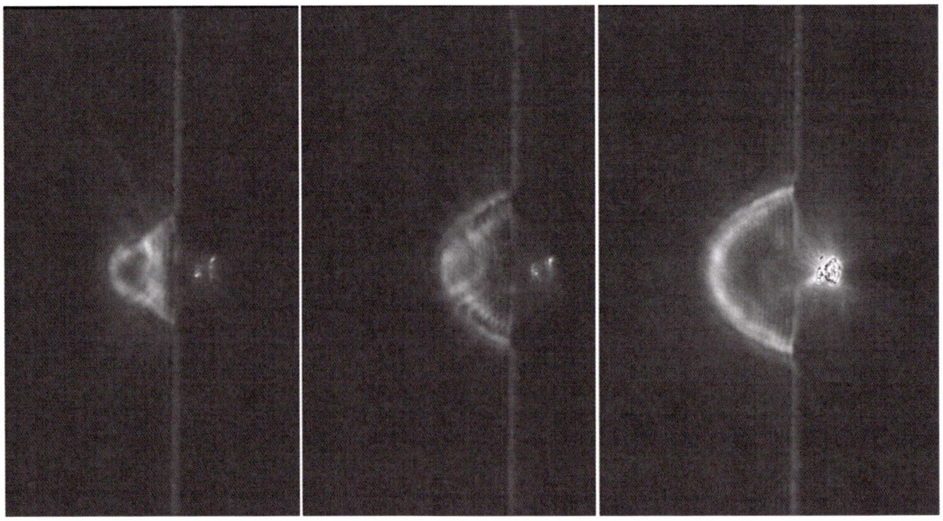

Fig. 2 Evolution of a laser-generated shock wave in air using schlieren. The 3 images correspond to delay t = 11.21 ns, 19.42 ns and 27.62 ns respectively. During this time the shock size (horizontal direction) increases from 177 to 192 μm, and finally 234 μm

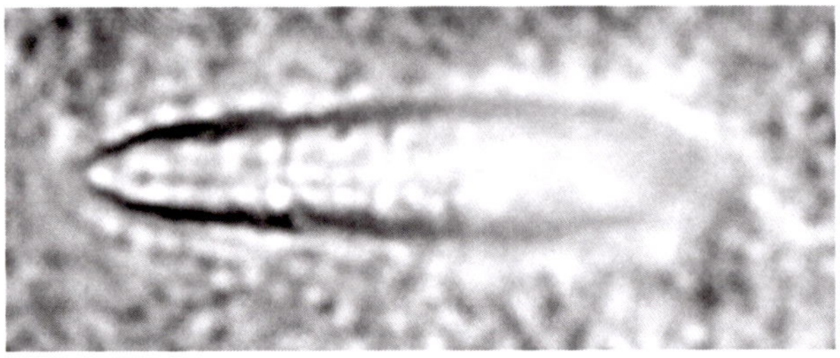

Fig. 3 Image of laser-induced plasma in air obtained using optical shadowgraphy. The probe beam crosses the plasma about 20 ns after the interaction of the pump beam has created the plasma. The size of the image is 200 × 460 μm

The drawback of shadowgraphy is the loss of spatial resolution. Indeed, the focal plane of the imaging lens realize the Fourier transform of the object to be imaged and the higher the spatial frequency, the bigger the distance from the focal point in the focal plane. Hence, placing a diaphragm in the focal plane means cutting the high-spatial frequencies (short spatial scales), which are essential in order to assure a detailed reconstruction of the image.

The schlieren technique is somewhat complementary to shadowgraphy. In this technique, we place an obstacle in the focal point of the imaging lens (see Fig. 4). Therefore, the refracted rays are allowed to go through while the un-refracted rays are cut. The strongly refracting zones are then materialized on the schlieren image by illuminated zones. Of course, schlieren allows maintain a good spatial resolution since it keeps the high spatial frequencies.

Figure 5 shows the comparison of shadowgraphy and schlieren in the case of the shock wave generated in air (as described about Fig. 2).

Finally, we close this paragraph by noticing that although shadowgraphy and schlieren are powerful diagnostic techniques, nevertheless they are mainly qualitative. As said before, one can attempt to do quantitative shadowgraphy or quantitative schlieren [29, 30] but indeed it is very difficult to retrieve the information on the density profile (this is not a surprise because indeed these diagnostics give information on the gradient of electron density or on its second derivative). An interesting development in this direction is however the development of the angular filter refractometry (AFR), which allows to characterize the

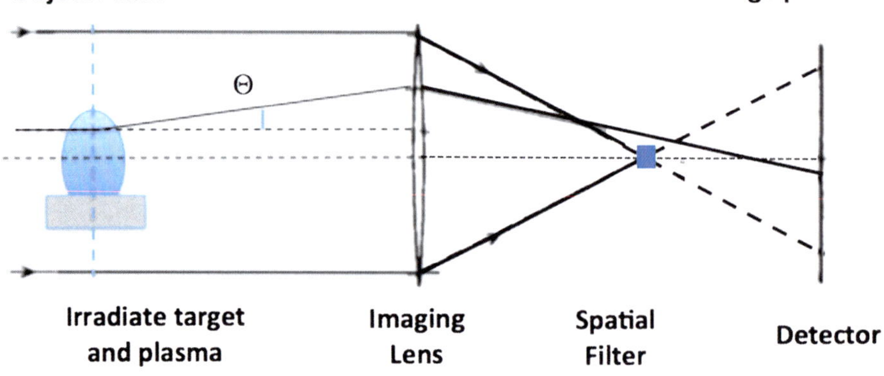

Fig. 4 In-principle scheme of schlieren with the obstacle placed in the focal point of the imaging lens

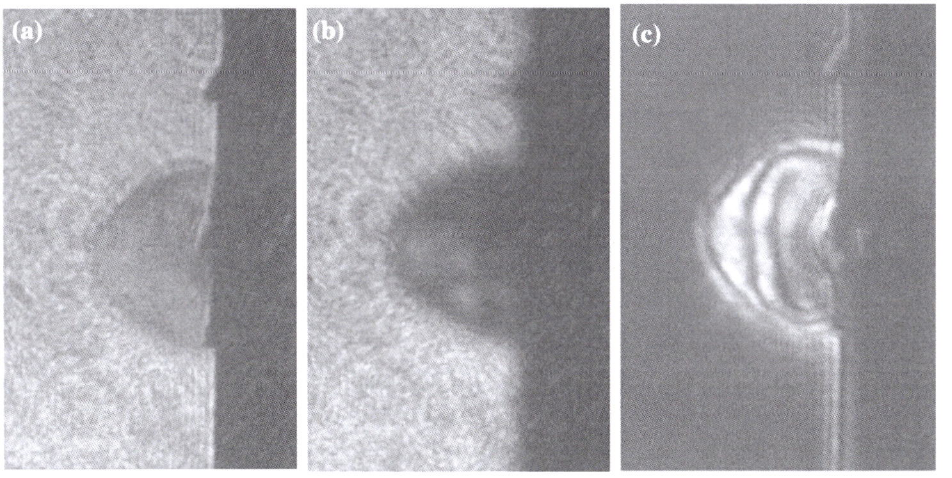

Fig. 5 Comparison of **a** conventional imaging, **b** shadowgraphy, and **c** schlieren techniques (same experimental configuration as in Fig. 2). The 3 images have been taken in identical conditions and with the same time delay

plasma density profile by producing a contour map of refraction angles [31]. This is accomplished by using angular filters (see Fig. 6) that block certain bands of refraction angles, producing shadows in the image plane. The plasma density is calculated from the measured refraction angles of a probe beam after passing through the plasma. The maximum measured density is limited by the f/# of the optical collection system as compared to the length of plasma in the direction of the probe and the magnitude of the transverse gradients (which determine the refraction angle through Eq. 2).

The AFR diagnostic uses an angular filter placed at the focus of the unrefracted probe beam (Fourier plane). The opaque regions of the angular filter block bands of refraction angles, resulting in shadows in the image plane. The diagnostic relies on the direct proportionality between the angle of refraction of a probe ray at the object plane and its radial location in the Fourier plane. This relation correlates the shadows produced by the angular filter with contours of constant refraction angle. For a single-lens imaging system and a collimated probe beam, it can be shown that a ray refracted at the object plane passes through the Fourier plane at a distance from the optical axis, r, which is equal to the focal length, f, of the collection lens times the refraction angle, Θ, regardless of its spatial origin in the object plane (assuming paraxial propagation). The direct proportionality of the spatial location of a ray to the amount of refraction allows for filtration of specific refraction angles in a deterministic manner. Depending on the value of the refraction angle, the probe rays will either be blocked by the opaque rings or transmitted between them, forming the light and dark bands in the image. By measuring the radial distance of each band, one can accurately calculate the refraction angles (again given by Eq. 2).

Chirped Optical Shadowgraphy

The technique of chirped pulse amplification (CPA) allows producing laser pulses with typical duration of hundreds ps to a few ns which are characterized by a chirp, i.e. a variation in time of the frequency. Normally this allows compressing the laser pulse in time inside a grating compressor, so to obtain short-duration very-high-intensity pulses. In the technique of chirped pulse shadowgraphy instead the beam is not compressed. The plasma to be probed is then illuminated with a chirped probe beam of relatively long duration. After the plasma, the beam is sent to an imaging spectrometer where the different wavelengths are dispersed in different direction. Since the different laser wavelengths arrive at different times on the plasma to be probed, we then have a univocal relation between time and wavelength.

In the imaging plane of the spectrometer we then get a 1D image resolved in time, as if we were operating with a

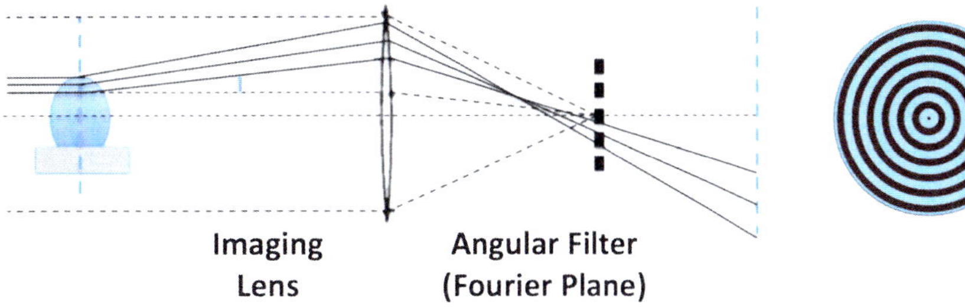

Fig. 6 Working scheme of the angular filter refractometry technique and example of filer used in the Focal plane (Fourier plane)

streak camera. However, the technique of chirped optical shadowgraphy allows getting a better time resolution.

This is a rather novel technique first introduced in [32]. In this experiment, optical shadowgraphy was to detect fast electron propagation in a gas medium. The region of the gas jet, which is ionized by the passage of fast electrons, creates a shadow in the recorded image. In [32] the chirped probe laser pulse was first compressed giving a 2D spatial resolution (classical shadowgraphy) and then, only partially compressed (i.e. still chirped) giving 1D time-resolved images (chirped shadowgraphy). In the first case, 2D snapshot images were obtained with the temporal resolution equal to the duration of the probe beam (< 350 fs). With this configuration, in relation to the propagation dynamics, several shots are needed with different delay between the main and probe beams. In the second case, only 1D images are obtained but the fast electron dynamics can be continuously followed in time over duration corresponding to the duration of the partially precompressed beam.

The scheme of the experimental set-up used in [32] is shown in Fig. 7. Fast electrons were created by a high-intensity laser on a thin foil (Ti) placed before a gas jet. Two gases (Ar and He) and different pressures were used.

Figure 8 shows shadowgrams obtained at different delays between the CPA beam and the probe. The cloud evolution can be followed and its average velocity between two different snapshots can be measured. However, as fast electron dynamics is inferred from several shots, the method is strongly affected by shot-to-shot fluctuations, greatly increasing error bars.

In chirped shadowgraphy the probe beam had longer pulse duration (τ = 19 ps) and was coupled to an imaging spectrometer. Here the gas medium (where the plasma is formed) is imaged on the entrance slit of the spectrometer, where the spectral axis through frequency dependence on time (chirp of the beam) corresponds to the temporal axis. The spatial resolution is obtained only in the direction parallel to the entrance slit. This diagnostic allows us to

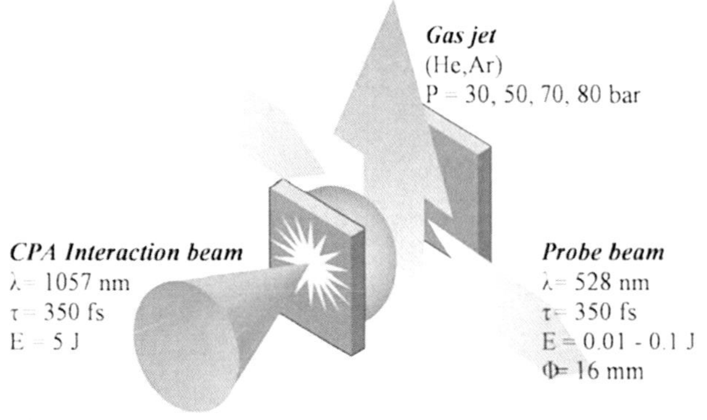

Fig. 7 Schematic experimental set-up. The distance between the foils is 1.2 mm. Thickness is 20 μm (Ti) and 15 μm (Al) (from [32])

measure on a single laser shot the variation of the ionization front velocity with time resolving early times. Figure 9 shows chirped shadowgrams obtained with Ar at different pressures. The images in Fig. 9 clearly show the following:

1. The propagation velocity and the cloud penetration distance increase with gas pressure. This result agrees with those obtained with the classical shadowgraphy. The dimension of the cloud at the end of the image corresponds to the dimensions observed in the classical shadowgraphy images.
2. The propagation velocity slows down in time. Figure 9 (bottom) (Ar 100 bar) clearly shows first a fast speed $v_{AB} \sim 0.67c$, then a deceleration to a speed $v_{BC} \sim 0.16c$. In addition, we can note a cloud penetration over a distance of the order of the millimeter. The propagation velocity in Fig. 9 (top) (Ar 20 bar) is lower ($v_{AB} \sim 0.12c$) and the final propagation is much less important ($\sim$ 300 μm).

Figure 10 shows the temporal evolution of the cloud dimension for various gas pressures, obtained from the chirped shadowgraphy images, and clearly shows a deceleration occurring at the very first moments of the cloud propagation. This phase is very difficult to be observed with classical shadowgraphy since it would require taking many images at delays close to each other.

Interferometry

Interferometry is based on the fact that the plasma has a refractive index and hence the light rays crossing it experiences a phase shift. As before, for an undercritical plasma the refraction index n of a plasma is real and

$$n = \sqrt{1 - \frac{\omega_p^2}{\omega^2}} = \sqrt{1 - \frac{n_e}{n_c}} < 1$$

The plasma (of typical size L) must be placed within an interferometric system, for instance in one of the two arms of a Mach–Zender interferometer, as shown in Fig. 11.

Then the phase shift $\Delta\Phi$ induced by a homogeneous plasma will be

$$\Delta\Phi = (k_{plasma} - k_o)L = k_o(n-1)L = k_o L\left(\sqrt{1 - \frac{n_e}{n_c}} - 1\right)$$

which, in the case of much undercritical plasmas, reduces to

$$\Delta\Phi \approx \frac{k_o L n_e}{2n_c} \qquad n_e \ll n_{cr}$$

We see therefore that interferometry can in this case directly provide the plasma density n_e. In the realistic case of non-homogeneous plasma, we have

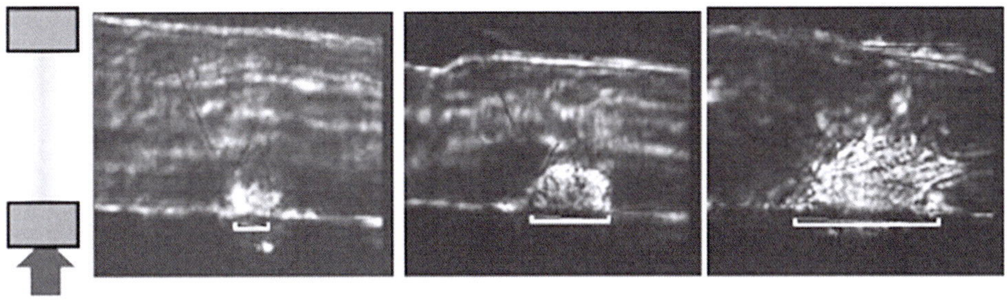

Fig. 8 Evolution of the electron cloud (He, 30 bar). From left: t = 10 ps, 16 ps, 50 ps. Bars correspond to 280, 540, and 1000 μm. The lower and upper dark parts in all images, respectively, represent the Ti and Al foils (from [33])

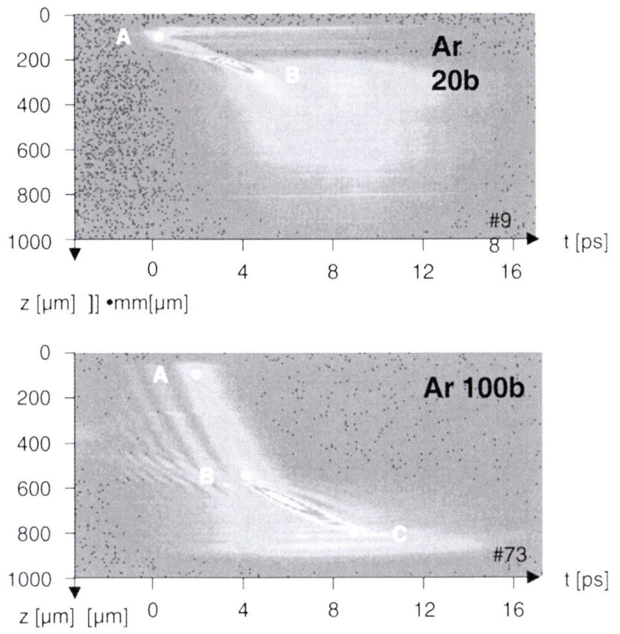

Fig. 9 Chirped shadowgraphy images obtained with a target of Al 15 μm and with a jet of argon to 20 bar (top) and to 100 bar (bottom). The laser intensity is 2×10^{19} W/cm^2. Time axis is the horizontal one and point A roughly represents the interaction time

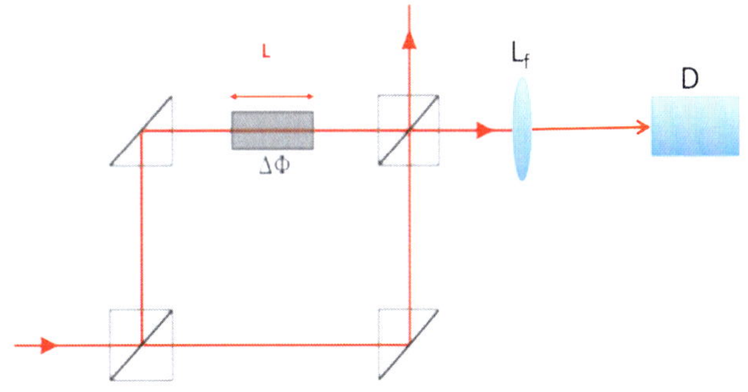

Fig. 11 Mach–Zender interferometric set-up used to measure the plasma refraction index. The lens L_f images the plasma on the detector D

$$\Delta\Phi \approx \frac{k_0}{2n_{cr}} \int n_e \, dl \qquad (n_e \ll n_{cr})$$

which means that interferometry measures the total electronic areal density along the line of sight. Defining the average electronic density as

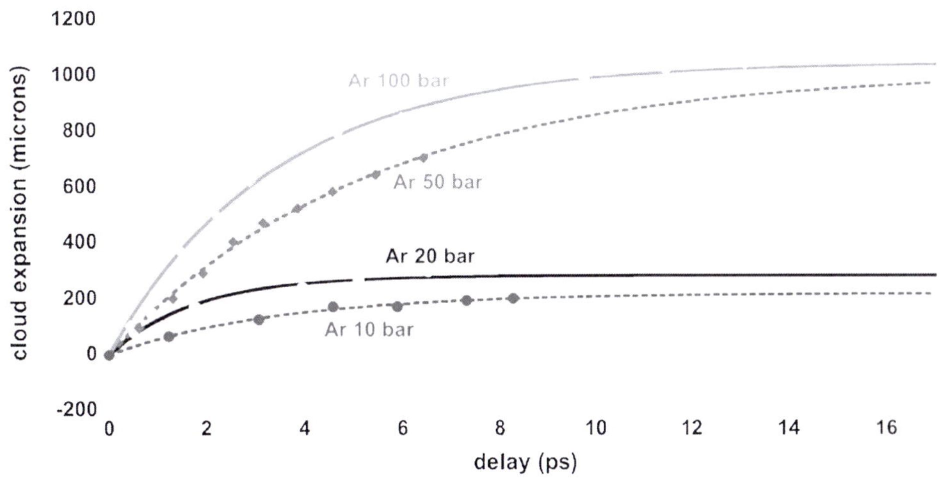

Fig. 10 Cloud size (top) and deduced velocity (bottom) versus time. Ar pressures are 10 bar (full circles), Ar 20 bar (empty circles), Ar 50 bar (full diamonds), Ar 100 bar (empty diamonds)

$$\langle n_e \rangle = \frac{1}{L} \int n_e \, dl$$

we see that

$$\frac{\Delta\Phi}{\pi} = \frac{L}{\lambda} \frac{\langle n_e \rangle}{n_{cr}} \tag{3}$$

or in other words, the fringe shift (phase shift in units of π) is proportional to the plasma length over the wavelength of the probing laser times the average density (defined over the line of sight) over the critical density.

In all practical situations, the last optical element in the Mach–Zender interferometer is slightly tilted in order to introduce fringes, i.e. a horizontal spatial carrier heterodyned frequency, with spatial wavelength λ_{het}. The carrier frequency is used to move the experimental information (i.e. the phase plane perturbation) away from the zero order in Fourier space, where all the low-frequency noise (such as spatial intensity fluctuations) is concentrated. In practice the presence of the fringes allows to more easily measure the phase displacement by simply observing the fringe shift.

The following figures show examples of interferometric images with laser-produced plasmas.

The first one (Fig. 12) represents a plasma created in air in the focal point of the focusing lens by a Nd:YAG laser with E $\approx$ 100 mJ, λ = 1.064 μm, $\tau \approx$ 40 ps [34]. The probing laser was a fraction of the main laser converted to second harmonic (532 nm) and sent across the plasma with a variable delay. Operating at two different laser wavelengths is advantageous because it allows (for instance by using interference filters) to strongly reduce the noise coming from the scattered pump light to the image formed by the probe beam. In addition, choosing a probe beam with shorter wavelengths, allows probing higher plasmas densities.

The second image (Fig. 13) shows an interferogram of plasma obtained, with the same laser focused onto a solid target placed in vacuum. In this case the plasma expands without creating a shock wave (since there is no air around it to be compressed) and the density profile smoothly changes from the solid density down to the critical density and below. Let's notice that the typical atomic density of air is of the order of $\approx 10^{19}$ cm^{-3}, therefore all the created plasma is undercritical and transparent to the probe beam: we see the fringes everywhere. Instead in the case of plasma from solid target, there is an overcritical region where the fringes disappear.

For sake of completeness, we recall here that one common method to retrieve the phase shift from the interferogram is to use Fourier transformation. This method was introduced early in the history of laser produced plasma [35] and then improved by using fast Fourier transform (FFT) techniques [36]. This methodology works very well when interferograms are only marginally affected by noise and reduction of fringe visibility. In order to produce accurate phase-shift maps when low-quality images are dealt with, more recently procedures for phase-shift map computation using ridge extraction in the continuous wavelet transform (CWT) framework were introduced [37]. Comparative analysis of the accuracy performances of CWT and of FFT approaches have shown that CWT-

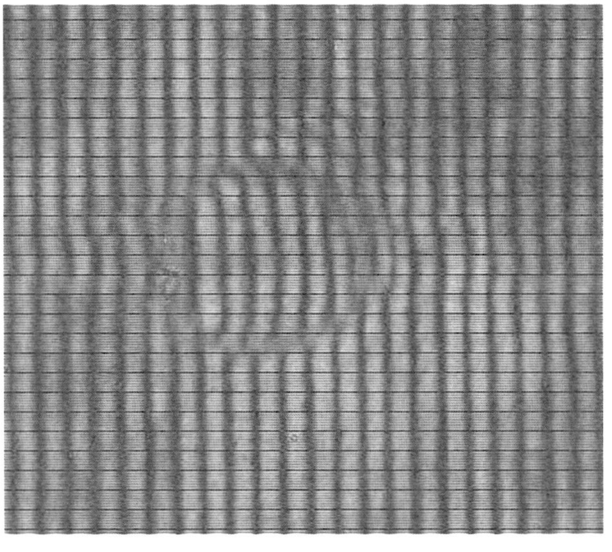

Fig. 12 Interferogram of a plasma created in air in the focus of the lens by a Nd:YAG laser with E $\approx$ 100 mJ, λ = 1.064 μm, $\tau \approx$ 40 ps [34]. The probe beam crosses the plasma 10 ns after the interaction of the pump beam has created the plasma. The size of the image is 200 × 170 μm

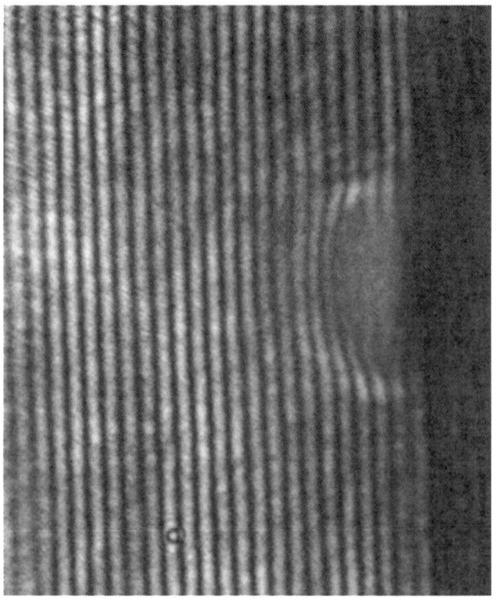

Fig. 13 Interferogram of a plasma expanding in vacuum created by focusing a ps-laser beam [34] onto an iron needle. The probe beam crosses the plasma 10 ns after the interaction of the pump beam. The size of the image is 300 × 360 μm

based tools can produce phase maps which are considerably less noisy and can better resolve local inhomogeneities.

These images immediately show some characteristics of the plasma (in particular the dimension). However, according to Eq. 3, the plasma density profile $n_e(x,y,z)$ is not directly obtained though interferometry but one can only obtain the average density over the line-of-sight $\langle n_e \rangle$. In order to get the detailed density profile, one needs to do additional assumptions.

Planar Quasi-Homogeneous Plasma

In some cases, one can assume that the plasma is quite uniform over the distance L. This is for instance the case of plasmas created on planar target by irradiation of a laser beam with a large focal spot (diameter $\approx$ L) smoother using PZP or KPP, which produces a top-hat distribution of intensity. In this case one can reasonably neglect the plasma edges with thickness d $\ll$ L. In addition, by limiting the observation time so that the expansion of the plasma is also $\ll$ L, one can reasonably neglect 2D effects and assume a 1D plasma expansion.

For instance, in the work in [1], the detector was a streak camera providing temporal resolution. Its slit was placed perpendicular to the target plane (i.e. parallel to the direction of plasma expansion) as shown in Fig. 14.

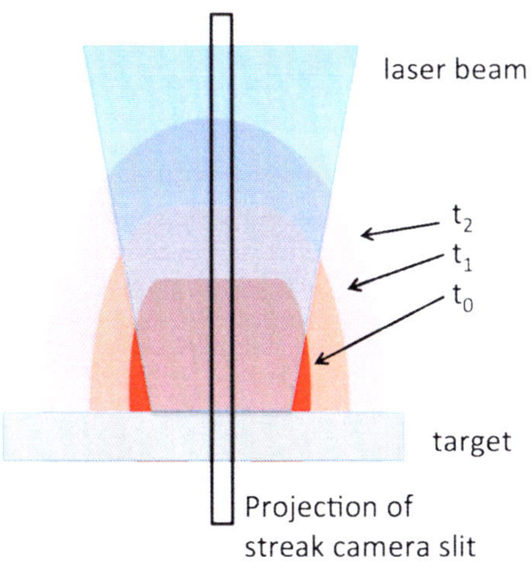

Fig. 14 Scheme of the experimental set-up used in [1]. The laser beam irradiates the solid target normally to the surface and the streak camera slit is perpendicular to the target surface and centered with respect to the focal spot. The laser beam was smoothed using Phase Zone Plate (PZP) thereby providing a top-hat irradiation profile and a "flat" plasma expansion (at least around the center of the focal spot and for short observation times). We schematically show the size of the plasma at different times t_0, t_1, t_2. Notice that at t_2 the plasma is bigger than the size of the laser focal spot and the expansion clearly becomes bidimensional even around the center of the focal spot

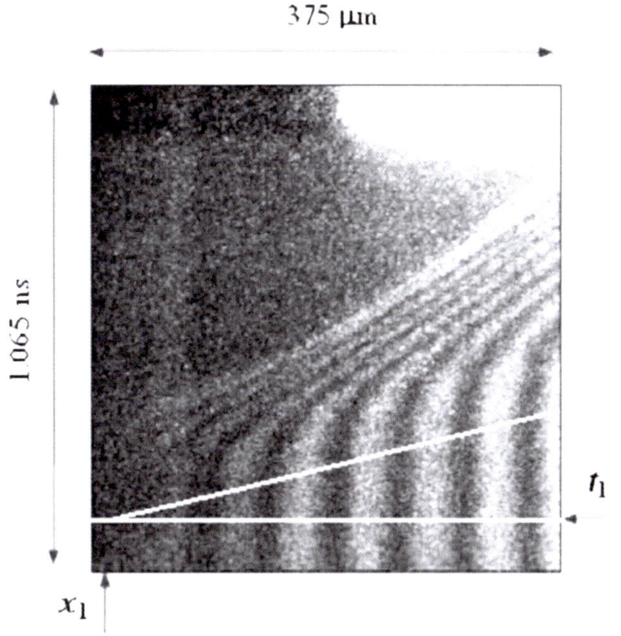

Fig. 15 Interferogram recorded with the streak camera on a planar CH_2 foil (shot at $E = 49.6$ J) Image size is 375 µm horizontal scale and 1.065 ns vertical scale. The vertical arrow shows the position x_1. The horizontal arrow roughly corresponds to the time t_1, at which the pump laser arrives on target. The white inclined line represents the ensemble of the points where the fringes begin to move (from [1])

Figure 15 shows a typical experimental result obtained using a Nd:YAG laser probe converted to 2ω with pulse duration longer than the desired observation window.

With all the assumptions, we have done, we can say that $n_e \approx \langle n_e \rangle$ and then we can simply apply Eq. 3 to retrieve electron density, which is here just a function of the distance from the planar foil and of time. Measuring the fringe shifts from Fig. 15 we get the plasma density profiles at different times, as shown in Fig. 16.

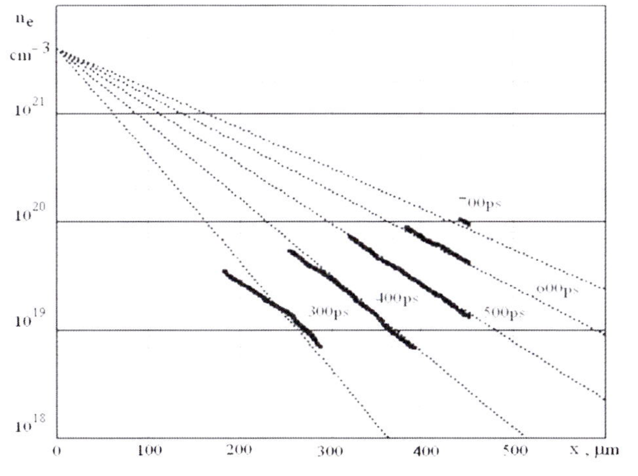

Fig. 16 Comparison of experimental profiles (obtained from the interferogram in Fig. 15) with the predictions of the analytic 1D model for a CH_2 target

The results in Fig. 16 show that in this case the density profiles agree quite well with the predictions of simple self-similar analytical models for isothermal plasma expansion, according to which

$$n_e(x,t) = n_{cr}\exp\left(-\frac{x - x_0}{c_s t}\right)$$

here c_s is the sound plasma velocity which is given in practical units by

$$c_s = 9.79 \times 10^5\sqrt{\frac{\gamma Z^* T_e}{m}}$$

where c_s is in cm/s, T_e in eV, m is the average ion mass in the plasma in atomic units and γ is the adiabatic constant. Isothermal conditions are here provided by the fact that the laser pulse duration is comparable to the observation window (therefore the plasma is continuously heated by the laser) and by the large free path of thermal electrons in the plasma corona, assuring efficient energy exchanges and temperature uniformization.

The graphs in Fig. 16 allow inferring the plasma electronic temperature T_e. From Fig. 16 we also see that the accessible density range is between 7×10^{18} and 10^{20} cm^{-3}. Of course, low densities are hardly detectable because the induced phase shift is very low (the refraction index being very close to 1). As for high densities, although the critical density for the probe beam is $\approx 3.9 \times 10^{21}$ cm^{-3}, in reality light rays which travel through the higher density (higher refraction index) regions of the plasma are strongly deviated and will fall out the collection solid angle of the lens L_f (remind Fig. 11 and also see later the paragraph on shadowgraphy). Increasing the optical aperture (by using for instance larger diameter optics) this effect can be reduced but the disadvantage is that the spherical aberrations of the optical system are increased. In case a large aperture optical system is attempted, complex lenses such as doublets or triplets should be used for an acceptable spatial resolution of a few microns. However this remains an unavoidable problem in interferometry, implying that usually only densities up to a maximum of $\approx n_{cr}/10$ can be probed (we notice here that indeed the same limitation applies to shadowgraphy and schlieren)

Plasmas with Cylindrical Symmetry

In many cases one can assume that the plasma has an axial (cylindrical) symmetry. This is for instance the case of a laser beam (normally with axial symmetry) focused perpendicularly on the surface of a planar target, or the case of a plasma generated in a gas. In all these cases, the mathematical Abel transform allows getting the detailed profile

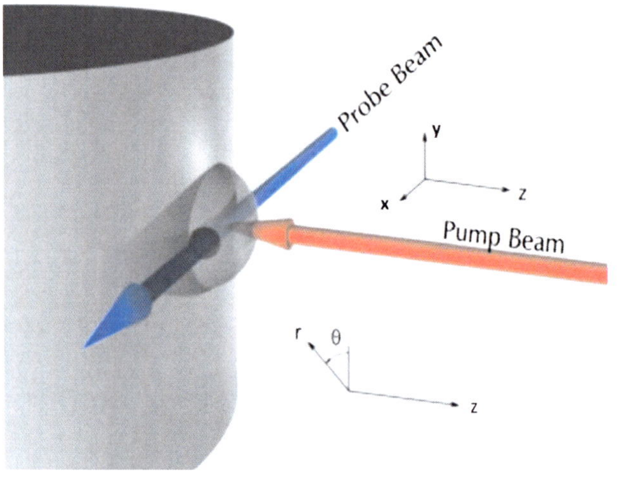

Fig. 17 Production ad probing of a plasma with cylindrical symmetry

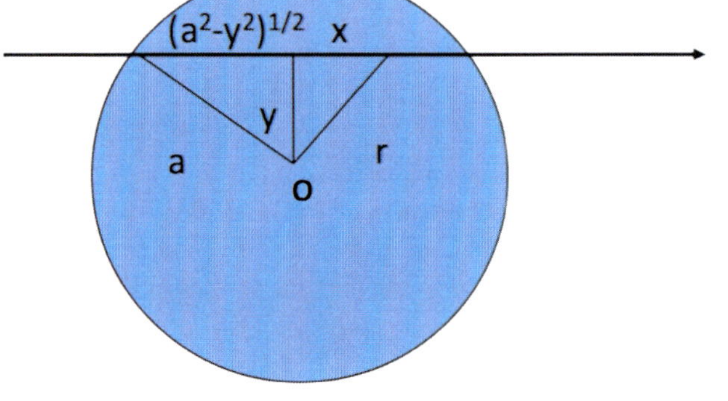

Fig. 18 Propagation of the probe laser beam inside the plasma of Fig. 16. Here a plane corresponding to a value of z has been selected. O is the center of symmetry, which also corresponds to the axis of the laser pump beam. The plasma radius is a

of electron density n_e as a function of spatial coordinates $n_e(r,z,t)$. See the experimental set-up schematically shown in Fig. 17 (notice again the use of needle as target)

Making a cut at level z, we have the situation described in Fig. 18.

Following the scheme in Fig. 18, the dephasing of the laser probe beam as it crosses the plasma at the position y, assuming as usual $n_e \ll n_{cr}$, is given by

$$\Delta\Phi(y) = \frac{k_0}{2n_{cr}}\int n_e dl = \frac{k_0}{2n_{cr}}\int_{-\sqrt{a^2-y^2}}^{\sqrt{a^2-y^2}} n_e(r)dx$$

Again this shows that dephasing depends on the areal plasma density (in units of electrons/m^2). If cylindrical symmetry does not exist, this quantity can still be calculated with relative good accuracy. When cylindrical symmetry exists the cubic density (electrons/m^3) can be extracted. By using the substitution

$$x = \sqrt{r^2 - x^2} \Rightarrow dx = -\frac{r\,dr}{\sqrt{r^2 - x^2}}$$

The dephasing $\Delta\Phi$ becomes

$$\Delta\Phi(y) = \frac{k_0}{2n_{cr}} \int n_e dl = \frac{k_0}{n_{cr}} \int_y^a \frac{n_e(r)}{\sqrt{r^2 - y^2}} dr$$

This is an integral equation of Volterra type, which has been studied by Abel. The dephasing function $\Delta\Phi(y)$ is the "Abel transform" of $n_e(r)$. If $n_e(a) = 0$, i.e. if the plasma has a finite size, one can invert it (Abel inversion) and get $n_e(r)$:

$$\frac{k_0}{n_{cr}} n_e(r) = \frac{1}{\pi} \int_r^a \frac{d}{dy}(\Delta\Phi(y)) \frac{dy}{\sqrt{y^2 - r^2}} dr$$

The "idea" behind Abel transform is shown in Fig. 19. One starts reconstructing the profile from the exterior (i.e. $y = a$), where one can assume that the value of density is uniform and the length of the (very short) plasma path is known. In this case $\Delta\Phi_1 = n_1 d_1$ and measuring $\Delta\Phi$ it is possible to deduce n_e. This information is then transported inside using the axial symmetry to the next path (next y value) where now it is possible to write $\Delta\Phi_2 = 2n_1 d_2 + n_1 d_3$ and so on…

Let's notice that in reality from a mathematical point of view the Abel transform can be applied also when the condition $n_e(a) = 0$ is not satisfied. It is sufficient that $\Delta\Phi$ drops to zero quicker that $1/r$.

The Abel transform is a powerful tool but it also has severe drawbacks. First of all, we have seen (Fig. 19) that the information on the electron density is reconstructed staring form the exterior. This means that the errors at the plasma edge will reflect up to the center of the plasma, even very severely.

Second, the mathematical operation of Abel inversion is very sensitive to noise. Small fluctuations in the data ($\Delta\Phi$) can imply very large fluctuations in the retrieved electron density n_e. This implies that before performing the Abel inversion it is usually needed to filter the data in order to reduce noise, but this also imply a loss of information.

Third, the Abel inversion requires perfect symmetry while experimental results usually show some asymmetries. So, before applying the Abel transform one must find the most likely symmetry axis and then either average the two half-images or duplicate one of the two halves to obtain a full dephasing map (which either correspond to filtering and losing information or to introducing spurious data).

As an application of the case of plasmas with cylindrical symmetry and of Abel transform let's consider the experimental results presented in [38]. In this work a Ti:Sapphire laser with energy of 60 mJ, and a duration of 150 fs was split in two parts, one to be used as pump (40 mJ) and the other acting as probe (20 mJ). The pump pulse was focused by a single lens with $f = 100$ cm on the target to a peak intensity on target of 4×10^{15} W/cm^2. The probe beam was frequency doubled by a 2 mm thick BBO crystal. The probe beam was then injected in a Mach–Zender interferometer and the interferogram was imaged on a 12-bit CCD.

Aluminum wires with 200 μm diameter were used as target, again in order to eliminate the problem of tilting with respect to the probe direction. Moreover, since the focal spot size was much smaller than the wire radius, the interaction area could practically be considered as a planar one.

The delay line has been varied to introduce delays ranging from $t = 0$ ps to $\Delta t = 100$ ps by steps of 3 ps. In Fig. 20, four interferograms show the plasma expansion at different times. The image resolution was about 1.2 μm

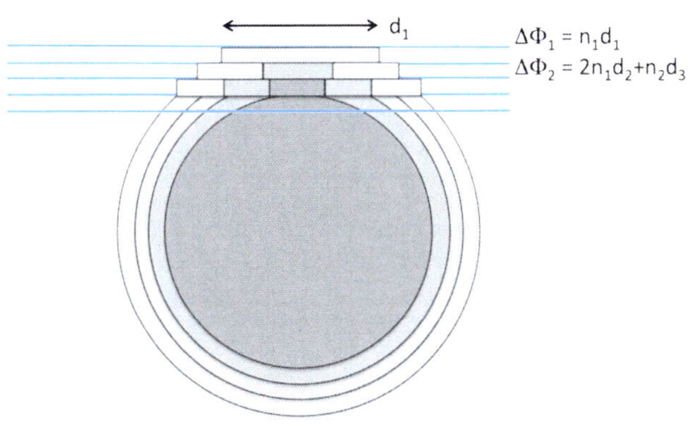

Fig. 19 Principle of Abel transform

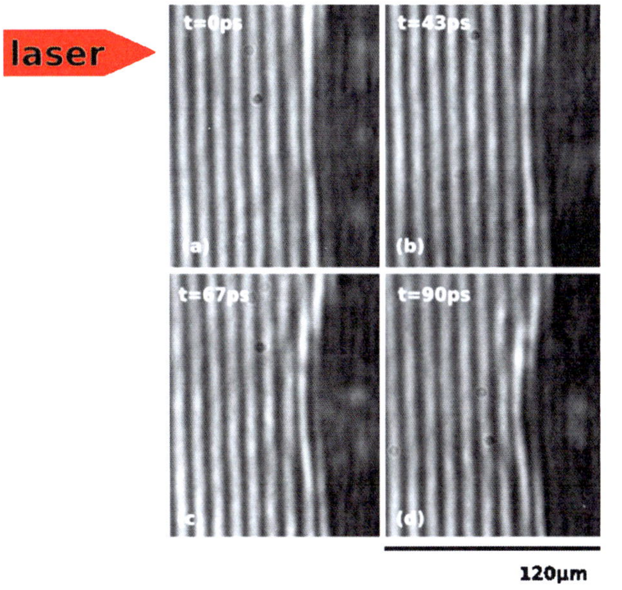

Fig. 20 Interferograms showing plasma expansion at four different delays: **a** 0 ps, **b** 43 ps, **c** 67 ps, and **d** 90 ps (from [35])

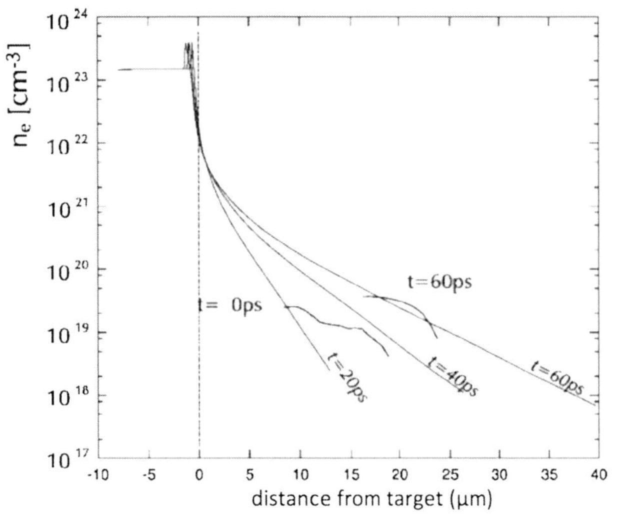

Fig. 21 Experimental plasma density profiles at 3 different delays: 0 ps, 20 ps, and 60 ps, and comparison with theoretical predictions (from [35])

given by diffraction limit and the interferometer interfringe was $\lambda_{het} \approx 10$ μm.

Some of the obtained density profiles are shown in Fig. 21 together with predictions from hydro simulations. Let's notice that unlike in the work previously presented (Ref. [1]), here the interferograms are recorder long after the end of the laser pulse (ps vs. fs) and hence the plasma expansion is rather adiabatic.

Polarimetry

The polarimetry diagnostic (Faraday rotation) was introduced in laser generated plasmas very early [4]. The application of the method for large density gradients dense plasmas accompanied by simultaneous shadowgraphy and interferometry is described in [39]. When a probe laser beam propagates in a magnetized plasma, the presence of the external magnetic field in the plasma create an anisotropy of the refractive index as [25, 40]:

$$n^2 = 1 + X\left(1 - i\nu/\omega - \frac{Y_\perp^2}{2(1 - X - i\nu/\omega)} \pm \sqrt{\frac{Y_\perp^2}{4(1 - X - i\nu/\omega)}}\right)$$

where $X = n_e^2/n_{cr}^2$, $Y = \omega_{ce}/\omega$, $Y_\perp = Y\sin\,\theta$. Here θ is the angle between the direction of propagation of the probe beam $\vec{k}$ and the direction of the magnetic field and ω_{ce} is the electron cyclotron frequency. The polarimetry diagnostics uses this anisotropy to retrieve information on the magnetic field inside of the plasma.

The most commonly used approach to measure magnetic fields is through the Faraday rotation. This is the case in which $\theta = 0$ and the polarization axis of the probe beam will change depending of the magnetic field strength and the plasma density. In the case of an underdense plasma with $X \ll 1$ and $Y \ll 1$, the analytical expression for the rotation of the polarization axis is

$$\Delta\Psi = \frac{1}{2}\int XY\cos\theta\frac{\omega}{c}dl \qquad (4)$$

Experimentally such rotation can be observed using a polarizing beam splitter like a Wollaston Cube, which allows the separation of the two components of polarization on an orthonormal basis $(\overrightarrow{e_x}, \overrightarrow{e_y})$. The rotation of polarization is then the ratio of the two projections of the electric field on the basis $(\overrightarrow{e_x}, \overrightarrow{e_y})$, i.e. it is $\Delta\Psi = \Psi_2 - \Psi_1 = E_{y2}/E_{x2} - E_{y1}/E_{x1}$. In the case of an external magnetic field of very high intensity, the rotation of polarization can become larger than $\pi/2$ defining the limits of this diagnostic. Generally, this diagnostic need to be used in combination with interferometry allowing to obtain the density profile, which, according to Eq. 4, is essential to retrieve the magnetic field map. This method to measure the magnetic field is well known and commonly used in astrophysics [41] or in the study of laser–plasma interactions [42]. When the magnetic field is purely orthogonal ($Y = Y_\perp$ or $Y_{//} = Y\cos\,\theta = 0$) Faraday rotation disappears.

When the magnetic field is not parallel to the direction of propagation of the probe beam, it is possible to use the Cotton–Mouton effect, which induces a modification of the ellipticity of the polarization during the propagation. In the case of an underdense plasma with $X \ll 1$ and $Y \ll 1$ this ellipticity modification is

$$\frac{b}{a} = \frac{1}{2}\int \frac{XY^2\sin^2\theta}{1 - Y^2}\sin 2\beta\frac{\omega}{c}dl$$

where b/a is the ratio of the minor and major axis of the ellipse defining the polarization of the probe beam (i.e. $b/$

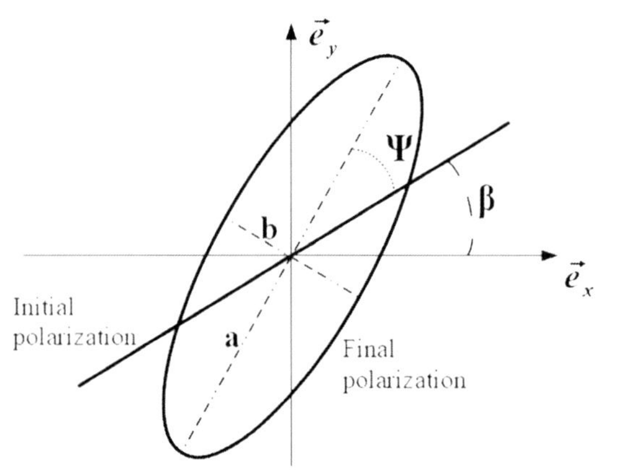

Fig. 22 The polarization ellipse and the polarization coordinate systems for a direction of propagation of the probe beam $\vec{k} = \vec{e_z}$

$a = 0$ for linear polarization beam and $b/a = 1$ for circular polarization) and β is the angle between the main polarization axis of the probe beam and $\vec{e_x}$, as described in Fig. 22.

If the magnetic field is only on the plane perpendicular to the direction of propagation of the probe beam, a simple polarizing cube can be used giving $b/a = E_y/E_x$ when the probe beam is initially linearly polarized along $\vec{e_x}$ (here E_y and E_x are the components of the electric field of the probe electromagnetic wave along $\vec{e_x}$ and $\vec{e_y}$. Caution needs to be used with this diagnostic method as the Cotton-Mouton effect is much weaker than the Faraday rotation.

If the magnetic field direction is unknown, one need to obtain the complete polarization of the laser as defined by the Stokes parameters [43]:

$$S = \begin{pmatrix} S_0 \\ S_1 \\ S_2 \\ S_3 \end{pmatrix} = \begin{pmatrix} E_x^2 + E_y^2 \\ E_x^2 - E_y^2 \\ E_a^2 - E_b^2 \\ E_r^2 - E_l^2 \end{pmatrix}$$

These parameters correspond to the projection of the laser intensity on 3 different basis $\overrightarrow{(e_x, e_y)}$, $\overrightarrow{(e_a, e_b)}$ oriented at 45° with respect to the previous one, and $\overrightarrow{(e_r, e_l)}$, the basis indicating left-had and right-hand rotations. Using the Stokes vector, we can retrieve the angle defining the polarization of the probe beam as:

$$\Psi = \frac{1}{2}\arctan\left(\frac{S_2}{S_1}\right)$$

$$\arctan\frac{b}{a} = \frac{1}{2}\arctan\left(\frac{S_3}{\sqrt{S_1^2 + S_2^2}}\right)$$

The easiest way to measure this parameter is to use a polarimeter measuring the intensity on the different basis show previously (Fig. 23). The intensity on the basis $\overrightarrow{(e_x, e_y)}$ is obtained using a polarizing beam splitter (i.e. a Wollaston in the figure). These two projections give the intensity S_0 of the probe beam. A polarizer at 45° extract the intensity on the direction $\vec{e_a}$, which allows calculating direction $\vec{e_b}$ as $E_b^2 = S_O - E_a^2$. Finally, the combination of a λ/4 wave plate and a polarimeter at 45° transmits only the intensity of the left-hand rotating component E_l wave (which allows calculating E_r as $E_r^2 = S_O - E_l^2$).

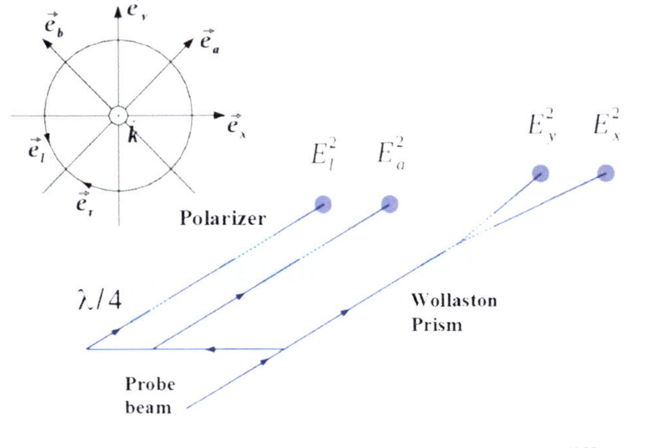

Fig. 23 Example of a two CCDs polarimeter. The three different basis $\overrightarrow{(e_x, e_y)}$, $\overrightarrow{(e_a, e_b)}$ and $\overrightarrow{(e_r, e_l)}$, are represented on the top-left corner

The Cotton–Mouton polarimetry method (measuring the Stokes parameters) was first introduced in high intensity laser solid target interaction experiments at CLF/RAL by Tatarakis et al. [44, 45]. More recently it has been successfully developed and applied by the research group at TIFR in India. They used a probe beam at normal incidence on the target and four different photodiodes measuring the 4 Stokes parameters with time resolution (but without spatial resolution) [46]. Later they improved the method using a CCD detector allowing obtaining spatially resolved images of the Stokes parameters in the magnetized plasma [47].

A limitation of the method is anyway that when the magnetic field is very strong, reaching MGauss intensities, Eq. 4 can no longer be used as the change of ellipticity becomes non-bijective. This means that different values of the magnetic field can produce the same final ellipticity.

In addition, as is the case of Faraday rotation, the polarimetry diagnostic does not allow alone to extract the magnetic field as the change of polarization also depends on the plasma density. Coupling to interferometry is therefore needed to estimate the magnetic field amplitude.

[For sake of completeness, we notice here that magnetic field in the high density region (near the critical density), which is not accessible with external probe pulses, can be estimated using self-generated harmonics in high intensity laser matter interactions. Tatarakis et al. [48] and Gopal et al. [49] measured the change of polarization produced by the Faraday effect on the self-generated harmonics of the laser allowing to get information about the magnetic field deeper inside the plasma, but without information about the path of the harmonics inside the plasma (i.e. the exact position of harmonics generation)].

As example of use of the polarimetry diagnostic, we refer to an experiment recently performed at the CELIA laboratory with the laser ECLIPSE, a Ti:Sapphire system delivering pulses of 30 fs and 200 mJ. These were focused on Al-coated SiO_2 targets to an intensity of 5×10^{17} W/cm^2 [47].

The laser beam was split in two part, one (95% of the energy) to be used as pump, and the other acting as the probe (2.5% of the energy). The pump pulse was focused by an off-axis parabola on the target at 45°. The probe beam was frequency doubled by a 1-mm thick BBO crystal

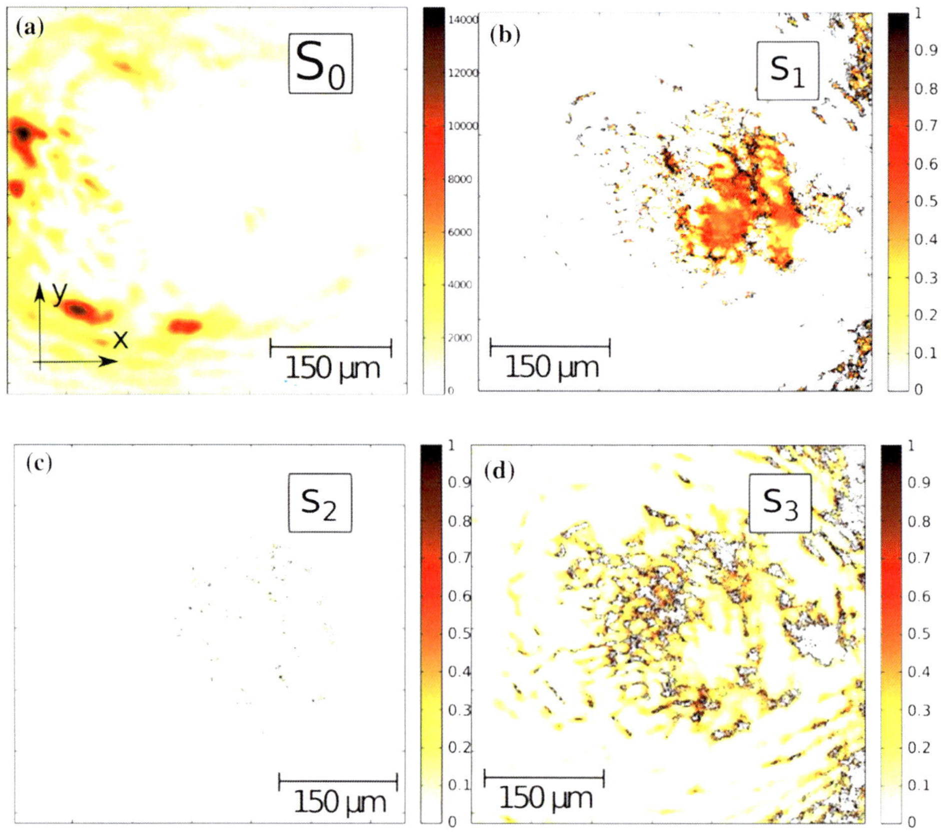

Fig. 24 Spatial mapping of the Stokes parameters of the reflected probe beam 1.5 ps after the interaction of the laser at 5×10^{17} W/cm^2 on an Al-coated fused silica target. **a** S_0, **b** $s_1 = S_1/S_0$, **c** $s_2 = S_2/S_0$, **d** $s_3 = S_3/S_0$

and focused at normal incidence by a lens with focal length f = 220 mm on a focal spot on target of about 300 μm. Frequency doubling was introduced with the dual purpose of letting the probe pulse propagate to higher density regions and of using a probe wavelength different from the pump, avoiding the problem introduced by the scattered pump light entering the interferometer beam path.

The probe beam was reflected by the plasma and collect by an imaging system (through the same lens used for focusing) with a 10 × magnification and then injected in the polarimeter which used two 12-bit charge coupled devices (CCD), each CCD measuring 2 of the 4 stokes parameters. The spatial resolution of the diagnostic was 10 μm. A delay line allowed introducing delays of the probe beam ranging from $\Delta t = -3$ ps to $\Delta t = 50$ ps in steps of 0.4 ps. Here Δt is the delay between the arrival on target of the main beam and of the probe beam (hence a negative Δt means that the probe beam arrives on target before the main pulse has struck it). The plasma density and scalelengths (which as explained before are needed to retrieve the value of magnetic field) were characterized suing interferometry in the same experiment.

Figure 24 shows the map of the four Stokes parameters, which are used to retrieve the ellipticity and rotation of the main polarization axis (Fig. 25). In this shot, we can observe that the rotation of the polarization axis is

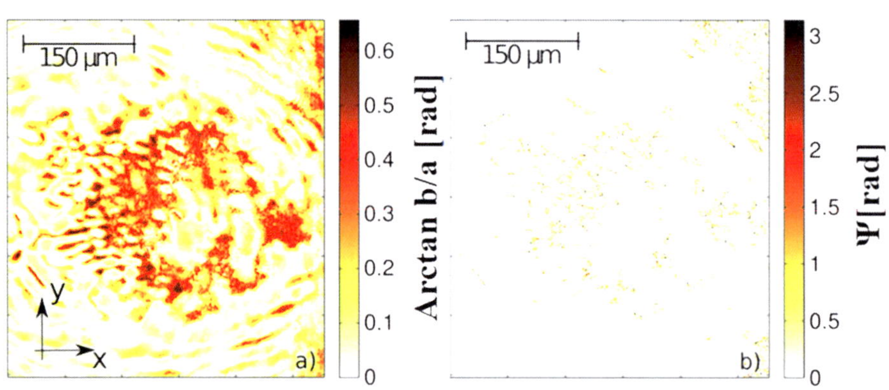

Fig. 25 Spatial mapping of angles **a** $arctan(b/a)$, **b** Ψ

negligible while its ellipticity is changing drastically. This implies the presence inside the plasma of a strong azimuthal magnetic field.

Due to the fact that the magnetic field strength is very large, the relation between magnetic field and elicit change is no longer bijective, as discussed earlier. Hence it is difficult to estimate the absolute value of the magnetic field. However, the comparison with simulation results, allowed to infer a value ~ 10 MegaGauss.

The polarimetry diagnostic allows measuring the magnetic field inside an underdense plasma with good temporal resolution (equal to the duration of the probe laser pulse) and good spatial resolution (define by the resolution of the imaging system).

Conclusions

This paper was addressed to give a few examples of the large variety of diagnostics needed to study laser-produced plasmas. We have focused on optical diagnostics for the undercritical region of the plasma: shadowgraphy, schlieren, interferometry, polarimetry showing some of the challenges related to measuring the detailed characteristics of small-size, fast-varying plasmas.

Acknowledgements This work has been carried out within the framework of the EUROfusion Enabling Research Projects: AWP17-ENR-IFE-CEA-01 «Preparation and Realization of European Shock Ignition Experiments» and AWP17-ENR-IFE-CEA-02 «StarkZee: Towards a universal benchmarked Stark-Zeeman code for spectroscopic diagnostics» and has received funding from the Euratom research and training programme 2014–2018 under grant agreement No. 633053. The views and opinions expressed herein do not necessarily reflect those of the European Commission. The work was also supported by the Competitiveness Program of NRNU MEPhI, Russia.

References

1. A. Aliverdiev, D. Batani, R. Dezulian, T. Vinci, A. Benuzzi-Mounaix, M. Koenig, V. Malka, Coronal hydrodynamics of laser-produced plasmas by interferometry. Phys. Rev. E. **78**, 046404 (2008)
2. K.L. Lancaster, J. Pasley, J.S. Green, D. Batani, S. Baton, R.G. Evans, L. Gizzi, R. Heathcote, C. Hernandez Gomez, M. Koenig, P. Koester, A. Morace, I. Musgrave, P.A. Norreys, F. Perez, J. Waugh, N. Woolsey, Temperature profiles derived from transverse optical shadowgraphy in ultra-intense laser plasma interactions at 6×10^{20} W/cm^2. Phys. Plasmas **16**, 056707 (2009)
3. M. Manclossi, D. Batani, D. Piazza, S. Baton, F. Amiranoff, M. Koenig, H. Popescu, P. Audebert, J.J. Santos, E. Martinolli, M. Rabec Le Gloahec, A. Antonicci, C. Rousseaux, M. Borghesi, C. Cecchetti, V. Malka, T. Hall, Optical shadowgraphy and proton imaging as diagnostics tools for fast electron propagation in ultra-high-intensity laser-matter interaction. Radiat. Eff. Defects Solids **160**, 575 (2005)
4. J.A. Stamper, B.H. Ripin, Faraday-rotation measurements of megagauss magnetic fields in laser-produced plasmas. Phys. Rev. Lett. **34**, 138 (1975)
5. F. Rosmej, A. Faenov, T. Pikuz, F. Flora, P. Di Lazzaro, S. Bollanti, N. Lisi, T. Letardi, A. Reale, L. Palladino, D. Batani, S. Bossi, A. Bernardinello, A. Scafati, L. Reale, A. Zigler, M. Fraenkel, E. Cowan, Inner-shell satellite transitions in dense short pulse plasmas. J. Quant. Spectrosc. Radiat. Transf. **58**, 859 (1997)
6. E. Martinolli, M. Koenig, J.M. Boudenne, E. Perelli-Cippo, D. Batani, T.A. Hall, Conical crystal spectrograph for high brightness X-ray Kalpha spectroscopy in subpicosecond laser-solid interaction. Rev. Sci. Instrum. **75**, 2024 (2004)
7. H. Nishimura, Y. Inubushi, Y. Okano, S. Fujioka, T. Kai, T. Kawamura, D. Batani, A. Morace, R. Redaelli, C. Fourment, J. Santos, G. Malka, A. Boscheron, A. Casner, M. Koenig, T. Nakamura, T. Johzaki, H. Nagatomo, K. Mima, X-ray polarization spectroscopy to study energy transport in ultra-high intensity laser produced plasmas. J. Phys. **112**, 022080 (2008)
8. M. Koenig, J.M. Boudenne, P. Legriel, T. Grandpierre, D. Batani, S. Bossi, S. Nicolella, R. Benattar, A computer driven crystal spectrometer with CCD detectors for X-ray spectroscopy of laser–plasmas. Rev. Sci. Instrum. **68**, 2387 (1997)
9. A. Stepanov, A. Starostin, V. Roerich, V. Makhrov, A. Faenov, A. Magunov, T. Pikuz, I. Skobelelev, F. Flora, S. Bollanti, P. Di Lazzaro, N. Lisi, T. Letardi, L. Palladino, A. Reale, D. Batani, S. Bossi, A. Bernardinello, A. Scafati, L. Reale, A. Osterheld, W. Goldstein, Modelling of the He-like magnesium spectral line radiation from the plasma created by XeCl and Nd-glass lasers. J. Quant. Spectrosc. Radiat. Transf. **58**, 937 (1997)
10. D. Batani, F. Bianconi, A. Giulietti, D. Giulietti, L. Nocera, Second harmonic polarisation and conversion efficiency in laser produced sparks. Opt. Commun. **70**, 38 (1989)
11. A. Giulietti, D. Giulietti, D. Batani, V. Biancalana, L. Gizzi, L. Nocera, E. Schifano, Spectroscopic evidence for sum frequency of forward and backscattered light in laser plasmas. Phys. Rev. Lett. **63**, 524 (1989)
12. D. Giulietti, V. Biancalana, D. Batani, A. Giulietti, L. Gizzi, L. Nocera, E. Schifano, Three half harmonic generation in laser plasma interaction: evidence for plasmon propagation. Il Nuovo Cimento **13 D**, 845 (1991)
13. L. Antonelli, D. Batani, A. Patria, O. Ciricosta, C. Cecchetti, P. Koester, L. Labate, A. Giulietti, L.A. Gizzi, A. Moretti, M. Richetta, L. Giuffrida, L. Torrisi, M. Kozlova, J. Nejdl, M. Sawicka, D. Margarone, B. Rus, G. Schurtz, X. Ribeyre, M. Lafon, C. Spindloe, T. O'Dell, Laser-plasma coupling in the shock-ignition intensity regime. Acta Tec. **56**, T57 (2011)
14. F. Pisani, M. Koenig, D. Batani, T. Hall, D. Desenne, J. Bruneau, C. Reverdin, Toroidal crystal spectrometer for time-resolved X-ray absorption diagnostic in dense plasmas. Rev. Sci. Instrum. **70**, 3314 (1999)
15. T. Hall, J. Al-Kuzee, A. Benuzzi, M. Koenig, J. Krishnan, N. Grandjouan, D. Batani, S. Bossi, S. Nicolella, Experimental observation of the shift and width of the aluminum K absorption edge in laser shock compressed plasmas. Europhys. Lett. **41**(5), 495–500 (1998)
16. D. Batani, L. Antonelli, S. Atzeni, J. Badziak, F. Baffigi, T. Chodukowski, F. Consoli, G. Cristoforetti, R. de Angelis, R. Dudzak, G. Folpini, L. Giuffrida, L.A. Gizzi, Z. Kalinowska, P. Koester, E. Krousky, M. Krus, L. Labate, T. Levato, Y. Maheut, G. Malka, D. Margarone, A. Marocchino, J. Nejdl, P. Nicolai, T. O'Dell, T. Pisarczyk, O. Renner, Y.J. Rhee, X. Ribeyre, M. Richetta, M. Rosinski, M. Sawicka, A. Schiavi, J. Skala, M. Smid, C. Spindloe, J. Ullschmied, A. Velyhan, T. Vinci, Generation of high pressure shocks relevant to the shock-ignition intensity regime. Phys. Plasmas **21**, 032710 (2014)

17. A. Ravasio, M. Koenig, S. Le Pape, A. Benuzzi-Mounaix, H.S. Park, C. Cecchetti, P. Patel, A. Schiavi, N. Ozaki, A. Mackinnon, B. Loupias, D. Batani, T. Boehly, M. Borghesi, R. Dezulian, E. Henry, M. Notley, S. Bandyopadhyay, R. Clarke, T. Vinci, Hard X-ray radiography for density measurement in shock compressed matter. Phys. Plasmas **15**, 060701 (2008)
18. A. Ravasio, S. Le Pape, A. Benuzzi-Mounaix, L. Romagnani, C. Cecchetti, D. Batani, T. Boehly, M. Borghesi, R. Dezulian, L. Gremillet, E. Henry, D. Hicks, B. Loupias, A. McKinnon, N. Ozaki, H.S. Park, P. Patel, A. Schiavi, T. Vinci, R. Clarke, M. Notley, S. Bandyopadhyay, M. Koenig, Proton radiography of a shock compressed target. Phys. Rev. E. **82**, 016407 (2010)
19. M. Koenig, B. Faral, J.M. Boudenne, D. Batani, S. Bossi, A. Benuzzi, Use of optical smoothing techniques for shock wave generation in laser produced plasmas. Phys. Rev. E Rapid Commun. **50**, R3314–R3317 (1994)
20. A. Benuzzi, W. Nazarov, M. Koenig, J. Krishnan, B. Faral, M. Temporal, D. Batani, L. Muller, F. Torsiello, T. Hall, N. Grandjouan, Dynamics of laser produced shocks in foam-solid targets. Phys. Plasmas Lett. **5**, 2827–2829 (1998)
21. D. Batani, A. Benuzzi, M. Koenig, I. Krasyuk, P. Pashinin, A. Semenov, I. Lomonosov, V. Fortov, Problems of measurement of dense plasma heating in laser shock wave compression. Plasma Phys. Controll. Fusion **41**, 93–103 (1999)
22. S. Skupsky, S. Kacenjar, Measuring fuel ρR for inertial fusion experiments using neutron elastic-scattering reactions. J. Appl. Phys. **52**, 2608 (1981)
23. J.A. Frenje, D.T. Casey, C.K. Li, J.R. Rygg, F.H. Séguin, R.D. Petrasso, V. Yu Glebov, D.D. Meyerhofer, T.C. Sangster, S. Hatchett, S. Haan, C. Cerjan, O. Landen, M. Moran, P. Song, D.C. Wilson, R.J. Leeper, First measurements of the absolute neutron spectrum using the magnetic recoil spectrometer at OMEGA. Rev. Sci. Instrum. **79**, 10E502 (2008)
24. T. Caillaud, O. Landoas, M. Briat, S. Kime, B. Rossé, I. Thfoin, J.L. Bourgade, L. Disdier, V.Y. Glebov, F.J. Marshall, T.C. Sangster, Development of the large neutron imaging system for inertial confinement fusion experiments. Rev. Sci. Instrum. **83**, 033502 (2012)
25. I.H. Hutchinson, *Principles of Plasma Diagnostics* (Cambridge University Press, Cambridge, 1987)
26. S. Minardi et al., Opt. Lett. **33**, 86 (2008)
27. M.M. Michaelis et al., Opt. Laser Technol. **23**(5), 283 (1991)
28. M.M. Michaelis, O. Willi, Opt. Commun. **36**(2), 153 (1981)
29. G. de Izarra, C. De Izarra, Quantitative shadowgraphy made easy. Eur. J. Phys. **33**(6), 1821 (2012)
30. G.S. Settles, *Schlieren and Shadowgraph Techniques* (Springer, Berlin, 2001)
31. D. Haberberger, S. Ivancic, S.X. Hu, R. Boni, M. Barczys, R.S. Craxton, D.H. Froula, Measurements of electron density profiles using an angular filter refractometer. Phys. Plasmas **21**, 056304 (2014)
32. D. Batani, S.D. Baton, M. Manclossi, D. Piazza, M. Koenig, A. Benuzzi-Mounaix, H. Popescu, C. Rousseaux, M. Borghesi, C. Cecchetti, A. Schiavi, Laser-driven fast electron dynamics in gaseous media under the influence of large electric fields. Phys. Plasmas **16**, 033104 (2009)
33. D. Batani, S.D. Baton, M. Manclossi, J.J. Santos, F. Amiranoff, M. Koenig, E. Martinolli, A. Antonicci, C. Rousseaux, M. Rabec Le Gloahec, T. Hall, V. Malka, T.E. Cowan, J. King, R. Freeman, Ultra-intense laser-produced fast electron propagation in gas jets. Phys. Rev. Lett. **94**, 055004 (2005)
34. A. Faenov, A. Magunov, T. Pikuz, D. Batani, G. Lucchini, F. Canova, M. Piselli, Bright, point X-ray source based on a commercial portable 40 ps Nd YAG laser system. Laser Part. Beams **22**, 373–379 (2004)
35. D.W. Sweeney, D.T. Attwood, L.W. Coleman, Interferometric probing of laser produced plasmas. Appl. Opt. **15**, 1126–1128 (1976)
36. K.A. Nugent, Interferogram analysis using an accurate fully automatic algorithm. Appl. Opt. **18**, 3101 (1985)
37. P. Tomassini, A. Giulietti, L.A. Gizzi, M. Galimberti, D. Giulietti, M. Borghesi, O. Willi, Analyzing laser plasma interferograms with a continuous wavelet transform ridge extraction technique. Appl. Opt. **40**(35), 6561 (2001)
38. A. Flacco, A. Guemnie-Tafo, R. Nuter, M. Veltcheva, D. Batani, E. Lefebvre, V. Malka, Characterization of a controlled plasma expansion in vacuum for laser driven ion acceleration. J. Appl. Phys. **104**, 103304 (2008)
39. M. Tatarakis et al., Phys. Plasmas **5**(3), 682 (1998)
40. C.B. Wharton, M.A. Heald, *Plasma Diagnostics with Microwaves* (Wiley, New York, 1965)
41. I.B. Zeldovich, A.A. Ruzmaikin, D.D. Sokolov, *The Fluid Mechanics of Astrophysics and Geophysics*, vol. 3 (Gordon and Breach Science Publishers, New York, 1983), p. 381
42. M. Borghesi, A.J. Mackinnon, R. Gaillard, O. Willi, A. Pukhov, J. Meyer-ter Vehn, Large quasistatic magnetic fields generated by a relativistically intense laser pulse propagating in a preionized plasma. Phys. Rev. Lett. **80**, 5137–5140 (1998)
43. S.E. Segre, A review of plasma polarimetry—theory and methods. Plasma Phys. Control. Fusion **41**(2), R57 (1999)
44. M. Tatarakis et al., Nature **415**, 280 (2002)
45. M. Tatarakis et al., Phys. Plasmas **9**(5), 2244 (2002)
46. A.S. Sandhu, A.K. Dharmadhikari, P.P. Rajeev, G.R. Kumar, S. Sengupta, A. Das, P.K. Kaw, Laser-generated ultrashort multi-megagauss magnetic pulses in plasmas. Phys. Rev. Lett. **89**, 225002 (2002)
47. S. Mondal, V. Narayanan, A.D. Lad, S. Ahmed, S. Sengupta, A. Das, Z.M. Sheng, P.K. Kaw, G. Ravindra Kumar, Measurement of hot electron transport in overdense plasma via self induced giant magnetic pulses. J. Phys. Conf. Ser. **44**, 022049 (2010)
48. M. Tatarakis, A. Gopal, I. Watts, F.N. Beg, A.E. Dangor, K. Krushelnick, U. Wagner, P.A. Norreys, E.L. Clark, M. Zepf, R.G. Evans, Measurements of ultrastrong magnetic fields during relativistic laser–plasma interactions. Phys. Plasmas **9**, 2244–2250 (2002)
49. A. Gopal, S. Minardi, M. Burza, G. Genoud, I. Tzianaki, A. Karmakar, P. Gibbon, M. Tatarakis, A. Persson, C.-G. Wahlström, Megagauss magnetic field generation by ultra-short pulses at relativistic intensities. Plasma Phys. Control. Fusion **55**, 035002 (2013)

Journal of Fusion Energy (2019) 38:315–329
https://doi.org/10.1007/s10894-018-0182-1

ORIGINAL RESEARCH

Fusion Neutrons: Tritium Breeding and Impact on Wall Materials and Components of Diagnostic Systems

Marek Rubel[1]

Published online: 1 September 2018

Abstract

A concise overview is given on the impact of fusion neutrons on various classes of materials applied in reactor technology: plasma-facing, structural and functional tested for tritium production and for diagnostic systems. Tritium breeding in the reactor blanket, fuel cycle and separation of hydrogen isotopes are described together with issues related to primary (tritium) and induced radioactivity. Neutron-induced damage and degradation of material properties are addressed. Material testing under neutron fluxes and safety issues associated with handling components in the radioactive environment are described. A comprehensive list of references to monographs and research papers is included to help navigation in literature.

Keywords Controlled fusion · Tritium · Radiation damage · Transmutation · Fuel cycle

Introduction

The driving force and the main goal of research on controlled thermonuclear fusion is to construct a power generating system based on the energy output of nuclear reactions between light isotopes. Processes under consideration involve deuterium (symbol D, d or ^{2}H), tritium (T, t or ^{3}H) and ^{3}He:

$$D + T \rightarrow \alpha(3.52\ \mathrm{MeV}) + n(14.06\ \mathrm{MeV}) \quad (1)$$

$$D + {}^{3}He \rightarrow \alpha(3.67\ \mathrm{MeV}) + H(14.69\ \mathrm{MeV}) \quad (2)$$

$$D + D \rightarrow T(1.01\ \mathrm{MeV}) + H(3.03\ \mathrm{MeV}) \quad (3a)$$

$$D + D \rightarrow {}^{3}He(0.82\ \mathrm{MeV}) + n(2.45\ \mathrm{MeV}) \quad (3b)$$

$$T + T \rightarrow {}^{4}He(3.77\ \mathrm{MeV}) + 2n(7.53\ \mathrm{MeV}) \quad (4)$$

The branching ratio of Reactions (3a) and (3b) is around one.

The energy release (Q value) and cross-sections (σ) of respective nuclear reactions indicate that the deuterium–tritium fusion, Reaction (1), is the best possible option [1, 2]. The reaction's cross-section (σ) is greater than that of other processes listed above. The maximum is around 70 keV (700,000,000 K) of deuteron energy but high D–T reactivity is reached already at 20 keV. Cross-section values of other processes are significantly lower and their maxima are above hundreds of keV, what—in other words—would require efficient particle heating and their confinement at the temperature of 2×10^{9} K or even higher. As a consequence, at present, all plans and efforts towards commercial fusion are concentrated on a reactor fuelled with deuterium and tritium. Two main schemes of fusing nuclei are: (1) magnetic confinement fusion (MCF) of plasma ignited and maintained by strong magnetic field of several tesla [3] and (2) inertial confinement (ICF) by intense laser [4] or ion beams [5]. Many scientific and technological issues and challenges are similar in all these approaches, especially when it comes to production of tritium fuel. In the following only the first scheme will be discussed, because it is most matured from the reactor technology point of view.

In every method of energy production, from hydro, via fossil materials, windmills to nuclear, the most important and indispensable factor is the fuel and its availability. Experiments in present-day controlled fusion devices are carried out mostly with easily available deuterium (Reactions 3a, 3b), but in the past experimental campaigns with

✉ Marek Rubel
rubel@kth.se

1 Royal Institute of Technology (KTH), 100 44 Stockholm, Sweden

D–T mixture (including 1:1 D–T operation) were performed at the Joint European Torus (JET) [6–8] and Tokamak Fusion Test Reactor (TFTR) [9]. The next D–T campaign in JET at the Culham Science Centre, UK is planned in year 2019. A step forward is expected to be done in the International Thermonuclear Experimental Reactor (ITER) constructed by seven partners to demonstrate high power performance operation and generation of tritium.

The aim of this lecture is to give an introduction to issues related to fusion neutrons which are energy carriers. They are indispensable for tritium breeding, but they also modify properties of wall materials causing radiation damage and transmutation. First the D–T reaction and its consequences are described. Then tritium breeding and fuel cycle are introduced and it is followed by the description of requirements for plasma-facing and reactor materials. Transmutation and radiation damage are presented in the next paragraph in which also the impact of neutrons on material performance is addressed. The work is concluded with remarks material testing and on safety issues associated with handling components in radioactive environment.

Deuterium–Tritium Fusion Reaction: Reactants and Consequences

In this paragraph all components of the D–T reaction are discussed because each of them plays an important and unique role. Deuterium and tritium are fuel species. The Q value of their reaction is 17.58 MeV of which, according to kinematics, 3.52 MeV is associated with fast alpha particle and 14.06 MeV is carried away by a fast neutron. A complete burn of 1 g of the equimolar D–T mixture corresponds to 1.2×10^{23} reactions producing energy of 67.9 GJ and 271.8 GJ carried by alphas and neutrons, respectively. Alpha particles are to transfer their energy to plasma thus enabling its efficient heating to achieve self-sustaining fusion. However, it also means that eventually the radiated power transferred from alphas to the fuel must be extracted by plasma-facing materials (PFM) and components (PFC). Fast neutrons are to deposit their energy in the absorber (reactor blanket) to facilitate heat exchange and transfer to electricity generating systems of a power plant and, simultaneously, for tritium production via nuclear reactions in the blanket.

The best way to exemplify the general power balance into consider a reactor of, for instance, 500 MW of fusion power. 80% of that (i.e. 400 MW) will be associated with neutrons and the rest (100 MW) with alphas. To the latter number one must add power injected by auxiliary heating systems, approximately 50 MW (neutral beam injection and resonance methods), is to be radiated by plasma and removed by PFC, i.e. 150 MW in total.

In summary, the reaction provides all necessary information regarding the direction how to proceed towards the main goal. However, it also indicates and signalises serious consequences and technological challenges related to power exhaust and radioactivity connected with the use of tritium and neutron-induced activation of plasma-surrounding structures. Neutrons are energy carriers, but—when passing through the surrounding wall components—they cause serious modification of structural (walls) and functional (diagnostics) materials. It is stressed here that issues related to power handling and exhaust, radiation damage and tritium breeding are universal for all confinement schemes realised for energy generation.

Tritium Breeding and Tritium Plant

Tritium is a low-energy β^- emitter:

$$T(\beta^-) \rightarrow {}^3\mathrm{He}\,(18.59\,\mathrm{keV},\ t_{1/2} = 12.323\,\mathrm{years}). \quad (5)$$

Resulting radioactivity of 1 g T equals to 9652 Ci (3.571×10^{14} Bq). In nature it is formed by the interaction of cosmic rays with atmospheric gases. The first step is the generation of neutrons which then interact with nitrogen:

$$^{14}\mathrm{N} + \mathrm{n} \rightarrow {}^{12}\mathrm{C} + \mathrm{T} - 4.3\,\mathrm{MeV} \quad (6)$$

$$^{14}\mathrm{N} + \mathrm{n} \rightarrow 3\,{}^4\mathrm{He} + \mathrm{T} - 11.5\,\mathrm{MeV}. \quad (7)$$

These processes yield annually around 150 PBq of tritium (1.5×10^{17} Bq, approximately 0.411 kg). The total inventory resulting from the global equilibrium (formation and decay) amounts to 2590 PBq ($\sim$ 7.2 kg). A small amount of tritium is also formed by the reaction of neutrons in rocks containing lithium (the main reaction is with ^{6}Li), boron, uranium and thorium [10]. In summary, tritium does not occur in nature in quantities to be considered as resources sufficient for reactor operation. According to the estimates one requires around 168 kg tritium per year for a reactor of 3 GW_{th} [2].

Tritium Breeding

The data shown above clearly indicate that tritium for a D–T reactor is to be produced specially for that purpose. It should be stressed that there are no tritium factories (and no such plans) to generate required huge amounts of tritium to supply a fusion plant. Tritium for many applications, including military, is generated nowadays in nuclear reactors as a minor fission product of ^{235}U, ^{233}U and ^{239}Pu (1 atom generated in 10^4 acts of fission) and in reactions of neutrons with deuterium (in heavy water nuclear reactors), lithium-6 or boron-10. Availability and production rates of T in CANDU reactors are discussed in [11]: around 100 g per reactor per year. Therefore, tritium for a fusion reactor must be produced on site, directly in the reactor blanket. Breeding based on reactions of neutrons with lithium is the choice. The element has two stable isotopes ^{6}Li and ^{7}Li with natural abundance of 7.59% and 92.41%, respectively. They are transmuted to tritium in following reactions:

$$^6\mathrm{Li} + \mathrm{n} \to \alpha + \mathrm{T} + 4.78\,\mathrm{MeV} \quad (E_{th} = 2.47\,\mathrm{MeV}), \tag{8}$$

$$^7\mathrm{Li} + \mathrm{n} \to \alpha + \mathrm{T} + \mathrm{n}' - 2.47\,\mathrm{MeV}, \tag{9}$$

$$^7\mathrm{Li} + \mathrm{n} \to 2\mathrm{T} + 2\mathrm{n}' - 10.3\,\mathrm{MeV}, \tag{10}$$

where E_{th} denotes the threshold energy of the process. All three reactions yield neutrons but physics of processes clearly indicates benefits of using ^{6}Li. The reaction cross-section increases with the decreasing neutron energy and, as obtained using JENDL-4.0 (Japanese Evaluated Nuclear Data Library), it reaches values exceeding 900 barn for thermal neutrons, while cross-sections of (^{7}Li,n) are much lower, as shown in Chapter 12 of Ref. [2]. Therefore, lithium heavily enriched ($> 80\%$) with ^{6}Li is a constituent of all candidate materials considered for the absorber. Several ceramics have been tested: Li_2ZrO_3, Li_8ZrO_6, Li_2TiO_3, $LiAlO_2$, Li_2SiO_3 and Li_4SiO_4. The other candidates are lithium oxide (Li_20), mixture of lithium and beryllium fluorides $(LiF)_2$–BeF_2 called Flibe (Li_2BeF_4) and Li17–Pb eutectic in the form of a liquid alloy or pebbles.

In the (^{6}Li, n) process one neutron produces only one tritium atom. To increase the efficiency of tritium breeding a neutron multiplier is to be added. It is mainly beryllium, but also lead (Pb) is considered in some concepts of tritium breeding blankets:

$$^9\mathrm{Be} + \mathrm{n} \to 2\alpha + 2\mathrm{n}' - 2.5\,\mathrm{MeV} \quad (E_{th} = 2.7\,\mathrm{MeV}), \tag{11}$$

$$^{208}\mathrm{Pb} + \mathrm{n} \to {}^{207}\mathrm{Pb} + 2\mathrm{n}' - 7.4\,\mathrm{MeV}. \tag{12}$$

There are also side reactions with low cross-sections leading to the production of tritium and helium and, eventually, to the accumulation of tritium in He bubbles formed in Be:

$$^9\mathrm{Be} + \mathrm{n} \to \alpha + {}^6\mathrm{He} \quad (E_{th} = 0.67\,\mathrm{MeV}) \tag{13}$$

$$^6\mathrm{He}\ (\beta^-) \to {}^6\mathrm{Li} \quad (E_\beta = 3.508\,\mathrm{MeV}, \quad t_{1/2} = 0.807\,\mathrm{s}) \tag{14}$$

$$^9\mathrm{Be} + \mathrm{n} \to {}^7\mathrm{Li} + \mathrm{T} - 10.40\,\mathrm{MeV} \quad (E_{th} = 11.6\,\mathrm{MeV}). \tag{15}$$

Several concepts for tritium breeding modules (TBM) are considered to be constructed by different ITER partners. The modules are to be consecutively tested on three ports (as currently proposed) once the reactor is ready to start full D–T operation. The list comprises water-cooled ceramic breeder (WCCB) [12], and series of helium-cooled modules: lithium–lead (HCLL) [13, 14], pebble bed (HCPB) [15], dual coolant lithium–lead (DCLL) [16] and ceramic breeders such as He-cooled (HCCB) [17] and lithium–lead ceramic (LLCB) [18, 19]. Some of those concepts [13, 14] are also being developed for a demonstration reactor of fusion power plant: DEMO. This huge effort focused on the development of blanket modules clearly shows that the success of fusion will depend on the breeding ratio which must be greater than one to supply enough fuel for the self-sustaining operation [20]. In calculations one takes into account expected breeding efficiency, the amount spent in burning plasma and tritium losses through natural radioactive decay, permeation through structures and long-term retention in PFC: the latter is discussed in "Plasma–Wall Interactions and Plasma-Facing Materials" section. The efficiency is strongly influenced by geometrical structure, breeder materials, neutron multipliers in the blanket system. The minimum requirement is the breeding ratio of 1.1. The estimates by Sawan and Abdou [21] have led to the value of 1.2. It must be stressed that the latter value is only a results of calculations which cannot be verified in any other system than the fusion reactor itself, as it requires the full fuel cycle to be in place. A simplified tritium cycle is shown in Fig. 1: neutrons originating from the D–T reaction interact in the blanket with lithium generating tritium which must be extracted to fuel the fusion process.

Availability of other reactants needed for fusion and tritium breeding, deuterium and lithium, does not present major difficulties. These are easily achievable reactants because they are found in nature in inexhaustible supply. Non-radioactive deuterium (natural abundance 0.015%) is extracted from sea water ($\sim$ 33.3 g D/m^3). Also lithium can be extracted from sea water (main assumed source containing 0.1–0.2 ppm of ^{7}Li + ^{6}Li) or obtained from common minerals such as lepidolite K (Li, Al)$_3$(Al, Si, Rb)$_4$ O_{10}(F, OH)$_2$, spodumene $LiAl(SiO_3)_2$, petalite $LiAlSi_4O_{10}$, amblygonite (Li, Na)$AlPO_4$(F, OH) and others.

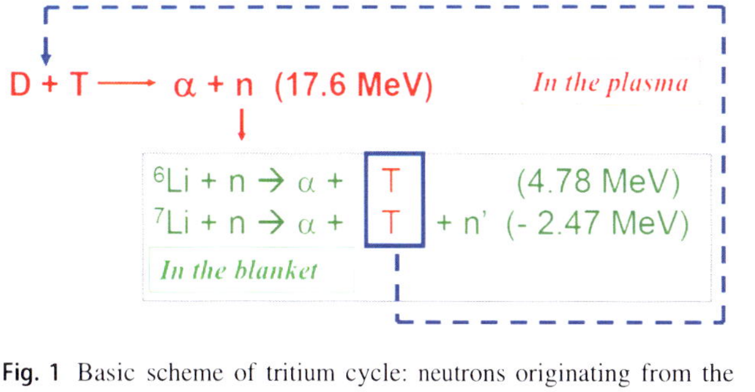

Fig. 1 Basic scheme of tritium cycle: neutrons originating from the D–T reaction interact in the blanket with lithium generating tritium which must be extracted to fuel the fusion process

Tritium Plant

All tritium from the reactor systems, both from the breeding blanket, exhausted from the torus (i.e. not used fuel pumped out from the reactor because only a small fraction of D–T mixture will be burnt in fusion reactions) and extracted from the cooling media (water and He), must be collected and handled by installations of the Tritium Plant before it can be used for plasma fuelling. Collected tritium must first undergo separation and purification from other gases used in a tokamak operation or also from various impurities in the vacuum system. Quantitative isotope separation is the next step.

Through the gas distribution system one supplies to the torus not only fuel, D_2, $D_2 + T_2$, T_2, but also light gases for wall cleaning (^4He and H_2) or ^3He if needed for resonance heating and heavier species for plasma edge cooling (i.e. impurity seeding) or massive gas injection, i.e. for disruption mitigation: N_2, Ne, Ar and possibly Kr. As a result, not spent fuel, other injected gases and volatile impurities are in the exhaust pumped out from the torus. Tritiated species occur in the form of gas molecules (I_2, where I denotes a mixture of hydrogen isotopes), water (I_2O), hydrocarbons (C_xI_y), carbon oxides (mainly monoxide CO), possibly ammonia (NI_3 if nitrogen is used for plasma edge cooling) and also mixtures containing I–O–C–N or other impurities forming volatile products. Carbon-containing compounds are mentioned here, because some carbon impurities will always be present even in a machine with all PFC made of metals, because of intrinsic impurities of in-vessel materials, a result of leaks and also contamination of components during the installation phase.

Separation methods comprise cryogenic distillation, condensation, electrolysis, diffusion via palladium (Pd) membranes, catalytic processes: oxidation of C_xI_y to CO or CO_2 and I_2O, decomposition of I_2O, C_xI_y, NI_3 and vapour stage exchange:

$$CO + I_2O \leftrightarrow CO_2 + I_2 \quad \text{[water gas shift]} \tag{16}$$

$$CI_4 + I_2O \leftrightarrow CO + 3I_2 \quad \text{[steam reforming]} \tag{17}$$

$$CI_4 \leftrightarrow C + 2I_2 \quad \text{[methane cracking]} \tag{18}$$

$$NI_3 \leftrightarrow N_2 + 3I_2 \quad \text{[ammonia cracking, Co-, Ni-based catalysts]} \tag{19}$$

$$2I_2O \leftrightarrow O_2 + 2I_2 \quad \text{[photocalalysis, electrolysis]}. \tag{20}$$

Catalytic cracking or decomposition of ammonia is the reverse reaction of the Haber–Bosch synthesis of ammonia.

Isotope Separation Station (ISS) with cryogenic installations is in the heart of the Tritium Plant. Differences in the boiling point of hydrogen isotopes, helium and remaining impurity gases allow for final purification and separation of species. Boiling points (T_B) of hydrogen isotopes are: 20.28 K (H_2); 22.13 K (HD), 23.67 (D_2) and 25.04 K (T_2). At these temperatures, helium isotopes (^3He and ^4He) are in the gaseous form, while other species, including neon (T_B = 27.104 K and melting at 24.56 K) had already been removed. A schematic flow diagram in an isotope separation facility based on the Tritium Separation Test Assembly (TSTA) operated until 1997 in Los Alamos National Laboratory, can be found in [22], while the information regarding the ITER fuel cycle have been presented by Mardoch [23]; it is also described in [2]. All tritium for introduction to the torus (freshly supplied to the reactor site and that leaving the ISS) must be stored in uranium or Co-Zr beds at low temperature. This is to ensure precise dosing and to avoid uncontrolled release of the radioactive gas. Also deuterium for the gas introduction system is stored in U-beds. The discharge of pure gases from the beds is realised at elevated temperature of about 720 K [7].

System for processing exhaust gases from the torus requires pumps fully compatible with tritium to ensure safe handling and closed gas loop for T_2 and D_2. They must be available from the very start of the D–D operation. The throughput is at the level 200 Pa m^3s^{-1} with 10% impurities. Testing of a prototype is ongoing also under tritium conditions [24].

Plasma–Wall Interactions and Plasma-Facing Materials

In a controlled thermonuclear device with magnetic confinement plasma–wall interactions (PWI) comprise all processes involved in the energy and mass exchange between the plasma and the surrounding wall. A scheme showing main processes of PWI is shown in Fig. 2, while a very comprehensive description of interactions and their consequences can be found in [25–27]. Energy leaves plasma in the form of electromagnetic radiation and kinetic energy of particles. Plasma-surrounding wall is irradiated by ions, charge-exchange neutrals, electrons, neutrons and photons originating from nuclear (γ) and electronic processes (X, UV). All of them modify material properties, from the very surface to the bulk. Simultaneously plasma is modified by impurities released by erosion processes from the wall.

Bombardment by particles originates from the fact that the discharge time is much longer than the particle confinement time ($t_{discharge} \gg \tau_p$). Some incoming particles are reflected and they instantly return to plasma, some remain in the wall for certain time before being released (this is recycling), but a fraction is trapped. The latter is referred to as long-term retention. Particle impact causes

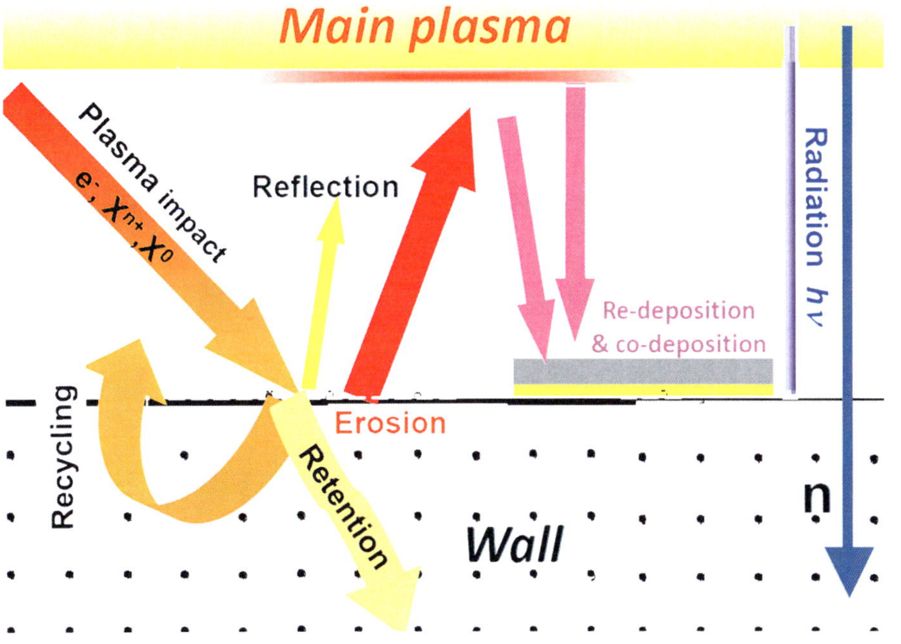

Fig. 2 Main processes in plasma–wall interactions

wall erosion. Species removed from the wall are ionised when they enter plasma and then they transported along magnetic field lines. Eventually, material removed from the wall, unless it is pumped out, must be deposited in another location either close or distant from the place of origin. In this process various plasma impurities are re-deposited together with fuel atoms thus forming mixed-material co-deposited layers. There is a complex chain of material migration: (a) erosion, (b) transport, (c) deposition being actually a co-deposition of fuel and impurity atoms, (d) re-erosion, (e) transport (f) re-deposition, etc. [27]. Micrographs in Fig. 3 show the appearance of such beryllium-rich (Fig. 3a) layers and the melt damage to a tungsten plate tested in a tokamak (Fig. 3b). Stratified Be-rich co-deposits are structurally very similar to carbon-rich layers observed in tokamaks with graphite walls. It is stressed here that this type of surface modification is not connected with neutrons, but the formation of mixed-material deposits will have an impact on the distribution of radioactivity (accumulation of T and also n-activated species) in a reactor. Also the change of tungsten PFC structure after melt events may eventually have an impact on mechanical properties of material under neutron irradiation.

The main erosion mechanisms are: physical sputtering, chemical erosion, melting and melt layer splashing, evaporation, arcing, photo- and electron-induced desorption. Neutron irradiation changes properties not only of plasma-facing, but also of structural, functional (e.g. T breeders and diagnostics) and other materials being in the neutron field. In summary, PWI processes are necessary, unavoidable and destructive. All these three aspects are to be considered in a list of requirements for different categories of fusion reactor materials.

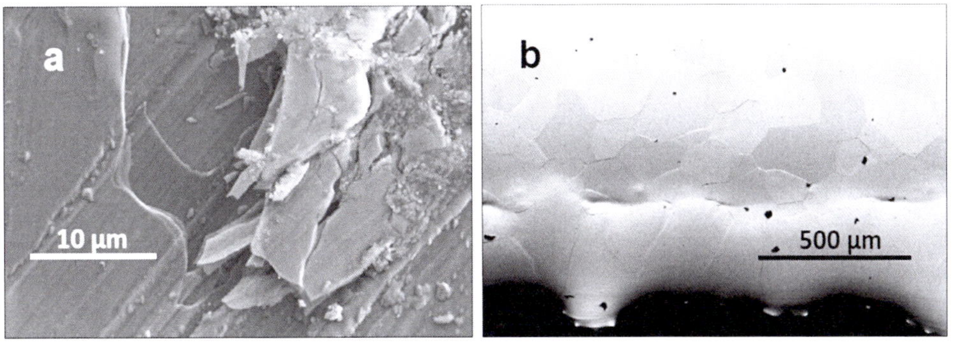

Fig. 3 Surface topography of materials exposed to plasma in tokamaks: **a** beryllium-rich co-deposit and **b** melt damage to a tungsten plate. Authors: Dr. E. Fortuna-Zalesna (Fig. 3a) and Dr. E. Wessel (Fig. 3b)

In-Vessel Materials and Components

Whichever product is manufactured, whether small or huge, it must possess some indispensable features to find a customer: functionality, stability, durability, reliability. The product must also comply with safety requirements and standards. These features are related to a broad range of factors including thorough design and selection of materials. A fusion reactor makes the requirements even more stringent. It is a nuclear device which must produce energy and—at the same time—must comply with safety requirements regarding the construction, operation, maintenance and decommissioning phases. It must also satisfy licensing criteria set by atomic agency authorities. The temperature gradients between the plasma and the surrounding wall are probably the greatest in the Universe and the operation is associated with intense nuclear radiation. Therefore, the technology for next-step devices with the D–T fuel presents challenges not encountered in present-day machines. This includes development and construction of components capable of reliable performance under exposure to intense particle fluxes and resulting high heat loads (above 10 MW m^{-2}) in highly radioactive environment. All starts with materials whose properties must satisfy several criteria. Compatibility with high vacuum for all in-vessel materials and high thermal conductivity of plasma-facing materials and components are the prerequisites. Others requirements for PFM comprise resilience to thermal shocks, high melting point, low-Z to reduce energy radiation losses by plasma contaminants, low neutron activation, low affinity towards chemical erosion resulting in the formation of volatile products, low sorption of hydrogen isotopes and high affinity to plasma impurity species (e.g. oxygen) towards the formation of stable non-volatile compounds. Another indispensable feature is the compatibility of PFM with the heat sink structure (mainly Cu alloyed with admixtures of chromium and zirconium) to ensure efficient and reliable active cooling. All these criteria together cannot be fully met by any single element, compound, alloy or composite material. Underlying physics and chemistry of erosion processes cannot be changed or eliminated, hence the goal of plasma edge engineering is to improve the control of interactions by optimizing the reactor operation scenarios and by the selection of the best-possible materials.

For over 2 decades, carbon, beryllium and tungsten were the major candidates for PFC. Advantages and drawbacks have been studied and discussed in detail [26–28]. Three materials were supposed to be used in ITER: beryllium on the main chamber wall, while tungsten and carbon in the form of fibre composites (CFC) in the divertor. The latter material was on the list because of not melting, high thermal conductivity (γ), excellent power handling capabilities and very low activation. However, high erosion rates by plasma and the formation of thick co-deposited layers (> 100 μm) rich with hydrogen isotopes (10–25 at.%) would lead to unacceptable levels of tritium inventory in the machine, especially in remote areas in the divertor. The full seriousness of the process and its detrimental impact on the reactor economy and safety was understood after the full D–T campaign in JET [7, 29–31]. As a result, a large-scale test in JET with a metal wall, beryllium and tungsten, was decided: JET with the ITER-Like Wall (JET-ILW) [32–34]. Distinctly reduced fuel inventory in the metal surrounding in comparison to the operation with carbon PFC in JET (JET-C) led to a major scientific and technological decision on resigning from using carbon in the ITER divertor [35]. This decision has also economical and safety aspects related to the motivated expectation of lowering the in-vessel inventory of tritium.

Radiation-Induced Effects

Neutrons pass the armour and structural materials of the blanket leading to the volumetric change of material composition by transmutation and to the damage of the crystalline lattice. This changes material properties and performance. A brief overview of major consequences for structural and diagnostic materials is presented in this section.

Transmutation

Transmutation is a conversion of one nucleus to another. It is the change in a nucleus structure, i.e. the formation of different isotope(s) or element(s) induced by bombarding the nucleus with particles or high-energy photons, (via Compton effect) which can induce nuclear reactions [36–41]. All isotopes undergo transmutation whose efficiency is related to the energy-dependent cross-section, $\sigma(E)$, of a given n-induced nuclear reaction.

Transmutation processes have serious consequences for all reactor materials, both structural and functional, and for reactor safety. This comprises the change of nuclear and also chemical composition of wall components which has a direct impact on thermomechanical properties: heat conductivity and capacity, and on linear expansion coefficient. It also induces radioactivity by transmuting stable nuclei to radioisotopes of different lifetime.

In the following, several examples will be considered for plasma-facing (Be, W and C) and structural materials (low-activation alloys). The transmutation of ^{9}Be (the only Be stable isotope) has already been partly addressed in reactions (11), (13) and (15). Another process is the activation of ^{9}Be to radioactive ^{10}Be, $t_{1/2} = 1.387 \times 10^{6}$ year. It decays by β^- emission to stable ^{10}B.

Carbon has two stable isotopes: ^{12}C (abundance 98.93%) and ^{13}C (1.07%). The efficiency of transmutation to a radioactive isotope ^{14}C (β^-, 0.156 MeV, $t_{1/2} = 5730$ year, specific activity 1.65×10^{11} Bq/g) in a fusion device would be very low because cross-sections for neutron capture by both stable isotopes are very low on the level of a few micro-barn (1b = 1×10^{-28} m^2).

Tungsten has four stable isotopes and one of a very long half lifetime: ^{180}W (natural abundance 0.12%, $t_{1/2} = 1.8 \times 10^{18}$ year), ^{182}W (26.50%), ^{183}W (14.31%), ^{184}W (30.64%), ^{186}W (28.43%). First stages of transmutation for the two heaviest and most abundant isotopes are:

$$^{184}\mathrm{W}(n,\gamma)^{185}\mathrm{W}(\beta^-) \rightarrow^{185}\mathrm{Re}(n,\gamma)^{186}\mathrm{Re}(\beta^-) \rightarrow {}^{186}\mathrm{Os} \qquad (21)$$

$$^{186}\mathrm{W}(n,\gamma)^{187}\mathrm{W}(\beta^-) \rightarrow^{187}\mathrm{Re}(n,\gamma)^{188}\mathrm{Re}(\beta^-) \rightarrow {}^{188}\mathrm{Os} \qquad (22a)$$

$$^{186}\mathrm{W}(n,\gamma)^{187}\mathrm{W}(\beta^-) \rightarrow^{187}\mathrm{Re}(n,2n)^{186}\mathrm{Re}(\beta^-) \rightarrow {}^{186}\mathrm{Os}. \qquad (22b)$$

In all cases the transmutation leads to the formation of ^{186}Os (stable isotope) which is then transmuted to other nuclei. Though the same intermediate isotope is produced in Reactions (21, 22a, 22b) the rates of processes are distinctly different because of differences in cross-section of respective steps. Temporal evolution of the composition change for tungsten and W alloys under ITER and fusion power plant conditions has been calculated by Gilbert and Sublet [37]. Besides the main products, i.e. Re and Os, there are some quantities of tantalum, hafnium, hydrogen, helium and, in the case of a power plant reactor also platinum and iridium. It is predicted that in ITER after 14 years of operation there will be 0.2 (at.%) of Re in W, while after 5 years of reactor operation Re and Os will constitute around 10 (at.%) of the composition. The new alloy will also contain H, He and a large number of other species, such as Ta, Hg, Pt. Calculations for transmutation of other elements are in [38–41]. One expects that the change of chemical composition would not be uniform through the bulk because cross-sections change with the neutron energy loss by stopping in the lattice. It should also be stressed that activation is to be taken into account in the case of plasma edge cooling, especially when the use of heavy gases, e.g. Kr, is considered [41].

Basic physics underlying the transmutation leading to the formation of radioactive nuclei cannot be overcome. The only way to minimize its effects is to use low-activation materials, i.e. materials containing elements of low cross-sections for transmutation or elements whose transmutation products are either non-radioactive or short-lived isotopes. It is clear that products and related radioactivity (i.e. energy spectrum and lifetime) strongly depend on the initial composition of the irradiated material. It also implies that not only major constituents should undergo low activation but also the quantity and quality of admixtures and impurities must be strictly controlled. For instance, while major constituents of low-activation vanadium alloys (V–3Ti–1Si) transmute to isotopes of short lifetime (^{46}Sc, ^{47}Sc, ^{51}Cr), the presence of even trace quantities of nickel impurities transmuted to ^{60}Co leads to a highly activated product ($t^{60}_{1/2}Co = 5.27$ year). The same remark regarding Ni may be applied to EUROFER-97, low-activation high-chromium alloy, developed for fusion reactor technology [42]. Fabrication of components containing low activation and high purity constituents is of great importance but also a reasonable approach is to be place taking into account limits in purification processes and their impact on the cost of production.

An important consequence of nuclear reactions is simultaneous formation of other transmutation products. In general, they belong to three categories:

1. Gaseous species such as hydrogen isotopes and helium: (n, p), (n, np), (n, d), (n, t) (n, α), (n, nα), (n, ^{3}He)
2. Gamma radiation: (n, γ), (n, n$'\gamma$)
3. Neutron breeding: (n, 2n), (n, 3n).

Particular attention should be given to processes of the first-group which generate hydrogen isotopes (H, D, T) or helium (^{3}He, ^{4}He). Gases are accumulated in the crystal lattice and they form bubbles and blisters. Bubbles are formed not only in the surface layer but also in the bulk. In a critical situation when the pressure of the accumulated gas overcomes a certain limit, blisters explode leading to the exfoliation; an example is shown in Fig. 4.

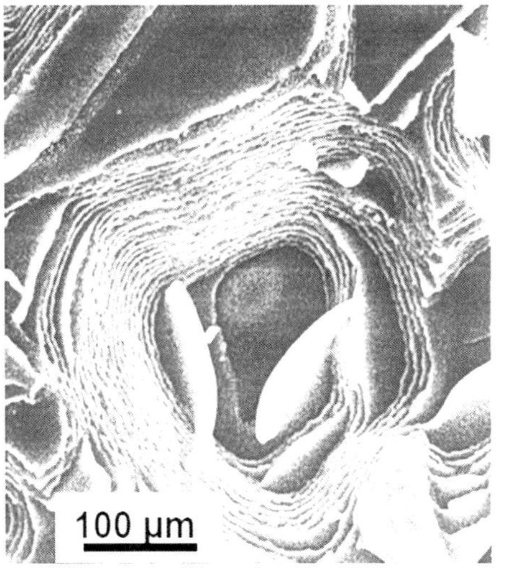

Fig. 4 Exfoliation of steel caused by high-dose irradiation

Radiation Damage

Energetic particles transfer their energy to the irradiated target through inelastic (electronic) and elastic (nuclear) interactions. Fundamental processes have been reviewed in [43, 44]. Inelastic energy losses lead to the ionisation and have a serious impact on electrical properties, while elastic stopping mean interactions with the lattice and—as a consequence—modification of the crystalline structure.

The measure of damage to a crystalline matter caused by bombardment with energetic particles is expressed in terms of so-called *displacement per atom* (dpa), i.e. the number of times each atom is dislodged from its place in the crystal by radiation. In other words, 1 dpa is equivalent to displacing all atoms once from their lattice sites. The cross-section for processes of neutron displacement damage is generally in the range from 1 to 10 barns. Damage depends on the fluence (total dose) and, in some cases, also on the neutron flux, though the latter is, however, is questionable. The dependence has been reported in earlier works [45, 46] but recent studies have not provided conclusive confirmation [47–49]. In carbon, beryllium and ceramic materials 1 dpa is produced by a neutron dose of around 1×10^{25} m^{-2}. Volumetric damage leads to the formation of dislocations, interstitials, voids and vacancies, in the crystal lattice. This results from the direct knock-on of atoms and/or ions from their sites. Knock-on atoms of sufficiently high energy initiate further displacements by producing collision cascades. Dislocation is defined as a line, plane or region in which there is a discontinuity in the regularity of the lattice. Voids and vacancies are the empty spaces formed by shifting atoms from their original sites. The distortion of crystals is further enhanced by the formation of gaseous products. As a consequence, all material properties are affected and deteriorated: chemical and physical such as thermo-mechanical, electrical conductivity. Phonon scattering on lattice defects has implications for thermal conductivity [50] leading to the decrease of that crucial parameter for materials of plasma-facing components and tritium breeding modules. The presence of voids and the related formation of gas bubbles leads to swelling. In the end effect the volume of metals [51–55] or ceramics is increased. On the contrary, carbon fibre composites (no bubble formation) shrink [56]. This in turn, leads to the significant drop in thermal conductivity, even by 70% from the original value [56]. 1 dpa typically results in 1% volume change. Swelling of a metal crystal changes drastically its mechanical properties causing hardening and resultant increased brittleness (embrittlement). Ductility is decreased by the agglomeration of He and H bubbles at grain boundaries and densification of dislocation network. The extent of damage is reduced at elevated temperatures due to annealing. However, for obvious reasons the temperature of materials and components cannot be increased indefinitely. Therefore, efforts in irradiation tests are focused on the definition of material operation limits (lifetime) at various neutron dose and irradiation temperature.

Impact on Materials of Diagnostic Systems

Diagnostic tools are essential for the safe and efficient functioning of a reactor: for the machine protection, guiding operation scenarios and to support advanced plasma research. The required and essential features of diagnostics are: reliability, repeatability, resistance to radiation, resistance to heat and—last but not least—maintainability. Thousands of components of diagnostic systems belong to several categories: mechanical supports, feedthroughs, seals, bolometer substrates, pressure gauges, windows, mirrors, neutron detectors, optical cables, connectors including mineral-insulated (MI) cables. Most of them, besides, certain mirrors (so-called *first mirrors* as plasma-facing components of optical diagnostic systems) will be protected from the direct plasma impact, but they still will be in the neutron field of diverse intensity. A list of materials used in diagnostics comprises hundreds of different substances. Many of them are insulators: quartz, silica, aluminium oxide (sapphire), aluminium nitride, beryllium oxide, boron nitride, alumino-silicates, diamond and many materials doped with boron. MI cables are indispensable for reliable operation and signal transmission from magnetic coils and thermocouples. There is a very broad program for development of radiation-hard materials for the use in fusion [57, 58] and many other disciplines where strong particle and electromagnetic radiation is generated [59]. Under extreme environment insulating properties of ceramics are changed and mostly deteriorated. The presence of voids and bubbles increases diffusion which is particularly dangerous in the case of radioactive tritium. The content of Table 1 summarizes several most pronounced processes and consequences of the irradiation. Surface effects listed in the last row originate from plasma–material interactions which are responsible for erosion and deposition. They are of particular importance in the case of first mirrors, therefore, a comprehensive test is carried out for ITER [60–67] and DEMO [68–70] both in tokamaks and under laboratory conditions.

Material Testing

Material testing under as realistic as possible conditions is a crucial procedure in the qualification process of materials and components in every technology. Mechanical integrity

Table 1 Radiation-induced effects in diagnostic materials

Process	Consequences
Radiation-induced (enhanced) conductivity (RIC)	Excitation of electrons into a conduction band
Radiation-enhanced absorption (RIA)	Light transmission loss
Radiation-induced electrical degradation (RIED)	Permanent enhancement of the volume conductivity caused by bulk radiation-induced defects
Radiation-enhanced diffusion (RED)	Increased tritium mobility in ceramic windows
Radiation-induced electro-motive force (RIEMF)	Induced voltage between centre and outer conductors of MI cables
Surface effects by sputtering, re-deposition, impurity segregation, (melting, evaporation, splashing)	Degraded specular reflectivity by erosion and deposition. Contamination of mirrors

and durability of items are absolutely essential. In the nuclear sector the lifetime of items is of particular importance for at least two reasons: safety of operation and complex procedures in maintenance and exchange of components. As pointed out in "In-Vessel Materials and Components" section, thermal conductivity of plasma-facing materials is one of the most important factors in selection of materials for PFC. There are numerous examples showing a pronounced decrease of that parameter not only after high-dose irradiation [56], but even after doses corresponding to 0.1–0.3 dpa [71]. This is turn will have a crucial impact on heat transfer from a PFC to a cooling structure.

In every branch of fusion science technology there are areas to be further developed and/or explored on the way to a power station. In the domain of materials, the list of main tasks also comprises:

(1) Advances in technology of high purity and radiation hard materials;
(2) Determination of fuel retention especially in plasm a-facing and structural materials of tritium breeding modules, to enable best-possible predictions of tritium inventory and its impact on a whole fuel cycle in a reactor.

It all calls for testing of selected materials and components under large doses of high energy neutrons to recognize the impact on electrical and thermo-mechanical properties. The crucial point is to produce a reactor-relevant level of damage (> 20 dpa per year) [72] including the modification of elemental composition. This can only be achieved by the irradiation with high fluxes or fast neutrons as they interact with the surface and the bulk of components. In other words, testing of radiation damage in reactor materials requires an efficient high-flux source of high-energy neutrons. There are advanced plans for the construction of the facility. The project will be presented in "Material Testing Under High-Energy Neutron Flux" section. Currently, irradiations are carried out in nuclear reactors. This has facilitated studies of the neutron impact on creep, ductility, thermal conductivity, retention of hydrogen isotopes [51–56, 71, 73]. The latter is examined also in ion-irradiated targets as they are easily available and distinctly cheaper than specimens from nuclear reactors.

Simulation of Neutron-Induced Effects by Ions: Impact on Test Mirrors

Damage can also be simulated by ion irradiation [74–77]. With energetic heavy ions very high damaged rates of tens of dpa are obtained, but the irradiation cannot lead to transmutation accompanied by generation of helium and hydrogen. Also the simulation of bulk effects is very limited as the implantation depth does not exceed a few micrometres. However, in one domain bulk effects do not play major role: testing of metallic mirrors which will be crucial components in all optical spectroscopy and imaging systems [78]. The test briefly described below was related to the simulation of neutron-induced effects that may degrade mirror performance in DEMO.

For the reflectivity of mirrors only the surface region is important: the optically active layer (OAL) calculated according to the the Beer–Lambert law. Light intensity penetrating a metal target falls exponentially with a decay constant known as absorption coefficient which depends on material and wavelength. In polycrystalline molybdenum, i.e. the material on which the test was performed, for the visible part of the spectrum it is in the range from 12 nm at the wavelength of 400 nm to approx. 22 nm at 800 nm; a detailed graph is shown in [68]. From the nuclear point of view, Mo is not an ideal material, as it has seven naturally occurring isotopes, three of which are measurably unstable with very long half-live times of more than 10^{14} years. Transmutation of Mo leads to long-lived radioactive isotopes of technetium (Tc), niobium (Nb), zirconium (Zr), molybdenum itself, hydrogen and helium from (n, p), (n, α) and (n, nα) reactions. It should be stated that all candidate materials for mirrors would be affected by neutrons in a reactor, i.e. undergo transmutation and activation. Moreover, there is no clear suggestion or indication regarding materials for mirrors in DEMO.

To simulate neutron-induced effects including both transmutation of Mo and generation of gaseous products the irradiation was done with 30 keV $^{98}Mo^{+}$, $^{93}Zr^{+}$ or $^{90}Nb^{+}$, 2 keV He^{+} and with 4 keV H_2^{+} corresponding to 2 keV for a monoatomic beam. Irradiation with the major transmutation product (technetium) could not be performed, because all Tc isotopes are radioactive. It would enormously increase cost of the test and would simultaneously limit the range of post-irradiation analyses. The selection of respective ion energies was based on predictive modelling using SRIM [79] in order to match the thickness of OAL, i.e. to deposit ions within that surface layer. All experimental details can be found in [68, 69]. The $^{98}Mo^{+}$ ions were used to produce damage, while the irradiation with zirconium or niobium was done, because both elements are formed as transmutation products in neutron-irradiated Mo. The amount of Zr and Nb would be at the level of 1×10^{14} cm^{-2} and 1×10^{13} cm^{-2} in the surface layer of 15 nm. The damage to the surface layer (so-called *first wall damage*) of Mo in a DEMO reactor is estimated at the level of 7 dpa per year and this would be accompanied by the presence of 45 appm of He, 479 appm of H [37, 40]. In the irradiation higher doses of H and He were used because these species, as charge exchange neutrals from plasma, may and will also interact with first mirrors.

Plots in Fig. 5a show the total reflectivity changes, in comparison to the initial reflectivity, of a Mo mirror following the irradiation with $^{98}Mo^{+}$ ions to the dose of 15×10^{14} cm^{-2} corresponding to 10 dpa (Fig. 5b). The diffuse reflectivity of the initial mirrors was 0.5% and increased to 1% after the irradiation. Therefore, total reflectivity represent specular component which a figure of merit for mirrors. In the visible range the reflectivity increases with fluency. It is related to the removal of molybdenum oxides from the surface [63]. The reflectivity in the near infrared range decreases by up to 5% due to the damage caused by implantation. The same result was obtained when the irradiation was performed at 573 K. Significant changes are caused by the irradiation with 3×10^{17} He/cm^2 as shown in Fig. 5b. Reflectivity decreases by approximately 20% with some differences dependent on the wavelength. Additional decrease was induced by irradiation with a H^{+} beam [69]. These changes are caused by the gas accumulation in the bubbles beneath the surface, up to the depth of 30–40 nm, thus modifying OAL. They are clearly visible on the cross-section shown in Fig. 6 for a He bombarded mirror.

It is perfectly well understood energy spectra of neutrons in the reactor (up to 14 MeV) and those of ions

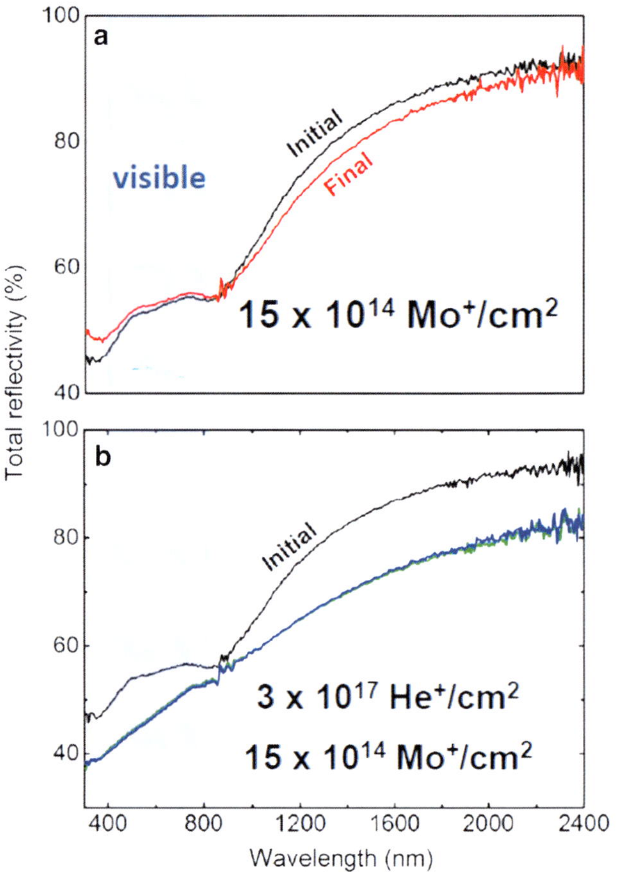

Fig. 5 The change of reflectivity in molybdenum test mirrors after irradiation with: **a** $^{98}Mo^{+}$ corresponding to 10 dpa; **b** both with He and $^{98}Mo^{+}$ beams

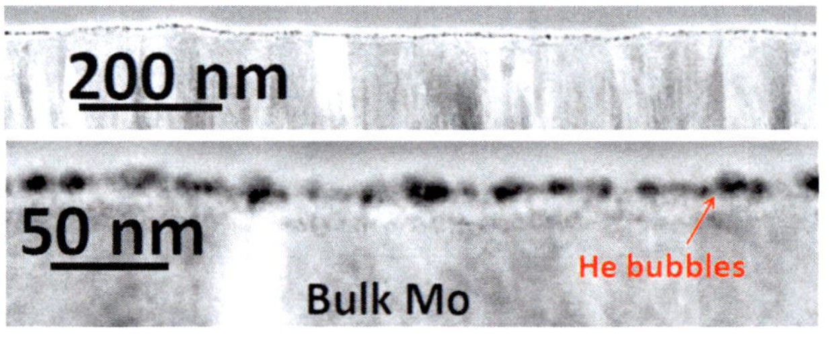

Fig. 6 STEM images of FIB-cut lamellae from the near-surface region of the Mo^{+} and He^{+} irradiated mirror. Author: Dr. Justyna Grzonka

(2–30 keV) are different. The modification and damage generated in mirrors under ion irradiation is not equal to that caused by neutrons, but the test under fully realistic conditions will become possible only in a real working fusion reactor, where all synergistic effects of material migration processes and neutron irradiation occur simultaneously. However, results of even such limited testing indicate that the major impact on the degradation of optical properties will be caused by H and He interaction, rather charge exchange species than those formed by transmutation.

Material Testing Under High-Energy Neutron Flux

High-flux testing requires a construction of an efficient test facility capable of simulating the neutron energy spectrum of a D–T fusion reactor. Over the years, different concepts of neutron sources have been proposed. Taking into account the required neutron flux and energy spectrum to simulate the irradiation conditions of plasma-facing components in a reactor the most promising and relevant sources are those based on the $^{7}Li(d, 2n)^{7}Be$, $^{6}Li(d, n)^{7}Be$ nuclear reactions (deuteron energy 30–40 MeV) generating neutrons with the energy spectrum peaking at around 14 MeV. Already in year 1976 a concept based on the deuteron irradiation of a lithium target was proposed [80]. In 1992 this approach using high-current deuterium beams was selected in year 1992 for the International Fusion Materials Irradiation Facility (IFMIF) [81] and in following years a conceptual design and technical specifications based scientific and engineering analyses were detailed and presented in year 2004 as a joint effort of EU, Japan, Russian Federation and the USA. In June 2007 the IFMIF-EVEDA project was established by the EU and Japan. EVEDA stands for Engineering Validation and Engineering Design Activities. There are three major components to be constructed, assembled and operated together to validate main technological challenges:

1. Accelerator facility,
2. Target facility (lithium loop),
3. Test facility (material irradiation chamber).

The project is based at the International Fusion Energy Research Centre (IFERC) in Rokkasho, Japan. Prototypes of three crucial facilities are either in operation or being manufactured. Respective modules of the accelerator, Linear IFMIF Prototype Accelerator (LIPAc), are manufactured mainly in EU (Belgium, France, Italy, Spain, some parts in Japan) and then shipped to IFERC, where the assembly takes place. The LIPAc is to prove the operation required for the first superconductive accelerating stage of IFMIF: 9 MeV, 125 mA of continuous wave D^{+} current resulting in a power of 1.125 MW that will validate the mastering of the space charge issues which are more relevant at low energies. Figure 7 shows the scheme of the LIPAc. All details about the program can be found in [82, 83].

The lithium target, EVEDA Lithium Test Loop (ELTL), was designed, assembled and operated at the Oarai Research and Development Center, Japan. Details about lithium technology for IFMIF are in [84], while the Li loop is described in [85, 86]. The aim was to prove the feasibility of technology, in particular to validate hydraulic stability of the Li target at a velocity up to 20 m s^{-1} under a vacuum condition of 10^{-3} Pa. A total of 2.5-t Li was placed in the ELTL chamber under clean controlled conditions to avoid any contamination by oxygen, nitrogen and water vapour. The 20 m high facility matched conditions required for the lithium loop exploitation at IFMIF: flow rate of 20 m s^{-1} at 300 °C, while IFMIF requires 15 ms^{-1} at 250 °C. The amplitude of the free surface stability is within +/1 mm at nominal flow conditions. The target of the ELTL is 100 mm wide, compared with the 260 mm of IFMIF, but with the same thickness of 25 ± 1 mm of the Li screen. Also the efficiency of liquid lithium purification from C, O, N and H has been successfully proven. In addition corrosion/erosion phenomena in steel conduits were studied and determined in the in LIFUS6 facility, a smaller Li loop operated at Brasimone, Italy [87].

High Flux Test Module (HFTM) is under fabrication at the Karlsruhe Institute of Technology (KIT), Germany. This is to house irradiated specimens at elevated temperatures in the range from 250 up to 550 °C with a maximum temperature spread of ± 3% within specimens inside a single capsule; 12 capsules are to be used. The cooling system is based on a fast flow of helium gas; it has been successfully validated in the HELOKA loop [88]. A test module to be placed in the medium flux region to assess the creep-fatigue behaviour under intense neutron and gamma irradiation was constructed and successfully validated at the Paul Scherrer Institute (PSI), Villigen, Switzerland.

In addition, remote handling studies for suitable replacement of the beam target have been carried out in Japan and Europe, together with the development in Japan of small specimens for fatigue, fracture toughness and crack growth. A set of small specimens for additional mechanical properties like tensile stress or creep has already been manufactured.

Successful operation of prototypes would mitigate risks for IFMIF which operation is planned to be based on 30–40 MeV from two continuous-wave linear accelerators with a total beam current of 2 × 125 mA. The neutron generation rate of some 10^{17} s^{-1} would result in a flux of some 10^{19} m^{-2} s^{-1} at the rear side of the target. Therefore, displacement damage (20 dpa per full power year) and transmutation products (He, H) in the irradiated material would match the neutron-induced effects anticipated in the fusion reactor environment. A drawing in Fig. 8 shows a scheme of the IFMIF assembly (length over 200 m) composed of the accelerator hall, lithium loop, test cell and a hall housing for facilities for post-irradiation examinations and for detritiation. The test cell will comprise a series of target chambers. The plan is install three chambers to allow

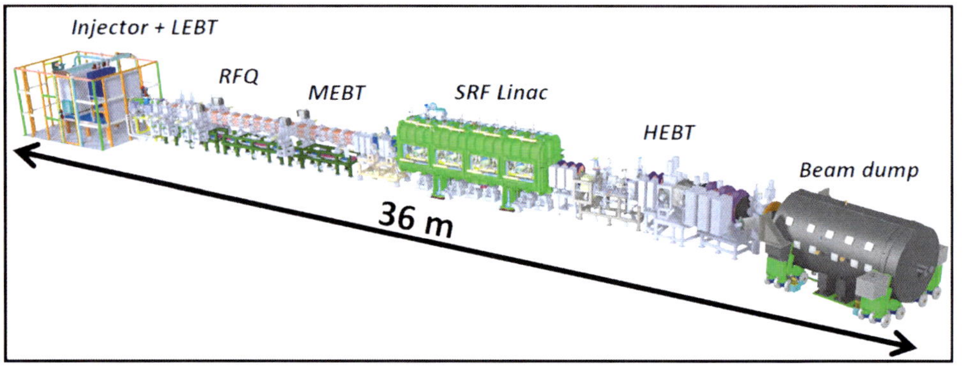

Fig. 7 A schematic drawing of LIPAc presently under installation and commissioning at IFERC. LEBT, low energy beam transport; RFQ, radio frequency quadrupole; MEBT, medium energy beam transport; SRF Linac, superconducting radio frequency linear accelerator; HEBT, high energy beam transport. *Source*: http://www.ifmif.org/?page_id=66

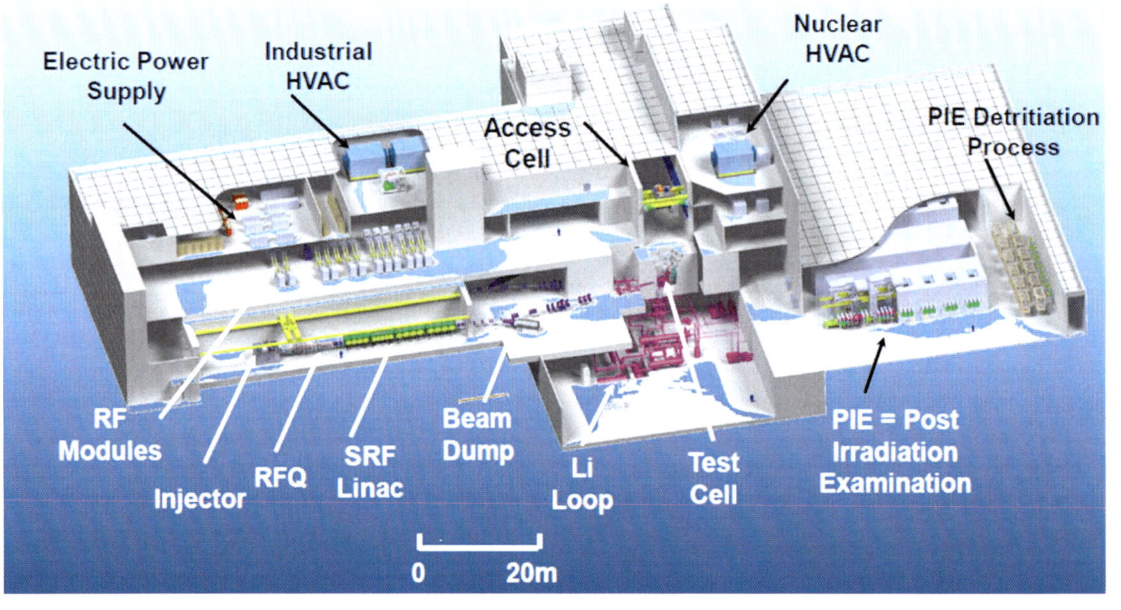

Fig. 8 A schematic plan of IFMIF. RFQ, Radio frequency quadrupole; SRF Linac, superconducting radio frequency linear accelerator; HVAC, Heating–Ventilation–Air Conditioning. Copyright ©Fusion for Energy (F4E)

for high flux irradiations in a volume of 0.5 L and lower dose irradiation in volumes of 6.0 and 7.5 L [70].

Concluding Remarks

Development and intense testing of materials and components belong to family of crucial elements for the progress in fusion science and technology. Having in mind scientific and technical challenges associated with the project, all efforts are to be made to ensure the best possible material selection, component manufacture and plasma edge engineering for a reactor-class machine. A reactor operated with the deuterium–tritium mixture is a nuclear device and—in addition to all machine construction and operation features—also radioactive aspects of fusion will undergo intense scrutiny. Proper testing and validation of material and component performance is essential to ensure their durability and integrity under high power loads in normal operation and in the case of off-normal events. Based on the best of our present knowledge we have to use and to develop tools, methods and materials best fitted for a steady–steady reactor operation. The entire dream and all hopes for commercial exploitation of controlled thermonuclear fusion for energy production depend on all aspects of power handling by PFC, efficient extraction of neutron energy and on efficient production, extraction and handling of tritium to fuel the reactor. In summary, it will depend on neutrons.

Acknowledgements This work has been carried out within the framework of the EUROfusion Consortium and has received funding from the Euratom research and training programme 2014–2018 under grant agreement No 633053. The views and opinions expressed herein do not necessarily reflect those of the European Commission.

References

1. H.-S. Bosch, G.M. Halle, Improved formulas for fusion cross-sections and thermal reactivities. Nucl. Fusion **32**, 611 (1992). https://doi.org/10.1088/0029-5515/32/4/I07/pdf
2. T. Tanabe (ed.), *Tritium: Fuel of fusion reactors* (Springer, Berlin, 2016)
3. J. Wesson, *Tokamaks*, 3rd edn. (Oxford Science Publications, Clarendon Press, Oxford, 2004)
4. E. Moses, Advances in inertial confinement fusion at the National Ignition Facility (NIF). Fusion Eng. Des. **85**, 983 (2010). https://doi.org/10.1016/j.fusengdes.2009.11.006
5. I. Hoffman, Review of accelerator driven heavy ion nuclear fusion. Matter Radiat. Extreme **3**, 1 (2018). https://doi.org/10.1016/j.mre.2017.12.001
6. P.-H. Rebut, Preliminary tritium experiment. Plasma Phys. Control. Fusion **34**, 1749 (1992). https://doi.org/10.1088/0741-3335/34/13/002
7. D. Stork et al., Systems for the safe operation of the JET tokamak with tritium. Fusion Eng. Des. **47**, 131 (1999). https://doi.org/10.1016/S0920-3796(99)00081-2
8. T.T.C. Jones et al., Technical and scientific aspects of the JET trace-tritium experimental campaign. Fusion Sci. Technol. **48**, 250 (2005). https://doi.org/10.13182/fst05-a922
9. C.H. Skinner, C.A. Gentile, J.C. Hosea, D. Mueller, J.P. Coad, G. Federici, R. Haange, Tritium experience in large tokamaks: application to ITER. Nucl. Fusion **39**, 271 (1999). https://doi.org/10.1088/0029-5515/39/2/410
10. J.N. Andrews, R.L.F. Kay, Natural production of tritium in permeable rocks. Nature **298**, 361 (1982). https://doi.org/10.1038/298361a0
11. M. Kovari, M. Coleman, I. Cristescu, R. Smith, Tritium resources available for fusion reactors. Nucl. Fusion **58**, 026010 (2018). https://doi.org/10.1088/1741-4326/aa9d25
12. M. Enoeda et al., R&D status of water cooled ceramic breeder blanket technology. Fusion Eng. Des. **89**, 1131 (2014). https://doi.org/10.1016/j.fusengdes.2014.01.035
13. G. Aiello et al., Development of the helium cooled lithium lead blanket for DEMO. Fusion Eng. Des. **89**, 1444 (2014). https://doi.org/10.1016/j.fusengdes.2013.12.036
14. J. Aubert et al., Status on DEMO helium cooled lithium lead breeding blanket thermo-mechanical analyses. Fusion Eng. Des. **109–111**, 991 (2016). https://doi.org/10.1016/j.fusengdes.2016.01.037
15. D. Carloni et al., Requirements for helium cooled pebble bed blanket and R&D activities. Fusion Eng. Des. **89**, 1341 (2014). https://doi.org/10.1016/j.fusengdes.2014.02.036
16. M.E. Sawan et al., Three-dimensional nuclear analysis for the US dual coolant lead lithium ITER test blanket module. Fusion Eng. Des. **85**, 1027 (2010). https://doi.org/10.1016/j.fusengdes.2010.01.002
17. K.M. Feng et al., New progress on design and R&D for solid breeder test blanket module in China. Fusion Eng. Des. **89**, 1119 (2014). https://doi.org/10.1016/j.fusengdes.2014.01.073
18. R. Battacharyay, Indian TBM team, Status of Indian LLCB TBM program and R&D activities. Fusion Eng. Des. **89**, 1107 (2014). https://doi.org/10.1016/j.fusengdes.2014.05.016
19. I.R. Kirillov et al., Results of lead–lithium ceramic breeder TBM conceptual design optimization. Fusion Eng. Des. **89**, 1421 (2014). https://doi.org/10.1016/j.fusengdes.2014.01.006
20. L.A. El-Guebaly, S. Malang, *Towards the Ultimate Goal of Tritium Self-Sufficiency*. http://fti.neep.wisc.edu/pdf/fdm1330.pdf
21. M.E. Sawan, M.A. Abdou, Physics and technology conditions for attaining tritium self-sufficiency for the DT fuel cycle. Fusion Eng. Des. **81**, 1131 (2006). https://doi.org/10.1016/j.fusengdes.2005.07.035
22. J.L. Anderson, P. LaMarche, Tritium activities in the United States. Fusion Technol. **28**, 479 (1995)
23. D.K. Murdoch et al., ITER fuel cycle development. Fusion Sci. Technol. **48**, 3 (2005)
24. Ch. Day, D.K. Murdoch, The ITER vacuum systems. J. Phys. Conf. Ser. **114**, 012013 (2008). https://doi.org/10.1088/1742-6596/114/1/012013
25. W.O. Hofer, J. Roth (eds.), *Physical processes of the interaction of fusion plasmas with solids* (Academic Press, New York, 1996)
26. G. Federici et al., Plasma-material interactions in current tokamaks and their implications for next step fusion reactors. Nucl. Fusion **41**, 1967 (2001). https://doi.org/10.1088/0029-5515/41/12/218
27. V. Philipps, P. Wienhold, A. Kirschner, M. Rubel, Erosion and redeposition of wall material in controlled fusion devices. Vacuum **67**, 399 (2002). https://doi.org/10.1016/S0042-207X(02)00238-5
28. J. Roth et al., Recent analysis of key plasma-wall interaction parameters for ITER. J. Nucl. Mater. **390–391**, 1 (2009). https://doi.org/10.1016/S0022-3115(00)00479-7
29. R.D. Penzhorn et al., Tritium depth profiles in graphite and carbon fibre composite material exposed to tokamak plasmas. J. Nucl. Mater. **288**, 170 (2001). https://doi.org/10.1016/S0022-3115(00)00705-4
30. J.P. Coad et al., Erosion/deposition issues in JET. J. Nucl. Mater. **290–293**, 224 (2001). https://doi.org/10.1016/S0022-3115(00)00479-7
31. M. Rubel et al., Beryllium and carbon films in JET following D–T operation. J. Nucl. Mater. **313–316**, 321 (2003). https://doi.org/10.1016/S0022-3115(02)01350-8
32. G.F. Matthews et al., ITER-like wall project overview. Phys. Scr. **T128**, 137 (2007). https://doi.org/10.1088/0031-8949/2007/T128/027
33. G.F. Matthews et al., JET ITER-like wall: overview of experimental program. Phys. Scr. **T145**, 014001 (2011). https://doi.org/10.1088/0031-8949/2007/T128/027
34. G.F. Matthews, Plasma operation with an all metal first-wall: comparison of an ITER-like wall with a carbon wall in JET. J. Nucl. Mater. **438**, S2 (2013). https://doi.org/10.1016/j.jnucmat.2013.01.282
35. S. Brezinsek, JET-EFDA Contributors, Plasma-surface interaction in the Be/W environment: conclusions drawn from the JET-ILW for ITER. J. Nucl. Mater. **463**, 11 (2015). https://doi.org/10.1016/j.jnucmat.2014.12.007
36. D. Li, K. Imasaki, K. Horikawa, S. Miyamoto, S. Amano, T. Mochizuki, Iodine transmutation through laser compton scattering gamma rays. J. Nucl. Sci. Technol. **46**, 831 (2012). https://doi.org/10.1080/18811248.2007.9711592
37. M.R. Gilbert, J-Ch. Sublet, Neutron-induced transmutation effects in W and W-alloys in a fusion environment. Nucl. Fusion **51**, 043005 (2011). https://doi.org/10.1088/0029-5515/51/4/043005

38. M. R. Gilbert, J.-C. Sublet, *Handbook of Activation, Transmutation, and Radiation Damage Properties of the Elements Simulated Using FISPACT-II & TENDL-2014; Magnetic Fusion Plants*. Technical report, CCFE-R(15)26, CCFE (2015)
39. M.R. Gilbert, J.-C. Sublet, *PKA Distributions of the Elements Simulated Using TENDL-2014*. Technical report CCFE-R(15)26-supplement, CCFE (2015)
40. M.R. Gilbert et al., An integrated model for materials in a fusion power plant: transmutation, gas production, and helium embrittlement under neutron irradiation. Nucl. Fusion **52**, 083019 (2012). https://doi.org/10.1088/0029-5515/52/8/083019
41. R.J. Walker, M.R. Gilbert et al., Neutron activation of impurity seeding gases within a DEMO environment. Fusion Eng. Des. **124**, 892 (2017). https://doi.org/10.1016/j.fusengdes.2017.01.057
42. J. Hoffmann et al., Improvement of reduced activation 9%Cr steels by ausforming. Nucl. Mater. Energy **6**, 12 (2016). https://doi.org/10.1016/j.nme.2015.12.001
43. N. Baluc, Materials for fusion power reactors. Plasma Phys. Control. Fusion **48**, B165 (2006). https://doi.org/10.1088/0741-3335/48/12B/S16
44. W. Yican, *Fusion Neutronics* (Springer, Berlin, 2017)
45. R.E. Larsen et al., Flux dependence of neutron enhanced diffusion in α-brass. Acta Metall. **12**, 1141 (1964). https://doi.org/10.1016/0001-6160(64)90093-8
46. S.R. MacEwen, The effect of neutron flux on dislocation climb. J. Nucl. Mater. **54**, 85 (1974). https://doi.org/10.1016/0022-3115(74)90079-8
47. D. Žontar et al., Time development and flux dependence of neutron-irradiation induced defects in silicon pad detectors. Nucl. Instrum Methods **A426**, 51 (1999). https://doi.org/10.1016/S0168-9002(98)01468-5
48. R. Stoller, The effect of neutron flux on radiation-induced embrittlement in reactor pressure vessel steels. J. ASTM Int. **1**, 1 (2004). https://doi.org/10.1520/JAI11355
49. F. Bergner et al., Flux dependence of cluster formation in neutron-irradiated weld material. J. Phys. Condens. Matter **20**, 104262 (2008). https://doi.org/10.1088/0953-8984/20/10/104262
50. M.V. Klein, Phonon scattering on lattice defects. Phys. Rev. **131**, 1500 (1963). https://doi.org/10.1103/PhysRev.131.1500
51. N. Sekimura, T. Okita, F.A. Garner, Influence of carbon addition on neutron-induced void swelling of Fe–15Cr–16Ni–025Ti model alloy. J. Nucl. Mater. **367–370**, 897 (2007). https://doi.org/10.1016/j.jnucmat.2007.03.065
52. I.A. Portnykh et al., The mechanism of stress influence on swelling of 20% cold-worked 16Cr15Ni2MoTiMnSi steel. J. Nucl. Mater. **367–370**, 925 (2007). https://doi.org/10.1016/j.jnucmat.2007.03.257
53. V.S. Neustroev, F.A. Garner, Severe embrittlement of neutron irradiated austenitic steels arising from high void swelling. J. Nucl. Mater. **386–388**, 157 (2009). https://doi.org/10.1016/j.jnucmat.2008.12.077
54. M. Rieth, A comprising steady-state creep model for the austenitic AISI L(N) steel. J. Nucl. Mater. **367–370**, 915 (2007). https://doi.org/10.1016/j.jnucmat.2007.03.062
55. S. Kondo, Y. Katoh, L.L. Snead, Cavity swelling and dislocation evolution in SiC at very high temperatures. J. Nucl. Mater. **386–388**, 222 (2009). https://doi.org/10.1016/j.jnucmat.2008.12.095
56. T.D. Burchell, Radiation damage in carbon–carbon composites: structure and property effects. Phys. Scr. **T64**, 17 (1996). https://doi.org/10.1088/0031-8949/1996/T64/002
57. G. Vayakis et al., Nuclear technology aspects of ITER vessel-mounted diagnostics. J. Nucl. Mater. **329–333**, 780 (2011). https://doi.org/10.1016/j.jnucmat.2011.01.081
58. M. Decreton, T. Shikama, E. Hodgson, Performance of functional materials and components in a fusion reactor: the issue of radiation effects in ceramics and glass materials for diagnostics. J. Nucl. Mater. **329–333**, 125 (2004). https://doi.org/10.1016/j.jnucmat.2004.04.012
59. A. Akindinov et al., Radiation hard ceramic RPC development. J. Phys. Conf. Ser. **798**, 012136 (2017). https://doi.org/10.1088/1742-6596/798/1/012136
60. M. Rubel et al., Mirror test for ITER at the JET tokamak: an overview of the programme. Rev. Sci. Instrum. **77**, 063501 (2006). http://link.aip.org/link/?RSINAK/77/063501/1
61. A. Litnovsky et al., Diagnostic mirrors for ITER: a material choice and the impact of erosion and deposition on their performance. J. Nucl. Mater. **363–365**, 1395 (2007). https://doi.org/10.1016/j.jnucmat.2007.01.281
62. M. Rubel et al., Overview of the second stage in the comprehensive mirrors test in JET. Phys. Scr. **T145**, 014070 (2011). https://doi.org/10.1088/0031-8949/2011/T145/014070
63. D. Ivanova et al., An overview of the comprehensive first mirrors test in JET with ITER-Like Wall. Phys. Scr. **T159**, 014011 (2014). https://doi.org/10.1088/0031-8949/2014/T159/014011
64. A. Garcia-Carrasco et al., Plasma impact on diagnostic mirrors in JET. Nucl. Mater. Energy **12**, 506–512 (2017). https://doi.org/10.1016/j.nme.2016.12.032
65. V. Voitsenya et al., Some problems of the material choice for the first mirrors of plasma diagnostics in a fusion reactor. Rev. Sci. Instrum. **70**, 790 (1999). https://doi.org/10.1063/1.1149402
66. M. Lipa et al., Analyses of metallic first mirror samples after long term plasma exposure in Tore Supra. Fusion Eng. Des. **81**, 221 (2006). https://doi.org/10.1016/j.fusengdes.2005.07.017
67. Y. Zhou et al., Study of first mirror exposure and protection in HL-2A tokamak. Fusion Eng. Des. **81**, 2823 (2006). https://doi.org/10.1016/j.fusengdes.2006.07.030
68. A. Garcia-Carrasco et al., Impact of helium implantation and ion-induced damage on reflectivity of molybdenum mirrors. Nucl. Instrum. Methods B **382**, 91–95 (2016). https://doi.org/10.1016/j.nimb.2016.02.065
69. M. Rubel et al., Metallic mirrors for plasma diagnosis in current and future reactors: tests for ITER and DEMO. Phys. Scr. **T170**, 014061 (2017). https://doi.org/10.1088/1402-4896/aa8e27
70. K. Ono et al., Effects of helium irradiation on degradation of optical properties of single and polycrystalline Mo mirrors for plasma diagnostics. Phys. Scr. **T138**, 014065 (2009). https://doi.org/10.1088/0031-8949/2009/T138/014065
71. J. Linke, Plasma facing materials and components for future fusion devices—development, characterization and performance under fusion specific loading conditions. Phys. Scr. **T123**, 45 (2006). https://doi.org/10.1088/0031-8949/2006/T123/006
72. J. Knaster, A. Möslang, T. Muroga, Materials reserach for fusion. Nat. Phys. **12**, 424 (2016). https://doi.org/10.1038/NPHYS3735
73. H. Fujita et al., Effect of neutron energy and fluence on deuterium retention behaviour in neutron irradiated tungsten. Phys. Scr. **T167**, 014068 (2016). https://doi.org/10.1088/0031-8949/T167/1/014068
74. O. Ogorodnikova et al., Annealing of radiation-induced damage in tungsten under and after irradiation with 20 MeV self-ions. J. Nucl. Mater. **451**, 379 (2014). https://doi.org/10.1016/j.jnucmat.2014.04.011
75. J. Grzonka et al., Electron microscopy observations of radiation damage in irradiated and annealed tungsten. Nucl. Instrum. Methods B **340**, 27 (2014). https://doi.org/10.1016/j.nimb.2014.07.043
76. S. Markelj et al., Deuterium retention in tungsten simultaneously damaged by high energy W ions and loaded by D atoms. Nucl. Mater. Energy **12**, 169 (2017). https://doi.org/10.1016/j.nme.2016.11.010
77. Z. Jiao, J. Michalicka, G.S. Was, Self-ion emulation of high dose neutron irradiated microstructure in stainless steels. J. Nucl.

Mater. **501**, 312 (2018). https://doi.org/10.1016/j.jnucmat.2018.01.054

78. A. Costley et al., Technological challenges of ITER diagnostics. Fusion Eng. Des. **74**, 109 (2005). https://doi.org/10.1016/j.fusengdes.2005.08.026
79. J.F. Ziegler et al., *The stopping and range of ions in solids*, 2nd edn. (Pergamon Press, New York, 1996)
80. P. Grand et al., An intense Li(d, n) neutron radiation test facility for controlled thermonuclear reactor materials testing. Nucl. Technol. **29**, 327 (1976)
81. A. Möslang et al., Suitability and feasibility of the International Fusion Materials Irradiation facility (IFMIF) for fusion materials studies. Nucl. Fusion **40**, 619 (2000). https://doi.org/10.1088/0029-5515/40/3Y/324
82. J. Knaster et al., Overview of IFMIF/EVEDA project. Nucl. Fusion **57**, 102016 (2017). https://doi.org/10.1088/1741-4326/aa6a6a
83. J. Knaster et al., IFMIF, the European–Japanese efforts under the Broader Approach agreement towards a Li(d, xn) neutron source: current status and future options. Nucl. Mater. Energy **9**, 46 (2016). https://doi.org/10.1016/j.nme.2016.04.012
84. J. Knaster, T. Kanemura, K. Kondo, An assessment of the evaporation and condensation phenomena of lithium during the operation of a Li(d, xn) fusion relevant neutron source. Heliyon **2**, e00199 (2016). https://doi.org/10.1016/j.heliyon.2016.e00199
85. H. Kondo et al., IFMIF/EVEDA lithium test loop: design and fabrication technology of target assembly as a key component. Nucl. Fusion **51**, 123008 (2011). https://doi.org/10.1088/0029-5515/51/12/123008
86. H. Kondo et al., Validation of liquid lithium target stability for an intense neutron source. Nucl. Fusion **57**, 066008 (2017). https://doi.org/10.1088/1741-4326/aa5fbd
87. J. Knaster, P. Favuzza, Assessment of corrosion phenomena in liquid lithium at $T < 873$ K. A Li(d, n) neutron source as case study. Fusion Eng. Des. **118**, 135 (2017). https://doi.org/10.1016/j.fusengdes.2017.03.063
88. G. Schlindwein et al., Mechanical testing of the IFMIF HFTM-DC prototype during operation in the HELOKA-LP helium loop. Fusion Eng. Des. **124**, 1077 (2017). https://doi.org/10.1016/j.fusengdes.2017.03.138

Journal of Fusion Energy (2019) 38:330–355
https://doi.org/10.1007/s10894-019-00213-9

ORIGINAL RESEARCH

Advanced Neutron Spectroscopy in Fusion Research

Göran Ericsson[1]

Published online: 27 February 2019

Abstract

This paper presents a review of the current state-of-the-art neutron spectroscopy in fusion research. The focus is on the fundamental nuclear physics and measurement principles. A brief introduction to relevant nuclear physics concepts is given and also a summary of the basic properties of neutron emission from a fusion plasma. Compact monitors/spectrometers like diamond, CLYC and the liquid scintillator are discussed. A longer section describes in some detail the more advanced, designed systems like those based on the thin-foil proton recoil and time-of-flight techniques. Examples of spectroscopy systems installed at JET and planned for ITER are given.

Keywords Fusion · Diagnostics · Neutron · Spectroscopy

Introduction

Neutrons are produced in most high performance magnetic confinement fusion research devices. Their spatial and energy distributions can be used to gain knowledge of the properties of the constituent fusion fuel [1]. On the most fundamental level, single-detector neutron monitors are used to estimate the total fusion power and energy. More advanced instruments, so called neutron cameras, give information on the neutron spatial distribution (in the poloidal plane), thereby providing information on the fusion power and the alpha particle birth profiles. In this paper we will focus on even more specialized measurements of the plasma neutron emission, namely neutron spectroscopy. We will present the basics of neutron measurements, in particular for spectroscopic use, give some examples of the measurement techniques employed and the information that can be gained from such measurements.

✉ Göran Ericsson
goran.ericsson@physics.uu.se

1 Department of Physics and Astronomy, Division of Applied Nuclear Physics, Uppsala University, Uppsala, Sweden

Some Pertinent Nuclear Physics

The neutron is a nucleon, i.e., a constituent of the atomic nucleus. As a free particle it is unstable but with a long lifetime of about 11 min (due to the properties of the weak force which governs its decay); thus the decay is not an issue for neutron measurements in fusion. Nucleons (protons, neutrons etc.) are influenced by the strong nuclear force, which is short ranged, acting on distances of femtometers [fm = 10^{-15} m] and thus only affects the properties within or close to the radius of the atomic nucleus. Unlike the proton, the neutron is electrically neutral and thus not affected by the electromagnetic force. This puts considerable restrictions on the types of measurement techniques that can be employed for its detection, as we will see later.

Atomic nuclei are systems of A bound nucleons: Z protons (each of charge $+e$) and N neutrons (charge 0) held together by the strong nuclear force. The nuclear mass number is defined as the sum of the number of protons and neutrons, $A = Z + N$. The proton number, Z, defines the chemical element; $Z = 1$ is Hydrogen, $Z = 2$ is Helium, $Z = 3$ is Lithium etc. For each element there exist a number of different variants, differing in the number of neutrons, N, they contain; these variants are called isotopes of the particular element.

The standard nuclear isotope notation is ${}^{A}_{Z}X_{N}$, where X is the chemical symbol (i.e., H for hydrogen, He for helium and so on).

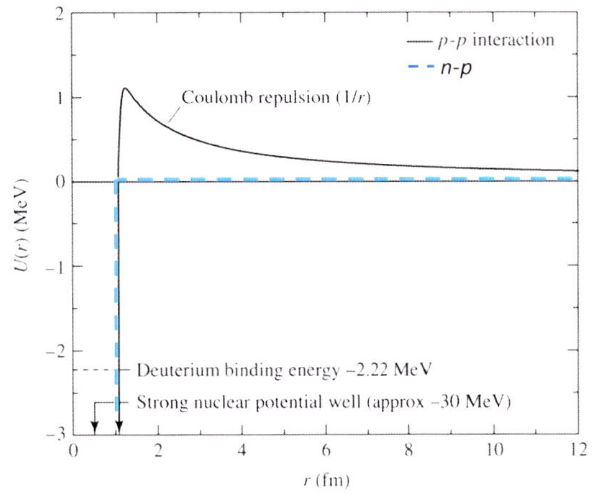

Fig. 1 The combined nuclear and Coulomb potential in the vicinity of a nucleus at r = 0

Unlike the short-ranged strong force, the electromagnetic force has infinite range. This is important when considering the possibility to initiate nuclear reactions, as at large distances, atomic nuclei (which always have a positive charge of +Z) repel each other through the Coulomb force. Only at the short distances of the nucleus, the strong force takes over and nuclear reactions can occur. For nuclear reaction purposes, the Coulomb repulsion potential can be estimated as:

$$V_c = (e^2/4\pi e_0)[\, Z_1 Z_2/(R_1 + R_2)] \tag{1}$$

where Z are the nuclear charges and R the "strong force radii" of the involved nuclei (where $R_i \approx R_x + 1$ [fm] is approximately the radius at which the strong force kicks in; R_x is the nuclear radius, which can be approximated as $R_x = 1.2 \cdot A^{1/3}$ [fm]).

Thus, for any nuclear reactions involving hydrogen to occur, such as fusion, the Coulomb energy at a distance of a few fm has to be overcome by the kinetic energy of the reacting particles. For deuterium and tritium the Coulomb potential is approximately $V_{c,dt} \sim 400$ keV [2] at the distance of the strong force. In reality, this is modified by quantum mechanical effects such that tunneling and resonant behavior and the onset of, for example, d + t fusion reactions is considerably lower. The situation is illustrated in Fig. 1, for p–p and n–p interactions. Note the lack of a Coulomb potential in n–p (n-nucleus) interactions. This fact is a crucial aspect of the possibility to use neutrons to induce fission reactions in heavy nuclei, such as Uranium and Plutonium.

A few other properties of particle interactions should be introduced for the purpose of the present discussion:

- *Cross sections, σ* this is a fundamental property of nuclear reactions, and is the effective area of the particles involved in a particular nuclear reaction. Due to the small areas involved, the unit barn [b] has been introduced, where 1 barn = 10^{-28} m^2 = 100 fm^2.
- *Nuclear binding energy, B* The strong force holds nuclei together in the nucleus, while the Coulomb repulsion (between protons) tends to tear it apart. The energy stored as binding energy in the nucleus is defined by the energy (mass) difference between the sum of the constituent nucleons and the nucleus at hand: $B = \Sigma(m_p) + \Sigma(m_n) - m_{AX}$. This comes about from the famous relation $E = mc^2$. Here the common notation to omit the c^2 in the formulas has been introduced; this notation will be used extensively below.
- *Nuclear reaction Q value* The Q value is the energy released or required in a nuclear reaction. It is defined as the sum of all particle masses before and after the reaction: $Q = \Sigma(m_{before}) - \Sigma(m_{after})$. Note that no reference to possible kinetic energy is involved. A reaction with a positive Q value means that energy is released in the reaction (as kinetic energy of the outgoing particles). A negative Q value, on the other hand, means that energy must be supplied in order for the reaction to occur. Also note that in all cases, the repulsive Coulomb potential must still be overcome in order to initiate the reaction, and this energy will be regained as kinetic energy of the outgoing particles; however, it is not part of the Q value definition.
- *Nuclear reactions* There are mainly two shorthand ways in which nuclear reactions are described: (i) a + B → C + d which means particles a and B react to produce particles C and d, alternatively written as (ii) B(a,d)C. The former notation is more common in reactions involving fundamental particles, while the latter is often used in nuclear reactions. In the latter notation, the B and C are taken to represent the involved (heavy) nuclei, while a, b are taken to be the lighter particles. Example: ^{6}Li(n,t)^{4}He is shorthand for the reaction where a neutron interacts with a Lithium-6 nucleus to produce a triton (t; ^{3}H) and an alpha particle (α; ^{4}He). Note that in strong nuclear reactions, the nucleon numbers are conserved, such that the number of neutrons and protons are the same before and after the reaction.
- *Nuclear excited states* In some nuclear reactions the resulting nucleus is left in a state with excess energy, an excited state. This is often noted by C^*, indicating that the nucleus C is not in its ground state, but carrying excess energy: B(a,d)C^*.
- *Nuclear decay* Nuclei in an excited state, X^*, will undergo decay to reach a stable state, for example by gamma (γ; electromagnetic) radiation to the so called ground state of the nucleus at hand, X_{gs}. However, depending on the level of excitation and the particular

nucleus, some decays occur through particle emission: $X^* \rightarrow Y + (n, p, \beta, \alpha)$, where the Y could still be in an excited state and undergo subsequent decay. Some heavy nuclei have no stable ground state. This is the case for all nuclei heavier than lead (Pb), although the lifetimes can sometimes be very long. For example, ^{238}U, which decays through α emission, has an estimated half life of 4.468 billion years. The decay modes of interest for the present discussion are:

- $X^* \rightarrow X_{gs} + \gamma$ (gamma decay; EM interaction; decay half life in the range fs–min)
- $X^* \rightarrow Y + \beta + \nu$ (beta decay β^-, β^+; nuclear transformation; weak interaction; decay half life in the range ms–hr)

Basic Neutron Detection Principles

A thorough review of radiation detection principles and practice is given in Ref. [3]. As mentioned previously, neutrons are electrically neutral, and cannot be detected through electromagnetic interactions. For charged particles, like protons and electrons, interactions with the electrons in the detector material produce moving, free electric charges. Collection and detection of these free charges is often the main method of particle detection. However, for neutrons, no free electrons are produced from traversing neutrons; direct nuclear reactions have to be utilized in order to produce such free charges in this case.

A special beneficial circumstance for neutron detection through nuclear reactions is that no Coulomb barrier prevents the neutron from interacting with a nucleus. Some of the useful nuclear reactions for neutron detection are:

(a) *Nuclear elastic scattering* X(n,n′)X′; where X′ is the nuclear recoil. These are "billiard ball" nuclear collisions/reactions, mediated by the strong nuclear force. "Elastic" in this context means that there is no internal nuclear reconfiguration of the nucleus X, i.e., it is in its ground state both before and after the reaction. Two-body kinematics governs the energies of the outgoing particles. A moving charged recoil nucleus can in this case create free electrons in the detector material which can be used for detecting the presence of the initial neutron.

- Examples: H(n,n)H(p_R), D(n,n)D, ^{4}He(n,n)^{4}He, ^{12}C(n,n)^{12}C (where p_R indicates a recoil proton).
- Detector systems utilizing this method are scintillators, thin-foil proton recoil and time-of-flight systems.

(b) *Nuclear inelastic scattering* X(n,n)X* followed by detection of the X* decay radiation. Several different decay modes of the resulting excited nucleus, X*, can be utilized γ, β, n, etc.

- Examples: ^{115}In(n,n)^{115}In^m; where m = * = metastable state, with a long half life.
- Detector types utilizing this method are for example activation foils.

(c) *Nuclear reactions* X(n,y)Z. These are reactions where two (or more) charged reaction products appear in the final state where y = p, d, α, pn, …

- Example: ^{12}C(n, α)^{9}Be, ^{3}He(n,p)^{3}H, ^{6}Li(n,t)α.
- Detector types: diamond semiconductor, ^{3}He tubes, Li glass scintillators.

(d) *Fission* X(n, x·n)Y*,Z*; heavy charged fission fragments

- Examples: X = ^{235}U, ^{238}U, Pu
- Detector types: Fission chambers (FC), Parallel plate avalanche counters (PPAC).

The Fusion Plasma as a Neutron Source: The Direct Emission

A more comprehensive summary of the concepts introduced here is given in for example Ref. [4]. Fusion energy research is focusing on plasmas with the hydrogen isotopes deuterium (D) and tritium (T) as fuel, where the intended fusion reaction is d + t. The attractive properties of this reaction are (i) its high energy release (17.6 MeV per reaction), (ii) its high cross section and (iii) its low threshold energy. Some of these properties are summarized in Fig. 2, showing the fundamental cross sections of relevant fusion reactions (in particular d + d, d + t, t + t) as function of ingoing particle energy and, more relevant for fusion plasmas, the reactivity $\langle\sigma v\rangle$ as a function of plasma thermal (kinetic) temperature.

In fusion experiments with D and T fuel the main fusion reactions (d + d, d + t, t + t) include the following neutron-producing channels (where the value in parenthesis is the neutron energy in the cold plasma limit):

$d + d \rightarrow {}^3He + n$ (2.45 MeV) (Branching ratio = 50%)
$d + t \rightarrow {}^4He + n$ (14.0 MeV) (Branching ratio = 100%)
$t + t \rightarrow {}^4He + n + n$ (0–8.8 MeV)

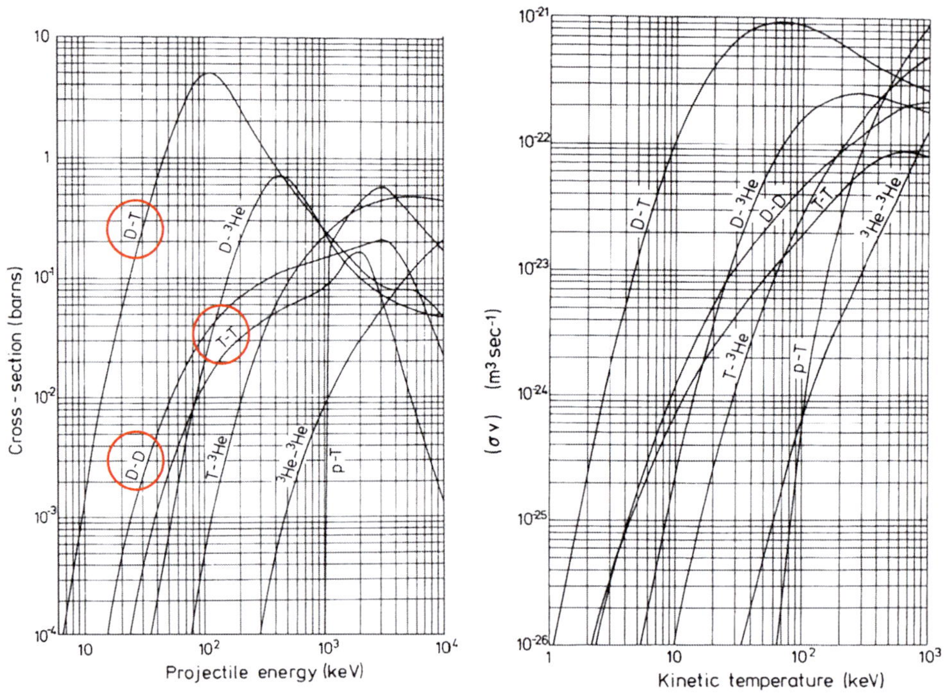

Fig. 2 (left) Cross sections of some nuclear fusion reactions relevant for fusion energy research. The three reactions most relevant for the present work are encircled. (right) The fusion reactivity $\langle\sigma v\rangle$ as function of the thermal plasma kinetic temperature ($k_B T_i$). From Ref. [5]

In addition, non-fuel nuclei of different kinds can also be involved in neutron producing reactions. For fusion plasmas, such "impurities" can be either particles deliberately injected, for example as part of the plasma heating (^{3}He), they can be the α particle "ash" from the fusion reactions, or unwanted elements that have entered the plasma from the surrounding walls and other structures (^{9}Be, ^{12}C):

$$d + \{^3He, ^4He, ^9Be, ^{12}C, \ldots\} \rightarrow n + X$$

However, in this presentation we will not discuss these latter reactions further, but concentrate on the emission properties of the main fusion reactions.

The neutron emission intensity is directly coupled to the fusion reaction rates in the plasma, since neutrons are produced in 50% of the d + d reactions and in 100% of the d + t reactions. The (local) fusion reaction rate can be written:

$$R = 1/(1 + \delta_{ab}) n_a n_b \langle\sigma v\rangle \,(m^{-3}s^{-1}) \tag{2}$$

Here, n_a, n_b are the number densities of the fusing ions, $\langle\sigma v\rangle$ is the reactivity, i.e., the integral over the velocity distributions of the two fusing ions and their relative velocity, weighted by the fusion cross section, σ:

$$\langle\sigma v\rangle = \iint f_a(\mathbf{v}_a) f_b(\mathbf{v}_b) \mathbf{v}_{rel} \sigma(\mathbf{v}_{rel}) d\mathbf{v}_a d\mathbf{v}_b \,(m^3 s^{-1}) \tag{3}$$

and δ is the Kronecker delta function: $\delta_{ab} = 1$ if a = b; $\delta_{ab} = 0$ if a ≠ b. This delta function must be introduced in order to avoid double counting in single (fuel) species plasmas.

The reactivity $\langle\sigma v\rangle$ can be evaluated for thermal plasmas as a function of plasma thermal (kinetic) temperature expressed in energy units as $k_B T_i$, where k_B is Bolzmann's constant and T_i is the plasma temperature in [K], as shown in Fig. 2(right).

For a plasma in thermal equilibrium the velocity distributions are given by Maxwell–Bolzmann functions:

$$f(v) = 4\pi v^2 (m/2\pi k_B T_i)^{3/2} \exp(-mv^2/2k_B T_i) \tag{4}$$

The neutron emission spectrum from such a thermal plasma can be calculated analytically (with some simplifications) and is very close to Gaussian in shape [6, 7]:

$$f(E_n)_{TH} \approx (\sigma_w (2\pi)^{1/2})^{-1} \exp(-(E_n - \langle E_n\rangle)^2/2\sigma_w^2) \tag{5}$$

where TH indicates thermal conditions, σ_w is the standard deviation and the energies are centered around $\langle E_n\rangle$ (see below). The width (σ_w) of the neutron energy distribution can be expressed in terms of the kinetic temperature as:

$$\sigma_w = (2 k_B T_i \langle E_n\rangle m_n/(m_1 + m_2))^{1/2} \approx C \,\mathrm{sqrt}(k_B T_i) \tag{6}$$

For d + d reactions the constant in front of the square root is $C_{dd} = 82.6$ [keV$^{-1/2}$]; for d + t it is $C_{dt} = 177$ [keV$^{-1/2}$]. Clearly, in thermal plasmas, or in situations where the thermal emission component can be clearly identified and measured, the width of the (thermal) neutron energy distribution gives an estimate of the plasma temperature, T_i. This is one of the primary physics parameters that neutron spectrometry can provide. It should be noted, however, that in an actual measurement, the information provided will be integrated along the instrument's field-of-view, and thus influenced by local plasma parameters such as density and temperature. Normally, the neutron emission will be heavily biased towards the core of the plasma.

It should be pointed out, that this thermal broadening also sets the "standard" for spectroscopic measurements, such that the resolution of the spectrometer should be matched to the expected thermal broadening of the spectra. The resolution should be good enough to resolve the thermal broadening, but it need not be considerably better than this. In most high-performance fusion experiments, such as JET, ASDEX, etc., plasma thermal temperatures from basic ohmic heating are around 2–3 keV; the corresponding broadening is then given by Eq. 6.

In the presence of non-Maxwellian components of the fuel velocity distribution function, the neutron energy spectrum will include components corresponding to these sources [7]. The origin of such non-thermal components could be the internal alpha heating or external heating sources, like NBI or RF. There is a non-trivial relationship between the non-thermal neutron energy spectrum and the underlying fuel velocity distribution and the connection is in general difficult to calculate analytically. Often Monte Carlo techniques are used in a forward modelling approach, where a model of the fuel ion velocity distribution is used to calculate the corresponding neutron energy distribution.

The underlying fuel velocity distribution can be taken from, for example, plasma modeling codes such as TRANSP or ASCOT. Alternatively, when full-scale plasma modeling is not available or too time consuming, more simplified models can be used where a number of distinct fuel components are used to represent the main processes in the plasma. These ion velocity components are then taken to represent ions in thermonuclear equilibrium, ions injected by the Neutral Beam Heating system, ions accelerated by the Radio-frequency heating system and (in particular in future high performance devices such as ITER and DEMO) ions affected, "knocked-on", by the high-energy alpha particles. In the latter three cases, the slowing down (through collisions with the thermal plasma) of the primary high-velocity ions must be calculated in order to gain an accurate representation of the component.

Fusion reactions involving ions from the different velocity components are then calculated, generating a number of distinct neutron emission components with characteristic energy distributions. These neutron energy components are for example (i) the thermal component, given by fusion reactions involving two ions from the bulk thermal distribution, (ii) the beam-thermal (or beam-target) component, involving one ion from the neutral beam slowing-down distribution and one from the thermal bulk,

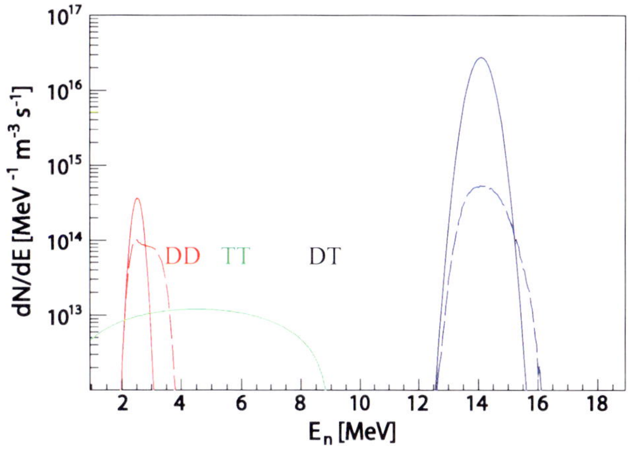

Fig. 3 Example of some of the primary neutron emission spectral components in a 50:50 DT plasma of typical ITER conditions; injection energy of the deuterium NBI is 1 MeV. Blue full line corresponds to thermal DT reactions, blue broken line to D beam-thermal T reactions, red full line to thermal DD reactions, red broken line to beam-thermal DD reactions and green full line to thermal TT reactions (Colour online)

(iii) the RF-thermal (RF-target) component, (iv) the beam–beam component etc. The particular conditions of a specific fusion plasma will determine which of these components are present on a significant level in the neutron spectrum [8, 9].

An example of some of the primary (i.e., directly from the fusion reactions) neutron emission components of a typical ITER plasma are shown in Fig. 3. Here, only thermal and beam-thermal components are shown. In addition, one can expect that also RF-thermal and alpha knock-on [10] components will be present in ITER. The calculation of the direct neutron flux at the position of the active detectors requires some computational tools. The calculation of the neutron energy distribution seen by a specific diagnostic (see further below) based on a specific fuel ion velocity distribution is done with codes like DRESS [11], GENESIS [12], ControlRoom [13]. Codes like LINE21 [14] take care of the sampling of the plasma volume and treat any intervening obstacles in the field of view of the diagnostic.

Fusion Neutrons: The Scattered Component

In neutron spectroscopic measurements at magnetically confined fusion facilities, the main source of information rests in the direct neutron emission as discussed above. In order to isolate the direct emission and to select the emission from a specific region of the plasma, collimated measurements are most often employed. The situation is schematically depicted in Fig. 4. In addition, more or less substantial neutron shielding is required around the active

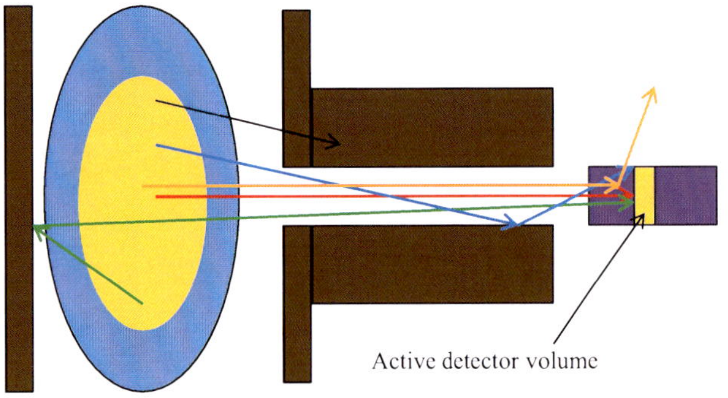

Fig. 4 Schematics of processes involved in neutron absorption, scattering, penetration and direct emission. Plasma in yellow, tokamak solid structures in dark brown (including instrument collimator), detector mechanics in purple, active detector volume in orange. Arrows illustrate possible neutron "histories": red is the direct flux at the detector, blue is (collimator) in-scattered neutron flux at the detector, green is (far wall) back-scattered neutron flux at the detector, orange is direct out-scattered or attenuated flux, black is neutrons stopped/absorbed in solid structures (Colour online)

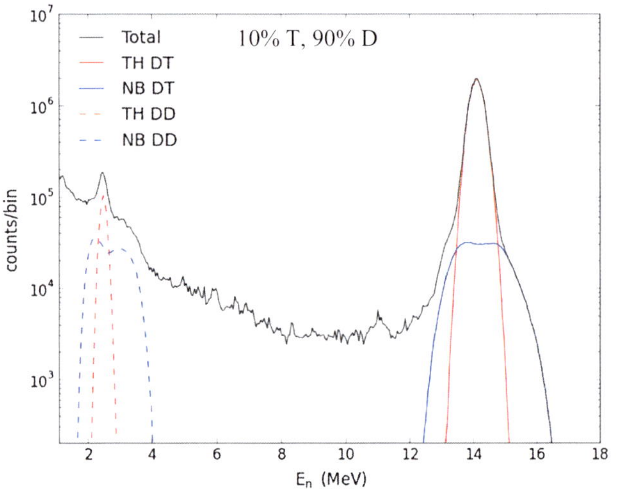

Fig. 5 Results from an MCNP simulation of direct and scattered neutron emission spectral components in a 90:10 DT plasma of typical ITER conditions; injection of NBI deuterium is at 1 MeV. Red full line corresponds to thermal DT reactions, Blue full line to beam-thermal DT reactions, red broken line to thermal DD reactions, blue broken line to beam-thermal DD reactions and black full line to the total neutron spectrum, including also scattered neutrons from dd and dt fusion reactions (Colour online)

detection system, in order to reduce the exposure to any detrimental background radiation at the measurement position. For fusion neutron measurements, this background is closely correlated in time and intensity with the primary fusion reactions and it is mainly composed of (capture) gammas and scattered neutrons (i.e., non-direct neutrons). Induced radioactivity and other sources normally play a minor role. These background components should be well understood both in order to design the diagnostic system and to interpret the data.

As seen in Fig. 4, a number of different processes contribute to the flux of scattered (low-energy) neutrons at the active detector. One of the most important contributions comes from far-wall backscattered neutrons. In this case, the region (area) of the tokamak internal wall in the direct field of view of the spectrometer system acts as the source. Note, however, that this area can be illuminated by neutrons from a large part of the plasma volume, thereby enhancing the effect. Other contributions come from neutrons scattered in materials close to the active detector, such as scattering in the collimator and other structures. These neutrons normally originate from a somewhat smaller region of the plasma, thereby reducing the effect of this contribution. For completeness, it is also important to consider any direct neutrons that are lost due to capture or out-scattering, in particular if absolute comparisons between models and measurements are done. The situation for gammas is to some degree similar to the neutrons, with the addition that absorbed neutrons often generate so called capture gammas, i.e., a gamma radiation due to the absorption of a neutron in a nucleus. The gammas will often appear in structures close to the active detector volume, and special care has to be taken in the design of the system to reduce this background component.

There exist several computer codes that calculate the neutron and gamma transport and thus can give an estimate of the background flux at a certain position in a specified geometry. Examples of such codes are MCNP [15], SERPENT [16] and TRIPOLI [17].

It is important to note that all scattered neutrons will be energy downgraded. The main processes are elastic and inelastic neutron scattering, $X(n,\acute{n})\acute{X}$and $X(n,\acute{n})X^*$. In both cases, the outgoing neutron, n', will be of lower energy than the one going into the reaction. This is illustrated in Fig. 5, which is similar to Fig. 3 but for a plasma with 10% T and 90% D. In this model calculation, done with MCNP, a significant level of scattered neutrons appears that is not part of the direct emission from the plasma.

As is clear from Fig. 5, scattered neutrons from the original d + t reactions (i.e., energy down-graded from 14 MeV) fill out the region below 14 MeV all the way to lowest energies. As can be understood from the figure, in situations of significant T relative density (n_T/n_{TOT-} $> 20\%$ or so), this background can act to obscure the signal from the fusion d + d reactions, making them hard or even impossible to discern.

The presence of any background in a measurement situation must of course be understood and controlled. In all diagnostic measurements for fusion, scattered neutrons and gammas will constitute a strong background source that will have to be carefully assessed both in the design of the diagnostic system and in the interpretation of the data.

Calculating the Neutron Energy in Fusion

In fusion, neutrons are emitted from the fusion reactions d + d → ^{3}He (0.8 MeV)+ n (2.5 MeV) and d + t → α (3.5 MeV)+ n (14.1 MeV). These are two body exothermic reactions with Q values equal to 3.27 MeV and 17.6 MeV for the DD and DT reaction, respectively. The kinetic energy carried by the neutron is determined by the kinematics of the reaction.

Consider the reaction d + d → ^{3}He + n. Using non-relativistic kinematics, the energy of the outgoing neutron can be written [9]:

$$E_n = \frac{1}{2}m_n V_{CM}^2 + \frac{m_{3He}}{m_{3He}+m_n}(Q+K) + V_{CM}\cos(\theta)\sqrt{\frac{2m_{3He}m_n}{m_{3He}+m_n}(Q+K)} \quad (7)$$

Here Q is the reaction Q value, Q_{dd} = 3.27 MeV; m_n and m_{3He} are the masses of the neutron and ^3He-particle, respectively; K is the relative kinetic energy, $K = 1/2\mu(v_{rel})^2$ with $v_{rel} = |\boldsymbol{v}_{d1} - \boldsymbol{v}_{d2}|$ and where μ is the reduced mass; $\boldsymbol{V}_{CM}$ is the center-of-mass velocity of the reactants (dd), $\boldsymbol{V}_{CM} = (m_d\boldsymbol{v}_{d1} + m_d\boldsymbol{v}_{d2})/(m_d + m_d)$; θ is the neutron CM scattering angle with respect to $\boldsymbol{V}_{CM}$.

In the case of dd fusion, the center of mass velocity simplifies to:

$$V_{CM} = \frac{m_d v_{d_1} + m_d v_{d_2}}{2m_d} + \frac{\mathbf{v}_{d_1} + \mathbf{v}_{d_2}}{2} \quad (8)$$

For the d(t,n) ^{4}He reaction, replace d, d, ^{3}He with d, t, ^{4}He, respectively, and Q = Q_{dt} = 17.6 MeV.

For reactants at rest the neutron energy would be constant and equal to $E_0 = m_{3He}/(m_{3He} + m_n)\ Q$ = 2.45 MeV and 14.0 MeV for the DD and DT reaction, respectively. When the reactants are not at rest the energy of the products is shifted by a quantity that depends on the reactant velocity (energy) and on the emission direction of the neutrons, through the term $V_{CM}{\cdot}\cos(\theta)$, which is responsible for the Doppler broadening effects in the neutron emission, and the K-factor which gives an additional small energy-dependence [7]. The emitted neutron energy spectrum is thus related to the reactant fuel ion velocity (energy) distributions and Eq. 7 is the basic relation used to interpret neutron spectroscopy measurements in fusion plasmas.

It is important to note that the term $V_{CM}\cos(\theta)$ in Eq. 7 introduces a directional dependence in E_n. For isotropic velocity distributions, f($\mathbf{v}_d$), the cos(θ) term averages to zero and the neutron energy distribution is the same in all directions. This is the case, for example, in a thermal plasma.

For non-isotropic f($\mathbf{v}_d$) the neutron energy distribution will depend on the measurement direction, i.e., on the diagnostic line-of-sight with respect to any preferred directions in the fuel ion velocity distributions. For measurements in magnetic confinement fusion such a preferred direction is the direction of the magnetic field. Certain populations of plasma fuel ions will have velocity distributions that depend on the pitch (v_{par}/v_{tot}), where v_{par} is the velocity component parallel to the (local) magnetic field). This situation applies to, for example, fuel ion populations due to external heating, such as those injected by Neutral Beam Injector (NBI) systems or accelerated by Ion Cyclotron Radio-frequency Heating (ICRH) systems. The cos(θ) dependence of E_n, together with the gyro motion of the ions (and electrons) around the magnetic field lines can create rather particular neutron energy spectra in certain directions. An example is shown in Fig. 6, for d + d reactions and a pitch equal to zero (i.e., NO parallel velocity, only perpendicular) and a number of different deuteron energies, measured at a viewing angle of 90° to the magnetic field. A very distinctive, quite broad, double-humped spectrum appears. Ions heated by ICRF can be modeled with such a process.

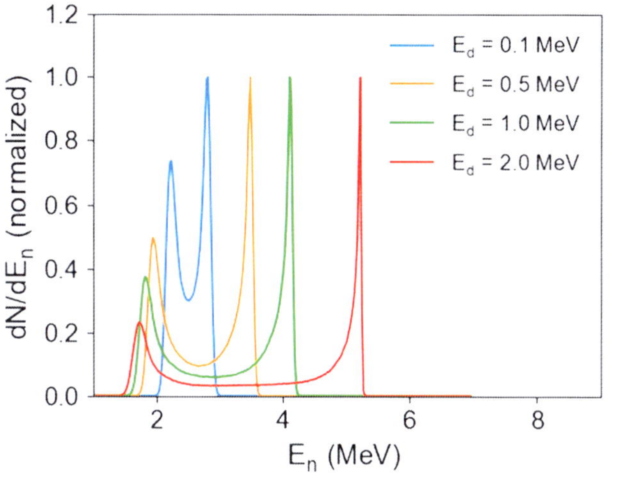

Fig. 6 Simulations of neutron energy distributions for a number of cases where high-energy (0.1–2.0 MeV) mono-energetic deuterons of pitch zero (i.e., with gyro motion completely perpendicular to the magnetic field) are made to interact with deuterons of a 5 keV bulk plasma. The emission is viewed by an instrument at an angle of 90° to the magnetic field

Role and Challenges for Fusion Neutron Spectrometry

The primary role of any fusion diagnostic system is obviously to provide time resolved information on relevant plasma/fuel ion parameters either for fundamental understanding of the fusion plasma physics and/or as part of the systems for machine control and protection. In the latter case, feed-back of results for active machine control should be provided on the ms time frame (or longer, depending on the parameter under study).

For neutron spectroscopy, the primary physics parameters/effects that can be provided are:

- The thermal fuel ion temperature, T_i. As discussed above, this quantity can be deduced from the width of the thermal neutron emission component.
- The fuel ion density and ratio n_T/n_D. This quantity can be deduced in two ways: (i) in plasmas with $n_T/n_D < 20\%$, by comparing the intensities of the thermal neutron emission components at 2.5 and 14 MeV, (ii) in plasmas of higher relative tritium contents by measuring and separating the thermal and beam-thermal components at 14 MeV. At ITER, neutron spectroscopy has been identified as the primary diagnostic to provide this quantity.
- The intensities of different neutron components, thereby providing quantities like thermal fuel ion fraction.
- More specialized information, like heating efficiencies and effects like RF tail temperatures, details on the fuel velocity distribution functions etc.
- A neutron spectrometer (system) can/should also be included as an extra sight line in any neutron camera system. In particular, with some complementary profile information from such a neutron camera, a well-characterized neutron spectrometer can provide an independent estimate of the total neutron rate (P_{fus}).

Some physics results based on analysis of neutron spectra can be found in for example Refs [14, 18–23]. Since neutron spectroscopic measurements are normally performed using collimated lines-of-sight (fields of view), the results provided will always be an integral measurement of the conditions in the volume covered within the field of view of the instrument. To disentangle more local effects, additional profile information (or assumptions) is required.

Some specific challenges for neutron diagnostics include [24]:

- The plasma constitutes an extended (> 100 m^3), continuous emission (min-hr) neutron source. This puts some requirements on the implementation of neutron diagnostic systems:
  - Collimated and well shielded measurements are normally required, often involving quite substantial radiation shielding installations.
  - The neutron flux at the detector(s) will be composed of both direct and scattered neutron contributions, and these have to be understood and handled.
  - Reliable, robust techniques should be used, as both the requirement of substantial radiation shielding and the desire to be close to the plasma make service and replacements cumbersome.
- The experimental conditions around the "reactor" are harsh, often involving:
  - High levels of neutron and gamma background radiation;
  - High temperatures, high B-fields, high-frequency EM noise interference;
- There are special requirements on the detectors in neutron diagnostic systems:
  - The harsh environments mean that detectors need to be robust and reliable under high temperature, high radiation, high B-field conditions or that these environmental effects are mitigated by appropriately designed interfacing, such as cooling, magnetic shielding etc.
  - The detectors are placed in a mixed field of neutron and gamma radiation and a separation of these signals is required for high quality results.
  - The system needs to resolve weak signatures in the neutron emission: good signal-to-background ratios are required. This is illustrated in Fig. 7, showing a simulation of the ITER measurement situation. Here it is seen that in order to access the alpha knock-on (direct imprint of the alpha heating in the neutron spectrum) a dynamic range of more than 4 orders-

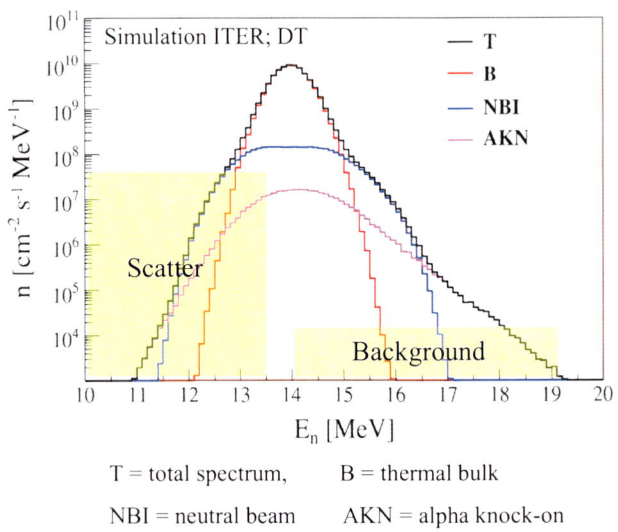

Fig. 7 Simulation of expected neutron emission spectral components in a 50:50 DT plasma of typical ITER conditions; injection energy of the deuterium NBI is 1 MeV. Red full line corresponds to thermal DT reactions, blue line to D beam-thermal T reactions, and magenta full line to the alpha knock-on effect. The estimated level of scattered neutrons and background are also indicated in yellow (Colour online). Adopted from Ref. [25]

of-magnitude is required. This is a very challenging experimental task.

- Results should be reported with a time resolution down to ms. For the type of counting experiments that are typical for neutron diagnostics this means a requirement to accept and acquire data of MHz signal rates. In other words, the count rate capability of the system has to be high, in the MHz region.
- For machine control and protection, information is required in real-time with down to ms time resolution: this means fast transfer rates from acquisition electronics to processing units as well as fast and robust data analysis.

- Interfacing can be an issue:
  - The size and weight of the neutron spectrometry systems are considerable.
  - The desire to place the systems close to the neutron source (i.e., the plasma), in order to maximize count rates, means that large and heavy components for radiation shielding are required.
  - Maintenance and replacement of components in positions close to the fusion source is a challenge.
  - A competition for "real estate" around the fusion device, where bulky neutron diagnostic systems, closely coupled and with "direct" lines-of-sight into the plasma, can be at a disadvantage.

Finally, it should be mentioned that the very high neutron rates expected in future high-performance fusion devices, such as ITER and DEMO, which are seen as problematic for many of today's plasma diagnostics, are instead a great opportunity for neutron diagnostics of all types.

Review of Neutron Spectrometry Techniques: Compact Detectors

There is a quite broad range of techniques and detectors that can (and have been) used for fusion neutron spectroscopy [26, 27]: they span from small compact single detectors/monitors of a few kg to very specialized instrument systems of considerable weight (10's of tons) and size (several m^3). Some of the more common techniques are schematically depicted in Fig. 8.

As can be seen in Fig. 8, many of the neutron detection techniques listed in "The Fusion Plasma as a Neutron Source: The Direct Emission" section have been tested and used in fusion. JET in particular has over the years had an ambitious program of neutron diagnostics, including, in particular, several different types of neutron spectrometers [26].

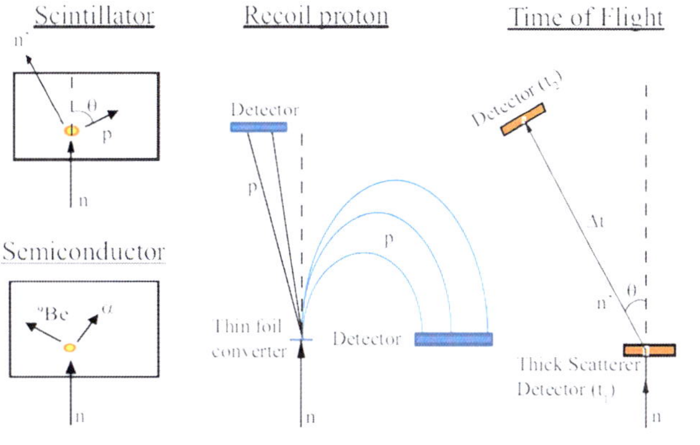

Fig. 8 Schematics of the main neutron spectroscopy techniques discussed in this paper. Top left illustrates use of a hydrogenous scintillator, bottom left use of a diamond semiconductor, middle shows the two variants of the thin-foil proton recoil technique; right the time-of-flight technique

Scintillators with a large hydrogen content are often employed as neutron counters and also, in a more limited capacity at least for fusion, as spectrometers. The main mechanism for detection in this case is (n, p) elastic scattering. This subject is only briefly introduced here and more extensively covered in another paper in these proceedings (M. Cecconello); it will not be discussed further here. Here, instead, we will focus on a few other neutron measurement and spectroscopy techniques.

Example 1: Semiconductor Detectors; Diamonds (C)

Diamond is a very interesting detector material for fusion diagnostics [28, 29], primarily for neutron counting and spectroscopy, for a number of reasons:

- There are a number of useful nuclear reaction channels with neutrons, some providing charged particles (only) in the final state. This provides a reasonable efficiency and the possibility for spectroscopy;
- It is mechanically a very robust material, and chemically benign (non-toxic, resistant and inert to many chemicals, etc.);
- It can withstand high temperatures;
- It is allegedly very radiation hard;
- There is rapid progress in fabrication of synthetic, single crystal diamonds suitable for neutron spectroscopy through the Chemical Vapor Deposit method; quality and size of samples is improving.

For low energy neutrons (i.e., for fusion neutrons in D plasmas), there are two main reaction channels open:

- $n + {}^{12}C \rightarrow n + {}^{12}C$ (elastic),
- $n + {}^{12}C \rightarrow {}^{13}C^*$ (n capture (γ))

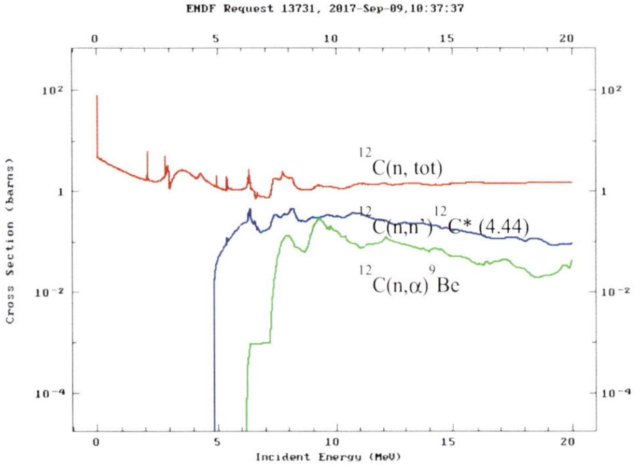

Fig. 9 Cross sections as function of neutron energy from the ENDF data library for some of the neutron induced reactions in diamond (^{12}C). Included are the total cross section (n,tot), the cross section for leaving the ^{12}C in its first excited state at 4.44 MeV and the cross section for the (n,a) reaction (suitable for neutron spectroscopy). Data from Ref. [30]

At higher neutron energies, from about $E_n > 5$ MeV, several new reaction channels open up (available for fusion neutrons in DT plasmas):

- n + ^{12}C → n′ + ^{12}C* (Q = − 4.4 MeV)
- n + ^{12}C → α + ^{9}Be (Q = − 5.7 MeV)
- n + ^{12}C → n′ + 3α (Q = − 7.3 MeV)

The cross sections for some of these reactions are shown in Fig. 9. As can be seen in the figure, the situation is rather complex, with many reaction channels open (depending on E_n), and in order to understand the response of a diamond detector one needs to know such cross sections in detail, both the excitation functions (dσ/dE; as shown in the figure) and the angular differential cross sections (dσ/dθ). In addition, detailed kinematics modeling of the reactions is required. A somewhat more complete table of the possible final state particles in n + ^{12}C reactions is given in Table 1. As shown, for fusion relevant neutron energies, $E_n < 20$ MeV, there are at least 8 open reaction channels.

Characterization of an actual diamond detector with mono-energetic 20 MeV neutrons is shown in Fig. 10.

Table 1 Some of the final state particles in n + ^{12}C reactions and their Q values

Products	*Q* (MeV)	Products	*Q* (MeV)
^{12}C + n	0	n′ + 3α	− 7.27
^{13}C + xg	4.95	^{12}B + p	− 12.59
^{12}C + n′ + g	− 4.44	^{11}B + d	− 13.73
^{9}Be + α_0	− 5.70	^{11}B + n+p	− 15.96

Here, α_0 denotes the situation where the ^{9}Be is left in its ground state. (Note that there are two columns of reaction products and corresponding Q values.)

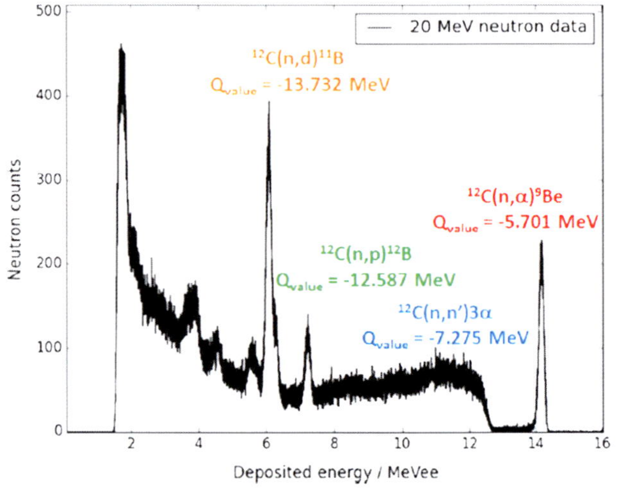

Fig. 10 Response of a diamond detector to mono-energetic neutrons of E_n = 20 MeV. Some of the structures in the spectrum are identified by their reaction channel and their Q value. Peaked structures typically correspond to charged-only final states, while broad structures correspond to final states with (at least) one neutral particle (here neutron). Adapted from Ref. [32]

Here, many of the reactions of Table 1 can be identified. In particular, the very favorable situation with the ^{12}C(n,α)^{9}Be reaction is evident. This reaction has the lowest Q value (Q = − 5.70 MeV) of all the channels with charge-only final states. The charged-only final state makes it well suited for spectroscopy, in particular of neutrons in DT plasmas, where the main direct emission is centered around E_n = 14 MeV. The energy of the final state particles, about 8.3 MeV for E_n = 14 MeV, is the highest for any charged-only channel, so the full-energy peak is well separated from the rest of the spectrum. Energy resolution of dE/E < 3% has been achieved with diamond detectors intended for use at fusion facilities [31].

Another beneficial property in this context is the fairly "flat" cross section for $E_n > 9$ MeV, simplifying the interpretation of data. A slight drawback of the diamond detector is the rather low ratio of good events in the (n,α) peak compared to the rest of the spectrum. As seen from Fig. 9, at E_n = 14 MeV this is on the level of a few percent. Even though the (n,α) events stand out clearly in the spectrum, all the other reaction channels will put a load on the detector and acquisition electronics, limiting the count rate capability (of good counts in the (n,α) peak) to below 100 kHz or so. This, in turn, will limit the achievable time resolution for physics results.

In order to analyze data from a diamond detector the response of the detector to neutrons in a broad range of energies is required. Such a response function, connecting the measured quantity of the instrument with the neutron energy, is often constructed from a mixture of data from measurements (such as those in Fig. 10) and simulations.

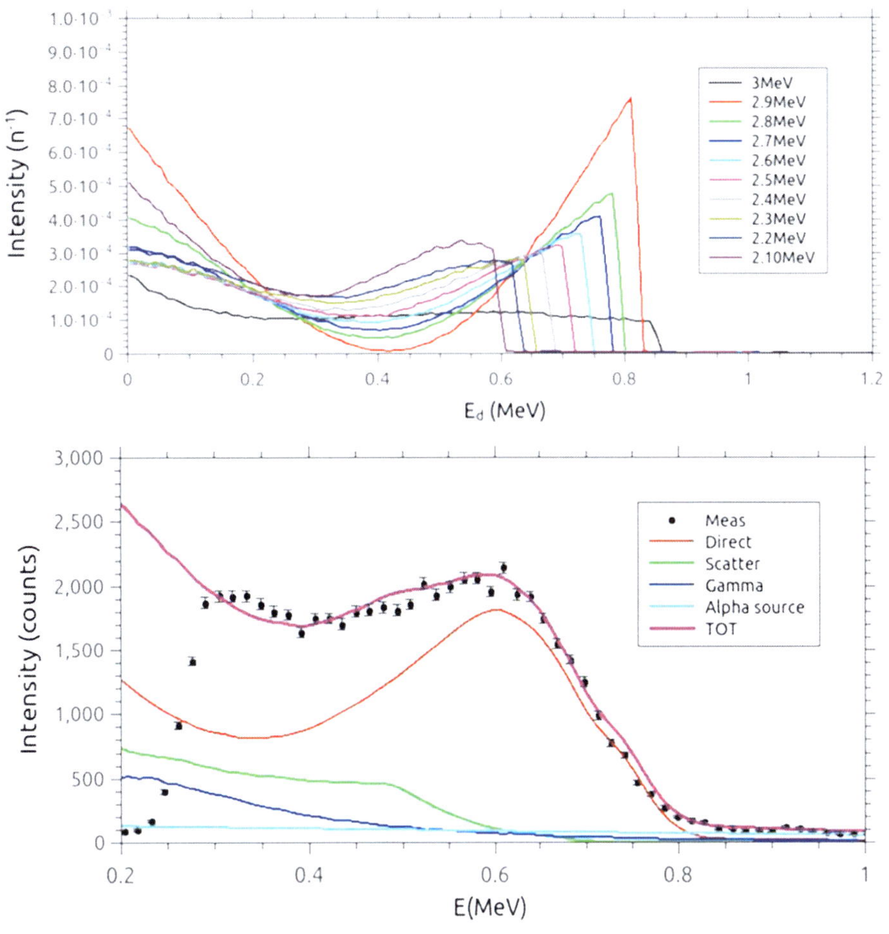

Fig. 11 (Top) Ideal response of a diamond detector to low energy mono-energetic neutrons in the range $2 < E_n < 3$ MeV through n-^{12}C elastic scattering. (Bottom) Experimental data from a diamond detector installed at JET (points with error bars). A model response has been fitted to the data, partly based on the calculated response functions in the top panel. The best fit is shown in magenta and includes components corresponding to Direct neutron emission (brown), Scattered neutrons (green), gamma rays (dark blue), and a calibration alpha source (light blue). From [33] (Colour online)

The simulation model is tuned to reproduce the experimental data, and is necessary in order to supplement the often limited number of measurements at different neutron energies. The measurements will put constraints on the modeling, and a good coverage of experimental data over the energy region of interest helps improve the modeling. The primary use of diamonds would be as neutron spectrometers in DT plasmas, which would suggest a detailed characterization of the response in the region $10 < E_{n\text{-}} < 20$ MeV. Such a characterization is described in Ref. [32].

Even though the charged-only channels, in particular the (n,α) reaction, are best suited for spectroscopy, it should be mentioned that also n-^{12}C elastic scattering can be used, in a manner similar to that in scintillators (where n–p elastic scattering is the primary reaction). This opens up for the use of diamonds also in D plasmas, something that has indeed been tested at JET [33]. Figure 11 shows results for a diamond used at JET in DD operations, where data from a set of 45 NBI heated discharges were summed to produce the pulse height spectrum. The calculated response for a number of mono-energies is shown in the top panel and the measured data, together with a component fit, at the bottom. A good agreement between modeling and data is obtained, indicating that physics results pertaining to the neutron spectrum can be extracted. An obvious drawback is the rather low neutron rate and efficiency of the detector, which makes it necessary to sum many discharges in order to obtain sufficient statistics to perform the analysis.

Diamond detectors are an active line of research and development in fusion neutron diagnostics. At JET, for example, two sets of diamond detectors are currently installed, one single crystal detector in a tangential line-of-sight, and a pixelated, 3 × 4 matrix, of 12 crystals in a radial, vertical LOS. The latter is shown in Fig. 12.

In summary, diamond detectors seem to offer a robust technique for neutron spectroscopy in particular for DT operations. However, the experimental use at high performance fusion devices is still somewhat limited, and some issues with diamond detectors remain to be studied:

- The low fraction of useful counts in the (n,α) peak, corresponding to a few % of the total load, means that the detector "efficiency" for useful counts is low. Thus, very high total count rates will be required and the count rate capability for neutrons and the influence of pile up at high rates must be studied.

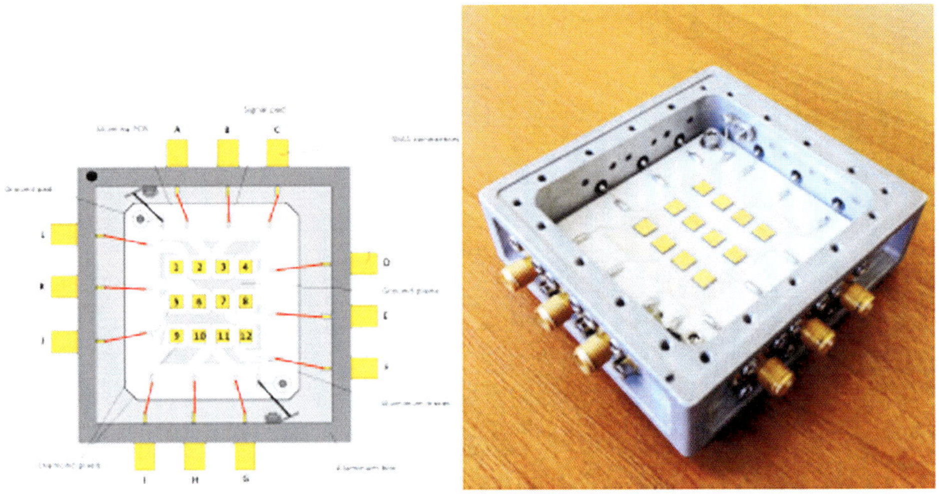

Fig. 12 (Left) Schematic design of the 3 × 4 pixel diamond detector installed in the Roof laboratory of JET. (Right) Photo of the diamond detector before installation. From [34]

- In diamonds produced with natural abundance carbon, about 1% of the nuclei will be ^{13}C. This means that the $^{13}C(n,\alpha)$ reaction can disturb measurements of weak neutron spectral components.
- A "polarization effect" has been observed, meaning a degradation of the output pulse height over time. The cause of this effect is at present not fully understood, and should be investigated further. The effect can be mitigated by reversing the detector bias after some time of exposure.
- The main production technique for artificial diamonds through the CVD process has limits when it comes to the dimensions of spectroscopy-grade crystals: at present the limit is a few tens of mm^2 in area, and a thickness up to 1 mm. Increased dimensions, in particular thickness, would be important to increase the efficiency of the detectors.
- The radiation hardness properties of diamond detector assemblies should be investigated further. Recent results have indicated issues, possibly related to the contacting of the detectors [35].

Example 2: In-Organic Scintillators for n Spectroscopy—CLYC: Cs_2LiYCl_6:Ce Crystals

A recent development with potential for neutron measurements, also fast neutron spectroscopy, at fusion facilities is the so called CLYC scintillator material [36, 37]. The chemical composition is Cs_2LiYCl_6:Ce, where the elements of interest for neutron induced reactions are mainly Li and Cl. Even though the potential for high-performance neutron spectroscopy might be somewhat limited, the material would offer interesting possibilities of measurements in mixed radiation fields of thermal and fast neutrons, in combination with a strong gamma component.

The reactions of interest for neutron measurements are:

- $^{6}Li(n,\alpha)^{3}H$ with a Q = 4.8 MeV, i.e., there is no threshold for this reaction, and it is therefore well suited for thermal neutron measurements. In principle, in view of the charged-only final state, this channel could be used for neutron spectroscopy, although the very strong thermal cross section tends to overshadow the effects of fast neutrons.
- $^{7}Li(n,\alpha n)^{3}H$ with a Q = − 2.5 MeV, i.e., a threshold reaction, thereby sensitive to counting of fast neutrons. However, the 3-body final state, involving a neutron, does not make this reaction channel suitable for spectroscopy.
- $^{35}Cl(n,p)^{35}S$ with a Q = 0.61 MeV. This charged-only reaction channel has a potential for neutron spectroscopy that could be explored.

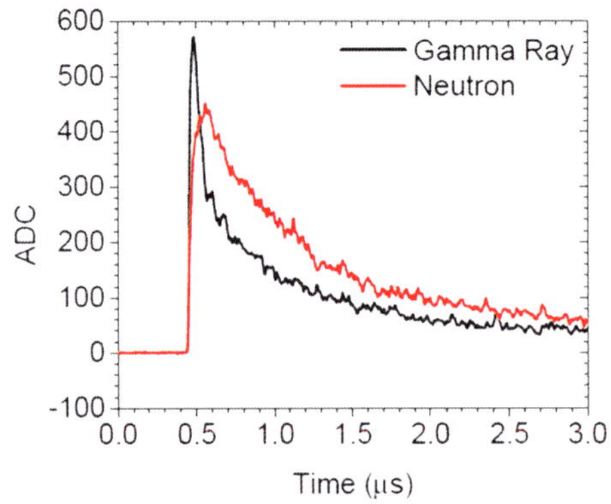

Fig. 13 Example of the difference in detector pulse shapes between neutron and gamma induced events in a CLYC detector

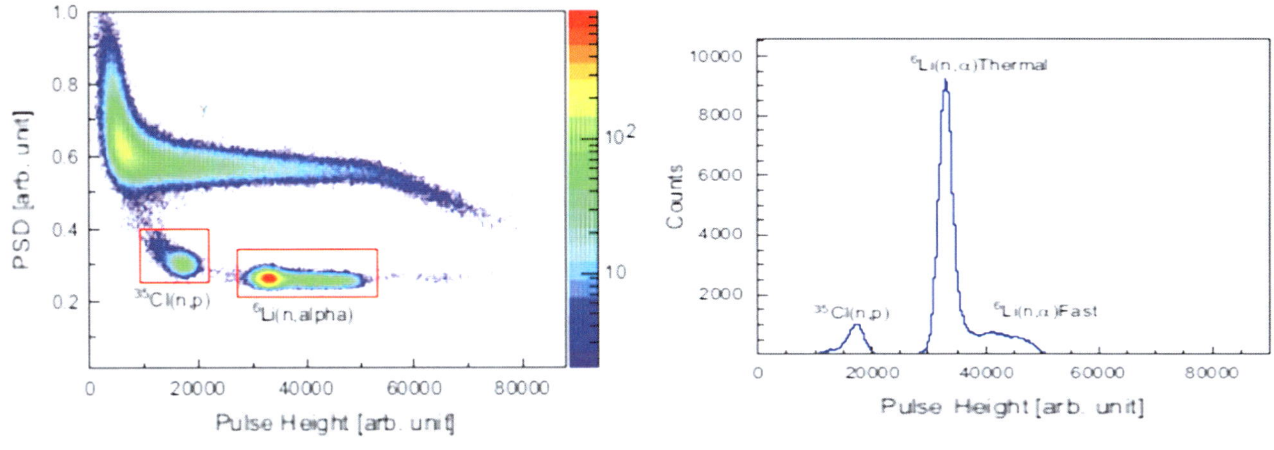

Fig. 14 Example of the use of a CLYC detector in a mixed field of gammas and neutrons. (left) Separation of events (reaction channels) through pulse shape discrimination (PSD). The neutron-induced reactions on ^{35}Cl and ^{6}Li are indicated. (right) Projection on the pulse height axis of the neutron-induced events in the red rectangles. From Ref. [37]

Some of the attractive properties of CLYC are:

- A good energy resolution of about dE/E ~ 4% for gammas at E_γ = 660 keV. This is worse than semi-conductor detectors, such as diamonds, but on par with many other recently developed in-organic scintillators, such as $LaBr_3$ and $CeBr_3$ based detectors.
- The pulse shape response to neutrons and gammas is quite different, as shown in Fig. 13, which opens up for excellent neutron-gamma event separation through Pulse Shape Discrimination analysis techniques.
- There is a possibility to tailor-suit the detector by selective enrichment of the Lithium isotopes:
  - enriching in ^{6}Li increases the sensitivity for thermal neutron counting
  - enriching in ^{7}Li improves the performance for fast neutron counting

Figure 14 illustrates the possibilities to separate different reaction channels by pulse shape analysis (Pulse Shape Discrimination, PSD). As can be seen, the neutron induced reactions are well separated from the gamma reactions.

Several issues remain for this material to be considered as a resource in fusion diagnostics:

- As shown in Fig. 13, the light decay constant is of the order μs, thereby restricting the count rate capability if pile up at high rates is to be avoided.
- In a mixed field of gammas and neutrons, as is often the case in fusion, the reactions involving neutrons are only a small fraction of the signals generated in the detector. The fraction of events useful in neutron counting or for neutron spectroscopy is small.
- The response of the detector to neutrons at different energies must be measured in detail, to investigate if the potential for spectroscopy can be fulfilled.
- The technique should be tested at a high-performance fusion facility to assess its potential in fusion neutron diagnostics.

Advanced Neutron Spectrometry: n–p Elastic Scattering

In this and the following sections we will focus on two techniques used in more specialized, designed neutron spectroscopy systems: the time-of-flight (TOF) technique and the thin foil proton recoil (TPR) technique. The design of these systems allows their performance to be adjusted to the needs of the specific measurement situation when it comes to the main system characteristics like energy resolution, detection efficiency and count rate capability. Still there are inherent limitations with the techniques that must be known and taken into account in any real implementation at a fusion device. As an introduction to this discussion we need to take a look at the basic reaction of neutron elastic scattering on hydrogen (protons), as this is the basis for the signals detected in the systems.

Elastic Scattering H(n,n)H

The basic reaction utilized in both the TPR and (forward) TOF techniques is neutron elastic scattering on hydrogen: $n + p \rightarrow n_R + p_R$, where R indicates the recoiling particles after the interaction. Neutron elastic (nuclear) scattering on hydrogen (i.e., protons in the target material) has several advantages:

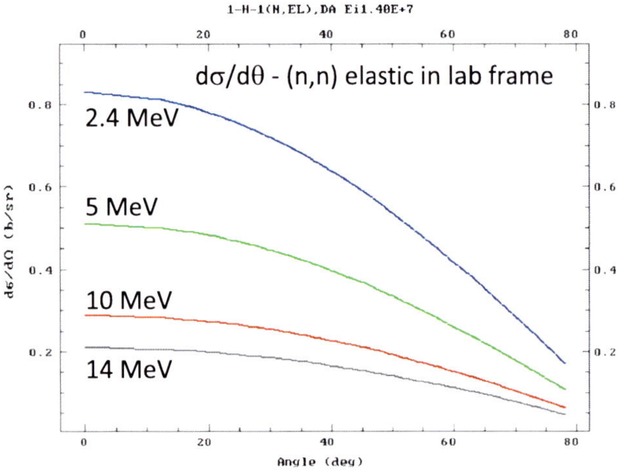

Fig. 15 The differential (dσ/dθ) cross section for elastic nuclear H(n,n)H scattering in the laboratory reference frame for a number of selected incoming neutron energies. Kinematically, the highest possible scattering angle is 90° (assuming $m_n = m_p$). Adapted from [30]

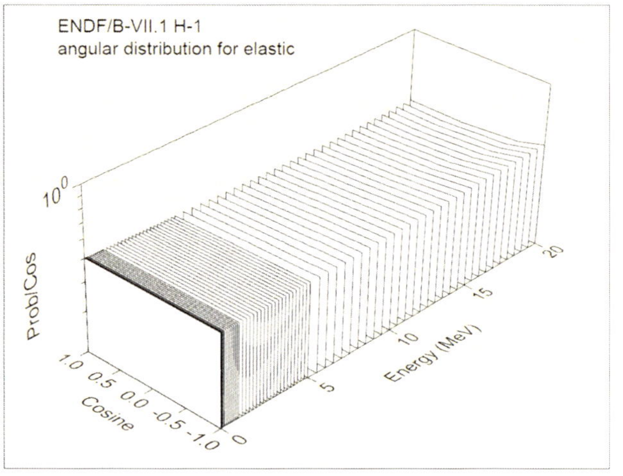

Fig. 16 The differential (dσ/d(cosθ)) cross section for elastic nuclear H(n,n)H scattering in the CM frame for $0 < E_n < 20$ MeV. From [38]

- The reaction has a high and well known (uncertainties of single%) differential cross section, both regarding dσ/dE and dσ/dθ, in fusion relevant neutron energies, $E_n < 20$ MeV. This opens up the possibility for good absolute measurements, if other experimental parameters are known to a similar level of precision. See Fig. 15.
- The differential cross section is close to isotropic in the Center-of-Mass system for $E_n < 20$ MeV, although a slight anisotropy in the backward (CM) direction starts to develop for $E_n > 10$ MeV. This means a corresponding slightly enhanced probability for protons scattered into the forward direction, and can be of relevance in some cases, in particular for absolute estimates. See Fig. 16 [30].
- There exist a range of suitable hydrogenous material, suitable both for active measurements (such as scintillators) and as self-supporting, passive scattering materials (plastic foils).
- The kinematics of n–p elastic scattering for E_n < 20 MeV is quite simple: no excited states of the outgoing particles exists and the neutron energy is well below any threshold for production of secondary particles.
- The kinematics of neutron elastic scattering on nuclei (Eq. 9) shows that scattering on Hydrogen gives the Hydrogen nucleus (proton) the highest possible charged recoil energy of any nucleus. For head on collisions ($\theta_p = 0$), practically the full neutron energy can be transferred to the recoil proton, $E_p = E_n$ (ignoring the small neutron–proton mass difference):

$$E_r = \frac{4A}{(1+A)^2}\left(\cos^2\theta\right)E_n \tag{9}$$

Here E_r is the energy of the recoiling nucleus, A is the atomic mass number of that nucleus, θ is the neutron scattering angle in the lab frame, with θ = 0 for head-on collisions.

One can identify three cases for using the p(n,n)p (elastic) reaction for neutron measurements:

1. Only look at the recoil protons: allow protons at ANY scattering angle; measure the energy of the recoil protons. This is used in most scintillators, like the liquid NE213 scintillator. (See paper by M.Cecconello in these proceedings.)
2. Only look at the recoil protons: passively allow protons only in a restricted angular interval; measure the energy of the recoil protons. This is used in the TPR technique, and has been implemented in the MPR/MPRu spectrometer at JET and suggested for ITER, as further discussed in Sect. 11 below.
3. Measure the recoil proton (energy and time) in an initial n–p scattering, follow the neutron, measure the proton (energy and time) in second n–p scattering: this is used in the TOF technique and implemented for example in the TOFOR spectrometer at JET.

In the following sections we will take a closer look at diagnostic systems based on points 2 and 3 above.

Time of Flight Systems

The neutron time-of-flight technique is based on a simple concept: measure the time t_{TOF} for the neutron to traverse a known distance L. From this, calculate the velocity, $v = L/$

t_{TOF}, and finally determine the (non-relativistic) neutron energy according to Eq. 10:

$$E_{n'} = \frac{1}{2}m_n v_{n'}^2 = \frac{1}{2}m_n\left(L\frac{}{t_{\mathrm{TOF}}}\right)^2 \tag{10}$$

Here, m_n is the neutron mass.

In some applications, a suitable time reference exists for the time measurement, like the RF system of an accelerator. In high-performance magnetic confinement fusion experiments, however, where plasma discharges can extend for several seconds, no such useful time reference exists. Instead, a neutron double scattering configuration is used, where the same neutron is detected in two (sets of) detectors, here called S1 for the first (start) detector and S2 for the second (stop) detector, and the distance and time between these detectors define the velocity of the scattered neutron. The principle of the technique is shown in Fig. 17(left). Note that in both S1 and S2 detectors, the actual measured signal is due to the recoils protons, according to the discussion in the previous section. In order to achieve sufficient (proton) signal amplitude in the elastic scattering in S1, the S2 coincidence detector is placed at some angle, α, with respect to the incoming neutrons from the plasma. The incoming neutrons are normally shaped into a "beam" by suitable collimation between the diagnostic and the plasma, as shown previously in Fig. 4.

In principle, one can envisage both forward scattering and backward scattering TOF systems. The latter would then depend on a primary scatter (in S1) on a nucleus heavier than hydrogen (proton). Such a possibility is offered by using a deuterated scintillator as S1. Such a solution is considered for the ITER High Resolution Neutron Spectrometer system presented later in this paper. However, in what follows we will mainly discuss the forward scattering TOF technique.

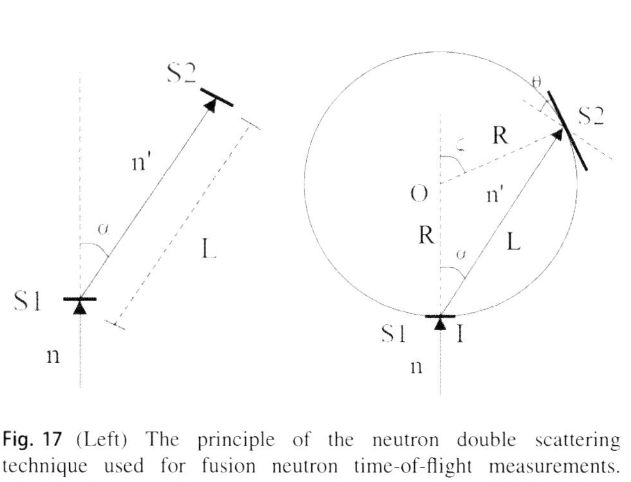

Fig. 17 (Left) The principle of the neutron double scattering technique used for fusion neutron time-of-flight measurements. (Right) Geometry for the concept of constant time-of-flight sphere. Also, the tilting of the S2 detector as used in the TOFOR implementation at JET is indicated. See text for details

In the double scattering configuration, the energy of the scattered neutron, $E_{n'}$, is related to that of the primary incident neutron, E_n, through the scattering angle α:

$$E_{n'} = E_n \cos^2(\alpha) \tag{11}$$

From geometric considerations (Fig. 17(right)), one can conclude that the distance L between any two points on a circle connecting the center of S1 and the center of S2 can be written:

$$L^2 = 2R^2(1 - \cos(\pi - 2\alpha)) = 4R^2\cos^2(\alpha). \tag{12}$$

Combining the three Eqs. 10, 11, 12 finally yields an expression for the energy of the primary, incoming neutron, E_n, that is independent of both the distance L and the scattering angle α, and only dependent on the (fixed) radius R and the measured time-of-flight:

$$E_n = 2m_n R^2/t_{TOF}^2 \tag{13}$$

This is the basis for the "constant time of flight sphere" concept, which states that under the assumption of elastic n–p scattering and as long as the secondary detectors, S2, are placed on a sphere of radius R, the energy of the incoming neutron is determined solely by the measured time-of-flight.

Equation 13 gives some important information on how we should design our TOF spectrometer. By differentiating the expression we obtain the expected uncertainty in the energy measurement, i.e., the energy resolution, as a function of the experimental parameters R and t_{TOF}, as given in Eq. 14:

$$\frac{dE_n}{E_n} = \sqrt{\left(2\frac{dR}{R}\right)^2 + \left(2\frac{dt}{t_{TOF}}\right)^2} \tag{14}$$

In the design of the spectrometer, one often strives for a certain maximum energy resolution, and an analysis of the contributions in Eq. 14 can help in that work.

Here it is important to note that the uncertainty, dR, in the radius R includes two components: first the uncertainty in the installed positions of the S1 and S2 detectors, and second, the additional uncertainty in the actual interaction positions within the finite thickness of the S1 and S2 detectors. While the former uncertainty is a common, systematic uncertainty for all measurements, the latter component is a true random uncertainty.

Also note that the dt used here means the uncertainty in the intrinsic time *determination*, and not the width of the measured time-of-flight peak (which is actually directly correlated to dE/E). This dt is given by uncertainties in the timing electronics, in the time pick-off method, in the statistical variations in the signal pulse shapes etc. In many cases dR ≫ dt and the expression for the energy resolution reduces to dE/E = 2dR/R, i.e., the resolution can be estimated from geometry alone. Furthermore, the (systematic) uncertainty in the determination of the positions of S1, S2 can be assumed to be of the order mm or better. This means that in many cases, it is the detector thickness of S1, S2 that will be the main contribution to the energy resolution.

As indicated above, when real measurement data is available, the energy resolution is given by the observed width and position of the measured time-of-flight distribution (peak), as given by $(dE/E)_{meas} = (2dt_{TOF}/t_{TOF})_{meas}$. Here dt_{TOF} is the full width at half maximum of the measured time-of-flight peak, and t_{TOF} the mean time-of-flight.

Several time-of-flight spectrometers have been used in fusion research over the year. Two recent examples are the TOFOR [39–41] and TOFED [42, 43] spectrometers, installed at JET and EAST, respectively. Here we will give a more detailed description of the TOFOR spectrometer [44].

The TOFOR spectrometer at JET was installed in 2005 and was built based on the following design drivers [adopted from Ref. 45]:

- Optimize the design for D plasma measurements ($E_n \approx 2.5$ MeV).
- Strive for an energy resolution corresponding to a $T_i \approx 4$ keV, i.e., $dE/E \approx 6.6\%$.
- The spectrometer should have a high count rate capability, with a maximum capacity of about $C_{cap} \approx 500$ kHz.
- It should be a simple, robust design, in particular including a large area S2 for high efficiency.

The design resulted in a system of compact dimensions, nicely fitting in the designated position in the JET roof laboratory, with $\alpha = 30$degr and $L = 1.22$ m, giving an estimated average t_{TOF} ($E_n = 2.45$ MeV) = 65 ns.

It was decided to base the design on fast plastic scintillator detectors coupled to photo-multiplier (PM) tubes. There were several reasons for this choice:

- Plastic scintillator is a hydrogenous material with suitable and well-known mechanical and detector properties. The composition is basically $(CH)_n$ thus providing ample scattering centers (protons) for (n,p) elastic scattering.
- Fast plastic means fast signal rise times which provide a good experimental situation for precise timing measurement.
- Plastic scintillators are robust, machineable, and have a good track record in nuclear and particle physics experiments.

TOFOR is composed of two sets of detectors, as depicted in Fig. 18(left):

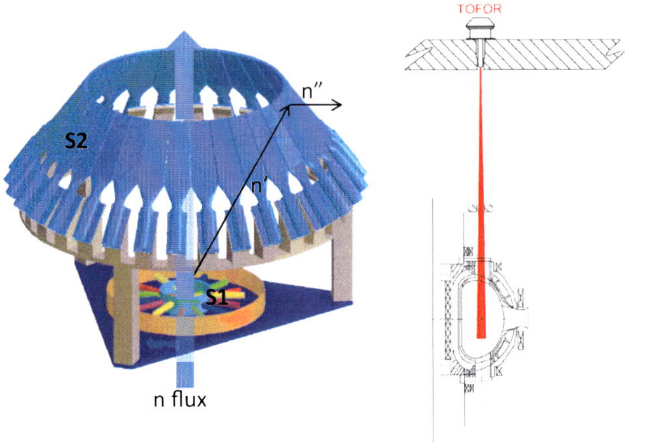

Fig. 18 (Left) Schematic drawing of TOFOR at JET. The neutron flux from the tokamak is shaped into a "beam" indicated by a broad, light blue arrow though the center of the figure. Primary neutrons, n, are scattered, n′, towards S2, where a second scattering might happen, giving a scattered neutron n″. The 5 scintillators of the S1 stack are placed in the center of the lower light blue ring. The 32 S2 detectors are indicated in medium blue, forming an "umbrella" symmetrically around the neutron beam. Dark blue, orange and brown are support structures. (Right) Geometry of the installation of TOFOR in the JET roof laboratory. The TOFOR is not to scale. The red cone indicates the instrumental field-of-view (Colour online)

i. A stack of five S1 "start" detectors, each a disk of radius $R_{S1} = 20$ mm and thickness $t_{S1} = 5$ mm, making a total material thickness of 25 mm and a geometrical thickness of 27 mm, allowing for 0.5 mm distance between the disks in the stack. Dividing the S1 into several detectors improves the C_{cap}, since pulse time properties, PM tube and acquisition electronics set a limit for the count rate in each individual detector channel at about 1 MHz.
ii. A set of 32 S2 "stop" detectors of "paddle" shape, each 350 mm long and 15 mm thick. The S2:s form an "umbrella" like structure placed symmetrically around the symmetry axis of the instrument, which coincides with the collimated neutron beam.

The installation at JET is schematically depicted in Fig. 18(right). TOFOR is installed in the roof laboratory at JET, about 19 m from the center of the plasma and with a line-of-sight going through the center of the plasma (note, the size of TOFOR is not to scale). The collimation is integrated in the roof laboratory floor which is about 2 m thick and also acts as radiation shield.

The geometry of the S1 and S2 detectors was determined in a performance optimization simulation study, where the thickness of the detectors was varied in order to find the combination with the highest possible efficiency for a selected energy resolution. The target energy resolution was set to 5.8% and the results from the optimization are shown in Fig. 19. Note that, in general, there is a

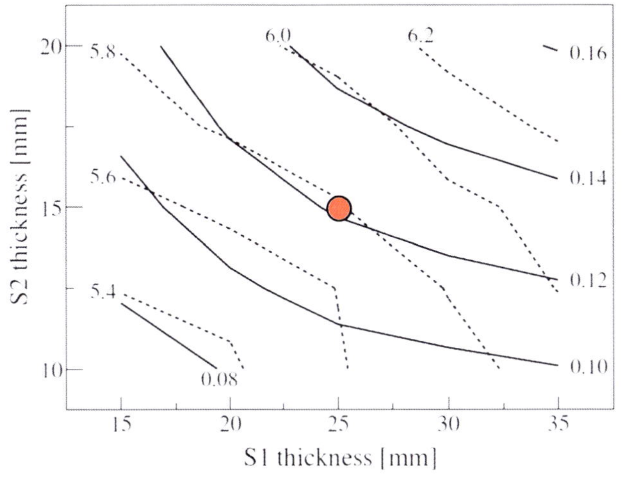

Fig. 19 Results from GEANT simulations of TOFOR's performance in a study to optimize the thickness of the S1 and S2 detectors. Full lines indicate efficiency contours in units of (cm^2), i.e., including the S1 detector area. Dashed lines indicate FWHM energy resolution contours in units of (%). The red dot indicates the selected geometry for the TOFOR design. Adapted from [39]

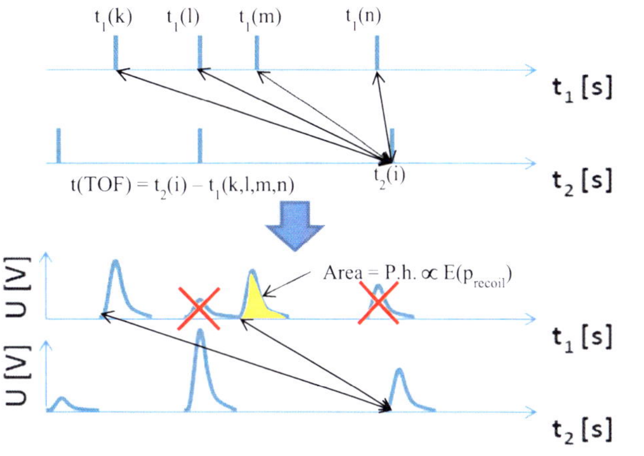

Fig. 20 Illustration of the way the time-of-flight is determined, also illustrating the are difference between original TOFOR time stamping electronics (top) and the upgrade ToFu system capability (bottom). With ToFu, correlation between time-of-flight and energy (pulse height) makes it possible to discard (red crosses) certain S1–S2 combinations as invalid (Colour online)

reciprocal relationship between resolution and efficiency, such that a higher efficiency leads to a worse resolution. The red dot indicates the selected optimal geometry with t_{S1} = 25 mm (5 × 5 mm) and t_{S2} = 15 mm. This combination actually has a slightly better resolution than the target (dE/E < 5.8%) for an efficiency of slightly higher than 0.12 cm^2 (area efficiency).

Since the S2 detectors are fairly large, and have their PM tubes attached at the lower end, they were tilted away from the constant time of flight sphere in a manner shown in Fig. 17(right). The tilt is such that neutrons at the upper tip, far from the PM tube, will reach the detector slightly before neutrons at the lower tip, close to the PM tube. This is to partly compensate for the slightly longer light transmission time through the length of the S2 detectors for neutrons interacting at the far end (with respect to the PM tube) compared to those interacting at the near end.

Combining the effects of the geometry, tilting of S2, and electronics, an overall energy resolution of about dE/E ≈ 7% was achieved for TOFOR for E_n = 2.5 MeV. Due to the choice to optimize TOFOR for D plasmas, the performance in DT plasmas (E_n = 14 MeV) will be somewhat reduced.

The original TOFOR installation included digital electronics that provided free-running time stamps (only) of all eligible events (i.e., events in the detectors above a set pulse-height discrimination threshold). As shown in Fig. 20(top), time-of-flight values were determined by selecting a signal in S2 (which is exposed to considerable lower rates than the in-beam S1) and calculating the time difference to all S1 signals in a specific time window, often taken to be ± 200 ns. In the example of the figure, four different S1 events are eligible to form a t_{TOF} with this specific S2 time stamp. Without pulse height information, all interactions in S1 (within the time window) are eligible as a valid S1–S2 coincidence, although in reality only one of the S1 interactions is actually induced by the same neutron that later interacted in S2. This makes the original TOFOR system sensitive to so called random (accidental) coincidences, i.e., t_{TOF} values formed between uncorrelated neutrons. The situation is particularly severe at high count rates, since the signal count rate (true coincidences) scales as $S_{true} \propto Y_n$, while the random background scales as $S_{false} \propto Y_n^2$, where Y_n is the neutron rate from the plasma. With the TOFOR type of electronics, at some neutron rate the false (random) signals will completely overshadow the true signals, making the system useless as a spectrometer.

The problem can be partly overcome by measuring correlated time and energy (pulse height) information. The original TOFOR electronics did not allow for this, and therefore an electronics upgrade program for TOFOR has been initiated, dubbed ToFu [46, 47]. The main element of the upgrade program is to replace the original TOFOR time stamping (only) electronics with state-of-the-art waveform digitizers (1 GHz sampling rate, 12–14 bit amplitude resolution, fast data transfers, etc.). This allows making more strict selection of eligible S1 "candidates" for each S2 signal, as shown in Fig. 20(bottom): some combinations of energy and time-of-flight simply do not comply with the requirement of elastic n–p scattering.

The ToFu upgrade program is well under way, and described in some detail in for example Ref. [47]. Figure 21 illustrates the new capabilities offered with ToFu.

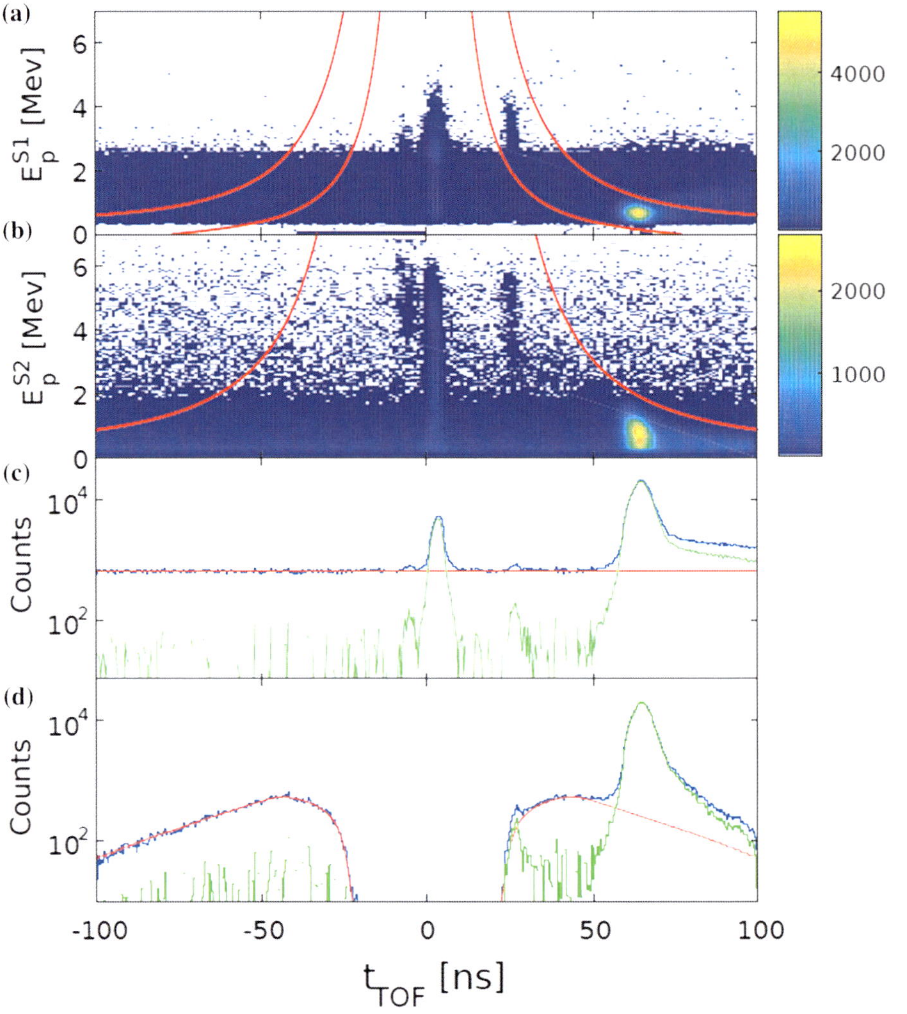

Fig. 21 Illustration of the improved background reduction offered with the ToFu electronics upgrade. **a** Correlation between t_{TOF} and energy deposition in the S1 detector, red lines indicate the allowed region for n–p elastic kinematics, **b** correlation between t_{TOF} and energy deposition in the S2, red line indicate the upper E_{S2} threshold; **c** projection of the t_{TOF} distribution in (**a**) (or **b**) as in the original TOFOR (no kinematic cuts), blue line is for full data set, red line is the estimated level of random coincidence background, green line is background subtracted data; **d** projection of the t_{TOF} distribution including kinematic cuts, blue line is all events that pass the cuts, red line is estimated random background, green line is background subtracted data. From [47] (Colour online)

Figure 21a, b show the possibility to plot the data in a two dimensional histogram, with energy versus time. The red lines in (a) indicate the kinematic correlations (energy-time) for eligible events in S1 directed towards the tips of the S2 detectors; events outside of these lines are NOT eligible single n–p elastic scattering events in S1 and can be discarded. In S2, panel (b), only an upper level can be set, indicated by the single red line. Still, all events above this line can be discarded. The process of applying these time-energy correlations is dubbed "kinematic cuts".

Figure 21 panels (c) and (d) show the projections of the recorded events in panels (a) and (b) without (panel c) and with (panel d) kinematic cuts. Note that panels (a) and (b) both include the same "coincident" events, so projections of panels (a) and (b) on the time-of-flight axis, without cuts, will be exactly the same, in each case giving the blue line in panel (c). Panel (c) illustrates the background subtraction possible with the original TOFOR electronics: the level of random coincidences for the data set under study was determined from a region on the negative time-of-flight side of the spectrum. This level was then subtracted from the data over the whole spectrum to find a representation of the true coincidences as shown in the green curve in panel (c) (or to be more precise, the level of random coincidences was included as a component in the physics analysis of the data).

With the new ToFu electronics, the kinematics cuts (red lines) are applied to panels (a) and (b) and only those events passing both cuts are plotted in panel (d) (blue line). The shape of the spectrum for times-of-flight formed with negative values can again be used (now mirrored around $t_{TOF} = 0$) to estimate the influence of the random background. The final selection of eligible events is shown in the green curve in panel (d).

There are a number of observations to be made when comparing the green curves of panels (c) and (d): (i) the gamma peak at $t_{TOF} = 4$ ns has completely disappeared in panel (d). This is as expected, since there should be no possibility that gamma interactions pass as single scattering elastic neutron events. The high t_{TOF} tail is much reduced in panel (d) compared to panel (c). This gives improved possibilities to understand the influence of scattered neutrons in all plasmas, and in particular improves the prospect of measuring 2.5 MeV neutrons in a strong background of 14 MeV neutrons, as will be the situation in future JET DT campaigns. Finally, but not so clear from this figure, the quality (in terms of statistical uncertainty) of the 14 MeV peak at 27 ns has improved considerably in panel (d) compared to panel (c).

The work on the ToFu upgrade is still ongoing, but it is already now clear that it will improve the TOFOR system data quality in both D and DT plasma operations.

Thin Foil Proton Recoil Systems

The thin foil proton recoil (TPR) technique is also based on elastic n–p scattering. Contrary to the TOF technique, in a TPR only a single elastic n–p scattering is used and no further measurement of the scattered neutron is done. The technique is based on measuring the energy of the scattered proton in a restricted angular range, preferably in the forward direction.

This leads to an important design restriction, namely that the primary proton "radiator" must be thin, in order to avoid a large and unknown proton energy loss. In practice, thin self-supporting CH_2 foils are used for the primary scattering. However, an advantage with the method is that these foils can be completely passive and any active signal measurements are done outside of the collimated neutron beam. This decoupling of the n–p scattering from the actual measurement facilitates the design and makes it, in principle, possible to shield the active detectors to any desired level. Thus, by careful instrumental design, this technique has the potential to provide excellent signal to background ratios, thereby allowing for measurements of very weak components in the neutron spectrum.

The two main implementations of the TPR technique are schematically shown in Fig. 22, where TPR here means the specific case with a proton energy-resolving detector placed at an angle θ out of the neutron beam. MPR stands for Magnetic Proton Recoil, and is further discussed below. The basic principle in both cases is: (i) neutrons from the fusion plasma are collimated into a neutron beam, (ii) the beam neutrons intersect a thin hydrogenous foil, (iii) some of the neutrons undergo elastic n–p scattering in the foil, (iv) the recoil proton is allowed to exit the foil and its energy is measured at some angle (in the forward direction).

There are several advantages of selecting the protons in the forward direction:

- The energy of the recoil proton is the highest possible, thereby facilitating its detection and separation from background.

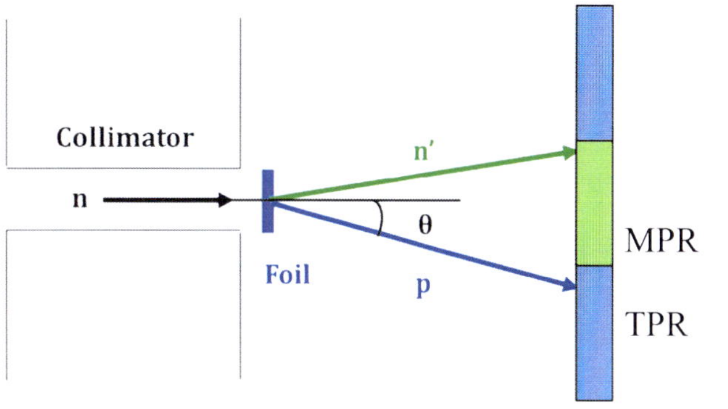

Fig. 22 Schematic illustration of the TPR/MPR neutron spectroscopy technique. TPR stands for Thin-foil Proton Recoil, MPR for Magnetic Proton Recoil

- The variation in proton energy is the smallest over a specified angular range (Fig. 23).
- The spread in proton energy loss for a specific foil thickness is minimized.

All the above points help to improve the performance and energy resolution of the system. The use of thin foils also improves the resolution, but decreases the efficiency. Thus, the TPR/MPR technique is best suited for high-performance DT operations where neutron rates are high, and the low efficiency is less of a limitation.

Since charged particles are tracked in the TPR/MPR, energy loss outside of the scattering foil and the active detectors should be avoided. Therefore, the space between the foil and the detector(s) should be evacuated. However, the required vacuum levels are moderate, and levels of the order $p < 10^{-3}$ mbar will normally suffice, putting rather relaxed demands on the vacuum system.

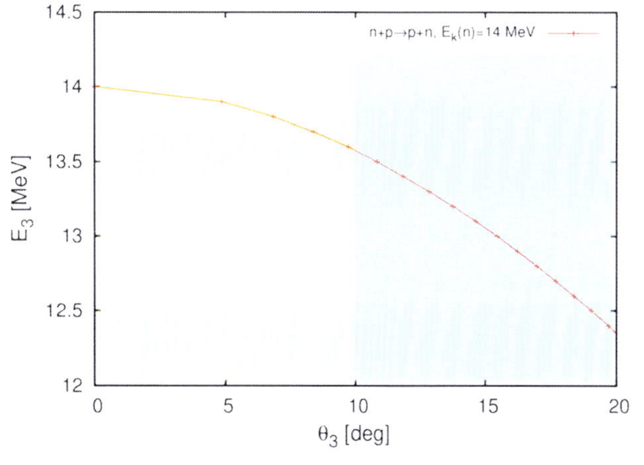

Fig. 23 Kinematics of elastic n–p scattering for incoming neutrons of E_n = 14 MeV. The recoil proton energy as function of proton angle is shown. The green region roughly corresponds to the situation in the MPR, the blue region to the TPR (Colour online). Adapted from [48]

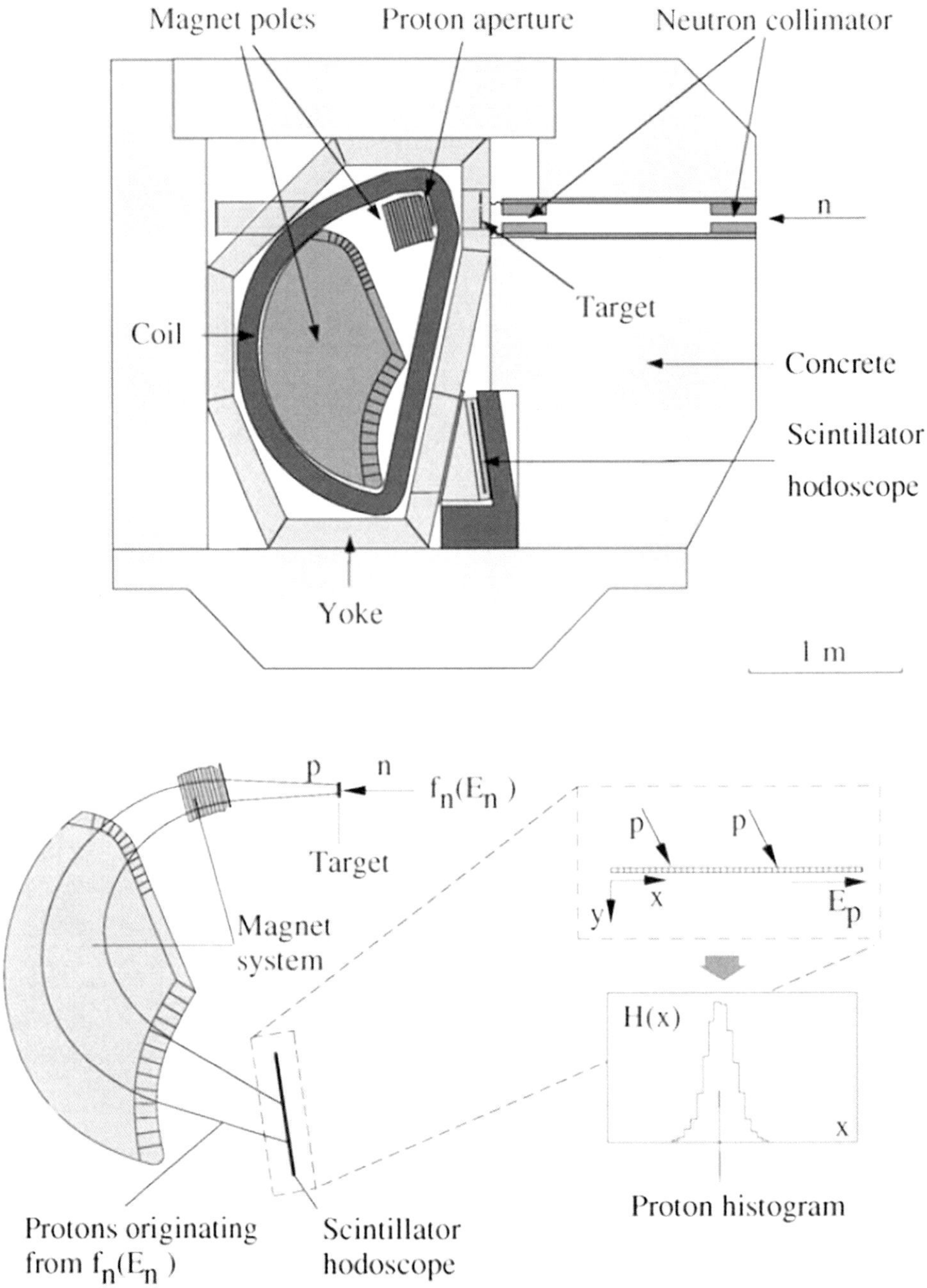

Fig. 24 Schematic drawings of the MPR/MPRu spectrometer at JET. (top) A cut side view through the center of the system, showing the concrete shield, the magnetic spectrometer system (split in two "dipoles"), the neutron collimator, the scintillator hodoscope etc. (bottom) A more detailed view of the magnetic and detector systems. From [54]

The MPR Spectrometer at JET

An instrument based on the MPR technique was installed at JET in the Torus Hall some 8 m from the plasma center in 1996 [49–51] and has provided a wealth of neutron spectrometry data. Some early results can be found in e.g. Refs [10, 49, 52, 53]. This technique employs a magnetic spectrometer to separate the selected forward-scattered protons according to their momentum (energy), as given by the cyclotron equation: $B\rho = p/q$, where B is the applied magnetic field, ρ is the bending radius, p is the particle momentum and q is the charge of the particle. This means an approximate bending radius of ρ=0.7 m for $E_p = 14$ MeV and B = 1 T. The MPR instrument at JET is depicted in Fig. 24. The magnetic spectrometer is of a traditional nuclear physics design, using ordinary magnetic materials. It is thus rather bulky (several m^3), heavy (about 20 tons) and requires considerable power to operate. The active detection system is a 32 element fast plastic scintillator hodoscope, placed approximately in the focal plane of the magnetic system about 1.5 m from the neutron beam (n–p scattering foil). This allows for a very high count rate capability of the spectrometer, of several MHz signal count rate, and very good signal-to-background ratios.

Due to its position in the JET Torus Hall, close to the plasma, a rather substantial radiation shield is required around the spectrometer. As seen in the figure, the instrument is enclosed in a concrete "bunker" weighing about 65 tons.

The MPR has been designed to provide very good resolution of 14 MeV neutrons, corresponding to $dE/E = 2.5\%$ for the optimized setting. There are three main contributions to the energy resolution of the MPR:

- Energy loss in the n–p scattering foil, dE_f. The scattering reaction can take place at any (random) point within the foil, and the outgoing recoil proton will have to traverse a distance in the foil, ranging approximately $0 < L < t_{foil}$ (somewhat adjusted for the outgoing proton direction). This inflicts an unknown energy loss to the proton.
- n–p scattering kinematics, dE_k. Due to the finite acceptance angle of outgoing protons, and the divergence of the neutron beam at the foil, n–p elastic scattering kinematics give a range of outgoing proton energies even for a mono-energetic incoming neutron.
- Spectrometer ion optics, dE_o. The finite area of the scattering foil (10 cm^2) and the properties of the magnetic system give a spread in proton positions at the hodoscope, even for mono-energetic neutrons. The finite width of the hodoscope detectors also add to this contribution.

In an optimized configuration, the three contributions described above should be matched and add about 200 keV each; the final resolution is obtained by adding the individual contributions in quadrature.

The MPR technique allows for a flexible system, and this has to some degree been implemented at JET. Thus, the magnetic spectrometer B-field can be varied in the range $0 < B < 1.4$ T, there are several scattering foils available in a carousel-like arrangement, and there are several apertures for the recoil protons to enter the magnetic system. This makes it possible to combine different settings in order to obtain the best performance for a specific measurement situation: thicker foils and larger proton apertures give higher efficiency, but poorer energy resolution. Alternatively, thin foils and small proton apertures improves the resolution at the expense of detection efficiency.

A major upgrade of the original MPR was undertaken in 2005 (MPRu) [51], where the original 37-scintillator hodoscope was replaced with a 32-element system composed of so called phoswich scintillators. These detectors combine two scintillator materials of slightly different timing properties, thereby allowing for a better background discrimination. In order to reduce the background further, a layer of Gadolinium-based paint has been applied to certain areas of the spectrometer. This should help to prevent some of the thermal neutrons from entering the region close to the detectors, thereby reducing the gamma background due to neutron-capture reactions at the detector location.

The MPR/MPRu operated successfully in the JET major DT campaign in 1997 and also in the more restricted trace tritium campaign in 2003. It will be the main 14-MeV neutron spectrometer in any upcoming JET DT campaigns, as those presently planned for 2019–2020. Some results from the MPR are given in Refs [54–56].

The Non-Magnetic TPR Technique

The principle of the non-magnetic thin foil proton recoil technique [57, 58] is shown in Fig. 25. As in all fusion neutron spectrometers, the neutron flux onto the instrument is shaped into a beam by suitable collimators. As in the MPR, the neutron beam intersects a thin hydrogenous foil where some neutrons undergo n–p elastic scattering. The recoil protons are measured in a (set of) energy resolving detector(s) placed some distance L away from the scattering foil. A suitable aperture in the detector allows the neutron beam to pass through without directly interacting with the detectors.

The system in the figure is equipped with two detectors for proton energy measurements, in a ΔE–E configuration. This two-detector configuration helps to reduce background, which mainly will be due to neutrons from the collimated beam, scattered towards the active detector surface. For fusion neutron applications (DT operations), the combined thickness of the two detectors should be sufficient to fully stop protons up to $E_{max} = 18$–20 MeV. The active detector area covers a proton scattering angle of $\theta \pm \Delta\theta$. In order to keep the resolution within the desired range, i.e., about 2.5% as in the MPR case, while

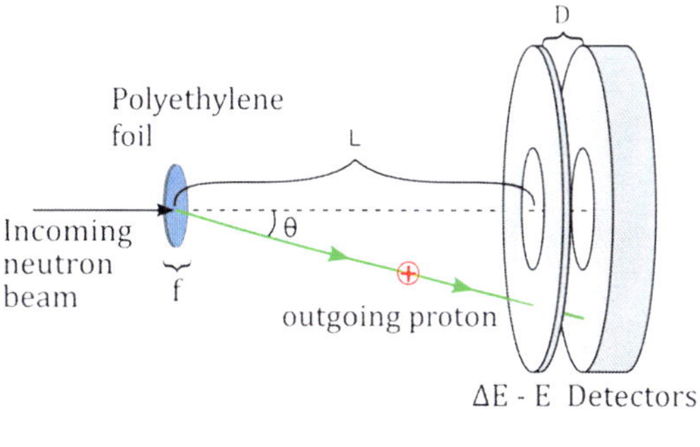

Fig. 25 Schematic drawings of a non-magnetic TPR system. Elastic n–p scattering takes place in the polyethylene (CH_2) foil, a set of two annular detectors to measure the energy of the outgoing recoil proton is placed at a distance L from the foil. The front detector is thin enough to allow protons above a certain minimum energy to pass through to the back detector which is thick enough to completely stop protons up to a certain maximum energy. From [59]

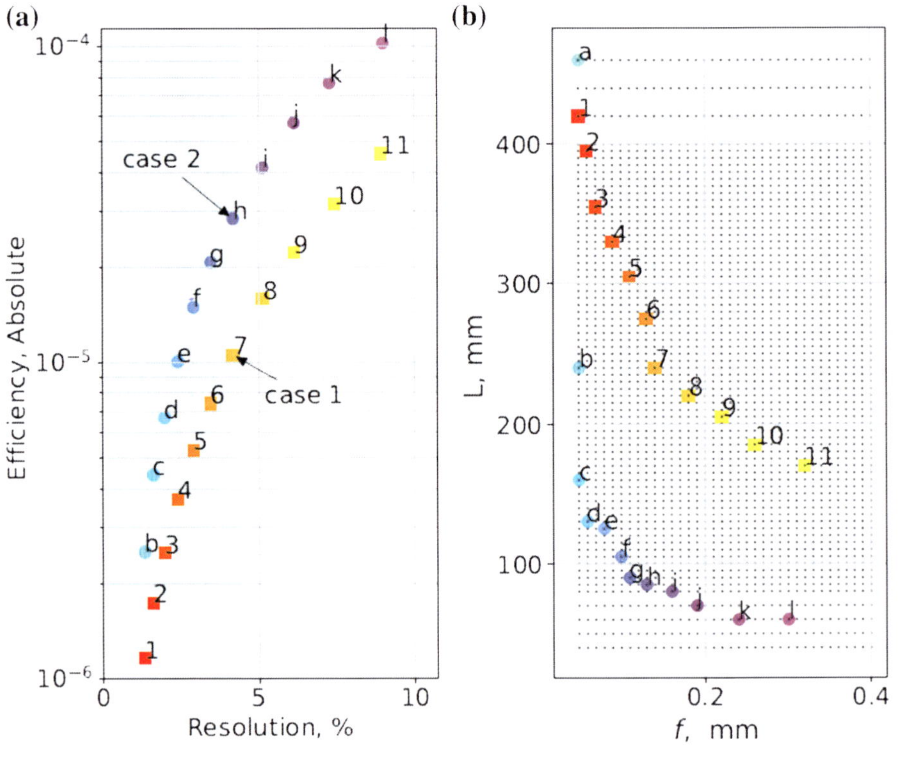

Fig. 26 Results of an optimization study of a Thin Foil Proton recoil system for ITER. Case 1 (numbered points) is for a non-segmented detector, case 2 (lettered points) for a segmented one. (Left) Scatter plot of the values of Resolution and Efficiency for the selected geometrical configurations. The blue shaded area indicates the acceptable range of resolution values. (Right) The geometry for the evaluated geometrical configurations in terms of foil thickness, f, and distance between foil and dE detector, L. From [59] (Colour online)

simultaneously providing an efficiency high enough to provide time-resolved measurements, the annular ΔE-E might require some radial segmentation. It is possible to increase the efficiency further by placing several TPR systems in tandem. This is possible since the very thin foils used for the n–p scattering leave the passing neutron beam almost undisturbed.

Possible choices for detectors for the TPR are high-performance scintillators (e.g., YAP) or semiconductor detectors. In particular segmented Si strip detectors are an attractive choice, since there is a large experience base with such detectors in the nuclear and high energy physics communities. The required total Si thickness for stopping neutrons of $E_n < 20$ MeV would be about 2 mm. Commercial, annular, segmented Si detectors of such dimensions exist ($t_{\Delta E} = 0.5$ mm, $t_E = 1.5$ mm) and are well-suited for a TPR application.

The resolution of the system will mainly depend on the thickness and radius of the foil, the angular range of the recoil protons and the intrinsic resolution of the energy-resolving detectors. As in the MPR case, a flexible system can be set up where a number of different pre-defined foils (of different thickness) can be selected, and the distance between foil and detectors varied.

In a study similar to that previously shown for the TOF technique, an optimization of the performance of the TPR can be performed [59]. In Fig. 26 the results from such a study are shown, where the efficiency (ε) and energy resolution were studied as function of the geometrical parameters of the system; here foil thickness, f, and distance between foil and detector, L. Two cases are shown, one where the detector segmentation has been used (Case 2; lettered results) and one without segmentation (Case 1; numbered points). Clearly the segmented solution gives a higher efficiency and more compact design, although at the cost of a more elaborate and expensive data acquisition system, in order to equip all segments with their own electronics read-out. Still, the performance ($\varepsilon = 3 \cdot 10^{-5}$, dE/E = 4.5%) and geometry of the system is quite attractive for neutron spectroscopy at high-performance devices such as ITER, although the system as shown here has not yet been tested at a real high-performance fusion facility.

Proposed High Resolution Neutron Spectrometer System for ITER

ITER has several neutron systems planned; some have been decided for full implementation while others are only developed to a conceptual level at this stage. Neutron spectroscopy on ITER falls in the latter category of so called "enabled" diagnostics, meaning there is work (and costs) allocated for a development up to a conceptual design but that full implementation is pending further decisions.

The main purpose of a High Resolution Neutron Spectrometry (HRNS) system [60] would be to measure the (core) fuel ion ratio (n_T/n_D) in ITER high-performance DT plasmas, which is a high-level requirement for process control; auxiliary measurements include the (core) fuel ion temperature, heating effects and contributions to the alpha heating physics. The system will be installed in the same port cell as the ITER Radial Neutron Camera (RNC) [61], some 12 m from the plasma center, so it could also be used as an extra channel for the RNC. The conceptual design has been concluded (end 2017) and reported to ITER. In order to fulfill the ITER requirements in dynamic range (in neutron flux), several spectrometers (systems) are required. The system also includes an element of redundancy and a system for measurement of the neutron emission in the initial D phase of operations. It is estimated that the proposed system will fulfill the ITER requirements on accuracy and time resolution in the measurements of the main parameter, n_T/n_D.

Several of the spectroscopy techniques discussed in the previous sections have been included into the conceptual design of the HRNS system for ITER. The design is schematically shown in Fig. 27. The system is placed in a port cell right behind the RNC and using one of the central collimated lines-of-sight of that system. The first system encountered by the incoming, collimated neutron beam is a "tandem" TPR system, where 3 (or more) detectors of the type discussed in the previous section are placed one after another. Due to the thin foils used for the n–p scattering in this system, most of the neutrons will pass unhindered (without any interaction) through the foils, and are thus available for use in subsequent sub-systems. The TPR system is placed in a dedicated vacuum chamber, with thin entrance and exit windows, to avoid too much energy loss of the recoil protons on their way between foil and detector. Due to its intrinsically low efficiency, the TPR system is intended for the highest power levels, P_{fus} = 500 MeV (down to say 100 MW). A diamond matrix is placed just downstream of the TPR system, to provide redundancy and cover some part of the dynamic range.

In order to provide some renewed collimation and reduce the background caused by neutrons scattered in the TPR + Diamond system, a dividing wall is proposed in front of the next section of the HRNS system.

The second section houses two time-of-flight neutron spectrometers, one of traditional forward scattering design (basically copied from the TOFOR installation at JET), and one of back-scatter design [62–64]. The two TOF systems are intertwined in order to save space. The back-scattering TOF system uses a deuterated scintillator as primary in-beam scatterer (D1). This enables neutron elastic scattering on deuterons, n-d, which both allows for and has a significant cross section in the backwards neutron direction. The kinematics of the reaction in the neutron energy range of interest in DT plasmas (i.e., 14 MeV ± 25%) makes it quite suitable for TOF operations: (i) the energy deposition (by deuterons) in the D1 is high (several MeV) and (ii) the subsequent neutron time-of-flight to the D2 (second scatterer) is well in the range for precision time measurements. The higher efficiency of a TOF system makes this back-TOF instrument useful in ITER operations in a lower P_{fus} range than the TPR, down to about 5–20 MW. In order to operate at higher P_{fus}, a local adjustable collimation must be installed in the wall between the two HRNS sections to reduce the incoming flux.

Both sections of the HRNS system are enclosed in local radiation shielding, mainly to protect the port cell from too high radiation levels due to the neutrons streaming through

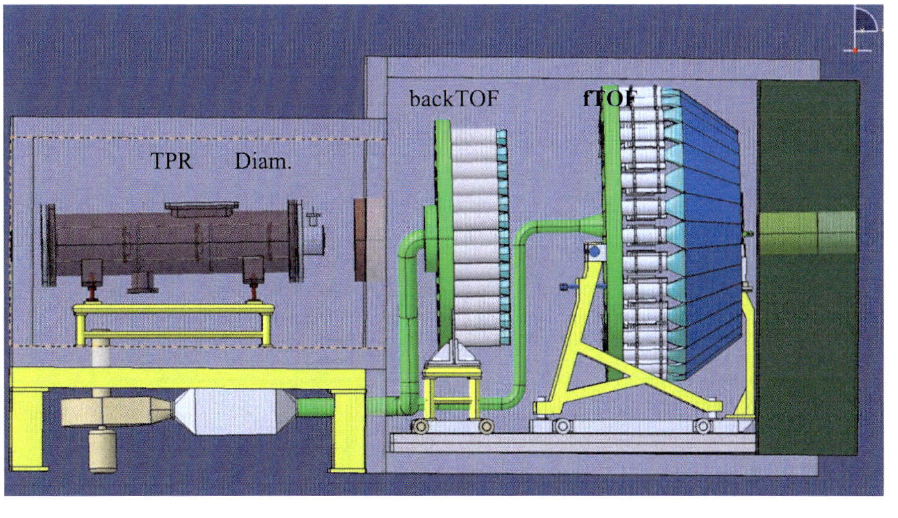

Fig. 27 Schematic view of the proposed, conceptual design for a High Resolution Neutron Spectrometry system for ITER. The collimated neutron beam from the plasma enters from the left, first detectors system is a tandem Thin-foil Proton Recoil (TPR) instrument, second system is a diamond matrix detector (Diam.), third is a back-scatter time-of-flight (backTOF), fourth is a forward scattering time-of-flight (fTOF). The system back wall is indicated in green, with a special neutron beam dump in army green. An air cooling system is shown in bright green, gray and light brown (Colour online)

the open collimator into the port cell. Particular attention must be put in the design of a beam dump for the remaining neutron beam at the far end of the system, both for radiation protection but also to protect the TOF systems from too high levels of unwanted background.

Tools for Spectrometer Development and Synthetic Diagnostics

As described in the introductory sections, the fusion neutron emission and its energy distribution can be a rich source of information about the state of the plasma, and in particular the state of the main plasma fuel ions. For the purpose of fusion energy research the details of the neutron energy distribution is not the primary concern; the important parameter is the state of the fuel ions. Therefore, in order to take full advantage of fusion neutron measurements, not only have the diagnostics to be well designed and characterized, there also has to be a suite of tools for interpretation of the acquired data. Furthermore, in recent fusion diagnostic work, the role of so called *synthetic diagnostics* has come to play an increasingly important role. This concept means that diagnostics that are to be implemented on a fusion device should be modeled in simulation codes to such a detail that the simulation models can be included into the larger tokamak modeling frameworks and provide realistic predictions of the data that would be produced in a real experiment. This approach fits very well into the thinking of *forward modeling* that is used in this section. In forward modeling, as applied here for fusion diagnostics, the starting point is the actual plasma conditions (fuel ion properties etc.) and these properties are then propagated through a number of modeling steps until a modeling estimate of the actual measurements is obtained. The result can then be compared to actual experimental data in an effort to obtain information about the plasma conditions. But the same processing chain can also be used on any proposed, not yet realized, diagnostic system in order to assess the performance of the system and to fine tune its design. This dual use of the forward modeling concept makes it particularly useful in fusion energy research.

This section will briefly discuss some of the most relevant tools in such a forward modeling approach.

The response of the neutron spectroscopy instrument to the incoming fusion neutron flux must be determined in detail. This includes both detailed understanding and modeling of the individual detectors used, as well as a characterization of the full instrumental response to fusion neutrons. The work results in the formulation of an "instrumental/detector response function". This response function could take the form of an analytical expression, but more often is represented by a response matrix. The determination of the response matrix often includes detector calibration by measurements, using radioactive sources (neutrons, electrons or gammas) or an accelerator and a step of particle/detector simulations in order to give a finer granularity and cover the full range of fusion-relevant neutron energies. Useful codes in this work are for example *GEANT* and *MCNP*.

Since neutron spectroscopy involves collimated lines-of-sight, *the field-of-view of the diagnostic* into the fusion plasma must be determined. The *LINE21* code has been developed to calculate the solid angle of each point in the plasma with respect to the detector. The calculation is done in a purely "optical" framework, where intervening structures are either considered fully transparent or completely opaque.

The *neutron transport* from the plasma to the diagnostic can be calculated with *MCNP, SERPENT* and similar neutron transport codes. The direct (un-scattered) flux is often quite straight-forward to determine, while the effect of scattered neutrons at the detector normally requires extensive modeling and calculations. Several processes are important for the scattered flux, as described in Sect. 5. The calculations can also include calculations of secondary emission of for example gammas, which can be of importance for background estimates.

The *properties of the neutron emission* from a certain point in the plasma can be calculated with codes like *DRESS, GENESIS, FPS*, etc. These properties include mainly the intensity and energy of the neutron emission in the direction of the diagnostic. These codes take as input the spatially resolved distribution functions of the plasma fuel ions (or other plasma populations).

Finally, the required *spatially resolved distribution functions of the plasma fuel ions* are obtained from codes such as *TRANSP, PION, ASCOT* etc., or a combination of these.

Implementation of a suite of codes such as these has been used in providing the physics results based on neutron spectroscopy measurements as presented in e.g. Refs [7, 40, 41, 65].

Summary, Outlook and Concluding Remarks

It has been shown in this review that fusion neutrons can provide a rich source of information about the state of the underlying plasma fuel ions. In order to exploit this opportunity, well designed, calibrated and characterized instruments are needed. This is especially true for the most challenging neutron measurements, namely those of the fusion neutron energy distribution with spectrometers.

Neutron measurements rely on nuclear detection techniques, and most of the available, developed techniques have been implemented at fusion devices. In particular JET has over the years provided a fruitful environment for neutron instrumentation developments.

The expected high rates of neutrons (and gammas) at future high-performance fusion devices, such as ITER and DEMO, pose a challenge to all diagnostics. However, for neutron measurements the higher neutron rates also constitute an advantage as this is the primary signal to be measured by the diagnostics.

Implementation of any diagnostic system at a high-performance device is challenging, and neutron diagnostics is no exception. Therefore it is important for the future that opportunities for test and prototype installations are given at today's devices. It is encouraging that such a program is pursued at several fusion machines, especially JET, in view of its planned DT campaign where rich opportunities of "learning by doing" should be available.

References

1. B. Wolle, Tokamak plasma diagnostics based on measured neutron signals. Phys. Rep. **312**, 1 (1999)
2. http://hyperphysics.phy-astr.gsu.edu/hbase/NucEne/coubar.html#c2. Accessed 8 July 2018
3. G.F. Knoll, *Radiation Detection and Measurement*, 4th edn. (Wiley, New York, 2010)
4. J. Wesson, *Tokamaks* (Oxford University Press, Oxford, 2011)
5. Kaye & Laby, Tables of Physical & Chemical Constants, National Physics Laboratory. http://www.kayelaby.npl.co.uk/atomic_and_nuclear_physics/4_7/4_7_4.html. Accessed 08 July 2018
6. G. Lehner, F. Pohl, Reaktionsneutronen als Hilfsmittel der Plasmadiagnostik. Zeitschrift für Physik **207**, 83 (1967)
7. L. Ballabio et al., Relativistic calculation of fusion product spectra for thermonuclear plasmas. Nucl. Fusion **36**, 1728 (1998)
8. J. Eriksson, in *Neutron Emission Spectrometry for Fusion Reactor Diagnosis: Method Development and Data Analysis*, PhD thesis, Digital Comprehensive Summaries of Uppsala Dissertations from the Faculty of Science and Technology, ISSN 1651-6214; 1244 (2015)
9. H. Brysk, Fusion neutron energies and spectra. Plasma Phys. **15**, 661 (1973)
10. J. Källne et al., Observation of the alpha particle "Knock-On" Neutron Emission from Magnetically Confined DT Fusion Plasmas. Phys. Rev. Lett. **85**, 1246 (2000)
11. J. Eriksson et al., Calculating fusion neutron energy spectra from arbitrary reactant distributions. Comput. Phys. Commun. **199**, 40 (2016)
12. M. Nocente, in *Neutron and gamma-ray emission spectroscopy as fast ion diagnostics in fusion plasmas*, PhD thesis, Università degli Studi di Milano-Bicocca (2012)
13. L. Ballabio, in *Calculation and Measurement of the Neutron Emission Spectrum due to Thermonuclear and Higher-Order Reactions in Tokamak Plasmas*, PhD thesis, Comprehensive Summaries of Uppsala Dissertations from the Faculty of Science and Technology, ISSN 1104-232X; 797 (2003)
14. F. Binda, *Liquid scintillators as neutron diagnostic tools for fusion plasmas: System characterization and data analysis*, PhD thesis, Digital Comprehensive Summaries of Uppsala Dissertations from the Faculty of Science and Technology, ISSN 1651-6214; 1646 (2018)
15. MCNP—A general-purpose Monte Carlo N-Particle code, https://mcnp.lanl.gov/. Accessed 08 July 2018
16. J. Leppänen et al., The Serpent Monte Carlo code: status, development and applications in 2013. Ann. Nucl. Energy **82**, 142 (2015)
17. J.C. Nimal, T. Vergnaud, TRIPOLI-3: a neutron/photon monte carlo transport code, in *Advanced monte carlo for radiation physics, particle transport simulation and applications*, ed. by A. Kling, F.J.C. Barão, M. Nakagawa, L. Távora, P. Vaz (Springer, Berlin, 2001)
18. G. Gorini et al., Feasibility of fuel density measurements in ITER using neutron spectrometery. Rev. Sci. Instrum. **68**, 581 (1997)
19. G. Gorini et al., Neutron spectrometry for plasma rotation. Rev. Sci. Instrum. **68**, 561 (1997)
20. C. Hellesen et al., Nucl. Fusion **53**, 113009 (2013)
21. J. Eriksson et al., Nucl. Fusion **55**, 123026 (2015)
22. M. Gatu Johnson et al., Plasma Phys. Control. Fusion **55** (2010) 085002
23. M. Gatu Johnson et al., Nucl. Fusion **50** (2010) 084020
24. N. Jarvis, Neutron measurement techniques for tokamak plasmas. Plasma Phys. Control. Fusion **36**, 209 (1994)
25. L. Ballabio et al., The alpha particle knock-on signature in the neutron emission of DT plasmas. Phys. Rev. E **55**, 3358 (1997)
26. N. Jarvis, Neutron spectrometry at JET. Rev. Sci. Instrum. **57**, 1717 (1986)
27. L. Giacomelli et al., Advanced neutron diagnostics for JET and ITER fusion experiments. Nucl. Fusion **45**, 1191 (2005)
28. A.V. Krasilnikov et al., Fusion neutronic source deuterium–tritium neutron spectrum measurements using natural diamond detectors. Rev. Sci. Instrum. **68**, 1720 (1997)
29. G. J. Schmid et al., A neutron sensor based on synthetic single crystal diamond. Lawrence Livermore National Laboratory, UCRL-CONF-200447, https://e-reports-ext.llnl.gov/pdf/301443.pdf. Accessed 08 July 2018
30. ENDF (Evaluated Nuclear Data File), https://www-nds.iaea.org/exfor/endf.htm. Accessed 08 July 2018
31. C. Cazzaniga et al., A diamond based neutron spectrometer for diagnostics of deuterium-tritium fusion plasmas. Rev. Sci. Instrum. **85**, 11E101 (2014)
32. M. Pillon et al., Experimental response functions of a single-crystal diamond detector for 5–20.5 MeV neutrons, Nucl. Instrum. Methods A **640**, 185 (2011)
33. C. Cazzaniga et al., Single crystal diamond detector measurements of deuterium-deuterium and deuterium-tritium neutrons in Joint European Torus fusion plasmas. Rev. Sci. Instrum. **85**, 043506 (2014)
34. D. Rigamonti, *Development of neutron and gamma-ray spectrometers for fusion plasma applications*, PhD thesis (open access), Universitá degli Studi di Milano—Bicocca (2018)

35. S. Baccaro et al., Radiation damage tests on diamond and scintillation detector components for the ITER radial neutron camera. IEEE Trans. Nucl. Sci. **65**, 2046 (2018)
36. N. D'Olympia et al., Pulse-shape analysis of CLYC for thermal neutrons, fast neutrons, and gamma-rays. Nucl. Instrum. Methods **A714**, 121 (2013)
37. A. Giaz et al., Fast neutron measurements with ^{7}Li and ^{6}Li enriched CLYC scintillators. Nucl. Instrum. Methods **A825**, 51 (2016)
38. Los Alamos National Laboratory, On-line resource, T-2 Nuclear Information Service, http://t2.lanl.gov/nis/data/endf/endfvii.1-n-pdf/h1.pdf. Accessed 09 July 2018
39. A. Hjalmarsson, *Development and Construction of a 2.5-MeV Neutron Time-of-Flight Spectrometer Optimized for Rate (TOFOR)*, PhD thesis, Digital Comprehensive Summaries of Uppsala Dissertations from the Faculty of Science and Technology, ISSN 1651-6214; 233 (2006)
40. M. Gatu Johnson, *Fusion plasma observations at JET with the TOFOR neutron spectrometer: instrumental challenges and physics results*, PhD thesis, Digital Comprehensive Summaries of Uppsala Dissertations from the Faculty of Science and Technology, ISSN 1651-6214; 715 (2010)
41. C. Hellesen, *Diagnosing fuel ions in fusion plasmas using neutron emission spectroscopy*", PhD thesis, Digital Comprehensive Summaries of Uppsala Dissertations from the Faculty of Science and Technology, ISSN 1651-6214; 716 (2010)
42. X. Zhang et al., Diagnosing NB plasmas on the EAST tokamak with new time-of-flight neutron spectrometer. Nucl. Fusion **54**, 104007 (2014)
43. X. Zhang et al., Second generation fusion neutron time-of-flight spectrometer at optimized rate for fully digitized data acquisition. Rev. Sci. Instrum. **85**, 043503 (2014)
44. M. Gatu Johnson et al., The 2.5-MeV neutron time-of-flight spectrometer TOFOR for experiments at JET. Nucl. Instrum. Methods **A591**, 417 (2008)
45. G. Gorini, J. Källne, High count rate time-of-flight spectrometer for DD fusion neutrons. Rev. Sci. Instrum. **63**, 4548 (1992)
46. M. Skiba et al., Fully digital data acquisition system for the neutron time-of-flight spectrometer TOFOR at JET. Rev. Sci. Instrum. **83**, 10D907 (2012)
47. M. Skiba, *Making TOFu: fusion plasma neutron emission spectrometry with a fully digital data acquisition system*, PhD thesis, Digital Comprehensive Summaries of Uppsala Dissertations from the Faculty of Science and Technology, ISSN 1651-6214; 1435 (2016)
48. On-line resource "Two-Body Kinematics Calculator and Plotter", http://skisickness.com/2010/04/relativistic-kinematics-calculator/. Accessed 09 July 2018
49. J. Källne et al., New neutron diagnostics with the Magnetic Proton Recoil spectrometer. Rev. Sci. Instrum. **70**, 1181 (1999)
50. G. Ericsson et al., Neutron emission spectroscopy at JET—results from the magnetic proton recoil spectrometer (invited). Rev. Sci. Instrum. **72**, 759 (2001)
51. E. Andersson Sundén et al., *The thin-foil magnetic proton recoil neutron spectrometer MPRu at JET*, Nucl. Instrum. Methods A610 (2009) 682
52. M. Tardocchi et al., Neutron spectrometery of radio-frequency heated deuterium-tritium plasmas. Rev. Sci. Instrum. **70**, 1171 (1999)
53. J. Frenje et al., Neutron emission Doppler-shift measurements in deuterium-tritium plasmas. Rev. Sci. Instrum. **70**, 1176 (1999)
54. H. Henriksson, *Neutron spectroscopy studies of heating effects in fusion plasmas*", PhD thesis, Comprehensive Summaries of Uppsala Dissertations from the Faculty of Science and Technology, ISSN 1104-232X; 861 (2003)
55. H. Sjöstrand, *Neutron spectroscopy: instrumentation and methods for fusion plasmas*, PhD thesis, Digital Comprehensive Summaries of Uppsala Dissertations from the Faculty of Science and Technology, ISSN 1651-6214; 556 (2008)
56. E. Andersson Sundén, *Neutron spectrometry techniques for fusion plasmas: instrumentation and performance*, PhD thesis, Digital Comprehensive Summaries of Uppsala Dissertations from the Faculty of Science and Technology, ISSN 1651-6214; 726 (2010)
57. S. Conroy et al., Neutron spectrometer for ITER using silicon detectors. Rev. Sci. Instrum. **79**, 10E508 (2008)
58. E. Andersson Sundén et al., Evaluation of neutron spectrometer techniques for ITER using synthetic data, Nucl. Instrum. Methods A **701** (2013) 62
59. B. Marcinkevicius et al., *A Thin Foil Proton Recoil Spectrometer for DT neutrons using annular silicon detectors*. 44th EPS Conference on Plasma Physics, (2017) http://ocs.ciemat.es/EPS2017PAP/pdf/P5.106.pdf. Accessed 09 July 2018
60. A. Wójcik-Gargula et al., *Calculations of the neutron-induced activity in air inside the cuboid 2 of the high resolution neutron spectrometer for ITER*, Henryk Niewodniczanskí Institute of Nuclear Physics, Polish Academy of Sciences, Report 2098/AP (2017), https://rifj.ifj.edu.pl/bitstream/handle/item/9/2098.pdf?sequence=1&isAllowed=y. Accessed 09 July 2018
61. D. Marocco et al., System level design and performance of the ITER radial neutron camera, in *Proceeding 26th IAEA Fusion Energy Conference, Kyoto Japan* (2016), https://nucleus.iaea.org/sites/fusionportal/Shared%20Documents/FEC%202016/fec2016-preprints/preprint0524.pdf. Accessed 09 July 2018
62. J. Källne et al., Measurements of the burn-up of fast ^{3}He and ^{3}H ions in deuterium plasmas. Phys. Scr. **T16**, 160 (1987)
63. N. Dzysiuk et al., Design of a backscatter 14-.MeV neutron time-of-flight spectrometer for experiments at ITER, in *AIP Conference Proceedings* 1612 (2014) 145
64. C. Hellesen et al., Conceptual design of a BackTOF neutron spectrometer for fuel ion ratio measurements at ITER. Nucl. Fusion **57**, 066021 (2017)
65. I. Wodniak, *Modelling and Measurements of MAST Neutron Emission*, PhD thesis, Digital Comprehensive Summaries of Uppsala Dissertations from the Faculty of Science and Technology, ISSN 1651-6214; 1364 (2016)

Journal of Fusion Energy (2019) 38:356–375
https://doi.org/10.1007/s10894-019-00212-w

ORIGINAL RESEARCH

Liquid Scintillators Neutron Response Function: A Tutorial

M. Cecconello[1]

Published online: 19 February 2019

Abstract
This tutorial is devoted to the understanding of the different components that are present in the neutron light output pulse height distribution of liquid scintillators in fusion relevant energy ranges. The basic mechanisms for the generation of the scintillation light are briefly discussed. The different elastic collision processed between the incident neutrons and the hydrogen and carbon atoms are described in terms of probability density functions and the overall response function as their convolution. The results from this analytical approach is then compared with those obtained from simplified and full Monte Carlo simulations. Edge effect, finite energy resolution, light output and transport and competing physical processes between neutron and carbon and hydrogen atoms and their impact on the response functions are discussed. Although the analytical treatment here presented allows only for a qualitative comparison with full Monte Carlo simulations it enables an understanding of the main features present in the response function and therefore provides the ground for the interpretation of more complex response functions such those measured in fusion plasmas. Although the main part of this tutorial is focused on the response function to mono-energetic 2.45 MeV neutrons a brief discussion is presented in case of broad neutron energy spectra and how these can be used to infer the underlying properties of fusion plasmas via the application of a forward modelling method.

Keywords Fusion reactions · Neutron · Liquid scintillator · Response function · Elastic scattering · Probability density function · Convolution · Energy resolution · Efficiency · Forward modelling

Introduction

The measurement of the neutron emission from deuterium-deuterium (DD) and deuterium-tritium (DT) fusion reactions is one of the most important methods of assessing the performance of present and future fusion reactors. Since a neutron is released for each fusion reaction occurring in the plasma, the measurement of neutron flux emitted from the plasma is directly correlated to the fusion power. The emitted neutrons' energy spectrum is characterized by two main components at 2.45 and 14.1 MeV from DD and DT reactions respectively.[1] The neutron energy spectrum is affected, among other things,[2] by the fusion plasma operating conditions. For example, the broadening of the Gaussian energy spectrum for 2.45 and 14.1 MeV neutrons is proportional to the square root of the plasma fuel ion temperature due to the relative velocity distribution of the reactants [1]. In addition, the neutron energy spectrum is affected by the different additional heating schemes that are normally employed in fusion devices such as neutral beam injection and radio-frequency heating. For example, in the case of neutral beam heating the emitted neutron energy spectrum is the combination of a thermal component due to the plasma fuel ions, a beam-thermal component from the reactions between beam and fuels ions and a beam-beam component. The relative intensity of these components is in turn affected by the plasma conditions. In present day devices, the beam-thermal component is dominating while in future fusion devices, the thermal

✉ M. Cecconello
marco.cecconello@physics.uu.se

1 Department of Physics and Astronomy, Uppsla University, Uppsala, Sweden

1 In fusion devices operated with DT fuel, due to their different cross-sections, the DT neutron emission dominates over the DD one while in D-only fuel devices DT reactions can be observed at a very low level where T is generated via one branch of the DD reaction.

2 Other parameters that affect the neutron production are the plasma density and the plasma effective charge but are not discussed here as they mainly affect the neutron yield and not the energy spectrum.

component will be dominant. In the case of radio-frequency heating, neutrons with energies above the DD and DT energies are observed due to fusion reaction from ions accelerated to energies of a few MeV. The measurement of the spatial and temporal evolution of the neutron emission from fusion plasmas in terms of its flux and energy spectrum can therefore provide important information on the plasma itself which can be used to optimize fusion power production.

Different diagnostics are used to measure the neutron emission from fusion plasmas [1, 2] but all rely on the conversion of the neutron into a charged particle that can then be detected. This conversion takes places either thanks to neutron induced nuclear reactions in which heavy charged fission fragments are produced[3] or via elastic scattering of neutrons with light atoms, typically hydrogen. In their simplest form, neutron diagnostics can just be used as counters where the measured counts are proportional to the number of neutrons emitted by the plasma and therefore to the fusion power. The proportionality constant is determined via the absolute calibration of such neutron counters and depends on several parameters such as the counter's efficiency and its position with respect to the neutron source (the plasma) and the fusion reactor [3, 4]. Neutron spectrometers are more sophisticated diagnostics in which the measured neutron energy spectrum is linked in a non-trivial way to the neutron source. Since the energy of fission fragments does not reflect the energy spectrum of the incident neutron, fission chambers can not be used as spectrometers, if one excludes the very crude spectroscopic capability offered by threshold reactions. For this reason, neutron spectrometers are all based on the conversion of the neutron into a light recoil particle via elastic scattering. Neutron spectrometers can be distinguished by the way in which the scattered neutron and the recoil particle are processed. In compact spectrometers, the recoil particle deposits its energy into the scattering medium which, depending on the material, can emit a pulse of scintillation light that is detected: in this case the scatterer itself acts as the detector. The light emission is then converted into a voltage signal and the voltage pulse height spectrum generated by the recoil particles is measured.[4] Since the voltage signal is proportional to the energy deposited by the recoil particle and this depends on the incident neutron energy, a detector based on scintillation material can, in principle, be used as a neutron spectrometer. Large spectrometers, instead, are based on the measurement of the scattered neutron or recoil particle in a detector other than the scatterer. For example, the recoil protons ejected by neutron scattering on a thin, hydrogen-rich foil can be detected in an array of detectors after they have been momentum and energy separated [5]. Alternatively, the time difference between the two scintillation events generated by the same incident neutron in two spatially separated scintillators provides the neutron time of flight and therefore its energy [6]. Recently, time-of-flight spectrometers are also taking advantage of the information on the amount of energy deposited by the recoil particles in the scatterer and in the detector to suppress the contribution from unwanted random coincidences that, especially in fusion devices, typically affect such instruments [7].

The often complex relation between the incident neutron energy spectrum and the output signal from the detector is referred to as the detector response function. Since the neutron energy spectrum at the detector's location is not mono-energetic and since the response function is usually dependent on the incident neutron energy, it is necessary to determine the response function for all the neutron energies of interest. The resulting set of response functions (one for each incident neutron energy) is referred to as the detector response function matrix.[5] It is the knowledge of this response function that allows to infer the characteristics of the neutron energy spectrum and ultimately of the plasma itself. Two different approaches can be used to relate the measured neutron energy spectrum to the fusion plasma source: (i) forward modelling and (ii) inversion algorithms. Forward modelling relies on the accurate modelling of all the processes that in the plasma affect the neutron emission, of the neutron transport from the source to the detector, of the conversion of the incident neutron field into recoil particles and eventually into the detector output signal [8]. Inversion algorithms make no assumption or modelling on the neutron source which instead is obtained by different least-squares minimization methods based on the knowledge of the detector response function [9]. Both methods have advantages and disadvantages but the one thing they have in common is the requirement for a very well characterized detector response function. Response functions for scintillator detectors are usually measured experimentally with well characterized neutron sources [10, 11] and interpreted with the help of dedicated Monte Carlo codes such as NRESP [12] or more general radiation transport codes such as MCNP [13] and MCNP-PoliMi [14] and GEANT4 [15] (which has the additional feature of simulating the scintillation light transport too). Very good agreement is found between measured and simulated

[3] Thermal neutron capture induced reaction in ^{235}U is a typical example.

[4] Pulse height spectra can be based on the peak amplitude or on the time integrated voltage signal.

[5] The response function for large spectrometers, which takes into account for example the effect of the time of flight geometry between scatterer and detector, is referred to as the instrument response function to distinguish it from the individual detectors' response functions.

neutron response functions for all the cited codes. For the purpose of this tutorial, NRESP will be used. In NRESP, the relevant cross-sections (and differential cross-sections) for neutron interactions with hydrogen and carbon in the energy range 0.02–20 MeV are included. The detector material composition and geometry including the liquid scintillator housing and in the optical window connecting to the photo-cathode are modelled. The deposited energy is then converted into a light pulse height distribution taking into account the finite detector energy resolution and experimentally measured light output functions for all the generated particles in the medium (recoils and α-s from nuclear reactions on carbon).

The interpretation of the measured or simulated response functions for liquid scintillator can be quite difficult if only the light output pulse height distribution is given. For a detailed understanding of the origin of the different features present in the response function it is necessary to analyse the contributions from the individual processes occurring in the liquid scintillator such as single and multiple elastic scattering, nuclear reactions and so on. This information can be obtained from the Monte Carlo codes discussed above but this is usually not trivial. In addition, although these codes can provide the expected light output pulse height spectrum, for example, for neutrons that have collided three times with protons, they do no provide an explanation for why it has that particular shape. The aim of this tutorial is to provide such an explanation by means of an analytical derivation of the expected light output pulse height distribution combined with a very simple Monte Carlo code used for its verification. The analytical approach is limited to a few simple cases but it provides nevertheless the basic understanding of how the real response function arises from a combination of multiple individual neutron interactions. This tutorial focuses in particular on the detailed explanation of the different contributions to the response function of a liquid scintillator to incident mono-energetic neutrons. As it will be shown, the interpretation of the response function even in this simple situation is far from trivial. The results obtained in this particular case are easily generalized for a broad neutron energy spectrum. The tutorial is therefore structured as follows. Section 2 is dedicated to a brief overview of the scintillation process, of the resulting recoil particle light output response function and of a simplified Monte Carlo code used for the study of multiple elastic scattering. Section 3 is devoted to the study of the light response function in the case of single elastic scattering processes. Section 4 is dedicated to the determination of the light response function in the case of multiple elastic scattering of a neutron with particles of the same species (for example, only with protons or only with carbon atoms). Section 5 discusses the mixed double scattering case in which the neutron scatters elastically with two different particles (for example first with a carbon atom and then with a hydrogen atom). In Sect. 6, the light response functions calculated with full Monte Carlo simulations are qualitatively interpreted with the help of the response functions derived in the previous two sections for each elastic scattering process. Section 7 briefly addresses some aspects affecting the light response function that have been neglected in the previous sections and elucidate the use of the response function matrix for the interpretation of the liquid scintillator response function when non mono-energetic neutron sources are present such as in the case of fusion reactions. In addition, the forward modelling method is also discussed. Final comments and remarks are presented in Sect. 8.

Neutron Response Function of Liquid Scintillators

Scintillators are among the most common type of detectors both for γ-rays and neutron radiation detection and operate on the principle of induced fluorescent light emission upon the interaction of the radiation within the material. A detailed description of the physical principles, material composition and application of scintillators can be found in [16]. For some scintillation materials, light emission depends on the type of incident radiation thus allowing the discrimination between γ-rays and neutrons, a feature that is essential in fusion application as neutron diagnostics often operates in mixed fields. Liquid scintillators based on organic materials such as benzene (C_6H_6), tuolene ($C_6H_5 \cdot CH_3$) and xylene ($C_6H_4 \cdot (CH_3)_2$), that is hydrogen-rich materials, belong to this category. The main interaction mechanism between γ-rays and liquid scintillators is Compton scattering with the electrons while neutrons interacts by elastic scattering with the hydrogen and carbon nuclei.[6] In both cases a recoil particle is generated (an electron for γ-rays and a proton or carbon nucleus for neutrons). Coulomb interactions between recoil particles and the organic scintillator molecules result in the conversion of the recoil kinetic energy into molecular excitation energy. Part of this excitation energy is then dissipated via thermal quenching and part via fluorescent light emission (commonly indicated as the light output L) in the UV region of the visible spectrum. Typical light pulses have a very fast rise time (few nanoseconds), intensities that depends both on the incident particle energy and type as shown in Fig. 1, and slow decay constants (between 20 and

[6] Inelastic scattering and nuclear reactions with carbon atom are also possible for neutron energies above 4 MeV and are discussed briefly later.

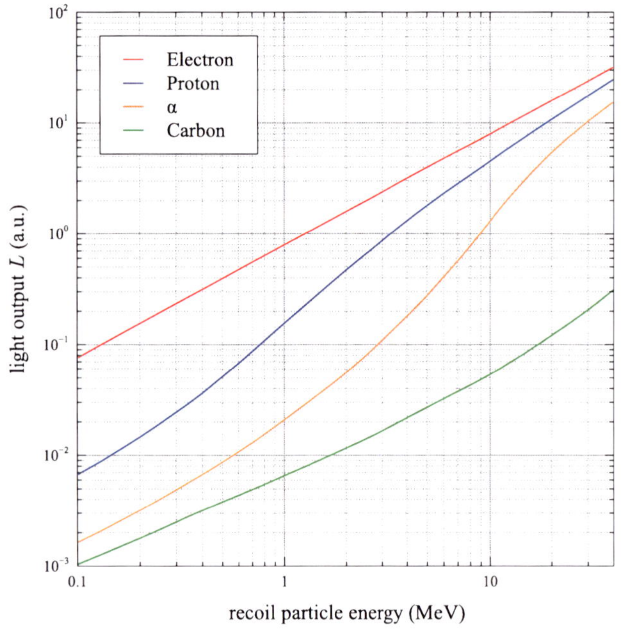

Fig. 1 Liquid scintillator light output L as a function the recoil particle energy. Data adapted from [19]

200 ns) that is different for different recoil particles [16]. Pulse shape analysis can then be used to discriminate γ-rays from neutrons: such detectors are therefore said to have Pulse Shape Discrimination (PSD) capabilities. The most common PSD techniques are the so-called *charge integration/separation* and the *zero-crossing* methods [17]. Recently, thanks to the digitization of the signal acquisition electronics with very high sampling rates (above 500 MSamples/s) it is possible to record and store individual light pulses and then reprocess them off-line enabling the optimization of the chosen PSD methods [18]: in such cases, the term Digital PSD (DPSD) is used.

The fluorescent light is then transported (either directly of via reflection with the liquid scintillator housing walls) to a photocathode as shown in Fig. 2: depending on the volume of the scintillator, significant attenuation of the fluorescent light in the scintillator medium itself can occur. The fluorescent light is then converted into electrons at the photocathode via the photoelectric effect and the initial few ejected photo-electrons are subsequently accelerated, focused and multiplied via secondary electron emission by multi-stage dynodes photomultiplier resulting in large gains (10^5–10^9). The conversion of photons into electrons and their multiplication is a process that, under the correct experimental conditions, is highly linear. Non-linearities occur for example if the photomultiplier is operated at high gains, at high counting rates and in presence of even weak magnetic fields. The electron current at the anode of the photomultiplier is converted into a voltage via a load resistor and the detector voltage output is fed into the acquisition system by co-axial cables (usually tens of meters in fusion experiments). Attenuation and distortion of the voltage signal in the cables will occur but these processes are linear and easily modelled. In present days fusion neutron diagnostics the detector voltage signal is digitized at very high sampling frequencies ranging from 250 MHz up to 4 GHz with high resolution from 10 to 14 bits. The ADC process can be considered linear if the integral an differential non-linearities are negligible which is often the case. Several conversion mechanism contribute therefore to the final signal that is measured for a single neutron interacting with the detector. If the liquid scintillator is to be used as a spectrometer it is then important that

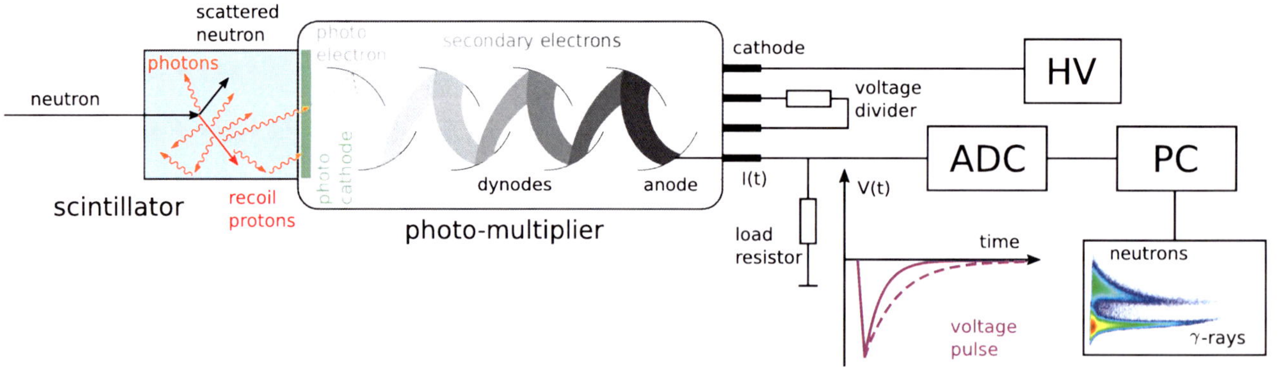

Fig. 2 Schematic drawing of a liquid scintillator detector and the conversion of the light output generated by a recoil proton after an elastic scattering with an incident neutron into a voltage signal. Photons emitted along the proton excitation track are converted into photo-electrons by the photo-cathode and then focused and multiplied by an arrangement of dynodes in multiple stages connected in series to the externally applied High Voltage (HV) via a voltage divider (the PMT internal wiring and de-coupling capacitors are not shown for clarity). The electron current collected at the anode is converted into a voltage signal by means of a load resistor which is fed into an Analog to Digital Converter (ADC) which is connected to a computer (PC) for its acquisition, storage and post-processing. Pulse shape discrimination between incident neutrons and γ-rays is carried out in the PC thanks to the different decay constant of the light pulse generated by the two different incident particles

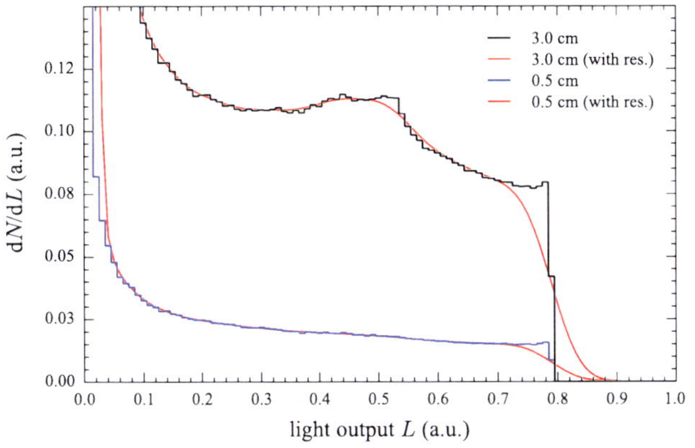

Fig. 3 Example of the light response function of a liquid scintillator to a mono-energetic neutron with energy of 2.45 MeV for a liquid scintillator of 6 cm diameter and two different thicknesses with and without the detector energy resolution effect included

a good linearity exists between the recorded signal and in the incident neutron energy. Under the correct experimental conditions all the above processes are linear but one: the fluorescent light output for recoil particles from neutron elastic collision is inherently non-linear as shown in Fig. 1. This non-linearity complicates significantly the neutron response function as discussed in detail in Sects. 3, 4 and 5.

Examples of the response function calculated with NRESP of two liquid scintillators with the same diameter (6 cm) and two different thickness (0.5 and 3.0 cm) for a beam of incident mono-energetic neutrons with an energy of 2.45 MeV are shown in Fig. 3. As can be seen, the response function exhibits a rich series of features some of which depends on the detector thickness. For the purpose of this tutorial a Simplified monTe cArlo neutron Response funcTion Simulator (STARTS) has been written to calculate the probability density function of the energy and light output distributions of multi-scattered neutrons and of all recoil protons and carbon nuclei in any possible combination. Contrary to the full Monte Carlo codes discussed above, STARTS is specifically aimed at the calculation of the pulse height spectra for specific types of elastic collisions. The following simplifications have been made in STARTS: (i) no cross-section dependence is included as the particles involved in the elastic scattering are selected by the user, (ii) all neutrons undergo a fixed number of elastic scattering defined by the user, (iii) the detector is considered infinite in size, (iv) recoil particles deposit all their energy in the detector, (v) the incident neutrons are mono-energetic and (vi) no energy resolution broadening is included in the calculation of the response function and (vii) the light output function used is taken from [12]. The first simplification implies that the relative intensity of the contribution to the response function from neutron-proton (np) and from neutron-carbon (nC) elastic collisions is neglected: this does not affect the shape of the response function. Simplifications (iii) and (iv) do not affect in any qualitative way the response function while simplification (vi) has an important effect which is however well known and easily understood. The impact of all these simplifications is briefly discussed in Sect. 3.

Response Function for Single Neutron Scattering

Consider an incident neutron with initial energy $E_{n,0}$ making an elastic collision with a proton, assumed to be at rest, as depicted schematically in panel (a) of Fig. 4. After the elastic collision the energies of the neutron and of the recoil protons will be $E_{n,1}$ and $E_{p,1}$ where the index "1" indicates that these quantities refer to the energy of the particles after the first collision. From classical mechanics, invoking the conservation of energy and linear momentum, it can be shown that in the case of a neutron colliding with a generic target particle these energies are given by:

$$\frac{E_{n,1}(\theta)}{E_{n,0}} = \frac{(1+\alpha)+(1-\alpha)\cos\theta}{2} \tag{1}$$

$$\frac{E_{t,1}(\theta)}{E_{n,0}} = \frac{(1-\alpha)(1-\cos\theta)}{2} \tag{2}$$

where the index t identifies the recoil target, θ is the scattering angle in the centre of mass reference system, $\alpha = (A-1)^2/(A+1)^2$ and $A = m_t/m_n$ that is the ratio between the target and neutron masses. It is useful, especially for the discussion in the following sections, to observe that $E_{n,1}/\alpha$ represent the maximum possible energy from which a neutron with energy $E_{n,1}$ could have originated in a single elastic scattering.

Under the assumption of an isotropic cross-section for the elastic scattering in the centre of mass, the scattering angle can assume any value between 0 and π with equal probability.[7] It can then be shown that the probability of the scattered neutron to have an energy in the interval $[E_{n,1}, E_{n,1}+dE_{n,1}]$ is given by:

$$p(E_{n,1})dE_{n,1} = \begin{cases} \dfrac{1}{1-\alpha}\dfrac{1}{E_{n,0}}dE_{n,1} & \text{if } E_{n,1} \in [\alpha E_{n,0}, E_{n,0}] \\ 0 & \text{otherwise} \end{cases} \tag{3}$$

where $p(E_{n,1})$ is the Probability Density Function (PDF) of the continuous random variable $E_{n,1}$. Note that $p(E_{n,1})$ is a

[7] This assumption is a good approximation for np elastic collision for neutrons of few MeV and slightly less accurate for the nC elastic collision case in which forward scattering is more favourable. For the sake of simplicity, this effect is here neglected as it makes a small difference to the calculations and results here discussed.

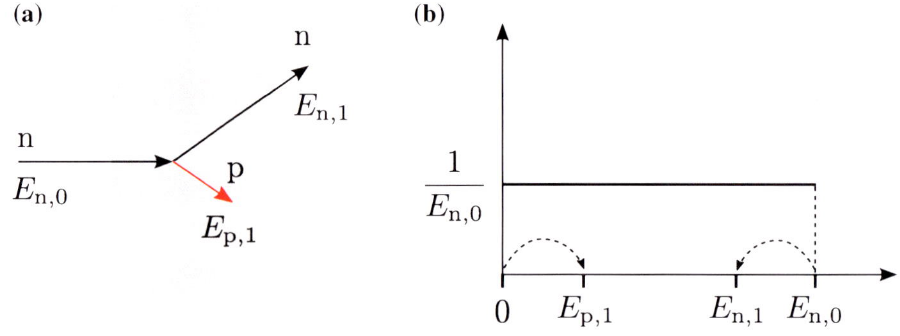

Fig. 4 **a** Depiction of an incoming neutron undergoing single elastic scattering resulting in a scattered neutron and a recoil proton; the shaded area indicates the liquid scintillator volume. **b** Probability density function for the energy of the scattered neutron

properly normalized PDF.[8] In fact, the total probability of the neutron having an energy in the interval $[\alpha E_{n,0}, E_{n,0}]$ after one collision is the sum (integral) of all the probabilities of the neutron having an energy in the interval $[E_{n,1}, E_{n,1} + dE_{n,1}]$ over all possible energies, which is equal to:

$$\int_{\alpha E_{n,0}}^{E_{n,0}} p(E_{n,1})dE_{n,1} = 1 \tag{4}$$

as it can be easily verified by inserting Eq. (3) into the above expression. note that Eq. (4) is the total probability law for mutually exclusive events for continuous random variables.

Similarly, the probability of the recoil target of having an energy in the interval $[E_{t,1}, E_{t,1} + dE_{t,1}]$ is given by:

$$p(E_{t,1})dE_{t,1} = \begin{cases} \frac{1}{1-\alpha}\frac{1}{E_{n,0}}dE_{t,1} & \text{if } E_{t,1} \in [0, (1-\alpha)E_{n,0}] \\ 0 & \text{otherwise} \end{cases}. \tag{5}$$

According to Eqs. (3) and (5), the probability of observing a scattered neutron and recoil target with energies in the range $[E, E + dE]$ is non-zero and constant within the appropriate energy interval as shown in panel (b) of Fig. 4. In the case of np elastic scattering, $\alpha = 0$ which implies that the recoil proton energy $E_{p,1}$ can assume any value between 0 MeV ("grazing" collision) and $E_{n,0}$ ("head-on" collision) with equal probability. In the particular case depicted in panel (b) of Fig. 4, the energy of the scattered neutron is $E_{n,1}$ and the energy of the recoil proton is $E_{p,1} = E_{n,0} - E_{n,1}$.

Since for mono-energetic neutrons with energy $E_{n,0}$ the recoil protons can equally assume energies in $[0, E_{n,0}]$ it follows that, in this simplified scenario, the response function of the detector is the one depicted in panel (b) of Fig. 4, that is a "box-like" function. As can be seen from Fig. 1, a recoil proton of energy $E_{p,1} = E_{n,0} = 2.45$ MeV would generate a light output yield of approximately $L \approx 0.8$ which is exactly the upper edge of the response functions shown in Fig. 3.

This simple "box-like" response function is modified by the non-linear dependence of the light output function on the recoil particle's energy. Following [16], if one assumes for the light output function the relation:

$$L(E_p) = kE_p^{\beta} \tag{6}$$

then:

$$E_p(L) = \left(\frac{L}{k}\right)^{1/\beta} \tag{7}$$

and therefore:

$$\frac{dE_p(L)}{dL} = \frac{1}{\beta L}\left(\frac{L}{k}\right)^{1/\beta} \tag{8}$$

where $k \approx 0.21$ MeV$^{-\beta}$ and $\beta \approx 3/2$ give a good approximation to the light output function for recoil protons shown in Fig. 1 up to 3 MeV. Equations (5) (with $\alpha = 0$) and (8) can be combined to obtain the light response function for the recoil protons:

$$\frac{dN}{dL} = \frac{dN}{dE_{p,1}}\frac{dE_{p,1}}{dL} = \frac{1}{E_{n,0}}\frac{1}{\beta L}\left(\frac{L}{k}\right)^{1/\beta} \tag{9}$$

which implies that the light output response function increases for low light yields. Figure 5 shows the light response function for the 3 cm thick detector calculated by NRESP, by Eq. (9) and by STARTS. As can be seen, the overall features at low light output yields ($L \lesssim 0.2$) and for $L \approx 0.8$ are well described qualitatively[9] but not in the range in between.

The "bump" seen in Figs. 3 and 5 at intermediate L values can not be explained as the result of the detector finite resolution, of edge effects for detectors of finite size

[8] This probability can also be thought as the probability of the neutron having that energy given that an elastic collision has occurred. The latter probability is linked to the detection efficiency as discussed in Sect. 7.2

[9] The term "qualitative" is here use to indicate that the absolute amplitude of the pulse height spectrum can not be calculated by STARTS and "ad-hoc" scaling factors are used instead.

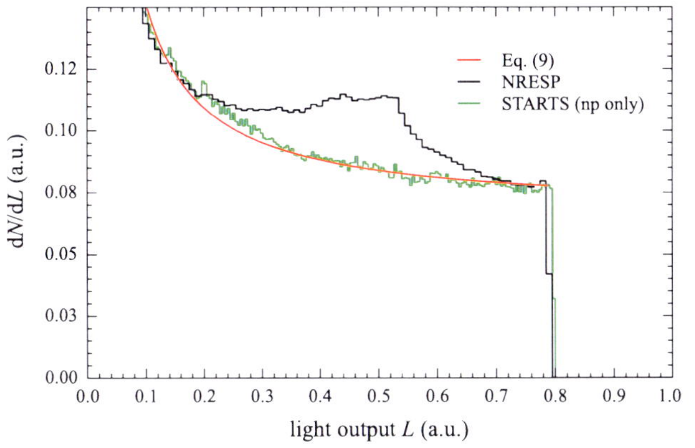

Fig. 5 Simulated light response function of a 3 cm thick liquid scintillator to a mono-energetic neutron with energy of 2.45 MeV using NRESP, STARTS and by Eq. (9)

nor as a consequence of the contribution from nC scattering. The effect of the detector finite energy resolution is mainly to "smear out" the sharp edge at the maximum light output as shown by the red curves in Fig. 3. For a detector of finite size it is possible for the recoil protons generated near the outer surface of the scintillator to escape after having deposited in the medium only a fraction of their energy. The mean free path λ of 2.45 MeV protons in liquid scintillators is of the order of $\lambda \lesssim 0.1$ mm. Even assuming that all the recoil protons generated within a distance λ from the outer surfaces escape, the overall light response function would be reduced in its amplitude by approximately 1 % but its shape would not be modified. The effect would be even smaller for recoil protons of lower energies. According to Eq. (2), the maximum energy for a recoil carbon, for which $\alpha \approx 0.716$, is $E_{c,1} = (1-\alpha)E_{n,0} \approx 0.7$ MeV. The corresponding light output is $L \approx 0.01$ (see Fig. 1) and therefore the contribution to the light response function from recoil carbon nuclei is confined to the very low end of the response function. On the scale used in Fig. 5 this is hardly visible.

It is possible to conclude therefore that the "box-like" response function for single np elastic scattering combined with the light output non-linearity together with the detector finite energy resolution describe quite well the simulated response function especially for the thin detector. However, for thick detectors this is not the case: in [16] this feature is described as the result of neutron double elastic scattering with protons in the scintillator. Section 4 is dedicated to the understanding of the origin and shape of this feature.

Response Function for Multiple Neutron Scattering

In the case of an incident neutron making two elastic collisions four different outcomes are possible as shown in Fig. 6: double elastic collision on two protons (track "c") or on two carbon nuclei (track "f") or mixed scattering first on a carbon nucleus and then on a proton (track "d") or the other way around (track "e"). As discussed in Sect. 3, the contribution to the total light output from recoil carbon nuclei is negligible and therefore the double scattering on carbon nuclei can also be neglected and is not further discussed here. In a similar fashion, the total light output in the case in which the neutron first collides with a proton and then with a carbon nucleus is almost equivalent to a single elastic scattering with a proton. Tracks of type "e" therefore contributes to the response function as single np events described in Sect. 3. The reponse function from double scattering on protons (track "c") is discussed in this section while track "d" (collision on carbon nucleus followed by collision on proton) is discussed in Sect. 5. The discussion of the neutron double elastic scattering on proton is divided in four parts. The probability density function for the energy of the neutron after two elastic collisions is derived in Sect. 4.1 while the probability density function for the energy of the second recoil target is derived in Sect. 4.2. The probability density function for the total deposited energy in two elastic collisions by the neutron is derived in Sect. 4.3 and finally the Probability Density Function (PDF) for the corresponding total light output is obtained in Sect. 4.4.

Doubly Scattered Neutron Energy Probability Density Function

The probability of a neutron with initial energy $E_{n,0}$ having energy in the range $[E_{n,2}, E_{n,2} + dE_{n,2}]$ after two elastic collisions given that after the first collision it had an energy in the range $[E_{n,1}, E_{n,1} + dE_{n,1}]$ is given by the conditional probability:

$$p(E_{n,2} \mid E_{n,1})dE_{n,2} = \frac{p(E_{n,2} \cap E_{n,1})}{p(E_{n,1})} dE_{n,2} \tag{10}$$

where $p(E_{n,2} \cap E_{n,1})$ is the PDF of the joint events resulting in the energies $E_{n,1}$ and $E_{n,2}$. Recalling Eq. (3), the probability $p(E_{n,2} \mid E_{n,1})dE_{n,2}$ is given by:

$$p(E_{n,2}|E_{n,1})dE_{n,2} = \begin{cases} \dfrac{1}{1-\alpha}\dfrac{1}{E_{n,1}} dE_{n,2} & \text{if } E_{n,2} \in [\alpha E_{n,1}, E_{n,1}] \\ 0 & \text{otherwise} \end{cases} \tag{11}$$

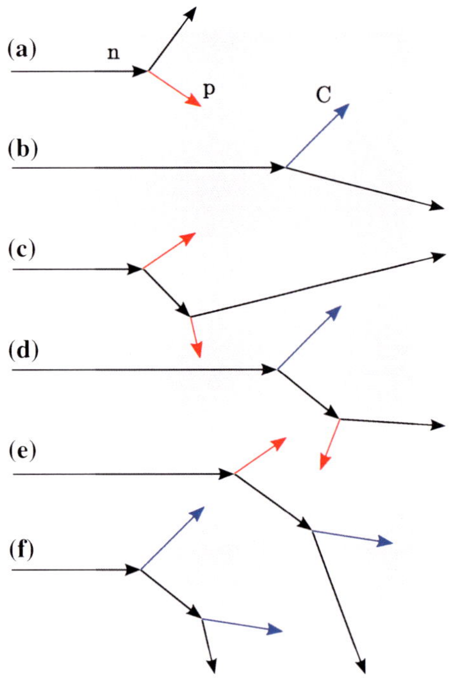

Fig. 6 Single and double elastic scattering events for a neutron incident on a liquid scintillator (gray shaded area). Tracks (a) and (b) are single scattering with a proton and carbon atom respectively. Tracks (c) to (d) represent the four possible collisions: np & np (c); nC & np (d); np & nC (e) and nC & nC (f)

which is equal to Eq. (3) with $E_{n,0}$ and $E_{n,1}$ replaced by $E_{n,1}$ and $E_{n,2}$ respectively.

The probability $p(E_{n,2})dE_{n,2}$ of a neutron with initial energy $E_{n,0}$ having energy in the range $[E_{n,2}, E_{n,2} + dE_{n,2}]$ after two elastic collisions regardless of which energy it had after the first collision can be obtained, using the law of the total probability, as the sum of the probabilities of all the possible combination of two collisions that could have resulted in $E_{n,2}$ being in such range. Each joint event has a probability given by $p(E_{n,2} \cap E_{n,1})dE_{n,2}$ which, using Eq. (10) and replacing $p(E_{n,1})$ with Eq. (3) in combination with Eq. (11), can be written as:

$$p(E_{n,2} \cap E_{n,1})dE_{n,2} = \frac{1}{(1-\alpha)^2}\frac{1}{E_{n,0}}\frac{1}{E_{n,1}}dE_{n,2}. \tag{12}$$

The total probability $p(E_{n,2})dE_{n,2}$ is then obtained by integrating over all possible energies $E_{n,1}$:

$$p(E_{n,2})dE_{n,2} = dE_{n,2}\int \frac{1}{(1-\alpha)^2}\frac{1}{E_{n,0}}\frac{1}{E_{n,1}}dE_{n,1}. \tag{13}$$

Note that $E_{n,2}$ ranges between the neutron initial energy $E_{n,0}$, corresponding to two subsequent "grazing" collisions, and the minimum energy $\alpha^2 E_{n,0}$ corresponding to two subsequent "head-on" collisions $(\theta = \pi)$. If $E_{n,2} \in [\alpha E_{n,0}, E_{n,0}]$, then $E_{n,1}$ can take any value between $E_{n,2}$ and $E_{n,0}$ (see panel (b) of Fig. 7) so that Eq. (13) results in:

$$p(E_{n,2})dE_{n,2} = \int_{E_{n,2}}^{E_{n,0}} \frac{1}{(1-\alpha)^2}\frac{1}{E_{n,0}}\frac{1}{E_{n,1}}dE_{n,1}dE_{n,2} \tag{14}$$

$$= \frac{1}{(1-\alpha)^2}\frac{1}{E_{n,0}}\ln\left(\frac{E_{n,0}}{E_{n,2}}\right)dE_{n,2}. \tag{15}$$

If $E_{n,2} \in [\alpha^2 E_{n,0}, \alpha E_{n,0}]$, then $E_{n,1}$ can range only between $\alpha E_{n,0}$ and $E_{n,2}/\alpha$ since in a single collision it is not possible for a neutron with initial energy $E_{n,1} \in [En,2/\alpha, E_{n,0}]$ to reach a final energy in the range $[\alpha^2 E_{n,0}, E_{n,2}]$ (see panel (c) of Fig. 7). In this case then, Eq. (13) gives:

$$p(E_{n,2})dE_{n,2} = \int_{\alpha E_{n,0}}^{E_{n,2}/\alpha} \frac{1}{(1-\alpha)^2}\frac{1}{E_{n,0}}\frac{1}{E_{n,1}}dE_{n,1}dE_{n,2} \tag{16}$$

$$= \frac{1}{(1-\alpha)^2}\frac{1}{E_{n,0}}\ln\left(\frac{E_{n,2}}{\alpha^2 E_{n,0}}\right)dE_{n,2}. \tag{17}$$

To summarize, the probability $p(E_{n,2})dE_{n,2}$ of a neutron with initial energy $E_{n,0}$ having energy in the range $[E_{n,2}, E_{n,2} + dE_{n,2}]$ after two elastic collisions is given by:

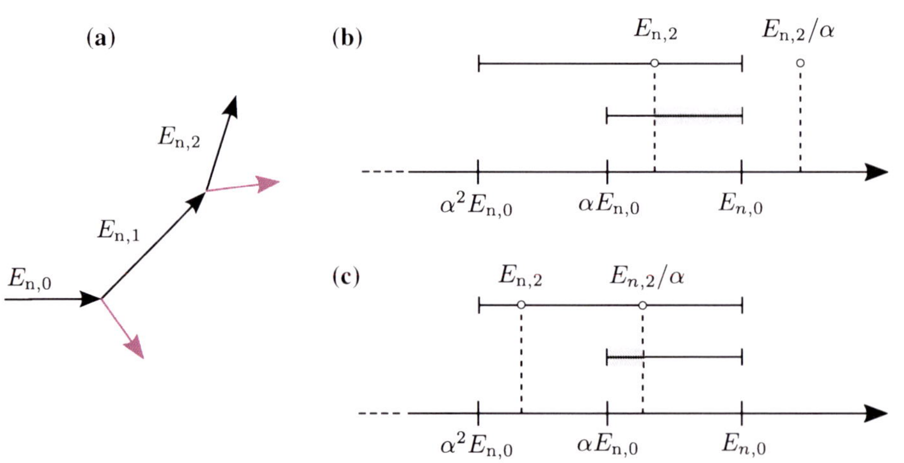

Fig. 7 A depiction of a doubly scattered neutron with energy $E_{n,0}$ from the same type of target nucleus (indicated by the magenta arrows) is shown in (a). b, c Show the range of possible energies that the neutron can have after one or two elastic scattering (short and long horizontal bars respectively). The shaded area indicates the range of possible energies for the neutron after one collision if after two collisions it has energy $E_{n,2}$

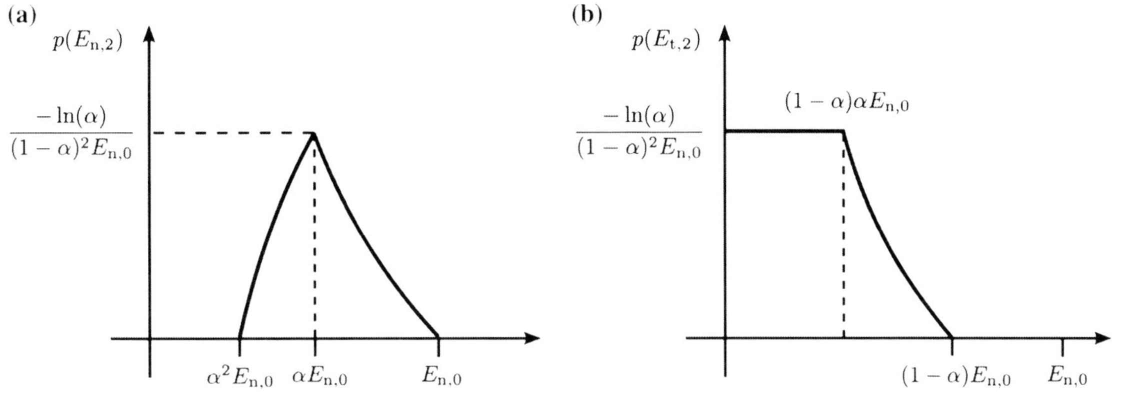

Fig. 8 Probability density functions for the twice scattered neutron (**a**) and for the second recoil target (**b**) for an incident neutron of energy $E_{n,0}$

$$p(E_{n,2})dE_{n,2} = \begin{cases} \frac{1}{(1-\alpha)^2}\frac{1}{E_{n,0}}\ln\left(\frac{E_{n,0}}{E_{n,2}}\right)dE_{n,2} & \text{if } E_{n,2} \in [\alpha E_{n,0}, E_{n,0}] \\ \frac{1}{(1-\alpha)^2}\frac{1}{E_{n,0}}\ln\left(\frac{E_{n,2}}{\alpha^2 E_{n,0}}\right)dE_{n,2} & \text{if } E_{n,2} \in [\alpha^2 E_{n,0}, \alpha E_{n,0}] \\ 0 & \text{otherwise} \end{cases} \tag{18}$$

and the corresponding PDF is shown in panel (a) of Fig. 8. In the particular case of the two targets being protons, $\alpha = 0$ and therefore $E_{n,1}$ and $E_{n,2}$ can both take any value in the interval $[0, E_{n,0}]$ which implies that Eq. (13) should be integrated in the interval:

$$p(E_{n,2})dE_{n,2} = dE_{n,2}\int_{E_{n,2}}^{E_{n,0}} \frac{1}{E_{n,0}}\frac{1}{E_{n,1}}dE_{n,1}. \tag{19}$$

resulting in:

$$p(E_{n,2})dE_{n,2} = \begin{cases} \frac{1}{E_{n,0}}\ln\left(\frac{E_{n,0}}{E_{n,2}}\right)dE_{n,2} & \text{if } E_{n,2} \in [0, E_{n,0}] \\ 0 & \text{otherwise.} \end{cases} \tag{20}$$

Second Recoil Target Energy Probability Density Function

Consider now the two recoils particles generated from the first and second elastic scattering with a single neutron with energy $E_{n,0}$. The probability for the first recoil to have an energy in the range $[E_{t,1}, E_{t,1} + dE_{t,1}]$ is given by Eq. (5) while the probability of observing the second recoil particle with an energy in the range $[E_{t,2}, E_{t,2} + dE_{t,2}]$ given that after the first collision the neutron has an energy $E_{n,1}$ is given by the conditional probability:

$$\begin{aligned} p(E_{t,2} \mid E_{n,1})dE_{t,2} &= \frac{p(E_{t,2} \cap E_{n,1})}{p(E_{n,1})}dE_{t,2} \\ &= \frac{1}{1-\alpha}\frac{1}{E_{n,1}}dE_{t,2}. \end{aligned} \tag{21}$$

The probability of observing the joint events in which the first scattered neutron has energy $E_{n,1}$ and collides with a second target transferring to it the energy $E_{t,2}$ is:

$$p(E_{t,2} \cap E_{n,1})dE_{t,2} = p(E_{t,2} \mid E_{n,1})p(E_{n,1})dE_{t,2}. \tag{22}$$

Substitution of the corresponding expressions for the PDFs results in:

$$p(E_{t,2} \cap E_{n,1})dE_{t,2} = \frac{1}{(1-\alpha)^2}\frac{1}{E_{n,1}}\frac{1}{E_{n,0}}dE_{n,1}dE_{t,2}. \tag{23}$$

The probability $p(E_{t,2})dE_{t,2}$ of observing a second recoil target with an energy in the range $[E_{t,2}, E_{t,2} + dE_{t,2}]$ is then obtained using the law of total probability, that is, by integrating Eq. (23) over all possible energies $E_{n,1}$ that could result in second recoil particle to be in such energy range. Note that the second recoil particle can have an energy in the interval $E_{t,2} \in [0, (1-\alpha)\alpha E_{n,0}]$ regardless of $E_{n,1}$, so the corresponding probability $p(E_{t,2})dE_{t,2}$ is:

$$p(E_{t,2})dE_{t,2} = \int_{\alpha E_{n,0}}^{E_{n,0}} \frac{1}{(1-\alpha)^2}\frac{1}{E_{n,1}}\frac{1}{E_{n,0}}dE_{n,1}dE_{t,2} \tag{24}$$

$$= \frac{1}{(1-\alpha)^2}\frac{1}{E_{n,0}}\ln\left(\frac{1}{\alpha}\right)dE_{t,2}. \tag{25}$$

If $E_{t,2} \in [(1-\alpha)\alpha E_{n,0}, (1-\alpha)E_{n,0}]$ then the incident neutron can only have energies in the interval $[E_{t,2}/(1-\alpha), E_{n,0}]$ so the corresponding probability $p(E_{t,2})dE_{t,2}$ is:

$$p(E_{t,2})dE_{t,2} = \int_{\frac{E_{t,2}}{1-\alpha}}^{E_{n,0}} \frac{1}{(1-\alpha)^2}\frac{1}{E_{p,1}}\frac{1}{E_{p,0}}dE_{p,1}dE_{t,2} \tag{26}$$

$$= \frac{1}{(1-\alpha)^2}\frac{1}{E_{n,0}}\ln\left[\frac{(1-\alpha)E_{n,0}}{E_{t,2}}\right]dE_{t,2}. \tag{27}$$

To summarize, the probability of the second recoil particle to have an energy in the interval $[E_{t,2}, E_{t,2} + dE_{t,2}]$ is given by:

$$p(E_{t,2})dE_{t,2} = \begin{cases} \frac{1}{(1-\alpha)^2}\frac{1}{E_{n,0}}\ln\left(\frac{1}{\alpha}\right)dE_{t,2} & \text{if } E_{t,2} \in [0, (1-\alpha)\alpha E_{n,0}] \\ \frac{1}{(1-\alpha)^2}\frac{1}{E_{n,0}}\ln\left[\frac{(1-\alpha)E_{n,0}}{E_{t,2}}\right]dE_{t,2} & \text{if } E_{t,2} \in [(1-\alpha)\alpha E_{n,0}, (1-\alpha)E_{n,0}] \\ 0 & \text{otherwise.} \end{cases} \tag{28}$$

The PDF $p(E_{t,2})$ is shown in panel (b) of Fig. 8. In the particular case of the two targets being protons, Eq. (28) reduces to:

$$p(E_{t,2})dE_{t,2} = \begin{cases} \frac{1}{E_{n,0}}\ln\left(\frac{E_{n,0}}{E_{t,2}}\right)dE_{t,2} & \text{if } E_{t,2} \in [0, E_{n,0}] \\ 0 & \text{otherwise.} \end{cases} \tag{29}$$

Figure 9 shows the probability density functions for the energy of two recoil protons and two recoil carbon nuclei calculated according to Eqs. (5) and (28) with $\alpha = 0$ and $\alpha = 0.716$ (panels (a) and (b) respectively) for $E_{n,0} = 2.45$ MeV and compared to STARTS calculations.

Total Deposited Energy for Doubly Scattered Neutrons

Most liquid scintillators are small in size compared to the distance travelled by neutrons with energies in the MeV range in the time required for the ADC to record a few samples (a few nanoseconds). If a neutron was to scatter twice within this time interval then the two recoil particles would be generated on a time scale so short that even present day fast data acquisition system would see the two collisions as a single event with an energy equal to the sum of the energy deposited by each individual recoil particle. As discussed at the beginning of Sect. 4, carbon contributes to the light response function only for $L \ll 0.1$ and therefore, even if the energy deposited can be a substantial fraction of the initial neutron energy, it is not discussed further.

Consider instead an incident neutron scattering elastically with two protons: the total energy deposited, in this case, is $E_d = E_{p,1} + E_{p,2}$. The probability density function $p(E_d)$ can be obtained by observing that if the first recoil particle deposits an energy $E_{p,1}$, then the probability of observing a deposited energy E_d is equal to the probability of the second recoil proton to have energy $E_{p,2}$, that is:

$$p(E_d \mid E_{p,1})dE_d = p(E_{p,2} \mid E_{p,1})dE_{p,2} = \frac{1}{E_{n,1}}dE_{p,2} = \frac{1}{E_{n,0} - E_{p,1}}dE_{p,2}. \tag{30}$$

The probability of observing $E_{p,1}$ and E_d is given by:

$$p(E_{p,1} \cap E_d)dE_d dE_{p,1} = p(E_{p,2} \mid E_{p,1})p(E_{p,1})dE_{p,2}dE_{p,1}. \tag{31}$$

Integration over all possible energies $E_{p,1}$, recalling that $p(E_{p,1}) = 1/E_{n,0}$, gives then the probability of observing E_d[10]:

$$p(E_d)dE_d = \int_0^{E_d} \frac{1}{E_{n,0}}\frac{1}{E_{n,0} - E_{p,1}}dE_{p,1}dE_{p,2} \tag{32}$$

$$= \frac{1}{E_{n,0}}\ln\left(\frac{E_{n,0}}{E_{n,0} - E_d}\right)dE_d. \tag{33}$$

Figure 10 shows the PDF for the total energy deposited by the two recoil protons from a neutron with $E_{n,0} = 2.45$ MeV calculated according to Eq. (33) and with the STARTS code.

Total Light Output for Doubly Scattered Neutrons

The non-linear relation between the proton recoil energy deposited in the scintillator and the emitted light output transforms non-linearly the two continuous random variables $E_{p,1}$ and $E_{p,2}$ into the two continuous random variables L_1 and L_2 thereby transforming non-linearly their outcome space as well. In particular, the outcome space for the total deposited energy by the two recoil protons given by:

$$E_{p,1} + E_{p,2} = E_d \tag{34}$$

becomes

$$L_1 + L_2 = L. \tag{35}$$

where if:

$$E_{p,1} \in [0, E_{n,0}] \Rightarrow L_1 \in [0, L(E_{n,0})] \tag{36}$$

$$E_{p,2} \in [0, E_{n,0} - E_{p,1}] \Rightarrow L_2 \in [0, L(E_{n,0} - E_{p,1})]. \tag{37}$$

Figure 11 shows the effect of this non-linear transformation of the probability density function $p(E_{p,2} \mid E_{p,1})$ into $p(L_2 \mid L_1)$ for $E_{n,0} = 2.45$ MeV calculated by STARTS. For

[10] Equation (32) is a special case of the general expression for the PDF of the sum of two continuous random variables x and y with PDFs p_x and p_y which is given by the convolution:

$$p_{x+y}(z) = p_x \otimes p_y = \int_{-\infty}^{\infty} p_x(\zeta)p_y(z - \zeta)d\zeta.$$

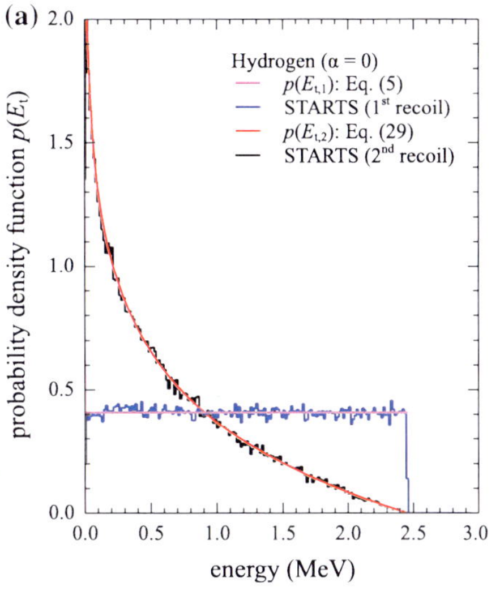

Fig. 9 Probability density functions for the energy of recoil particles for an incident neutron of initial energy of 2.45 MeV undergoing elastic collisions with two particles of the same species: two protons (**a**) and two carbon atoms (**b**)

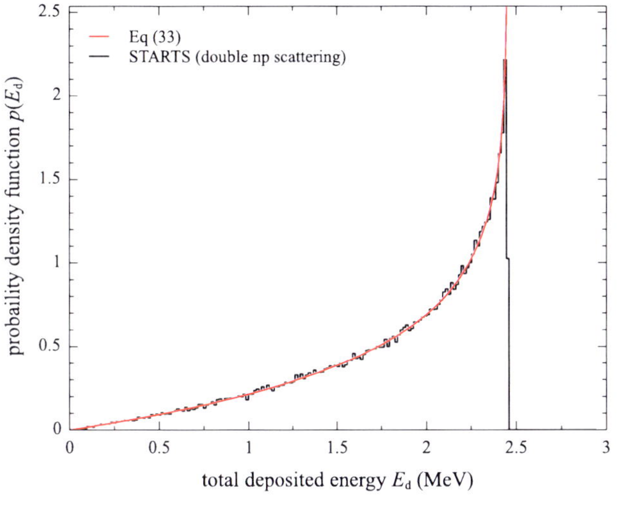

Fig. 10 Simulated (red line) and theoretical (dashed black line) probability density function of the total deposited energy by two recoil protons for a neutron with initial energy of 2.45 MeV (Color figure online)

"head-on" collisions, the neutron energy is transferred only to one proton and the light output is $L(E_p) \approx 0.8$. For $\theta < \pi$, resulting for example in $E_{p,1} = 0.425$ MeV, then $E_{p,2} \in [0, 2.025]$ MeV (see left panel of Fig. 11). The recoil proton with energy $E_{p,1} = 0.425$ MeV will result in a light pulse $L_1 = 0.05$ and therefore $L_2 \in [0, 0.60]$ (see right panel of Fig. 11). The maximum light output in this case is $L = 0.65$ for a total deposited energy of 2.45 MeV. The insert on the right panel of Fig. 11 shows $p(L_2 \mid L_1 = 0.05)$ calculated by STARTS.

Note however that the observable quantity is not the total dep osited energy E_d but the light output $L = L_1 + L_2$ where L_1 and L_2 must fulfil the condition:

$$E_p(L_1) + E_p(L_2) \leq E_{n,0}. \tag{38}$$

The probability density function for the total light output p_L is then calculated as the convolution $p_{L_1} \otimes p_{L_2}$[11]. The first term, p_{L_1}, can be written in terms of $p(E_{p,1})$ observing that the probability $p(E_{p,1})dE_p$ of a recoil proton to have an energy in the range $[E_{p,1}, E_{p,1} + dE_{p,1}]$ is equal to the probability $p(L_1)dL$ that the corresponding light output is in the range $[L_1, L_1 + dL_1]$ from which follows[12]:

$$p(L_1) = p(E_{p,1})\frac{dE_p(L_1)}{dL_1}. \tag{39}$$

Replacing $p(E_{p,1})$ with its corresponding expression [see Eq. (5)], Eq. (39) becomes:

$$p(L_1) = \frac{1}{E_{n,0}}\frac{dE_p(L_1)}{dL_1}. \tag{40}$$

In a similar fashion, the PDF for the conditional probability $p(L_2 \mid L_1)dL_2$ is then:

$$p(L_2 \mid L_1) = \frac{1}{E_{n,0} - E_p(L_1)}\frac{dE_p(L_2)}{dL_2} \tag{41}$$

and therefore, the probability density function $p(L)$ of observing a total light output L resulting from two recoil protons with light pulses L_1 and L_2 is:

$$p(L) = \int_0^L \frac{1}{E_{n,0}}\frac{1}{E_{n,0} - E_p(L_1)}\frac{dE_p(L_1)}{dL_1}\frac{dE_p(L_2)}{dL_2}dL_1. \tag{42}$$

The PDFs $p(L_1)$ and $p(L_2 \mid L_1)$ are non-zero only if L_1 and L_2 satisfy the condition given in Eq. (38). This implies that

[11] See footnote 10.

[12] The quantity (dE_p/dL) can be calculated numerically from a tabulated data set of $\{E_p, L\}$ values or analytically if a functional dependence is given as, for example, in Eq. (6). In this tutorial, (dE_p/dL) is calculated numerically using the light output function shown in Fig. 1.

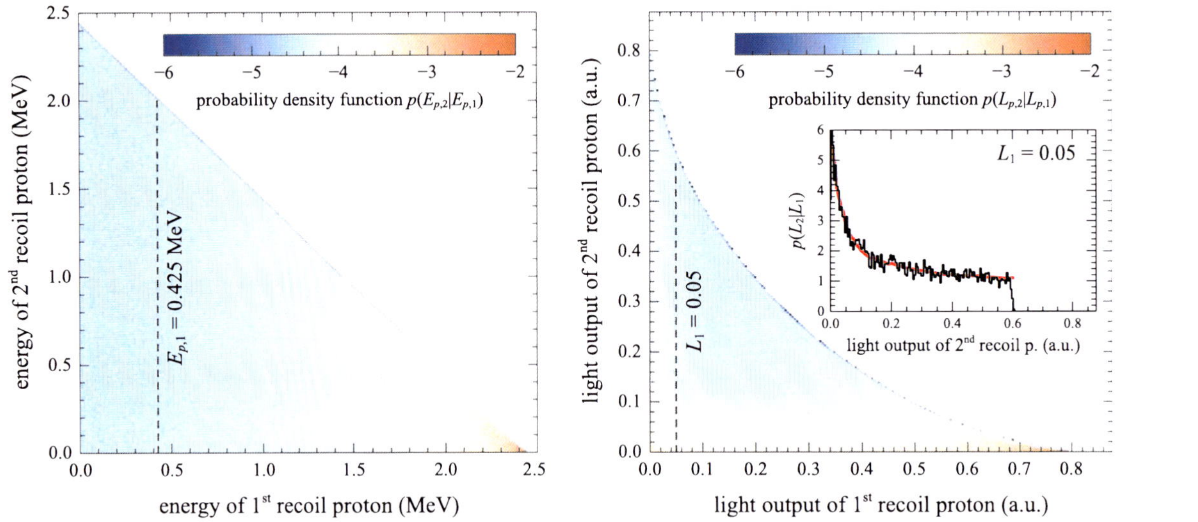

Fig. 11 Left panel: PDF $p(E_{\mathrm{p},2} \mid E_{\mathrm{p},1})$ for the two recoil protons scattered elastically by a neutron with initial energy of 2.45 MeV: the vertical dashed line represents $p(E_{\mathrm{p},2} \mid E_{\mathrm{p},1} = 0.245\ \mathrm{MeV})$. Right panel: PFD $p(L_2 \mid L_1)$ corresponding to the one based on the energy outcome space shown on the left panel: the vertical dashed line represent the range of possible light outputs L_2 given that for $E_{\mathrm{p},1} = 0.425$ MeV, $L_1 = 0.05$. The corresponding simulated (black line) and theoretical (red line) PDFs are explicitly shown in the insert (Color figure online)

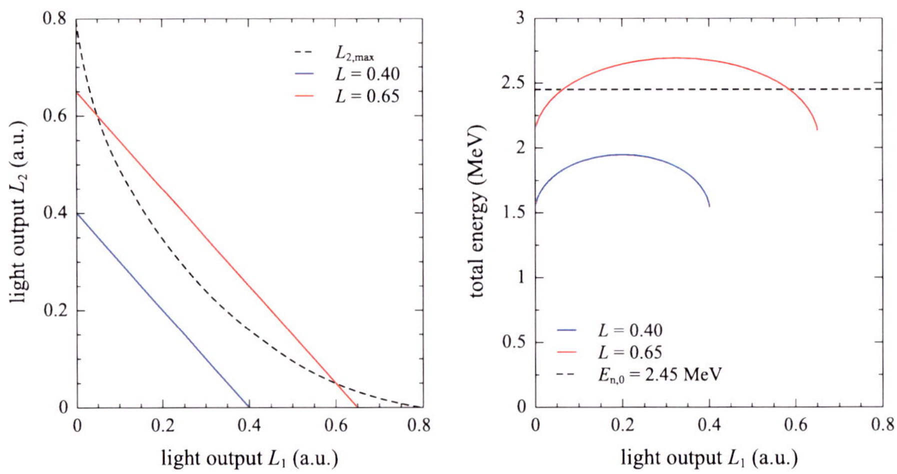

Fig. 12 Left panel: maximum possible light output L_2 as a function of L_1 for $E_{\mathrm{n},0} = 2.45$ MeV. The red and blue lines are examples of possible observable light output from two recoil protons. Right panel: total deposited energy as a function of L_1 corresponding to the two possible observable light output shown on the left panel (Color figure online)

for a given $E_{\mathrm{n},0}$ and L_1 there is a maximum value $L_{2,\max}$ for which this condition is satisfied and is given by:

$$L_{2,\max}(L_1) = L[E_{\mathrm{n},0} - E_{\mathrm{p}}(L_1)]. \tag{43}$$

The dependence of $L_{2,\max}$ on L_1 is shown by the dashed black line in the left panel of Fig. reffig:EconsL: the region of possible values of L_1 and L_2 is the one below the curve $L_{2,\max}(L_1)$. The curve $L_{2,\max}(L_1)$ can be interpreted as all the possible combinations of L_1 and L_2 values for which $E_{\mathrm{d}} = E_{\mathrm{n},0}$.

As a result, the regions where $p(L_1)$ and $p(L_2 \mid L_1)$ are non-zero depends on both L and $E_{\mathrm{n},0}$. Consider in fact first the case where the observed total light output is $L_{\mathrm{obs}} = 0.4$: such a light output could be obtained by any combination of L_1 and L_2 values related by Eq. (35) (see the blue line in the left panel of Fig. 12). The corresponding deposited energy is shown by the blue line in the right panel of Fig. 12. Since in this case $E_{\mathrm{d}} < E_{\mathrm{n},0}$, then $p(L_1)$ and $p(L_2)$ are non-zero for all $L_1 \in [0, L_{\mathrm{obs}}]$. Conversely, consider now the case in which the observed total light output is $L_{\mathrm{obs}} = 0.65$ (see red line in the left panel of Fig. 12). In this case, $p(L_1)$ and $p(L_2)$ are non-zero only if $L_1 \in [0, L_{1,a}] \cup [L_{1,b}, L_{\mathrm{obs}}]$ where $L_{1,a}$ and $L_{1,b}$ are the light outputs for which $L_{\mathrm{obs}} = L_{2,\max}(L_1)$. For $L_1 \in [L_{1,a}, L_{1,b}]$ it turns out that $E_{\mathrm{p}}(L_1) + E_{\mathrm{p}}(L_2) > E_{\mathrm{n},0}$ (as shown by the red line on the right panel of Fig. 12) which is of course not physically possible. It is clear then, that as L increases so does the integral $p(L)$ until $L = L_{\mathrm{B},1}$ with:

$$L_{\mathrm{B},1} = L_1^* + L_{2,\max}(L_1^*) \tag{44}$$

Fig. 13 Evaluation of the PDF $p(L)$ for the total light output L for $L = 0.25$, $L = 0.50$ and $L = 0.65$

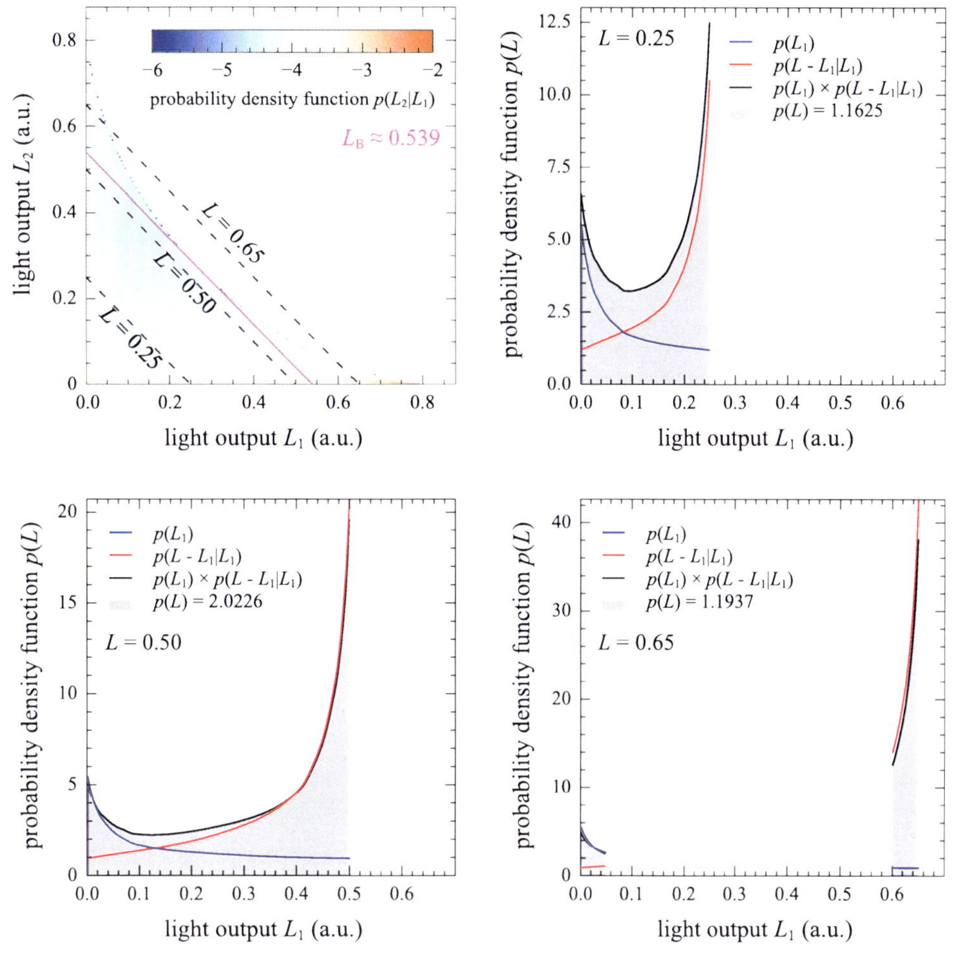

where L_1^* is the point of tangency, i.e.:

$$\left.\frac{dL_{2,\max}(L_1)}{dL_1}\right|_{L_1=L_1^*} = -1. \tag{45}$$

The PDF $p(L)$ reaches its maximum for $L = L_{B,1}$ and goes to zero as $L \to 0$ since the regions where the PDF $p(L_1)$ and $p(L_2)$ are non-zero become vanishing small[13]. The way in which the convolution integral in Eq. (42) is calculated as a function of L is elucidated in Fig. 13 which shows how the integrand depends on L_1 for three different values of the total light output L. Figure 14 shows instead the PDF $p(L)$ given by Eq. (42) for all possible values of the total light output, that is for $L \in [0, L_M]$, compared with the results from STARTS.

The predicted contribution to the total light output response function due to a neutron of a given initial energy making two elastic scattering with protons shown in Fig. 14 can be individuated in the NRESP response function shown in Fig. 5: the sharp knee for $L \approx 0.539$ can now be understood in terms of the maximum energy repartition between the two recoil protons. A closer comparison between Figs. 5 and 14 reveals however that, for $L < L_{B,1}$, $p(L)$ does not drop as predicted by Eq. (42) which indicates that double scattering on protons is not sufficient to reproduce NRESP response function. For this to happen, it is necessary to consider the contribution from triple np scattering. In a fashion similar to what has been done for the double scattering case, it is possible to write the PDF for the light output for the third recoil proton as:

$$p(L_3) = \frac{1}{E_{n,0} - E_p(L_1) - E_p(L_2)} \frac{dE_p(L_3)}{dL_3}. \tag{46}$$

The probability density function of the sums $L = L_1 + L_2 + L_3$ is then calculated as the convolution $p_L = p_{L_1} \otimes p_{L_2} \otimes p_{L_3}$:

$$p(L) = \int_0^L p(L_1) \int_0^{L-L_1} p(L_2) p(L - L_1 - L_2) dL_2 dL1. \tag{47}$$

and is shown in Fig. 15 together with the one calculated by STARTS. As can be seen, $p(L)$ does not drop much for $L \in [L_{B,2}, L_{B,1}]$ where $L_{B,2}$ corresponds to situation in

[13] In the case of the light output function used in this tutorial and for $E_{n,0} = 2.45$ MeV, $L_1^* \approx 0.2697$ and $L_{B,1} \approx 0.5389$.

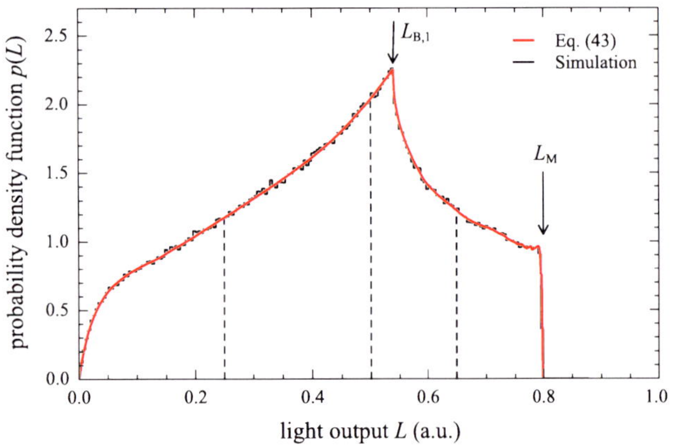

Fig. 14 Probability density function for the light output corresponding to the total energy deposited in the scintillator when a neutron with an energy of 2.45 MeV scatters elastically with two protons: simplified simulation (black) and expected (red). The vertical dashed lines indicate the values of the total light output shown in Fig. 13 (Color figure online)

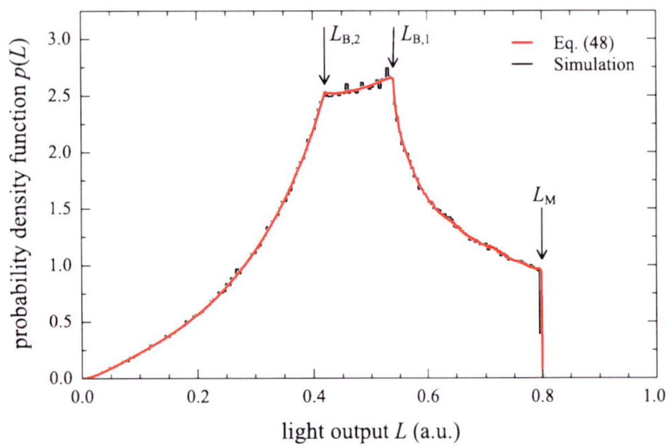

Fig. 15 Probability density function for the light output corresponding to the total energy deposited in the scintillator when a neutron with an energy of 2.45 MeV scatters elastically with three protons: simplified simulation (black) and expected (red) (Color figure online)

which $E_p(L_1) + E_p(L_2) + E_p(L_3) = E_{n,0}$. It is clear from the results shown in Figs. 5, 14 and 15 that a linear combination of the response functions from one, two and three recoil protons can reproduce NRESP output. This is however postponed until Sect. 6 as the last important component to the total light response function, that is the one arising from recoil protons from a neutron that has undergone a prior elastic collision with a carbon atom (track "d" of Fig. 6), is discussed in the next section.

Response Function for Neutron Scattering with Different Targets

From the discussion in Sects. 3 and 4.2 it is clear that, even if the light output from the recoil carbon makes a negligible contribution to the response function, the energy repartition between the recoil and scattered particles will affect the light output of the scattered proton in the second elastic collision. The corresponding energy and light output PDFs can be derived by generalizing to the case where an incident neutron with initial energy $E_{n,0}$ makes two elastic collisions with atoms characterized by A_1 and A_2 such that $A_1 > A_2$ (and therefore $\alpha_1 > \alpha_2$). The probability of observing a neutron with energy between $[E_{n,1}, E_{n,1} + dE_{n,1}]$ after the scattering with A_1 is:

$$p(E_{n,1})dE_{n,1} = \frac{1}{1-\alpha_1}\frac{1}{E_{n,0}}dE_{n,1}. \tag{48}$$

The probability of the twice scattered neutron to have an energy in the range $[E_{n,2}, E_{n,2} + dE_{n,2}]$ after the scattering with A_2 is obtained by applying the law of total probability to the joint event $p(E_{n,2} \cap E_{n,1}) = p(E_{n,2} \mid E_{n,1})p(E_{n,1})$ where the integral is carried out over all possible $E_{n,1}$:

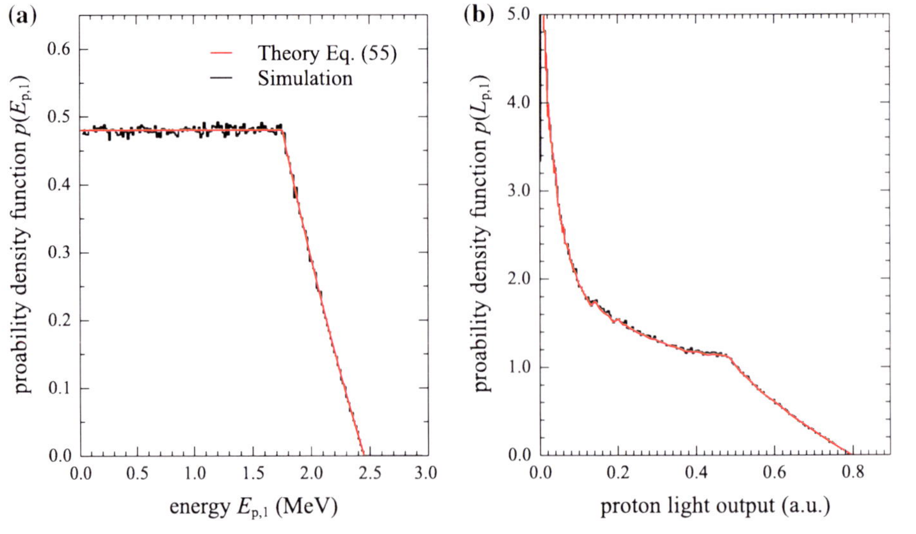

Fig. 16 Simulated (black line) and theoretical (red line) probability density function for the energy (**a**) and the light output (**b**) for the recoil proton from an elastic scattering with a neutron that has previously undergone a collision with a carbon atom (Color figure online)

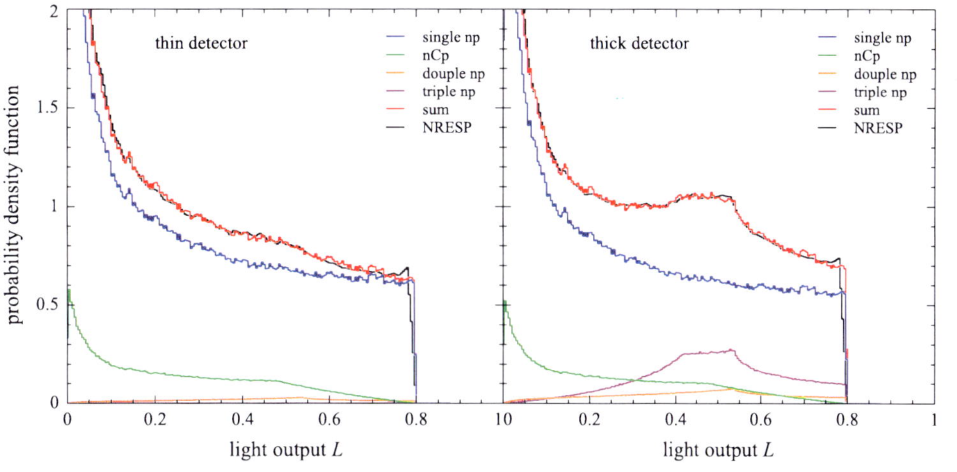

Fig. 17 Comparison between the light output response function of a "thin" (left panel) and a "thick" (right panel) liquid scintillator to a mono-energetic neutron with energy of 2.45 MeV calculated by NRESP (black line) and the predicted one (red line) from a linear regression of the theoretical light output response function for single (blue line), double (orange line) and triple (dark magenta) np scattering events together with the nCp scattering contribution (green line) (Color figure online)

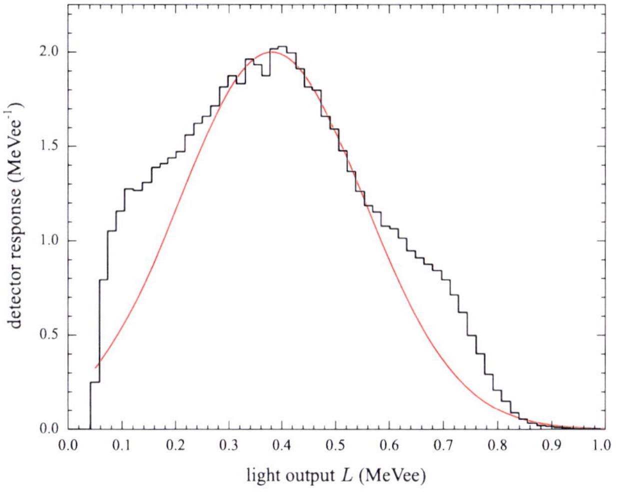

Fig. 18 An example of the central limit theorem: multiple np scattering giving rise to a normal-like distribution contribution (red line) to the response function for 2.45 MeV neutrons into a large EJ-309 liquid scintillator (black line): data from [11] acquired with a non-zero threshold (response function goes to zero $L \lesssim 0.1$) (Color figure online)

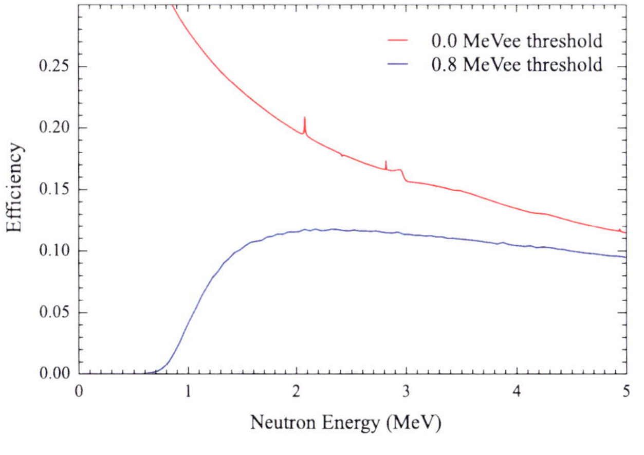

Fig. 19 Liquid scintillator efficiency calculated by NRESP for a 1.5 cm thick scintillator with and without an acquisition threshold

$$p(E_{n,2})dE_{n,2} = \frac{1}{1-\alpha_1}\frac{1}{1-\alpha_2}\frac{dE_{n,2}}{E_{n,0}}\int\frac{1}{E_{n,1}}dE_{n,1} \tag{49}$$

The integration limits above depends on $E_{n,2}$ possible ranges. In particular

$$\text{if}\begin{cases} E_{n,2}\in[\alpha_1E_{n,0},E_{n,0}] & \Rightarrow E_{n,1}\in[E_{n,2},E_{n,0}]\\ E_{n,2}\in[\alpha_2E_{n,0},\alpha_1E_{n,0}] & \Rightarrow E_{n,1}\in[\alpha_1E_{n,0},E_{n,0}]\\ E_{n,2}\in[\alpha_1\alpha_2E_{n,0},\alpha_2E_{n,0}] & \Rightarrow E_{n,1}\in[\alpha_1E_{n,0},E_{n,2}/\alpha_2]. \end{cases} \tag{50}$$

then:

$$p(E_{n,2})dE_{n,2} = \frac{1}{1-\alpha_1}\frac{1}{1-\alpha_2}\frac{dE_{n,2}}{E_{n,0}} \times\begin{cases} \ln\left(\frac{E_{n,0}}{E_{n,2}}\right) & \text{if } E_{n,2}\in[\alpha_1E_{n,0},E_{n,0}]\\ \ln\left(\frac{1}{\alpha_1}\right) & \text{if } E_{n,2}\in[\alpha_2E_{n,0},\alpha_1E_{n,0}]\\ \ln\left(\frac{E_{n,2}}{\alpha_1\alpha_2E_{n,0}}\right) & \text{if } E_{n,2}\in[\alpha_1\alpha_2E_{n,0},\alpha_2E_{n,0}]\\ 0 & \text{otherwise.} \end{cases} \tag{51}$$

In the case $\alpha_1 > 0$ and $\alpha_2 = 0$, the PDF $p(E_{n,2})$ then becomes:

$$p(E_{n,2})dE_{n,2} = \frac{1}{1-\alpha_1}\frac{dE_{n,2}}{E_{n,0}} \times\begin{cases} \ln\left(\frac{E_{n,0}}{E_{n,2}}\right) & \text{if } E_{n,2}\in[\alpha_1E_{n,0},E_{n,0}]\\ \ln\left(\frac{1}{\alpha_1}\right) & \text{if } E_{n,2}\in[0,\alpha_1E_{n,0}]\\ 0 & \text{otherwise.} \end{cases} \tag{52}$$

The probability density function for the energy of the recoil target A_1 is given by (5) with α replaced by α_1 if $E_{A_1,1}\in[0,(1-\alpha_1)E_{n,0}]$ and zero everywhere else. The PDF for the energy of recoil target A_2 is given by Eq. (49) but the integration limits are now given by:

$$\text{if}\begin{cases} E_{A_2,1} \in [0, (1-\alpha_2)\alpha_1 E_{\mathrm{n},0}] \\ \quad \Rightarrow E_{\mathrm{n},1} \in [\alpha_1 E_{\mathrm{n},0}, E_{\mathrm{n},0}] \\ E_{A_2,1} \in [(1-\alpha_2)\alpha_1 E_{\mathrm{n},0}, (1-\alpha_2)E_{\mathrm{n},0}] \\ \quad \Rightarrow E_{\mathrm{n},1} \in \left[\dfrac{E_{A_2,1}}{1-\alpha_2}, E_{\mathrm{n},0}\right]. \end{cases} \tag{53}$$

In this case, $(1-\alpha_2)E_{\mathrm{n},0}$ corresponds to A_2's maximum possible energy if the neutron has lost no energy in the collisions with A_1 (grazing collision) while $(1-\alpha_2)\alpha_1 E_{\mathrm{n},0}$ is A_2's maximum possible energy if the neutron has lost the maximum energy possible in a "head-on" collision with A_1. Integration of Eq. (49) with $E_{\mathrm{n},1}$ in the intervals specified in (53) gives the probability for the second recoil target to have an energy in $[E_{A_2,1}, E_{A_2,1} + dE_{A_2,1}]$:

$$p(E_{A_2,1})dE_{A_2,1} = \frac{1}{1-\alpha_1}\frac{1}{1-\alpha_2}\frac{dE_{A_2,1}}{E_{\mathrm{n},0}} \times \begin{cases} \ln\left(\dfrac{1}{\alpha_1}\right) & \text{if } E_{A_2,1} \in [0, (1-\alpha_2)\alpha_1 E_{\mathrm{n},0}] \\ \ln\left[\dfrac{E_{\mathrm{n},0}}{E_{A_2,1}}(1-\alpha_2)\right] & \text{if } E_{A_2,1} \in [(1-\alpha_2)\alpha_1 E_{\mathrm{n},0}, (1-\alpha_2)E_{\mathrm{n},0}] \\ 0 & \text{otherwise.} \end{cases} \tag{54}$$

In the case $\alpha_1 > 0$ and $\alpha_2 = 0$, the probability of observing the recoil proton with energy in the range $[E_{\mathrm{p},1}, E_{\mathrm{p},1} + dE_{\mathrm{p},1}]$ is therefore:

$$p(E_{\mathrm{p},1})dE_{\mathrm{p},1} = \frac{1}{1-\alpha_1}\frac{dE_{\mathrm{p},1}}{E_{\mathrm{n},0}} \times \begin{cases} \ln\left(\dfrac{1}{\alpha_1}\right) & \text{if } E_{\mathrm{p},1} \in [0, \alpha_1 E_{\mathrm{n},0}] \\ \ln\left(\dfrac{E_{\mathrm{n},0}}{E_{\mathrm{p},1}}\right) & \text{if } E_{\mathrm{p},1} \in [\alpha_1 E_{\mathrm{n},0}, E_{\mathrm{n},0}] \\ 0 & \text{otherwise.} \end{cases} \tag{55}$$

Panel (a) of Fig. 16 shows the PDF for the energy of the recoil proton as a function of $E_{\mathrm{p},1}$ calculated by Eq. (55) together with STARTS results. The corresponding PDF for the light output is given by applying Eqs. (9)–(55) and is shown in panel (b).

Comparison with Full Monte Carlo Simulations

It is now possible to proceed to the qualitative comparison between a combination of the theoretical PDFs derived in the previous sections with the light output response

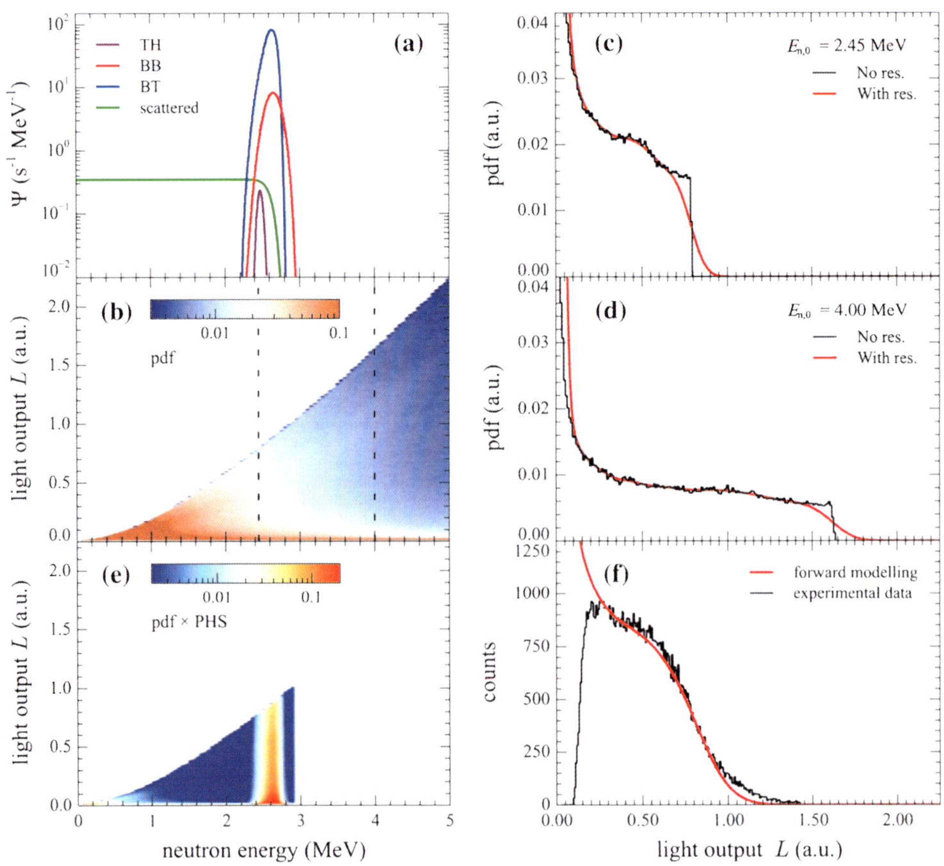

Fig. 20 Example of forward modelling. **a** Shows the individual components of the neutron energy spectrum for a neutral beam heated DD fusion plasmas at the detector in its different components: THermal (TH), Beam-Beam (BB) and Beam-Thermal (BT) assuming a 5 % contribution from scattered neutrons. **b** Shows light output response function matrix: the light output response function for two specific energies (2.45 and 4.00 MeV) indicated by the vertical dashed black lines are shown in (**c**) and (**d**) with and without the effect of the finite energy resolution. **e** Shows the folding of the energy spectrum in (**a**) with the response function matrix in (**b**). Finally, in (**f**) the linear combination of the folded response functions for all neutron energies shown in (**e**) are compared with the experimentally measured light output pulse height spectrum

function obtained by full Monte Carlo simulation such as NRESP. In this context, a full Monte Carlo simulation refers to a simulation in which the proper cross-sections of the relevant processes are taken into account, the density of hydrogen and carbon atoms in the scintillation material are specified and a realistic geometrical model of the detector is used. Without this level of details, neither STARTS nor the analytical approach can properly estimate the relative importance (amplitude) of the different components constituting the light output response function although one can reasonably assume that the role of multiple np scattering becomes more important as the size of the detector increases and that collisions with carbon nuclei followed by collisions on hydrogen are always present given that the cross-sections for these processes are of the same order of magnitude. The weights of the light output response functions corresponding to the different elastic scattering processes are obtained by a simple multiple linear regression where the dependent variable is the overall response function calculated by NRESP. The relative amplitude of the different components so obtained has just an indicative (qualitative) nature and for a proper estimate of these weights one should exclusively use full Monte Carlo codes.

With these words of caution well present in mind, the comparison between the NRESP light output response function and the one from the linear regression based on STARTS calculations is shown in Fig. 17 for both the "thin" and "thick" detectors. As can be seen, the light response function for the "thin" detector can be well matched neglecting the contribution from triple np scattering while this component is essential to match the response function for the "thick detector".

As an example of how the analytical approach can be used to make predictions regarding the expected response function of a liquid scintillator, consider the case of a scintillator with a very large active volume. It is obvious that multiple elastic scattering of the incident neutron on protons should make a large contribution to the total light output response functions. Observing that the response function for multiple np scattering can be calculated by multiple convolutions of the same probability density function, although each weighted by a different factor which depends on the scattered neutron energy before each collision, and using the central limit theorem[14] then it is possible to predict that expected response function should approximate a normal probability density function. This is indeed the case as shown in Fig. 18 where the detector response function of a 12.7-by-12.7 cm cylindrical EJ-309 liquid scintillator measured experimentally [11] is compared with a normal distribution.

Having now described, albeit qualitatively, the building blocks of the light response function of a liquid scintillator in the simplest case, it is now possible to tackle its interpretation in presence of those effects neglected so far and in more realistic scenarios such as those encountered in fusion devices.

Response Functions in Realistic Scenarios

In this section additional aspects affecting the response function of liquid scintillators that have been neglected in the previous sections are discussed. These can be divided in two categoties: those aspects that are intrinsic to the scintillator properties which are discussed in Sects. 7.1 and 7.2 and those which depend on the incident neutron energy spectrum, discussed in Sect. 7.3.

Detector Properties Affecting the Light Output Response function

The response function discussed so far has been limited to the case of 2.45 MeV mono-energetic incident neutrons for which the predominant interaction mechanism is elastic scattering on hydrogen and carbon nuclei. For neutrons of higher energies the following competing processes occur: inelastic nC scattering (for $E_n > 4.8$ MeV) and the nuclear reactions $^{12}C(n,\alpha)^9Be$ (for $E_n > 7.4$ MeV) and $^{12}C(n,n')^9Be^*$ (for $E_n > 10$ MeV). At these high energies, nuclear reactions induced by the neutron interacting with the detector material surrounding the liquid scintillator cell are also possible. Inelastic scattering results in the scattered neutron and recoil particles to have a lower energy than in the case of elastic scattering. The light output from these recoils particles have smaller amplitude and will be distributed continuously from zero to the maximum light output possible. The light output from α-s of few MeV is much higher than that of recoil carbons of similar energy but still lower by a factor of about 10 compared to the one for recoils protons (see Fig. 1). The contribution from α-s appear as an additional "edge" at the low end part of the light output response function superimposed to the "box-like" response for protons. Indicating with L_p the light output resulting from a "head-on" collision of a high energy neutron with a proton and with L_α the light output produced by an α particle then $L_\alpha/L_p \approx 0.1$ for $E_{n,0} = 15$ MeV [10]. Due to the complexity of these competing events as well as the strong angular dependence of their cross section, full Monte Carlo simulations are

[14] The central limit theorem holds only if both the average value and variance of the random variable exist. The random variable in this case is the total light output L for which the existence of its average value and variance is guaranteed.

indispensable for a correct modelling of experimental response functions at neutron energies above few MeVs.

The transport of the scintillation light from its point of generation inside the active detector volume to the PMT becomes important for large detectors or detectors with one dimension much larger than the other. This is due to the fact that the scintillator itself is a self-absorbing medium and therefore the same deposited energy will result in two light pulses of different amplitude depending on the location of the recoil particle generation with respect to the PMT. The effect of light transport and self-attenuation is usually negligible for small volume liquid scintillators (a few centimetres). For larger scintillators, self-attenuation results in light pulses being distributed at lower light output than expected. For example, in the case of incident mono-energetic neutrons onto a long cylindrical scintillator with a PMT at one end of the cylinder, self-absorption will result in the "smearing" of the "head-on" collision edge towards lower amplitudes thus worsening the detector energy resolution. If liquid scintillators are to be used for neutron spectroscopy it is therefore essential that they are of limited dimensions. Codes such GEANT4 [15] and MCNP-PoliMI [14] are capable of treating properly the transport of photons and should be used for scintillators of shape more complex than the standard cylindrical one and for large volumes.

Two additional quantities that affect the light response function are the light output as a function of the deposited energy by different recoil particles and the energy resolution function. The light output function depends by the size and geometry of the scintillator and affects the relative position of the different features in the light output pulse height spectrum and in particular the position of the "edge" corresponding to the deposition of the full neutron energy in a single elastic scattering with a proton ("head-on" collision). The energy resolution dependency on the neutron energy results in the smearing of the response function and it is mostly predominant at the location of the "head-on edge". The energy resolution is affected in turn by the geometry and size of the scintillator and typically ranges between 5 and 30 % in the neutron energy range 0–3 MeV. For a discussion on the different methods for measuring the light output and energy resolution functions the reader is referred to [10, 11] the references therein. Since the "head-on edge" region is the one generated by neutrons that have not undergone any collision, the spectral information on the neutron source is mainly contained in this region and therefore the determination of the light output and energy resolution functions is very important especially at the energy of interest (for example at 2.45 and 14.1 MeV for fusion plasmas).

Detector Efficiency

The efficiency of a liquid scintillator to a mono-energetic neutron with energy $E_{n,0}$ for single elastic scattering on hydrogen is given by [16]:

$$\epsilon(E_{n,0}) = \frac{n_H\sigma_H(E_{n,0})}{n_H\sigma_H(E_{n,0}) + n_C\sigma_C(E_{n,0})} \left\{1 - e^{-\left[n_H\sigma_H(E_{n,0}) + n_C\sigma_C(E_{n,0})\right]d}\right\} \tag{56}$$

where d is the detector thickness, n_H and n_C are the number density of hydrogen and carbon atoms respectively, and σ_H and σ_C are the neutron scattering cross-section on hydrogen and carbon. The exponential terms represents the fraction of neutrons of energy $E_{n,0}$ passing through a scintillator of thickness d (i.e. the uncollided neutrons) and therfore the terms within curly brackets represents the fraction of neutrons interacting in the scintillation material.

The detector efficiency can be calculated using full Monte Carlo codes as the integral of the pulse height spectrum: the efficiency of a 1.5 cm thick scintillator as a function of the incident neutron energy is shown in Fig. 19. Since in reality, pulse height spectra are measured with an acquisition threshold, as shown in Fig. 18, the lower limit of the integration must be accurately determined. This is typically done by calibrating the detector light output against the energy deposited by recoil electron using standard γ-rays calibration sources (such as ^{22}Na, ^{137}Cs and ^{207}Bi) and then converting the experimental light output threshold into a deposited energy threshold using the corresponding light output function [20, 21]. The light output so calibrated is given in units of MeV electron-equivalent (MeVee) as shown in Fig. 18. In this case, the light output threshold is approximately 0.1 MeVee resulting in a deposited energy threshold of approximately 0.8 MeVee for recoil protons (see Fig. 1). In particular, the efficiency and the scintillator cross-sectional area determine the number of counts per seconds that are measured for a given incident neutron flux.

As shown in Fig. 19, the probability of a neutron to interact with the scintillator increases as its energy decreases. As a results, neutrons that have undergone (multiple) scattering before reaching the detector can make a significant contribution to the pulse height spectrum. Neutrons that have collided first with the scintillation material will have an increased probability of making multiple collisions especially for thick detectors as the detection efficiency increases. It is clear then that the dependency of the efficiency on the neutron energy has a non-trivial effect on the light response function both in shape and amplitude. Full Monte Carlo codes ensure that all these effects are included in the response function.

Neutron Energy Spectra in Fusion Devices and Forward Modelling

In fusion devices, liquid scintillators are typically located inside thick neutron shield and view the neutron source (the plasma) via long collimators and the energy spectrum of the neutrons at the detector are anything but mono-energetic. A typical fusion neutron energy spectrum contains spectral components from (i) DD and DT thermal emission (approximately gaussian with a width which depends on the plasma ion temperature), (ii) fusion reactions involving ions with energies much larger than the thermal ions produced by additional heating and (iii) scattered neutrons, that is, neutrons that before their first interaction with the scintillator have already suffered one or multiple collisions with the material surrounding it (from the fusion device itself, to the neutron diagnostic in which the detector is installed). In such a situation, the interpretation of the experimental response function requires the accurate estimate of the neutron flux and energy spectrum at the detector location. In turns this requires accurate plasma physics models for the calculation of the DD and DT neutron sources and of the neutron emission along the line of sight by which the neutron source is seen by the detector to provide the direct (not collided) neutron flux and energy spectrum at the detector. In addition, Monte Carlo neutron transport simulations are required to estimate the scattered neutron flux and its energy spectrum given a specific neutron source and line of sight. The resulting neutron energy spectrum at the detector will contains all these spectral components at different levels.

An example of the different components of the neutron spectrum at a detector location in a DD plasma with neutral beam heating is shown in panel (a) of Fig. 20 which can be thought of as the neutron emissivity multiplied by the detector area and the solid angle. The response function for each incident neutron energy is then calculated resulting in the response function matrix in which, as shown in panel (b), the columns contain the light output response function at for a particular neutron incident energy. The response functions for incident neutron with energies 2.45 and 4.00 MeV are shown in panels (c) and (d): each response function is then folded with the energy resolution function. The neutron energy spectrum can then be thought as a set of weights that multiply the response function matrix for each incident neutron energy resulting in the weighted response function matrix shown in panel (e). Adding together the contribution from all neutron energies present in the neutron energy spectrum is then equivalent to the column-wise sum of the elements of the weighted response function matrix: the result in shown in panel (f) together with experimentally measured light output pulse height spectrum integrated over a given time interval and multiplied by the detector efficiency. The comparison shown in panel (f) highlights one final aspect of the forward modelling, that is, the conversion of the matrix of normalized response functions multiplied by the absolute neutron energy spectrum into absolute counts. The geometry of the line of sight is already included in the neutron energy spectrum shown in panel (a) so this is achieved by multiplying the neutron energy spectrum by the integration time used to calculate the measured pulse height spectrum in panel (f) and by the detector efficiency.

In this case, the agreement is not perfect. Several reasons could be responsible for this. One possibility is that the original neutron spectrum might not have been properly estimated because one or more plasma parameters that are used in the modelling codes are measured or estimated incorrectly. Among such quantities are the fuel ion and the fast ion (from the additional heating) distribution functions in space and energy, the plasma temperature and density. If one assumes that the detector response function is well known, then such discrepancy can be used to improve the models used for the description of the plasma. This method of proceeding in the modelling of the measured light pulse height spectra is therefore called "forward modelling" since, as the name implies, it starts from the source of the neutrons and then propagates the resulting neutron field taking into account all the aspects affecting the neutron transport from the source to the detector (including realistic geometries and material compositions of the fusion device and of the diagnostic) and then converts the neutron field at the detector into the light output pulse height spectrum via the response function matrix. It is clear therefore how crucial it is that the liquid scintillator response function is determined as accurately as possible if information on the plasma itself is required from the neutron measurements from liquid scintillators.

Summary

The interpretation of the neutron response function of liquid scintillators can be rather subtle as discussed in this tutorial even in the simplest case imaginable of a mono-energetic neutron source. This case, which can be approximately obtained in neutron calibration facilities, however is very useful to understand the different processes that contributes to the neutron light response function which can then easily generalized to the case of fusion neutron sources. Response functions can be easily obtained with Monte Carlo codes as discussed here but their use and the interpretation of the calculated results can be quite complex and as shown in this tutorial not strictly necessary for a qualitative understanding of the response function.

Monte Carlo codes are however indispensable whenever a quantitative comparison with experimental measurements is crucial. This is particularly true in the case where one wants to infer absolute quantities from measurements based on the response of liquid scintillators from fusion plasmas such for example the absolute neutron yield (which requires an accurate determination of the detector efficiency) or the underlying fuel ions distribution functions from the measured neutron energy spectrum. In such cases, forward modelling is a robust although quite complex method that enables an understanding of the underlying plasma properties and therefore can provide useful feedback information on the theory and tools used to model the physics of fusion plasmas. This, in turns, requires a very good overall characterization of the response function of liquid scintillators. It is hoped that this tutorial provides an useful introduction to the understanding of the response functions obtained by full Monte Carlo codes.

Acknowledgements The author wish to thanks S. Conroy and A. Hjalmarsson at the Department of Physic and Astronomy at Uppsala University for useful discussions. This work was funded by the Swedish Research Council VR research Grant 2015-03869 and the European Union's Horizon 2020 research and innovation programme under Grant Agreement Number 633053. The views and opinions expressed herein do not necessarily reflect those of the European Commission.

References

1. O.N. Jarvis, Neutron measurement techniques for tokamak plasmas. Plasma Phys. Control. Fusion **36**(2), 209 (1994)
2. B. Wolle, Tokamak plasma diagnostics based on measured neutron signals. Phys. Rep. **312**(1), 1 (1999)
3. D.B. Syme et al., Fusion yield measurements on JET and their calibration. Fusion Eng. Des. **89**(11), 2766–2775 (2014)
4. P. Batistoni et al., 14 MeV calibration of JET neutron detector-sphase 1: calibration and characterization of the neutron source. Nuclear Fusion. **1;58**(2), 026012 (2018)
5. E. Andersson Sundén et al., The thin-foil magnetic proton recoil neutron spectrometer MPRu at JET. Nucl. Instrum. Methods Phys. Res.A **610**(3), 682–699 (2009)
6. M. Gatu Johnson et al., The 2.5-MeV neutron time-of-flight spectrometer TOFOR for experiments at JET. Nucl. Instrum. Methods Phys. Res.A **591**(2), 417–430 (2008)
7. M. Skiba et al., Kinematic background discrimination methods using a fully digital data acquisition system for TOFOR. Nucl. Instrum. Methods Phys. Res.A **838**, 82–88 (2016)
8. F. Binda et al., Forward fitting of experimental data from a NE213 neutron detector installed with the magnetic proton recoil upgraded spectrometer at JET. Rev. Sci. Instrum. **85**, 11E123 (2014)
9. A. Zimbal et al., Compact NE213 neutron spectrometer with high energy resolution for fusion applications. Rev. Sci. Instrum. **75**(10), 3553–3555 (2004)
10. H. Klein et al., Neutron and photon spectrometry with liquid scintillation detectors in mixed fields. Nucl. Instrum. Methods Phys. Res.A **476**, 132–142 (2002)
11. A. Enqvist et al., Neutron light output response and resolution functions in EJ-309 liquid scintillation detectors. Nucl. Instrum. Methods Phys. Res.A **715**, 79–86 (2013)
12. Dietze G. et al., Monte Carlo Codes for the calculation of the neutron response functions and detection efficiencies for NE 213 scintillation detectors, PTB-ND-22, Braunshweig, (1982)
13. T. Goorley et al., Initial MCNP6 release overview. Nucl. Technol. **180**, 298–315 (2012)
14. S.A. Pozzi et al., MCNP-PoliMi: a Monte-Carlo code for correlation measurements. Nucl. Instrum. Methods Phys. Res. A **513**, 550–558 (2003)
15. S. Agostinelli et al., Geant4–a simulation toolkit. Nucl. Instrum. Methods Phys. Res. A **506**, 250–303 (2003)
16. G.F. Knoll, *Radiation Detection and Measurement*, vol. 864 (Wiley, New York, 2010)
17. D. Wolski et al., Comparison of n-γ discrimination by zero-crossing and digital charge comparison methods. Nucl. Instrum. Methods Phys. Res. **360**(3), 584–592 (1995)
18. B. Esposito et al., Digital pulse shape discrimination in organic scintillators for fusion applications. Nucl. Instrum. Methods Phys. Res. **518**(1–2), 626–628 (2004)
19. V.V. Verbinski et al., Calibration of an organic scintillator for neutron spectrometry. Nucl. Instrum. Methods **65**(1), 8–25 (1968)
20. J.W. Fowler et al., Efficiency calibration of scintillation detectors in the neutron energy range 1.5–25 MeV by the associated particle technique. Nucl. Instrum. Methods **175**(2–3), 449–463 (1980)
21. G. Dietze, H. Klein, Gamma-calibration of NE 213 scintillation counters. Nucl. Instrum. Methods Phys. Res. **193**(3), 549–556 (1982)

Journal of Fusion Energy (2019) 38:376–385
https://doi.org/10.1007/s10894-018-0183-0

ORIGINAL RESEARCH

Review of Neutron Diagnostics Based on Fission Reactions Induced by Fusion Neutrons

Urszula Woźnicka[1]

Published online: 14 August 2018

Abstract

Neutrons are an exceptional diagnostic signature in magnetic fusion experiments with high temperature plasma. A part of neutron detection methods relies on detecting nuclear reactions products, initiated by neutrons. Among different possibilities, the neutron-induced fission reactions can be used. These methods use the specific properties of the radioactive decay phenomena induced by neutrons of heavy elements like uranium or thorium. The neutron-induced fission reactions produce energetic fission fragments that can be detected in a special type of ionization chambers that has one of electrode covered with fissionable material. A fission reaction may be also adapted to neutron detection by activation method. A dedicated target-sample made from fissionable isotope (like ^{235}U, ^{238}U, ^{232}Th), irradiated by neutrons, is the source of prompt and delayed neutrons. The decay curve of delayed neutrons can be registered almost immediately after irradiation by fusion neutrons, with a delay resulting exclusively from the time of transport of the samples to the measuring set-up. The yield of a fusion neutron source can be calculated from that measurement. Fission phenomena and fissionable materials are applicable in plasma neutron diagnostics. The lecture gives the introduction to the nuclear fission process and the particular attention is done to the phenomenon and physics of delayed neutrons which are generated during the fission reaction. Examples of delayed neutrons activation set-up and fission chambers dedicated for tokamaks are presented.

Keywords Thermonuclear fusion diagnostics · Fission reactions · Delayed neutrons · Neutron activation method · Fission chambers

Introduction

Neutrons resulting from the thermonuclear fusion of hydrogen isotopes (deuterium and/or tritium, DD, DT) have properties characteristic for a given nuclear (fusion) reaction. For this reason, neutron measurements are a valuable measure of the course of nuclear reactions in plasma. The principal objective of neutron diagnostics is to characterize the fusion plasma under study as a neutron source. In deuterium plasma or deuterium–tritium mixture plasma mainly 2.5 and 14 MeV neutrons are generated from the nuclear reactions $^{2}H(^{2}H,n)^{3}He$ and $^{3}H(^{2}H,n)^{4}He$. Some neutrons can be generated from nuclear reactions between light ions (p,^{2}H,^{3}He) and plasma impurities: light plasma ions accelerated to high energies (> 1 MeV) produce neutrons over a broad energy range. Neutrons can be generated also from photonuclear reactions, as an effect of an acceleration of plasma electrons at high energies (> 10 MeV) which generate further (γ, n) reactions.

Neutron measurement methods that are possible in fusion reactors experiments (tokamaks, etc.) are restricted by the difficult access to the plasma vessel and a harsh environment caused by strong mixed radiation fields, strong magnetic fields and currents, powerful high frequency wave generators and power supply, heat loads, etc.

Diagnostics of fusion neutrons provide an indicator of plasma properties [1], like fusion power, ion temperature profile, deuterium and tritium energy spectra, incidence of fast particles, neutron source profile and fluence, fuel ion ratio (tritium-*to*-deuterium) in plasma core, etc. In particular, neutron data will be instrumental for the real-time control of fusion power in future reactors. Detection systems are built on the basis of neutron cameras, monitors,

✉ Urszula Woźnicka
Urszula.Woznicka@ifj.edu.pl

1 Institute of Nuclear Physics, Polish Academy of Sciences, Radzikowskiego 152, Kraków 31-342, Poland

spectrometers as well as fission chambers and neutron activation systems.

Since the neutrons are electrically neutral particles they have usually to be converted into charged ones to be detected. Generally, a neutron detector must be equipped with a converter (to convert neutron to detectable radiation) and with a conventional radiation detector (e.g. gaseous, semiconductor, scintillation). Two basic types of neutron interaction with matter are suitable for this conversion: neutron scattering and absorption. Neutron absorption reaction is a common method of neutron detection. This mainly bases on radiation capture and nuclear fission. The neutron is absorbed by target material (converter) emitting secondary particles such as protons, alpha or beta particles, photons or fission fragments. Some reactions are threshold reactions requiring a minimum energy of neutrons. The most common isotopes for the neutron converting material are presented in Table 1 with the corresponding cross-sections and abundances in the natural elements [2].

Although fissile isotopes have significantly lower fission cross-sections compared to the absorption cross-sections of other isotopes used as neutron converters (Table 1), they have other unique properties that can be used to detect neutrons. The fission fragments are heavy charged particles. That fission fragments create a high output signal, because they deposit a large amount of energy in a detector sensitive volume. This allows an easy discrimination of the background radiation, e.g. gamma ray.

- Uranium-lined ion chambers (*fission chambers*) are common neutron sensors employed to monitor intense sources of neutrons e.g. in fusion devices.

The method bases on registering specific products of fission reactions—*delayed neutrons*. The rate of *decay of delayed neutrons* can be registered by conventional neutron counters and next recalculated to the primary neutron flux which reached the fission sample. The method has application for neutron sources of very high fluxes.

- Samples of selected uranium isotopes can be used in *activation methods* to determine a neutron fluence or an energy range of the neutron flux, using selected threshold reactions.

Nuclear Fission Reaction and Fission Chamber for Neutrons Detection

In nuclear physics and nuclear chemistry, nuclear fission refers to either a nuclear reaction or a radioactive decay process in which the nucleus of an atom splits into smaller parts (lighter nuclei), often producing free neutrons and photons. Fission reaction is mainly induced by neutrons but also spontaneous radioactive decay (not requiring a neutron) is also referred to as fission, and occurs especially in very high mass number isotopes (See, e.g. [3–5]).

Some nuclides are capable of undergoing fission after capturing a neutron and the fission cross section on that reaction depends on the incident neutron energy (Table 1 and Fig. 1). Nuclides like ^{235}U, ^{233}U, and ^{239}Pu are fissioned with thermal neutrons, e.g. the fission cross section of ^{235}U is 575 b for incident neutron energy 0.025 eV and strongly goes down with neutron energy because the binding energy resulting from the absorption of a neutron is greater than the critical energy required for fission. Among them only ^{235}U is present in natural uranium (0.71%). By a contrast, the binding energy released by ^{238}U (with 99% of abundance) which absorbed a thermal neutron is less than the critical energy, so the neutron must be faster and having higher energy to realize a fission reaction. For ^{238}U the characteristic threshold of incident neutron energy is about 1 MeV and achieves value about 1 b for neutron energy (2.45–14) MeV. Figure 1 presents the fission cross sections in the function of neutron incident energy for ^{238}U, ^{235}U and for ^{232}Th.That last isotope characterizes by the sharp energy threshold on 500 keV and fission cross section in the neutron energy range (2.45–14) MeV is about 0.5 b. The presence of the energy threshold is a valuable feature

Table 1 Most common isotopes use as neutron absorbers applied in neutron detectors

Neutron induced reaction	Capture/fission cross section σ (b)	Abundance of the isotope in the natural element (%)
$^{10}_{5}B(n,\alpha)^{7}_{3}Li$	3837(thermal neutrons)	19.9
$^{3}_{2}He(n,p)^{3}_{1}H$	5333(thermal neutrons)	0.0137
$^{6}_{3}Li(n,\alpha)^{3}_{1}H$	940 (thermal neutrons)	7.5
$^{113}_{48}Cd(n,\gamma)^{114}_{48}Cd$	20,615 (thermal neutrons)	12.22
$^{235}_{92}U(n, \text{fission})$	~ 582 (thermal neutrons)	0.72
$^{238}_{92}U(n, \text{fission})$	~ 1 (2.45–14) MeV	99.27
$^{232}_{90}Th(n, \text{fission})$	~ 0.5 (2.45–14) MeV	100

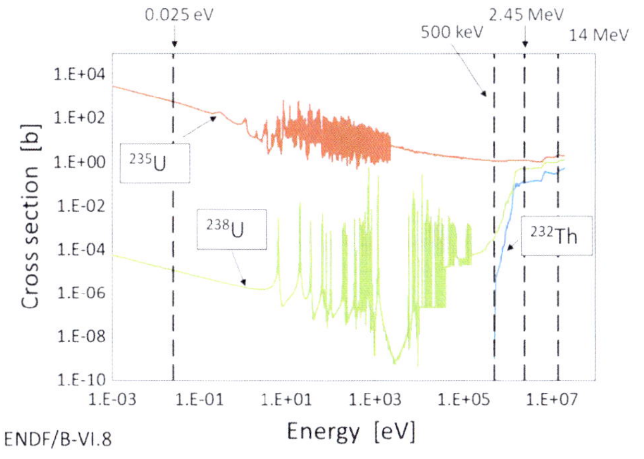

Fig. 1 Fission cross-sections for ^{235}U, ^{238}U and ^{232}Th from ENDF/B-VI.08 library at 300 K [6]. Dashed lines indicate characteristic energy levels of neutrons: thermal neutrons (0.025 eV), energy threshold for ^{232}Th (500 keV), neutron energy for DD reaction (2.45 MeV), and neutron energy for DT reaction (14 MeV)

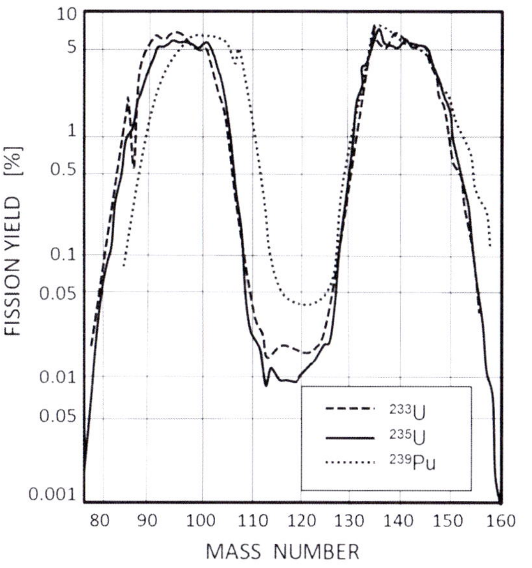

Fig. 2 Characteristic course of the two-hump distribution curve of fission fragments (after exposure with thermal neutrons) with a visible shift for ^{239}Pu in relation to ^{233}U and ^{235}U

for the detection of fast neutrons. Both ^{238}U and ^{235}U isotopes are characterized by the presence of multiple resonant lines in the area below 500 keV. This should be taken into account when designing systems for fast neutron measurements with these isotopes and provide adequate shields against neutrons from the resonant region.

Any combination of lighter nuclei can be produced during neutron induced fission reaction, but the most common event is a slightly unequal fission in which one daughter nucleus has a mass of about 90 to 100 u (u—atomic mass unit) and the other the remaining 130–140 u [4]. Unequal fissions are energetically more favorable because it allows one product to be closer to the energetic minimum near mass 60 u (only a quarter of the average fissionable mass), while the other nucleus with mass 135 u is still not far out of the range of the most tightly bound nuclei. Distribution of the mass of products depends on the energy of incident neutrons and on the kind of the parent nucleus. The much known two-hump curves of mass distributions of fission fragments after exposure of ^{233}U ^{235}U and ^{139}Pu with thermal neutrons are presented in Fig. 2. It is noteworthy that the mass distributions differ slightly for different isotopes. These differences can be important in the design of fission detectors in which the signal is formed by ionized fission products.

The neutron induced fission reactions can be directly used for a neutron detection via fission products. They carry high kinetic energy, about 160 MeV. Thank to this, the fission detector is characterized by a very strong signal originating from fission products and weak disturbances from any competitive reactions or other events generating the measurement background.

The most common form of fission detector [7] is a cylindrical ionization chamber which consists of a pair of electrodes, one of this is coated with the fissile deposit, e.g. ^{235}U, ^{238}U or ^{242}Pu. The detector is filled with the inert gas, mainly with argon and small other additions. Fission chambers can operate in pulse and/or current mode depending on the dynamic range of measurement. The signal of a fission chamber depends primarily on the fissile deposit thickness and the geometric conditions under which the fission fragments are collected. In the case of deposits that are very thin compared with fragments range, a double-humped fission fragments energy spectrum is observed as the detector response with maxima at approx. 70 and 100 MeV, with a shape analogous to that shown in Fig. 2. If the deposit is made thicker to enhance detection efficiency, the energy loss of fragments within the deposit will reduce the average fragment energy and distort the shape of measured distribution.

Fission chambers are robust neutron detectors (can operate under high temperatures and under electromagnetic loads). Normally highly enriched ^{235}U is used for the coating of fission chamber electrodes, which makes fission chamber sensitive mainly to thermal neutrons. Another coatings of threshold fissionable isotopes, e.g. ^{238}U can be used to detect fast neutrons. It is expected that large devices dedicated to thermonuclear fusion research, such as ITER [1, 8] will be equipped with fission detectors to measure of neutron yield of DD od DT fusion plasma. Knowledge of neutron source strength is essential for measurement of a total fusion power generated by tokamak. Fission detectors will be the important diagnostic for ITER operation and the neutron emission rate measured by fission chambers will be used as a feedback parameter for fusion output control in ITER.

For example, a set of fission chambers were used for neutron flux measurements in TFTR [9] tokamak fusion plasma experiments. Both types of fissionable materials threshold (^{238}U, ^{232}Th) and without threshold (^{235}U) were used in the detectors for neutron flux measurement. ^{235}U fission chambers were covered by neutron moderators to improve sensitivity and detector response function. The different mass of fissionable material was used to cover the measurement possibilities for a wide range of neutron yield (from 10^{10} to $> 10^{19}$ n/s). The ^{235}U amount in two most sensitive fission chambers was 1.3 g. The other two fission chambers with ^{235}U contained 0.01 g of fissile material with $\sim$ 130 times lesser sensitivity than most sensitive detectors. Two thresholds fission chambers contained 1.3 and 0.3 g ^{238}U and their sensitivities were differing in 4 times than each other. With these detectors the measurements of neutron strength uncertainty have been obtained on the level of $\pm$ 13%.

Also on JT-60U [10] and JET [11] tokamaks the time-resolved neutron yield has been measured with fission neutron chambers. A wide range of the neutron flux, typically 9 or 10 decades, has been measured. ^{235}U detector is used in the low and medium neutron yield discharges and the ^{238}U detector is used in the high neutron yield discharges such as high-power NB heating experiments.

Delayed Neutrons of Nuclear Fission

The fission products are usually unstable due to being relatively neutron-rich nuclei and many of them are subject to beta decay. This releases additional energy in the form of beta particles, antineutrinos and gamma rays. In some cases, available energy in a highly excited state is enough to emit neutrons (delayed neutrons) instead of gamma emissions.

For example, ^{235}U absorbs thermal neutron and forms ^{236}U nucleus which fissions into two large fission fragments (Eq. 1). There are several hundreds of ways for the ^{236}U compound nucleus to fission. These fragments emit, on average, two or three free neutrons (in average 2.47), called *prompt* neutrons, for example:

$$^{235}_{92}U + n \rightarrow ^{236}_{92}U^* \rightarrow ^{90}_{36}Kr \rightarrow ^{143}_{56}Ba + 3n \tag{1}$$

The primary fission fragments (^{90}Kr and ^{143}Ba) have excess of neutrons as compared to the stable nuclei with the same atomic number. In most cases they undergo several successive β^- transitions to adjust their neutron/proton ratio. The fragment-pair above decays as follows:

$$^{90}_{36}Kr \xrightarrow[33s]{\beta^-} ^{90}_{37}Rb \xrightarrow[2.7min]{\beta^-} ^{90}_{38}Sr \xrightarrow[28y]{\beta^-} ^{90}_{39}Y \xrightarrow[64h]{\beta^-} ^{90}_{40}Zr \text{ (stable)} \tag{2a}$$

$$^{143}_{56}Ba \xrightarrow[30s]{\beta^-} ^{143}_{57}La \xrightarrow[12min]{\beta^-} ^{143}_{58}Ce \xrightarrow[33h]{\beta^-} ^{143}_{59}Pr \xrightarrow[13.7d]{\beta^-} ^{143}_{60}Nd \text{ (stable)} \tag{2b}$$

Some fission fragments occasionally undergo a stage of radioactive decay that yields a new nucleus in an excited state that emits an additional neutron, called a delayed neutron, to get to ground state, e.g.:

$$^{87}_{35}Br_{52} \xrightarrow{\beta} ^{87}_{36}Kr^*_{51} \rightarrow ^{86}_{36}Kr_{50} + n \tag{3}$$

These fission fragments which emit the neutron are called delayed-neutron precursor atoms. More than 99% of *prompt* neutrons are emitted very shortly (within 10^{-12} s) after the fission. Delayed neutrons are emitted long after the fission (sometimes several minutes later) with a yield of about 1%. The decay chain is summarized as follows:

$$^{A}_{Z}X_N \xrightarrow{\beta} ^{A}_{Z+1}Y^*_{N-1} \rightarrow ^{A}_{Z+1}Y_{N-2} + n(\textit{delayed neutron}) \tag{4}$$

The X nucleus is a delayed-neutron *precursor*, and the Y nucleus is a delayed-neutron *emitter*. Obviously, for these neutrons the 'delay time' is determined by the half-life of the *precursor* nucleus X, which can be quite large, since the β^- decay is governed by the weak interaction. The main precursors of delayed neutrons during fission reaction are collected in Table 2 with the corresponding half-lives and abundances [12].

Properties and Parameters of Delayed Neutrons

An interesting consequence of the decay chain is that the excitation energy of the delayed-neutron emitter nucleus is usually much lower than the excitation energies of the direct fission fragments. Therefore, the average energy of the delayed neutrons is also smaller (300 $\div$ 600 keV) than that of the prompt neutrons (2 MeV in average).

Examples of energy distributions of the delayed neutrons emitted after fission in ^{232}Th, ^{238}U and ^{235}U calculated using MCNP [13, 14] code are presented in Fig. 3. The simulations have been performed for flat cylinders (2R = 1.8 cm, H = 0.25 cm) of the above pure isotopes, irradiated by the 2.45 MeV neutrons. Delayed neutrons emitted after 10 s of the irradiation have been counted. The current of neutrons through the target surface has been obtained, containing neutrons of energy from 10^{-10} MeV to 2.5 MeV. The "Normalized emission" label in the plots means the neutron current through the target surface in a given neutron energy interval normalized to the total

Table 2 The main fission fragments (*precursors*) of delayed neutrons

Precursor	Half-life of the precursor nucleus (s)	Precursor abundance
Br-87	55.6	0.033
I-137	24.5	0.178
Br-88	16.3	0.111
I-138	6.46	0.046
Rb-93	5.93	0.024
Br-89	4.38	0.101
Rb-94	2.76	0.162
I-139	2.30	0.046
As-85	2.08	0.107
Y-98 m	2.00	0.088
Kr-93	1.29	0.0048
Cs-144	1.00	0.0070
I-140	0.86	0.0052
Br-91	0.542	0.017
Rb-95	0.384	0.049
Rb-96	0.203	0.017
Rb-97	0.170	0.0052

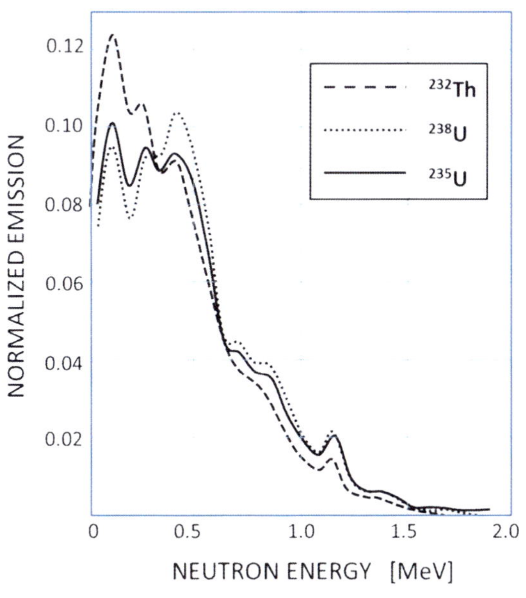

Fig. 3 Energy distributions of the delayed neutrons emitted after fissions in selected isotopes

current (in the whole energy range) per one starting source neutron.

More than 66 delayed neutron precursor nuclei are identified so far. Their half-lives range between 0.12 and 78 s, therefore their delayed neutrons appear with considerably differing delay times. The time after which the delayed neutron appears is determined by the life of the precursor. Observation of delayed neutrons as a function of time shows their multi-exponential nature.

Among variety of possible mathematical representations of the delayed neutrons decay, a six-group model is commonly used [15, 16]. Tables 3 and 4 present parameters of 6 group model of delayed neutrons after irradiation of ^{238}U and ^{232}Th by neutrons with energy of about 2 MeV [12]. The corresponding decay curves and its distribution into 6 exponents with different decay constants are shown in Fig. 4a, b. It is evident that the curves for different isotopes slightly differ from each other (see the summary shown in the Fig. 4c for ^{238}U and ^{232}Th).

A general decay curve describing the number of particles (neutrons) after time t is an exponential function:

$$N(t) = N(0)e^{-\lambda t} \tag{5}$$

where $N(0)$ is the number of particles in $t = 0$ and λ is the decay constant.

In order to obtain the decay rate of particles, one should differentiate Eq. (5) over time:

$$\frac{dN(t)}{dt} = -\lambda N(0)e^{-\lambda t} \tag{6}$$

Based on the above relationship the decay rate $R(t)$ of delayed neutrons can be written as:

$$R(t) = \frac{dN}{dt} = C\sum_{i=1}^{n} \beta_i \lambda_i e^{-\lambda_i t} \tag{7}$$

where β_i—contribution of delayed neutrons for the ith group of precursors, λ_i—decay constant of the ith group of delayed neutrons, n—number of groups, C—constant.

Table 3 Parameters of 6 group model of delayed neutrons after irradiation of ^{238}U by neutrons with energy of about 2 MeV. Number of neutrons per one fission event, $\nu = 2.805$, number of delayed neutrons per one fission event, $\nu_d = 0.0460$

Number of a group i	Time of a half-life $(T_{1/2})_i$ (s)	Decay constant λ_i (1/s) $\lambda_i = \ln(2)/(T_{1/2})_i$	Relative abundance β_i
1	52.38 ± 1.29	0.01323 ± 0.00033	0.013 ± 0.003
2	21.58 ± 0.39	0.03212 ± 0.00043	0.137 ± 0.005
3	5.00 ± 0.19	0.1386 ± 0.00537	0.162 ± 0.016
4	1.93 ± 0.07	0.359 ± 0.013	0.388 ± 0.007
5	0.49 ± 0.23	1.415 ± 0.066	0.225 ± 0.008
6	0.172 ± 0.009	4.03 ± 0.21	0.075 ± 0.003

Table 4 Parameters of 6 group model of delayed neutrons after irradiation of ^{232}Th by neutrons with energy of about 2 MeV. Number of neutrons per one fission event, $\nu = 2.616$, number of delayed neutrons per one fission event, $\nu_d = 0.0531$

Number of a group i	Time of a half-life $(T_{1/2})_i$ (s)	Decay constant λ_i (1/s) $\lambda_i = \ln(2)/(T_{1/2})_i$	Relative abundance β_i
1	56.03 ± 0.95	0.01237 ± 0.00021	0.034 ± 0.002
2	20.75 ± 0.66	0.0334 ± 0.0011	0.150 ± 0.005
3	5.74 ± 0.24	0.1208 ± 0.0050	0.155 ± 0.021
4	2.16 ± 0.08	0.3209 ± 0.012	0.446 ± 0.015
5	0.571 ± 0.042	1.2139 ± 0.089	0.172 ± 0.013
6	0.211 ± 0.019	3.285 ± 0.30	0.043 ± 0.006

Activation Method of Neutron Measurements Based on Delayed Neutrons

The activation method using delayed neutrons has been successfully applied in large fusion devices [11, 13, 17–19]. This variant of the activation method is implemented on JET together with the standard activation method based on gamma-spectrometry. The measurement set-up consists of a steal vessel with cadmium shield filled with polythene moderator (40 cm in diameter). The central hole is used to enter the activated sample through the pneumatic rabbit system end. The six ^{3}He proportional counters (one-inch diameter, 15 cm sensitive length) surround the hole (at a radial distance of 8 cm), forming a hexagonal array. [19].

Unlike the indium used in gamma-spectrometry based activation method, the fissionable samples can be irradiated in each plasma discharge, one after another, due to short half-lives of delayed neutron precursors. In the system at JET [19], the sample of fission material is positioned in the irradiation end before the discharge and removed few seconds after the discharge is terminated. The rabbit system transfers the activates sample to the measurement set-up in 15 s. The counting of the detectors starts together with discharge start and sampled at 500 ms intervals, after sample arrival the neutron signal is registered. Such a system gives information about the integral number of delayed neutrons, and the time decay curve was not analyzed. In the treatment of the ^{238}U measurements, the 18% contribution of the small ^{235}U contamination (0.4%) has been taken into account. The detection efficiency has been established to be (14.6 ± 0.28)% for ^{238}U.

In the Institute of Nuclear Physics PAN, an experimental set-up for delayed neutron measurements was built and dedicated to activation methods applicable for fusion devices. A measuring chamber was designed, optimized with use of Monte Carlo simulations, and constructed [13, 20–22]. The device (called DET-12) is equipped with the set of twelve neutron detectors and the relevant system of data acquisition and digitizer allowing to determine the time decay curve of delayed neutrons. Figure 5 shows a general view of the DET-12 device and its components are specified in Fig. 6. The construction and dimensions of the individual components have been optimized in order to obtain the best performance of the device and the best possible signal-to-noise ratio. The background has been eliminated both from the scattered neutrons inside the device and from neutrons coming from the outside.

In the measuring system of the DET-12 device, the GaGe Octopus 8240CompuScope digitizer card was used. (http://www.gage-applied.com/digitizers/12-bit/pci/compuscope-octopus-82xx.htm, 22.05.2018). The card acts as a

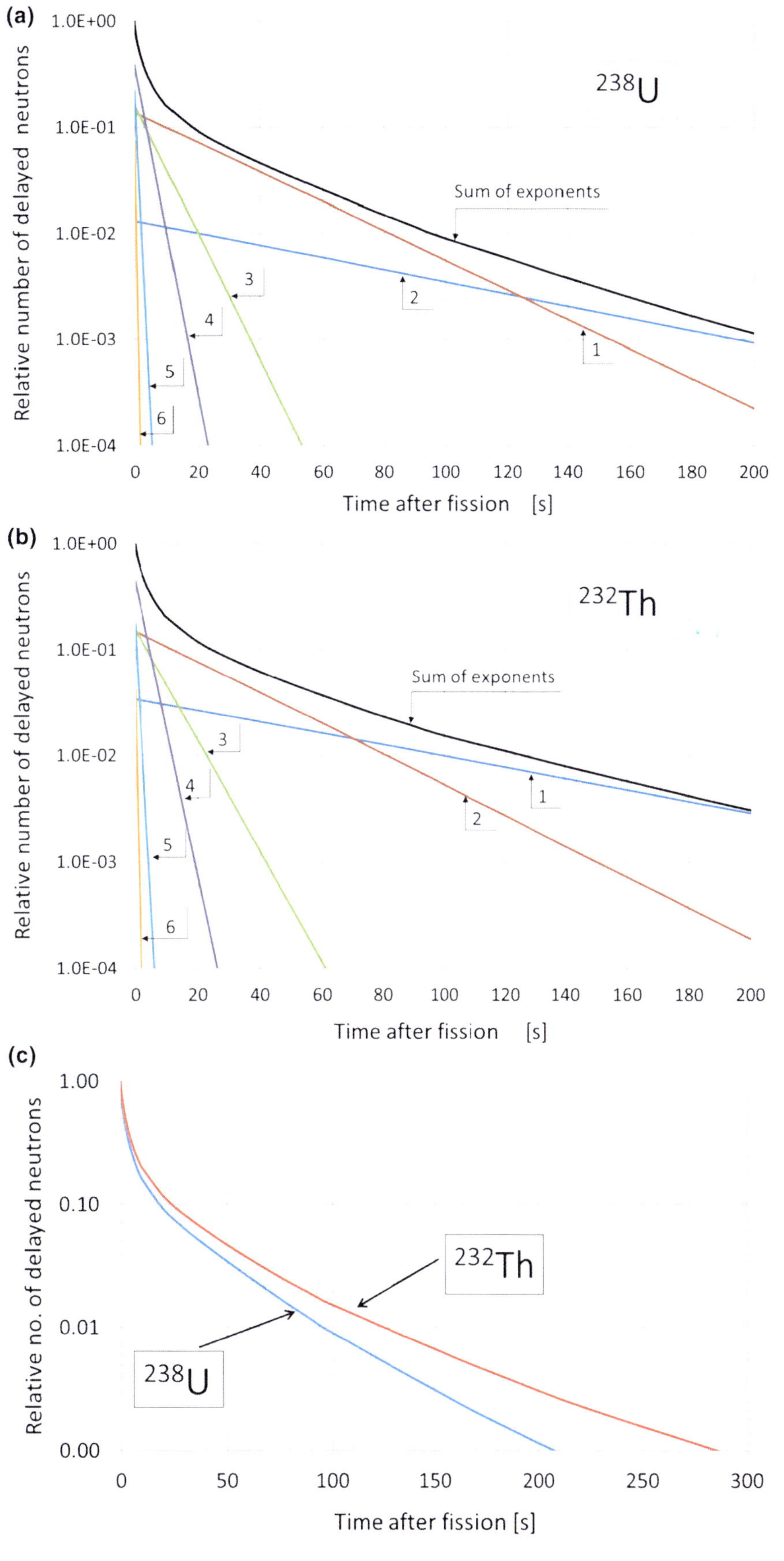

Fig. 4 Time decay curves of delayed neutrons from ^{238}U (**a**) and ^{232}Th (**b**) fission by 2 MeV neutrons. Decay curves from each group (1–6) of delayed neutrons are marked. **c** Total decay curves of delayed neutrons for ^{238}U (**a**) and ^{232}Th are presented together for the visualization of differences in decay rates

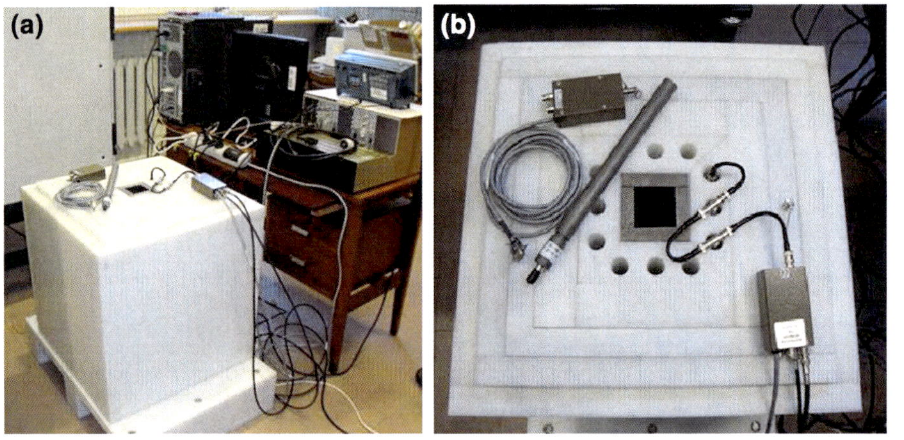

Fig. 5 General view of the DET-12 device for delayed neutrons measurements from fissionable samples. **a** Sample and detectors container (white), external electronics (on the right), **b** Container without the top cover

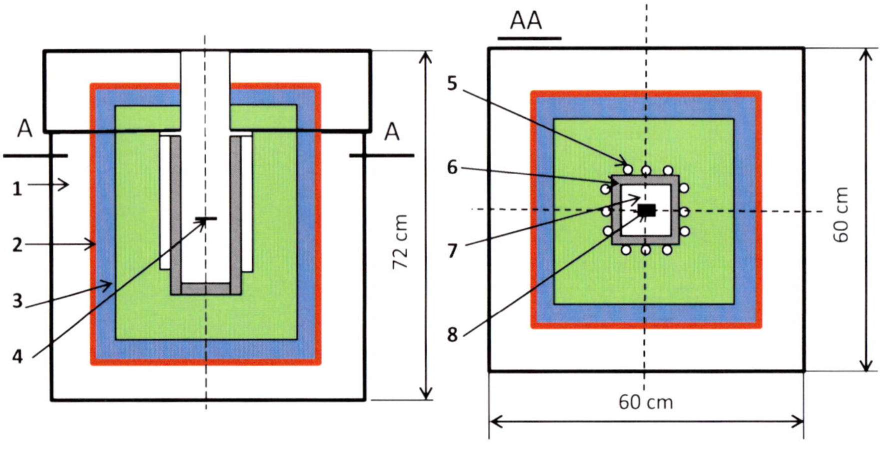

Fig. 6 Schematic cross section of the DET-12 setup for the delayed neutron detection (layers not in scale). 1 Borated polyethylene, 2 Cadmium layer, 3 Boron Carbide, 4 Polyethylene, 5 Detector holder, 6 Bismuth, 7 empty space (air), 8 Irradiated sample position

recorder of all impulses from helium detectors. It allows to record the shape of pulses, register the amplitude and the time in which they were recorded (the so-called "time stamp", allowing to arrange the pulse according to its time of appearance from the moment of registration). Built-in extended memory allows to record all pulses with maximum sampling up to several hundred seconds from the start of the digitizer. This allows later to accurately analyze the pulses in terms of their appearance time and the ability to assign a given pulse to the time channel [23].

Time decay curve of delayed neutrons generated in the target-sample one can measure in time analyzer with Δt channel width. Number of counts in channel *k* one calculate from Eq. (7):

$$N(t_k) = C\sum_{i=1}^{n}\beta_i\lambda_i\int_{t_k}^{t_k+\Delta t}e^{-\lambda_i t}dt \tag{8}$$

this gives

$$N(t_k) = C\sum_{i=1}^{n}\beta_i\left(1-e^{-\lambda_i\Delta t}\right)e^{-\lambda_i t_k} \tag{9}$$

where $C = \varepsilon N_0\ P$, ε—detector efficiency, N_0—number of primary neutrons reaching the sample-target, [n/cm^2], *P*—material parameter: product of macroscopic fission cross section, Σ_f, and number of delayed neutrons per one fission event, ν_d.

For the target-sample which is a single isotope (e.g. ^{238}U or ^{232}Th) material parameter is:

$$P = \nu_d\sigma_f\frac{N_A}{A}m \tag{10}$$

where σ_f—microscopic fission cross section (cm^2), N_A—Avogadro number, A—atomic number, *m*—mass of the sample-target (g)

The purpose of the measurement is to determine the number of neutrons falling on a sample. This number can be estimated from the number of counts in a single channel of the formula (9):

$$N_0 = \frac{N(t_k)}{\varepsilon P\sum_{i=1}^{n}\beta_i\left(1-e^{-\lambda_i\Delta t}\right)e^{-\lambda_i t_k}} \tag{11}$$

The above-described constant *C* corresponds to the ideal analytical solution describing the decay curve of

Fig. 7 An example of the calculated delayed neutrons decay curve (Eq. 9) measured in time intervals $\Delta t = 1$ s for ^{238}U sample

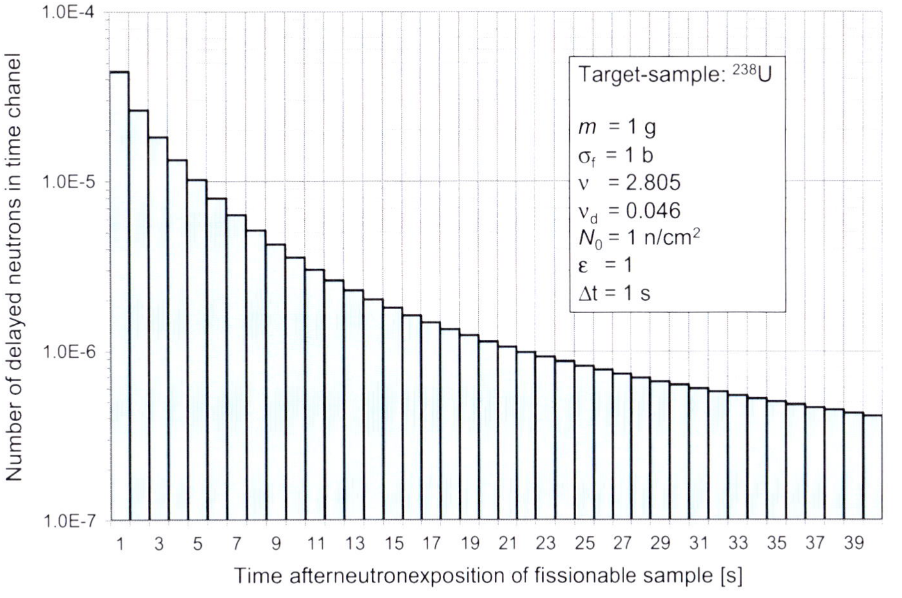

the delayed neutron flux. In fact, the calibration constant C depends not only on the efficiency of detection, the macroscopic cross-section on the split, the size and detailed elemental composition of the target, but also on the entire experimental environment. For a specific application of the DET-12 system, it will be necessary to carry out detailed Monte Carlo numerical modeling that reproduces the measurement conditions at the fusion device, supplemented by calibration measurements. The latter can be carried out in laboratory conditions using an isotopic source. Detailed procedures related to the adaptation of DET-12 to the actual measurement conditions at the fusion device are presented in paper [23].

An example of the calculated decay curve registered in $\Delta t = 1$ s channels obtained using the above formulae is presented in Fig. 7. A rapid drop in counts over time means that the measurement of delayed neutrons should be performed as soon as possible after exposure. This effect is a disadvantage (very fast drop in count statistics) and advantage (immediate experimental result, the registration of the decay curve takes few minutes at most) of the activation method using delayed neutrons. Target, which is irradiated with a primary neutron flux, is placed as close as possible to the tokamak's chamber and after the exposure has to be transported by pneumatic mail to the place of measurement. From this it follows that a certain delay between the measurement time and the exposure time is inevitable. On the other hand, the final result of the measurement is obtained very quickly after irradiation. This, compared to the classical activation method, is an advantage, as often isotopes used in classical activation method require a long cooling time before the final spectrometer gamma measurement.

Summary

Fission chambers and the activation method which uses *delayed neutrons* as measurement methods for determination of the yield of a fusion neutron source are successfully applied (and will be applied) in large fusion devices. The main feature of fission chambers are very small dimensions, a wide range of neutron yield which can be measured and extremely low backgrounds from gamma rays and from other competing reactions. The neutron delayed activation method can be a good method of detecting very high neutron fusion fluxes due to the relatively small number of delayed neutrons which are registered. Compared to the classical activation method, the final measurement result (fusion neutron fluence) can be obtained in a shorter time after the target has been activated. Targets for the delayed neutron method must be measured as soon as possible after irradiation. In the case of classical activation, it is often necessary to cool the target in order to extinguish the high activity of short-lived isotopes, which extends the total measurement time. Knowledge of the physics of the heavy elements fission phenomenon is undoubtedly very helpful in the selection of the right measurement way both in the case of fission chambers and of the activation method that use the presence of delayed neutrons.

Acknowledgements Lecture on Summer School of Plasma Diagnostics: PhDiaFusion 2017. 'Neutron and gamma diagnostics for fusion plasma'. 11-15 September 2017, Podlesice, Poland. https://phdia2017.ifj.edu.pl/. The work was performed as part of the Institute of Nuclear Physics, Polish Academy of Sciences Task Number S61414: 'High-temperature plasma diagnostics'. Thank you very much to my colleague Eng. Andrzej Igielski for helpful discussions and comments during the preparation of the article.

References

1. A.V. Krasilnikov, M. Sasao, YuA Kaschuck, T. Nishitani, P. Batistoni, V.S. Zaveryaev, S. Popovichev, T. Iguchi, O.N. Jarvis, J. Källne, C.L. Fiore, A.L. Roquemore, W.W. Heidbrink, R. Fisher, G. Gorini, D.V. Prosvirin, AYu. Tsutskikh, A.J.H. Donné, A.E. Costley, C.I. Walker, Nucl. Fusion **45**, 1503 (2005). https://doi.org/10.1088/0029-5515/45/12/005
2. S.F. Mughabghab, *Atlas of Neutron Resonances*, 5th edn. (Elsevier, Amsterdam, 2006)
3. K.H. Beckurts, K. Wirtz, *Neutron Physics* (Springer, Berlin, 1964). ISBN 978-3-642-87614-1
4. A.M. Weinberg, E.P. Wigner, *The Physical Theory of Neutron Chain Reactors* (The University of Chicago Press, Chicago, 1958)
5. G.D. James, J.E. Lynn, A. Michaudon, *Nuclear Fission and Neutron-induced Fission Cross-sections* (Pergamon Press, Oxford, 1981). ISBN 9781483189765
6. ENDF-6 Formats Manual Data Formats and Procedures for the Evaluated Nuclear Data Files ENDF/B-VI and ENDF/B-VII and https://www-nds.iaea.org/exfor/endf.htm.
7. G.F. Knoll, *Radiation Detection and Measurement*, 3rd edn. (Wiley, Hoboken, 2000)
8. M. Ishikawa, T. Kondoh, T. Nishitani, Y. Kusama, J. Plasma Fusion Res. Ser. **8**, 334–337 (2009)
9. H.W. Hendel, R.W. Palladino, Cris W. Barnes, M. Diesso, J.S. Felt, D.L. Jassby, L.C. Johnson, L.P. Ku, Q.P. Liu, R.W. Motley, H.B. Murphy, J. Murphy, E.B. Nieschmidt, J.A. Roberts, T. Saito, J.D. Strachan, R.J. Waszazak, K.M. Young, Rev. Sci. Instrum. **61**, 1900 (1990). https://doi.org/10.1063/1.1141115
10. T. Hayashi, T. Nishitani, A.M. Sukegawa, M. Ishikawa, K. Shinohara, Rev. Sci. Instrum. **79**, 10E506 (2008)
11. O.N. Jarvis, Plasma Phys. Control. Fusion **36**, 209 (1994). https://doi.org/10.1088/0741-3335/36/2/002
12. Delayed Neutron Data for the Major Actinides. A report by the Working Party on International Evaluation Co-operation of the NEA Nuclear Science Committee, NEA/WPEC-6, OECD PUBLICATIONS. https://www.oecd-nea.org/science/wpec/volume6/volume6.pdf. Accessed 2002
13. B. Bieńkowska, K. Drozdowicz, B. Gabańska, A. Igielski, R. Prokopowicz, U. Wiącek, U. Woźnicka, Physical assumptions for a design of the DET-12 chamber for detection of delayed neutrons. Rept. IFJ No. 2074/AP, Kraków, 2014. https://www.ifj.edu.pl/publ/reports/2014/2074.pdf?lang=pl. Accessed 2002
14. X-5 Monte Carlo Team, MCNP: a general Monte Carlo N-particle transport code. Version 5, Rept. LA-UR-03-1987, Los Alamos National Laboratory, 2003
15. G.R. Keepin, T.F. Wiemett, R.K. Zeigler, Phys. Rev. **107**(4), 1044 (1957)
16. G.R. Keepin, *Physics of Nuclear Reactors* (Addison-Wesley Publ. Co., Massachusetts, 1965)
17. M.T. Swinhoe, O.N. Jarvis, Nucl. Instrum. Meth. Phys. Res. **221**(2), 460 (1984). https://doi.org/10.1016/0167-5087(84)90020-6
18. H. Tourwe, A. Fabry, P. D'hondt, Delayed neutron counting system for JET plasma neutron yields diagnostics, in *Reactor Dosimetry*, ed. by J.P. Genthon, H. Röttger (Springer, Dordrecht, 1985), pp. 309–315. https://doi.org/10.1007/978-94-009-5378-9_30
19. P. van Belle, O.N. Jarvis, G. Sadler, S. de Leeuw, P. D'Hondt, M. Pillon, Rev. Sci. Instr. **61**, 3178 (1990). https://doi.org/10.1063/1.1141679
20. G. Tracz, B. Bieńkowska, K. Drozdowicz: Detection of delayed neutrons from neutron activation of fissionable substance samples. Monte Carlo modelling of response of the DET-12 device. Rept. IFJ No. 2061/PN, Kraków, 2013. https://www.ifj.edu.pl/publ/reports/2014/2061.pdf?lang=pl. Accessed 2002
21. B. Bieńkowska, K. Drozdowicz, B. Gabańska, A. Igielski, W. Janik, A. Kurowski, G. Tracz, U. Wiącek, J. Dankowski, Technical design and operation tests of the DET-12 device for detection of delayed neutrons. Rept. IFJ No. 2075/AP, Kraków, 2014. https://www.ifj.edu.pl/publ/reports/2014/2075.pdf?lang=pl. Accessed 2002
22. A. Kurowski, B. Bieńkowska, K. Drozdowicz, B. Gabańska, A. Igielski, R. Prokopowicz, G. Tracz, U. Wiącek, U. Woźnicka, J. Fusion Energ. **37**, 120–123 (2018). https://doi.org/10.1007/s10894-018-0156-3
23. U. Wiącek, U. Woźnicka, J. Fusion Energ. **37**, 120–123 (2018). https://doi.org/10.1007/s10894-018-0156-3
24. A. Kurowski, The method of measuring neutron fluxes from the synthesis reaction in thermonuclear reactors by detecting delayed neutrons from the activation of fissile materials. Design and construction of the measurement system. Ph.D. thesis (in Polish), Institute of Nuclear Physics, Polish Academy of Sciences, Kraków, 2018

Journal of Fusion Energy (2019) 38:386–393
https://doi.org/10.1007/s10894-018-0195-9

ORIGINAL RESEARCH

Neutron Diagnostics for Tokamak Plasma: From a Plasma Diagnostician Perspective

J. Bielecki[1] · A. Kurowski[1]

Published online: 8 October 2018

Abstract
Neutron measurement is considered to be one of the major diagnostic methods in tokamak plasma. The neutron diagnostics provide key information on plasma physics, machine protection and control issues. In this article, the commonly used techniques of neutron emission diagnostics for tokamak plasma are presented. The requirements and limitations imposed on the diagnostic systems are discussed. Several basic techniques of total neutron yield and neutron spectrum measurements such as neutron counters, neutron activation technique, time-of-flight spectroscopy or magnetic proton recoil spectroscopy are described in a practical way, from a plasma diagnostician point of view. Primary physical quantities that can be derived from those measurements are pointed out and discussed.

Keywords Neutron diagnostics · Plasma · Tokamak · Plasma tomography · Neutron counters · Neutron activation · Time of flight spectroscopy · Plasma focus

The Importance of Neutron Diagnostics

In order to obtain a comprehensive insight into the processes taking place in the tokamak fusion plasma it is necessary to simultaneously determine several plasma parameters. Preferably it should be done with temporal and spatial resolutions smaller than the typical time and length scales of the investigated phenomena. Due to the extreme heat, no physical measurement devices can be placed directly inside the plasma. However, since the thermonuclear plasma is a source of electromagnetic and particle radiation, the plasma diagnostics rely on measurements of the emission, transmission or scattering of particles or electromagnetic radiation. Fusion plasma diagnostics is an extensive subject itself, and constitutes a large and important part of fusion experiments. In this article, we focus on neutron diagnostics which play a major role in plasma research on tokamak machines. The subject is presented in a pragmatic way from a plasma diagnostician point of view.

✉ J. Bielecki
jakub.bielecki@ifj.edu.pl

[1] Institute of Nuclear Physics Polish Academy of Sciences (IFJ PAN), 31-342 Kraków, Poland

The role of a plasma neutron diagnostician is to characterize the plasma under study as a source of neutrons [1]. To do that, tokamak devices are equipped with several neutron diagnostics that vary in terms of purpose of application and technical parameters. In fact, measurements of the neutron emission have been carried out since the beginning of thermonuclear fusion research.

In the most promising of the hydrogen fusion reactions, when a mixture of deuterium and tritium is used, about 80% of fusion energy yield is carried by the neutron. Thus, the number of produced neutrons is a direct measure of the progress toward the achievement of controlled thermonuclear reactor. Nevertheless, this is not the only reason why the neutron measurements are of special interest for tokamak plasma. Neutron diagnostics play also a crucial role in study of physical phenomena occurring in tokamak plasmas and will play a prominent role in the control of the next-step thermonuclear power plants. Indeed, since neutrons originate in the fusion reaction, they carry information on the properties of the ions taking part in the reaction. As neutrons do not have the electric charge, they can easily escape from the tokamak magnetic field and be registered by a diagnostic system. The important information on ion species partaking in or affecting the fusion process can be extracted from the neutron source strength or the neutron energy spectrum. Study of the neutron spectrum can provide information on ion temperature and ion density. Neutron emission can be also useful to monitor plasma position, its extend and shape. This requires measurements of spatial emission profiles, so that neutron emissivity distribution can be reconstructed using special tomographic methods.

A lot of effort has been devoted to develop reliable methods of neutron measurement and to improve the accuracy of existing experimental techniques of neutron measurements. A part of neutron measurement techniques that are commonly used in nuclear fusion devices are based on concepts originally developed for fission reactors and experimental neutron physics. However, significant differences between the techniques that are directly applicable for fusion research and those used in conventional fission reactors are defined by the requirements and limitations on the neutron measurements in the thermonuclear plasma environment. This subject will be discussed in "Requirements and Limitations of Neutron Diagnostics" section.

In the currently operational tokamaks the reactions of the primary interest are: $D(d,n)^3He$, $D(d,p)T$, and $D(t,n)^4He$. The first two reactions are almost equally probable branches of fusion reaction involving deuterium fuel. The neutrons produced in deuterium fusion reactions have energy of $E_{n0} = 2.45$ MeV while the neutrons from deuterium–tritium reaction have energy of $E_{n0} = \sim 14.1$ MeV. This allows us to focus mostly on this energy ranges. However, the observation of the neutron emission in tokamak experiments is not uniquely related to the production in thermonuclear reactions. The accelerated ions ($^3He,p,d,t$) of energy greater than 1 MeV are present in the plasma. They can interact with plasma impurities e.g. with Be intrinsically present in the plasma due to plasma interaction with Be walls of modern tokamaks. This leads to emission of energetic neutrons from reactions: $^9Be(^3He,n)^{11}C$, $^9Be(\alpha,n)^{12}C$, $^9Be(p,n)^9B$, $^9Be(d,n)^{10}B$, $^9Be(t,n)^{11}B$ [2]. Moreover, photoneutrons can be produced when high energy runaway electrons strike the vacuum vessel structures depositing a considerable amount of energy in a few tens of milliseconds [3]. These high-energy electrons are generated by radio-frequency heating e.g. electron cyclotron resonance heating (ECRH). Non-fusion neutrons are also produced in disintegration of deuterons by these energetic electrons in the reaction $D(e',en)H$ or by bremsstrahlung photons in the reaction $D(\gamma,n)H$. These photoneutrons can be also detected by tokamak neutron diagnostics. Finally, the neutron diagnostics should be able to register scattered fusion neutrons of lower energy. In order to accomplish all the required measurements, modern tokamaks are equipped with numerous complementary neutron diagnostics.

Requirements and Limitations of Neutron Diagnostics

The conditions in which neutron diagnostics have to operate are very severe. This is mostly due to a wide range of plasma temperature (from several eV up to tens of keV) and harsh radiation environment (high fluxes of neutrons, gamma and X-ray radiation). Thus, neutron diagnostics suffer from radiation damage and can be exposed to thermal and mechanical stress. Since the observed neutron flux can vary by a few orders of magnitude, the neutron detectors should have high dynamic range. Another aspect that should be taken into consideration is the magnetic field. If a neutron diagnostic uses magnetic field, an appropriate shielding has to be designed and constructed. The diagnostic magnetic field must not disrupt the tokamak magnetic field and vice versa, the tokamak field should not interfere with the diagnostic. For example, magnetic proton recoil (MPR) neutron spectrometer at Joint European Torus (JET) needs a heavy shielding against the tokamak powerful magnetic fields to avoid interference [4]. An additional challenge in design of plasma neutron diagnostics is a limited access to the plasma. Usually, the space for diagnostics is severely limited and only a small number of ports is available. The complications in precise spectroscopic measurements are also related to the electromagnetic noise from the tokamak high-power auxiliary systems (e.g. Neutral Beam Injection—NBI, Ion Cyclotron Resonance Heating—ICRH or lasers of the Light Detection and Ranging—LIDAR system). This electromagnetic noise can interfere with a neutron diagnostic detection chain and significantly decrease signal-to-noise ratio.

The requirement on spatial, temporal and energy resolution are mostly defined by the physical phenomena to be studied. For the regular neutron yield monitoring, the temporal resolution of 10 ms is usually sufficient. However, investigations of transport properties or magnetohydrodynamics (MHD) instabilities require increased temporal and spatial resolution. In general, a trade-off between high temporal and spatial resolution and the accuracy of the measurements in terms of statistics and signal-to-noise ratio is necessary.

Another important requirement for a neutron diagnostic is its proper characterization and calibration. Due to the nature of the neutron transport, the problems in absolute calibration of neutron diagnostics are not trivial and this task is very often supported by Monte Carlo calculations of neutron transport. Low-activity radioactive sources are used for quick checks, intershot and overnight calibrations of neutron diagnostics. These sources are usually embedded into a tokamak diagnostic near the detector. More precise, but also more laborious, procedures involve in-vessel calibrations where a strong standardized radioisotope neutron source (e.g. ^{252}Cf) [5] or a neutron generator [6] is inserted and moved around the vacuum vessel to map out the volume occupied by the plasma.

Physical Quantities Determined from Neutron Diagnostics Measurements

Fusion Reaction Rate

Fusion reaction rate (i.e. number of fusion reactions per unit volume per unit time) is given as:

$$r = \frac{n_i n_j}{1 + \delta_{ij}} \langle \sigma v \rangle, \tag{1}$$

where n_i is the density of ith species, δ_{ij} is the Kronecker delta. Assuming that the reactants have a distribution of velocities (e.g. a thermal distribution), $\langle \sigma v \rangle$ is the average of the fusion cross section σ over the relative velocities v, called the reactivity.

The fusion reaction rate is usually determined indirectly from the neutron flux measured by neutron counters or the activation technique. This indirect approach means that other plasma parameters such as density, temperature, impurities concentration along with the discharge parameters have to be known. Then, a transport code that includes a physical model of the interactions (e.g. beam-plasma interactions, or radio frequency wave interactions with a plasma), is used to calculate the fusion reaction rate in an iterative manner. One of the most commonly used transport codes is TRANSP [7].

Plasma Ion Temperature

The ion temperature can be determined from the Doppler broadening measured by a neutron spectrometer. Fusion reactions are two-body exothermic reactions with Q values 3.3 MeV and 17.6 MeV for DD and DT, respectively. Since the reactants are not at rest, energy of the products is shifted by a quantity that depends on the reactant energy and on the emission direction of the neutrons.

Assuming that the reactants have thermal distribution i.e. ion populations are in equilibrium with a Maxwellian velocity distribution of ion temperature T_i, the resulting neutron energy spectrum is nearly of the Gaussian shape [8]:

$$f(E_n) = \frac{1}{W\sqrt{\pi}} exp\left(-\frac{(E_n - E_{n0} - \Delta E_{th} - \Delta E_r)^2}{W^2}\right), \quad W = \left(\frac{4m_n \langle E_n \rangle T_i}{m_n + m_\alpha}\right), \tag{2}$$

where m_n and m_α denote the mass of the neutron and the mass of the alpha particle, respectively. In Eq. 2, E_n is the measured neutron energy and E_{n0} is the characteristic neutron emission energy which applies in the limit $T_i \to 0$. If $T_i > 0$, the distribution is shifted as expressed by ΔE_{th} parameter. Moreover, finite plasma rotation velocity introduces an additional shift expressed by ΔE_r (see "Toroidal Rotation Velocity" section).

The thermal Doppler width (FWHM) is proportional to $\sqrt{T_i}$:

$$FWHM = 2W\sqrt{ln2}. \tag{3}$$

Equation 3 may be stated in practical units as: $FWHM = 82.2\sqrt{T_i}$ and $FWHM = 177.2\sqrt{T_i}$, for DD and DT reaction, respectively, where both neutron energies and ion temperatures are measured in keV.

Toroidal Rotation Velocity

It is foreseen that the neutron spectroscopy can be potentially used for determination of the plasma toroidal rotation velocity V_t in the core of burning plasma (e.g. in the ITER tokamak). The plasma rotation velocity V_t can be, in principle, calculated from the difference spectrum obtained from simultaneous measurements of neutrons emitted in two opposite tangential directions using spectrometers with collimated lines of sight (LoS), as shown in Fig. 1. Such a technique is an alternative to determination of V_t from the Doppler shift of characteristic photon emission lines in the X-ray region.

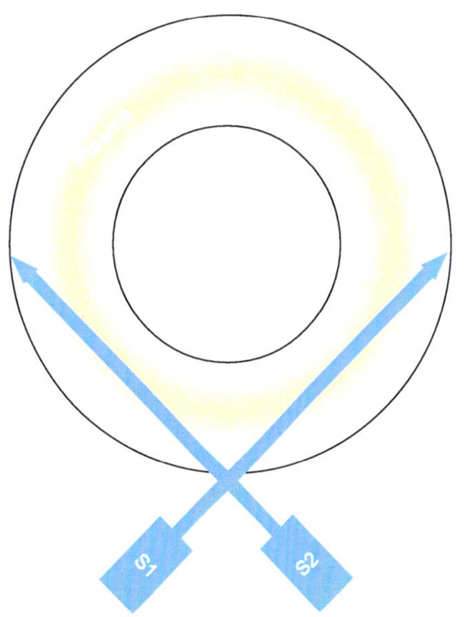

Fig. 1 A schematic layout of two neutron spectrometers for plasma rotation velocity determination. The information on plasma rotation velocity can be derived by subtracting spectra measured in two tangential LoS

Assuming the conditions of thermonuclear fusion plasma with the Maxwellian ion distribution, the energy distribution of the neutron emission from $D(t,n)^4He$ reaction is described by Eq. 2. The energy shift due to plasma rotation is given as $\Delta E_R = 0.54V_t$, where ΔE_R is expressed in keV and V_t in km/s. In ITER, predicted plasma rotation velocity will be in the range of $1\,\text{km/s} < V_t < 50\,\text{km/s}$.

Determination of V_t using a single spectrometer is very challenging due to required accuracy of fusion neutron spectrum measurement. Therefore, it has been proposed [9] to use a pair of spectrometers that view the same plasma volume from opposing directions, i.e. co- and counter-tangential LoS as shown in Fig. 1. A difference spectrum constructed based on the data obtained from both spectrometers can be used to derive V_t. There are two advantages of such an approach. Firstly, the energy difference, for a given V_t is doubled. Secondly, the precision depends now to a lesser degree on the absolute values of the system but is transferred to the relative one relating the two spectrometers to each other. Thus, the accuracy of V_t determination is shifted from the absolute calibration of the spectrometers to the relative calibration of the two spectrometers to each other.

The main limitation of the technique, that relies on collimated neutron measurements, is the fact that the measured spectra are integrated over the LoS. Thus, the measured spectrum consists of contributions from plasma volumes with different neutron emissivity, temperature and rotation velocity. It is however a sensible assumption that most of neutrons come from the hot core of the plasma. Therefore, in the first approximation, the observed shift ΔE_R is mostly related to plasma rotation velocity in the hot plasma core.

Neutron Diagnostics

Neutron Counters

During the fusion experiments with auxiliary heating, the neutron yield can change by 2–3 orders of magnitude. Thus, it is necessary to use neutron detectors with a wide dynamic range. Moreover, such a detector should have a fast response time, since fast ions slowing down time is ~ 100 ms. The detector should be also immune to hard X-rays and gamma-rays. Most commonly used neutron detectors in fusion plasma studies are: BF_3 proportional counters, ^{3}He proportional counters and ^{235}U and ^{238}U fission chambers.

The BF_3 and ^{3}He counters use $^{10}B(n,\alpha)^7Li$ and $^3He(n,p)T$ reactions, respectively. The ^{253}U and ^{238}U fission chambers utilize the $^{235}U(n, fission)$ or $^{238}U(n, fission)$ reactions releasing up to 200 MeV of energy. Since these detectors are more sensitive to low-energy neutrons, they are often embedded into a moderator to slow down fast neutrons.

A fission chamber is basically an ionization chamber with the electrodes coated with fissile material: ^{235}U or ^{238}U. The ionization of the filling gas is caused by fission fragments. Since $^{238}U(n, fission)$ reaction has a threshold of ~ 1 MeV, a chamber based on this isotope is sensitive only to fast neutrons. As a common practice, ^{235}U and ^{238}U fission chambers are installed in pairs around the fusion devices to monitor the neutron flux. Figure 2 shows a schematic view of a fission chambers pair.

Neutron Activation Systems

Time-integrated measurements of the total neutron yield are routinely preformed using the neutron activation system. In this case, samples of materials (usually in the form of foils) with the well-known neutron reaction cross-sections are used. These samples are transferred to irradiation points located near the plasma. The important quantities for selection of the activation foils include: reaction cross-section, reaction threshold energy, daughter nuclides half-life time and characteristic energies. Table 1 summarizes the most widely used reactions for neutron activation measurements at modern tokamaks. In order to deduce the total neutron emission rate in the whole plasma, based on the neutron flux at the irradiation point, it is necessary to rely on the precise Monte Carlo modelling of neutron transport in the fusion machine.

The primary goal of the activation measurements is to provide a reliable measure of fusion energy production over a wide dynamic range. Neutron fluxes from DD and DT reactions are relatively easily measured using several foils with different threshold energy reactions. A full neutron energy spectrum can be retrieved by using a multi-foil activation technique. In this case, several samples with different reaction energy thresholds are used and the spectrum unfolding must be performed using a dedicated computer code. The activation measurements enable also determination of the triton burn-up ratio (for DD plasma) and D/T fuel ratio n_d/n_t in DT plasmas.

Time of Flight Spectrometry

In the inertial confinement experiments, the neutron emission occurs only for a very short time interval (typically hundreds of picoseconds for laser fusion, nanoseconds in plasma focus (PF) devices). The most convenient way to obtain the neutron energy spectrum is thus to use time-resolved neutron detector located far from the plasma. The arrival time of neutrons depends on their velocity and thus on their energy. This is the principle of the time-of

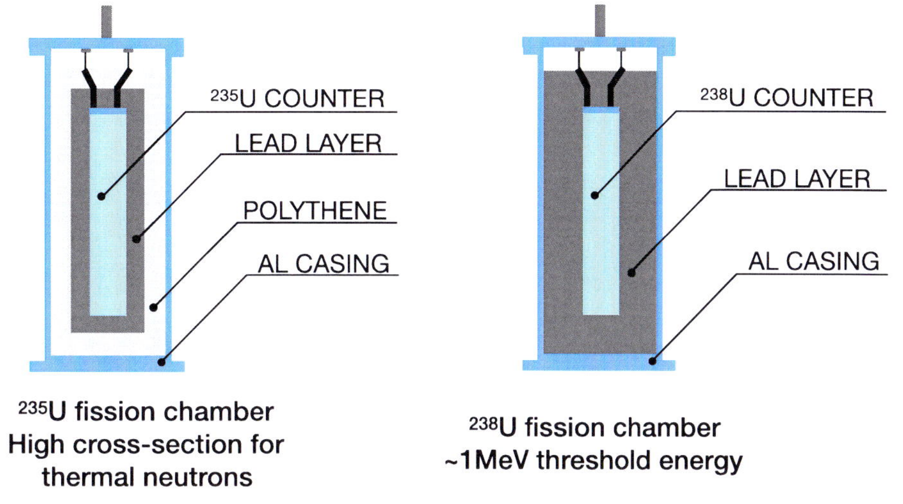

Fig. 2 A schematic view of a pair of ^{235}U and ^{238}U fission chambers routinely used as neutron flux monitors

Table 1 Commonly used reactions for neutron activation measurements

Reaction	Reaction threshold energy (MeV)	Half-life	Fuel type
^{58}Ni(n,p)^{58}Co	1	70.82 days	DD
^{64}Zn(n,p)^{64}Cu	1.8	12.7 h	
^{115}In(n,n)^{115m}In	0.5	4.5 h	
^{232}Th(n,f)F.P.	1.2	1 min	
^{27}Al(n,p)^{27}Mg	2.6	9.5 min	DT
^{28}Si(n,p)^{28}Al	5	2.3 min	
^{56}Fe(n,p)^{56}Mn	5	2.6 h	
^{63}Cu(n,2n)^{62}Cu	10.9	9.7 min	

flight (ToF) technique that allows for obtaining the neutron energy spectrum from the time-dependant neutron intensity.

Let us focus at first on the emission from a plasma focus device. The device emits a short pulse of hard X-rays and neutrons. In Fig. 3, a time trace of detected radiation emitted from a plasma-focus device operated with deuterium plasma is shown. As a detector, a fast scintillation probe (UFNSP-1, ACS laboratory, Poland) was used and the distance between the plasma and the detector was 5 m. The first peak is associated with the detection of X-rays, while the second peak, separated by ~ 210 ns is due to neutron detection. If the distance between the plasma source and the scintillation probe is known, the energy of neutrons can be simply calculated from the ToF difference between X-rays and neutrons. This is schematically

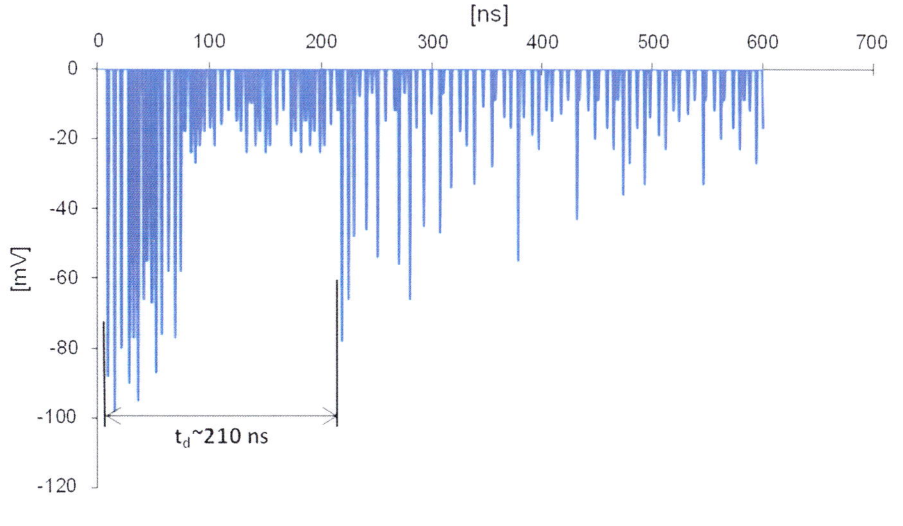

Fig. 3 Separation of the X-ray and neutron signals in ToF method used in pulsed sources registered by ultra-fast scintillation probe

illustrated in Fig. 4 where the ToF for photons and neutrons of 2 MeV, 2.5 MeV and 3 MeV energy is plotted as a function of distance d.

For plasma discharges longer than nanoseconds, ToF method becomes a bit more complicated. Measured ToF of neutrons has to be associated with proton recoil technique on thin foils or thin scintillators. In Fig. 5, measured neutron n incoming on a thin plastic scintillator $S1$ gives *start* pulse of recoiled proton according to the scattering process:

$$n + p \rightarrow p_R + n' \tag{4}$$

Second *stop* pulse signal is given by $S2$ scintillator. ToF of the recoiled neutron n' is measured on a known distance L between detectors $S1$ and $S2$. Energy of measured neutron n, determined by nonrelativistic kinematics is independent on the scattering angle θ and given by the following formula [10]:

$$E_n = 2m_n r^2 / t_{TOF}^2, \tag{5}$$

where r is radius of the time-of-flight sphere (typically $\sim$ 1 m) shown if Fig. 5 and t_{TOF} is the flight time of the scattered neutron.

Spatially Resolved Neutron Measurements: Neutron Camera

By using several detectors with different LoS located in a common poloidal plasma cross-section, it is possible to measure line-integrated neutron emissivity. The detectors are often arranged in several cameras viewing plasma from different directions. Each LoS is equipped with one or more neutron detectors, typically based on liquid or plastic scintillators (e.g. at JET [11]) or diamond detector (foreseen for ITER). Each detection channel must be properly shielded in order to avoid registration of the back-scattered neutrons. Collimation of neutrons is much more challenging then collimation of light. This is because of the long penetration distance of neutrons in essentially all materials. This imposes that neutron collimators are usually large structures. Ideally, each detector should function as a neutron spectrometer allowing determination of the neutron energy spectrum. In practice however, often the spectrometric capabilities are not available and detectors work simply as counters.

The line-integrated measurements along a number of chords enable to reconstruct the spatial distribution of the neutron emissivity in the poloidal cross-section of the plasma. This is possible by using so-called tomographic

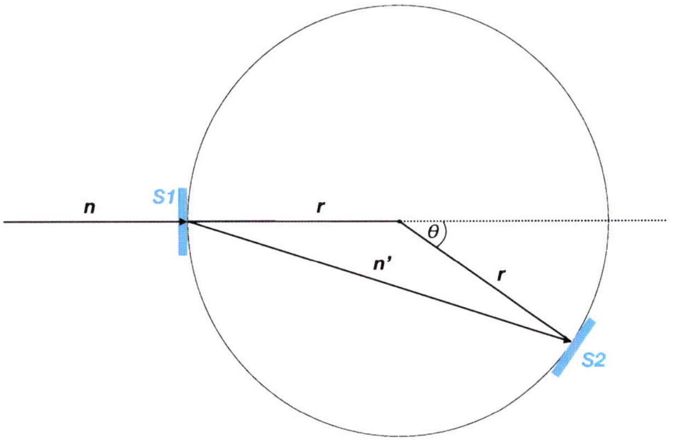

Fig. 5 Principle of ToF method for neutron measurement with long time plasma discharges

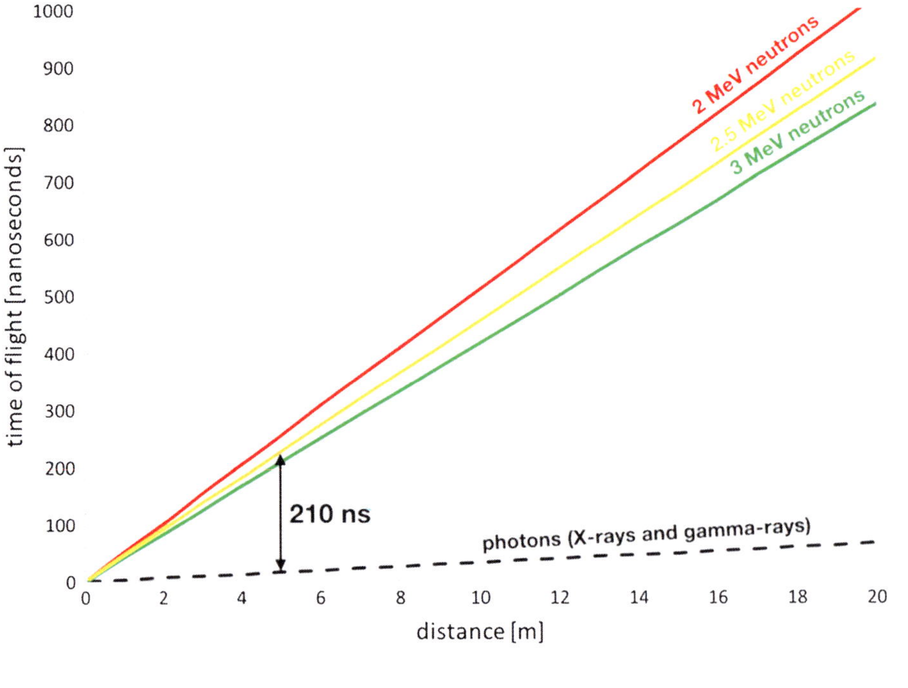

Fig. 4 Time of flight for X-rays and 2 MeV, 2.5 MeV and 3 MeV neutrons as a function of distance d

inversion methods. This task is however not trivial, since the reconstruction is an ill-posed problem. Several approaches to the tomographic inversion for fusion devices have been developed. The most common ones are based on Tikhonov regularization [12], minimization of the Fisher information [13] or maximum entropy method [14]. The other methods include neural networks [15] or genetic algorithms [16].

The reconstructed distribution of 14.1 MeV neutron emissivity provides information on fusion reactivity and alpha particle birth profile. Moreover, the information on the spatial distribution of neutron emission can be used to determine and control plasma shape and position.

Magnetic Proton Recoil Spectrometry

Magnetic proton recoil (MPR) system is based on n–p elastic scattering and magnetic dispersion. A simple schematic diagram of MPR spectrometer is showed in Fig. 6. A collimated beam of neutrons is converted to protons through n–p elastic scattering in the thin polyethylene (CH_2) foil. The relationship between the recoil proton energy E_p and the neutron energy E_n is given by [17]:

$$E_p = E_n \cos^2\theta_{np}, \tag{6}$$

where θ_{np} is the scattering angle between neutron and the recoiled proton. The recoil protons emitted in the forward direction are pre-selected by the proton collimator and the entrance gap of the dipole magnet. The forward recoil protons have a maximum n–p cross-section and a minimum uncertainty in the kinematic energy spread, accordingly enhancing the neutron detection efficiency and energy resolution. The protons are passively spatially separated in the magnetic field according to their momentum as described by [17]:

$$\rho = \frac{p}{qB} = \frac{\sqrt{2mE_p}}{qB}, \tag{7}$$

where ρ is the bending radius, p is the proton momentum, q is the charge, B is the magnetic field, and m is the mass. Thus, the distribution of recoil proton energies can be converted into a spatial distribution of protons on the scintillators array. Since the bending radius is inversely proportional to the magnitude of magnetic field, to minimize the dimension of MPR spectrometer it is necessary to improve the dipole magnetic field. On the other hand, however, the stronger magnetic field of the diagnostic, the more shielding is required to avoid the interference with the tokamak magnetic field.

Energy resolution of the MPR spectrometer at FWHM of the recoil proton energy distribution results mainly from energy loss ΔE_f in the CH_2 foil, the kinematic energy broadening ΔE_k and the ion-optical aberrations of the magnet ΔE_m. Thus, the total energy resolution ΔE_t can be estimated as [17]:

$$\Delta E_t = \sqrt{\Delta E_f^2 + \Delta E_k^2 + \Delta E_m^2}. \tag{8}$$

The neutron detection efficiency of the MPR spectrometer can be expressed by [17]:

$$\varepsilon(E_n) = \frac{d\sigma_{np}}{d\Omega} \times N_H \times \Omega_S \times T_P, \tag{9}$$

where $d\sigma_{np}/d\Omega$ is the differential cross section in the laboratory system for elastic scattering. The cross section depends on the incident neutron energy and the scattering angle. N_H is the number of protons per unit area in the CH_2 foil, Ω_s is the solid angle defined by the proton collimator and entrance gap of the dipole and T_p is the magnet transmission.

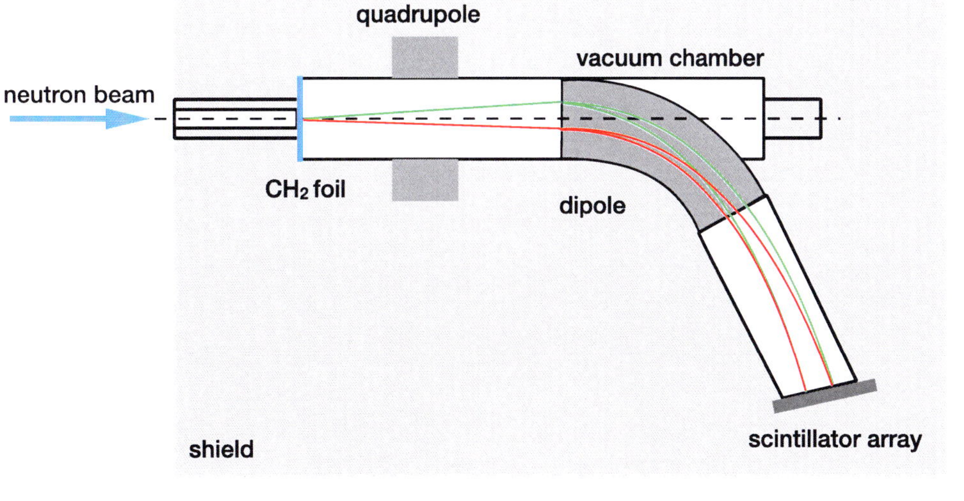

Fig. 6 Schematic layout of a MPR spectrometer. Figure based on [17]

Summary

Neutron diagnostics play a crucial role in tokamak plasma physics. They provide key information on the machine performance but also allow studying physical phenomena such as particle transport or MHD activities. Time-integrated measurements e.g. using neutron activation spectroscopy provide reliable assessment of the neutron yield and thus the fusion power. Time-resolved neutron measurements enable to study the neutron flux evolution and investigations of plasma MHD activities. Spatially-resolved measurements provided by neutron cameras enable to determine the spatial distribution of neutron emissivity and alpha particle birth profile. Finally, neutron spectroscopy performed with ToF, MPR or liquid scintillator spectrometers provides an insight into physical properties of plasma such as ion temperature or plasma rotation velocity.

In the era of bigger tokamaks such as ITER the approach to neutron diagnostics must be revised. The high fluxes of neutrons will impose the requirements on a new design of detection systems and necessity to use radiation-hardened electronics. Among many plasma parameters to be measured in ITER, the neutron diagnostics will allow evaluation of the fusion power which plays a key role for machine protection as well as for plasma optimization and achieving the fusion gain factor $\mathbb{Q}$, related to the reactor performance. The requirements on ITER neutron diagnostics define the neutron emissivity measurements within 10% accuracy, with a temporal resolution of 1 ms and spatial resolution of a tenth of the minor plasma radius, i.e. 200 mm. The approach to the in-vessel calibration of the neutron diagnostics should be also revised. Indeed, 10% of accuracy in the assessment of the foreseen ITER fusion power is equivalent to more that 3 times fusion power achieved at JET tokamak. This indicates that alternative approaches to the calibration should be considered.

Acknowledgements The authors gratefully acknowledge the Polish National Science Centre Grant (NCN G61527) which facilitated research presented in the paper.

References

1. O. Jarvis, Plasma Phys. Control. Fusion **36**, 209 (1994)
2. M. Gatu Johnson et al., Plasma Phys Control. Fusion **52**, 085002 (2010)
3. O. Jarvis, G. Sadler, J. Thompson, Nucl. Fusion **28**, 198 (1988)
4. E. Andersson Sundén et al., Nucl. Instrum. Methods Phys. Res. A **610**, 682–699 (2009)
5. D. Syme, Fusion Eng. Des. **89**, 2766–2775 (2014)
6. P. Batistoni et al., 14 MeV calibration of JET neutron detectors—phase 2: in-vessel calibration. Nucl. Fusion **58**, 106016 (2018)
7. B. Grierson et al., Fusion Sci. Technol. (2018). https://doi.org/10.1080/15361055.2017.1398585
8. H. Brysk, Plasma Phys. **15**, 611 (1973)
9. G. Gorini, Rev. Sci. Instrum. **68**, 561 (1997)
10. M. Gatu Johnson et al., Nucl. Instrum. Methods Phys. Res. A **591**, 417 (2008)
11. M. Anton et al., Plasma Phys. Control. Fusion **38**, 1849–1878 (1996)
12. J. Bielecki et al., Rev. Sci. Instrum. **86**, 093505 (2015)
13. A. Jardin et al., Phys. Scr. **91**, 044007 (2016)
14. J. Kim et al., Rev. Sci. Instrum. **77**, 10F513 (2006)
15. E. Ronchi et al., Nucl. Instrum. Methods Phys. Res. A **613**, 295–303 (2010)
16. J. Bielecki, Fusion Eng. Des. **127**, 160–167 (2018)
17. Zhang et al., Rev. Sci. Instrum. **86**, 125115 (2015)

Journal of Fusion Energy (2019) 38:394–405
https://doi.org/10.1007/s10894-019-00216-6

ORIGINAL RESEARCH

Development of Microwave Imaging Diagnostics for WEST Tokamak

R. Sabot[1] · J. -C. Giacalone[1] · Y. Nam[1] · A. Berne[1] · C. Brun[1] · D. Elbèze[1] · F. Faisse[1] · L. Gargiulo[1] · M. Kim[2] · W. Lee[2] · P. Lotte[1] · H. K. Park[3] · B. Santraine[1] · G. Yun[4] · WEST team

Published online: 21 March 2019

Abstract
WEST is an upgrade of the Tore Supra tokamak to test ITER like divertor elements over very long pulses. For the studies of MHD instabilities affected by tungsten impurities, two microwave imaging diagnostics have been developed to obtain density and temperature fluctuation maps. The core reflectometer was refurbished and now measures the core density profiles in few microseconds. Measurements of several thousand profiles in sequence with microsecond dead time (burst mode) will be available in 2019 enabling the reconstruction of density maps with unprecedented time resolution. The major issue for the implementation of an electron cyclotron emission imaging diagnostic was the heat radiation during long pulses on the large vacuum window and the first optics. Two metallic in-vessel mirrors inside the personnel access tube solved this issue. The first mirror can withstand the heat radiation without water cooling, the second mirror redirects the beam toward the personnel access port. The oblique and small view angle of this access eases also the thermomechanical constraints on the vacuum window. As the space between the port flange and the tokamak access lobby was tight, a compact optical enclosure with vertical aligned optics was designed. To ensure a good reproducibility of the mirrors and the optical enclosure positions after a maintenance period and thus avoid re-alignment, different mechanical systems are used to manipulate the mirrors and the optical enclosure and to fine tune their position and their orientation. The first channel covers the low field side and will provide 192 pixel images of temperature fluctuations with 1–2 microsecond time resolution. The installation of the in-vessel mirrors is scheduled during March 2019. Diagnostic qualification tests will be done before the restart of the experiments late May 2019.

Keywords Plasma diagnostics · Microwave imaging diagnostics · Electron cyclotron emission imaging diagnostic · ECEI · Reflectometry

Introduction

The large gradients required to ignite fusion reactions inside a magnetic confined plasma can destabilize numerous instabilities from large scale modes widening up to a fraction of the plasma dimension to microscale modes at Larmor radius scale. Some instabilities are ubiquitous, others are periodic, whereas onset of others leads to plasma termination. Because these instabilities govern the stability and the performances of a plasma, they have been a major topic for fusion research since the beginning of the nuclear fusion quest. Since their first development in the 90's, microwave imaging instruments have been installed on many devices to obtain a direct visualisation of the instability spatial structures and their temporal evolution [1, 2].

First images were obtained using 1D measurements and the plasma rotation to reconstruct a map of the electron temperature. This technique first developed for soft X-rays [3], was applied to electron cyclotron emission measurements to reconstruct 2D images during sawtooth on JET [4] and TFTR [5, 6]. In the 90's, electron cyclotron emission imaging (ECEI) were developed to get direct images of the temperature fluctuations [1, 7]. It has proven extremely powerful for understanding the physics of sawtooth crash

✉ R. Sabot
roland.sabot@cea.fr

1 CEA, IRFM, 13108 Saint Paul-lez-Durance, France

2 National Fusion Research Institute, Daejeon 34133, Korea

3 Ulsan National Institute of Science and Technology, Ulsan 44919, Korea

4 Department of Physics, Pohang University of Science and Technology, Pohang 37673, Korea

[8, 9], the structure of edge localized modes [10] or fast particles induced instabilities [11]. Microwave reflectometry imaging (MIR) have also been developed to measure density fluctuations. MIR systems are usually install in parallel to ECEI sharing the same front-end optics. However, MIR diagnostics are much more difficult to operate and are thus installed on only few devices.

For the WEST [Tungsten (W) Environment in Steady-state Tokamak] tokamak, two microwave diagnostic will provide images of plasma instabilities. The core reflectometer has been upgraded to measure the density profiles in few microseconds, allowing the reconstruction of core density with very high spatial and temporal resolution. Using the rotation technique, 2D images can be reconstructed [12]. An ECEI system developed through UNIST-WEST collaboration will image the temperature fluctuations. This diagnostic will be commissioned by the end of 2018 and first images are expected for the 2019 experimental campaign.

WEST is a transformation of the Tore Supra tokamak plasma in an X-point divertor configuration to test ITER divertor technology during long pulse operation [13]. Internal coils were installed to obtain flexible equilibria: lower, upper or double X-point configuration. The carbon fibers composite (CFC) bottom limiter was replaced by a divertor made with a mix of actively cooled ITER like tungsten elements and tungsten coated graphite tiles. These uncooled tiles restrict the plasma duration to typically $\sim$ 10 s at high power. The upper divertor is made of water cooled tungsten coated copper tiles. WEST first plasma was obtained in 2016, the divertor configuration was commissioned in 2017 and the lower hybrid (LH) and ion cyclotron resonance heating (ICRH) systems started operation in 2018. Additional heating power up to 5 MW and plasma duration over 30 s on the upper divertor at reduced power were obtained during the 2018 campaign. The lower divertor with water cooled ITER like monoblocks will be completed in 2020 (WEST phase 2) allowing long pulse operation, up to 1000 s with up to 10 MW of additional heating power.

This paper describes the development of two WEST microwave imaging diagnostics. First, the upgrade reflectometer is presented. The ECEI diagnostic is then described with a focus on the innovative approaches to the integration of the ECEI optics in WEST. Although the diagnostic relies on the ECEI principles, the large optics required by ECEI force an implementation inside the personnel access port, which induced new constraints on the optics design compared to standard ECEI. These diagnostics will complement the set of diagnostics dedicated to fluctuations measurements [14] and the new magnetic diagnostic sets [15]. Simultaneous images of the temperature and the density fluctuations induced by MHD instabilities and turbulence will help to better understand the role of impurities and in particular tungsten on instability thresholds and their evolution during long pulses.

The Upgraded Core Reflectometer

Based on radar principle, reflectometry diagnostics measure the phase of a microwave beam after reflection of a cutoff layer inside the plasma [16, 17]. As the probing wave frequency increases, the reflecting layer moves deeper inside the plasma. The density profile is reconstructed from the variation of the phase with the probing frequency using an inversion method. In high magnetic field tokamaks, X-mode reflectometers on the outer side can reach the plasma center even for flat or hollow profiles as the upper cut-off frequency F_R increases with magnetic field:

$$F_R = \frac{1}{2}\left(\sqrt{4F_p^2 + F_{ec}^2} + F_{ec}\right) \quad (1)$$

where F_p is the plasma frequency and F_{ec} the electron cyclotron frequency.

Density Map Reconstruction Principle

The method developed by Nagayama et al. [5] was applied to the core density measurement to obtain 2D maps of the core density in Tore Supra [18]. The principle is similar to the Magnetic Resonance Imaging, but instead of rotating the detector around the scanned body, the plasma rotation makes the plasma to spin in front of the reflectometer. Compared to ECE, reflectometry offers the advantages of a local measurement with almost continuous radial resolution (few millimetres) and a very high sensitivity ($< 1\%$).

Assuming a rigid body rotation at ω_θ, a density profile measurement done at $t = t_i$ in the equatorial plane ($\theta = 0$) gives also the density profile at an angle $\theta_i = \omega_\theta(t_0 - t_i)$. However, as plasma is not stationary, one must take into account the evolution of the density profile to perform reconstructions at the same time for all poloidal angles. To correct this evolution, the density $n_e(\theta_i, t_0, \rho)$ at $t = t_0$, at the poloidal angle θ_i and the normalized radius ρ is obtained by linear interpolation between backward measurement made at $t = t_0 - \theta/\omega$ and time forward measurement.

When the density profile is measured on both side of the magnetic axis, the time forward measurement corresponds to the measurement made half turn latter at $t = t_0 + (\pi - \theta)/\omega_\theta$:

$$n_e(\theta_i, t_0, \rho) = \frac{(\pi - \theta_i) \cdot n_e(0, t_0 - \theta_i/\omega_\theta, \rho) + \theta_i \cdot n_e(0, t_0 + (\pi - \theta_i)/\omega_\theta, \rho)}{\pi} \quad (2)$$

The density profiles must be processed to produce smooth images on a constant map (ρ_g, θ_g). The map poloidal angle resolution $\Delta\theta_g$ is set the profile measurement period Δt_p and the rotation ω_θ:

$$\Delta\theta_g > \omega_\theta \Delta t_p \quad (3)$$

Using a plasma equilibrium, the density profiles $n_e(R, t)$ are evaluated on the normalized radius grid ρ_g.The density profile is then resampled at $\frac{\Delta\theta_g}{\omega_\theta}$ times to get an integer number of profiles N_p per rotation period. When the time resolution is high enough, high frequency variation can be low pass filtered before resampling. The density maps on the grid (ρ_g, θ_g) are then obtained from relation 2. To reduce discontinuity in the maps around $\theta = \frac{\pm\pi}{2}$ and benefit from measurements on the low and high field side, an odd number of poloidal angle per disc half is chosen $N_p = 2(2n + 1)$. This reconstruction technique is still valid with some minor modifications when the rotation speed is slowly evolving in time or along the radius. This reconstruction method requires fast profile measurements: the duration of the measurement must be much shorter than the plasma rotation period which should also be much shorter than the characteristic time of the instability or the profile evolution.

On Tore Supra, the core reflectometer measured profiles in 40–100 μs, depending on plasma parameters, with 5 μs dead time between two profiles. For the image shown on Fig. 1, the profiles were measured in 45 μs with a dead time of 5 μs. The core rotation duration was around 900 μs (1.1 kHz), thus roughly 18 profiles were measured per rotation period ensuring a good angular sampling ($\Delta\theta \sim 20°$). In this discharge, the sawtooth period lasted around 40 ms, 4440 density profiles were measured consecutively in 222 ms covering 5 sawtooth periods.

Figure 1 shows two examples of density structures during sawtooth: a internal ring and a weaker external ring which appear few milliseconds after the crash and a ring that can be observed 10 ms later. Both structures were recovered and interpreted with the XTOR-2F code [19]. During the sawtooth crash, the magnetic surfaces are reconnected and the dense hot core is expulsed outside the core. This expulsed density is then poloidal homogenised by poloidal flows to form an external ring. These flows also reinject a fraction of the expulsed density inside the core in a crescent shape structure which is latter poloidal homogenised in a more internal ring [20]. These structures are not observed on the temperature because the temperature follows the magnetic surface reconnection while the density is transported by poloidal flows, the energy transport along the field lines being much faster than particle transport.

Upgrading the Core Reflectometer Channels

Profile Measurements

The core reflectometer is built on the original microwave setup which was developed to achieve short profile measurements [21, 22]. The core reflectometer has two independent channels (see Fig. 2). The first one runs with a Voltage Controlled Oscillator (VCO) and a frequency synthesizer to measure density profiles or density fluctuations during the same discharge [23]. The second channel runs only with a frequency synthesiser. It can probe another zone of the plasma or frequencies close to the first channel ones to perform radial correlations.

The wave is generated by a low frequency source and frequency multipliers are used to reach the D-band. The source signal is divided into a probing and a reference signal and a single sideband modulation shifts the frequency for heterodyne detection [24]. The diagnostic upgrade was an opportunity to swap the frequency multipliers. The multiplication order for the profile/fluctuation channel is now 8 with a 2nd harmonic mixer. The

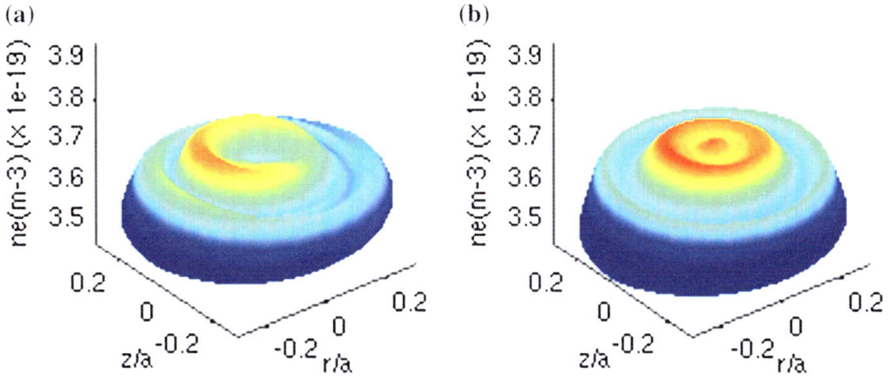

Fig. 1 Examples of fine density structures after a sawtooth crash (TS shot 44,634): **a** two (one internal and a weaker external) crescents just observed after the crash (1–2 ms), **b** 10 ms latter the crescents are homogenized in two rings

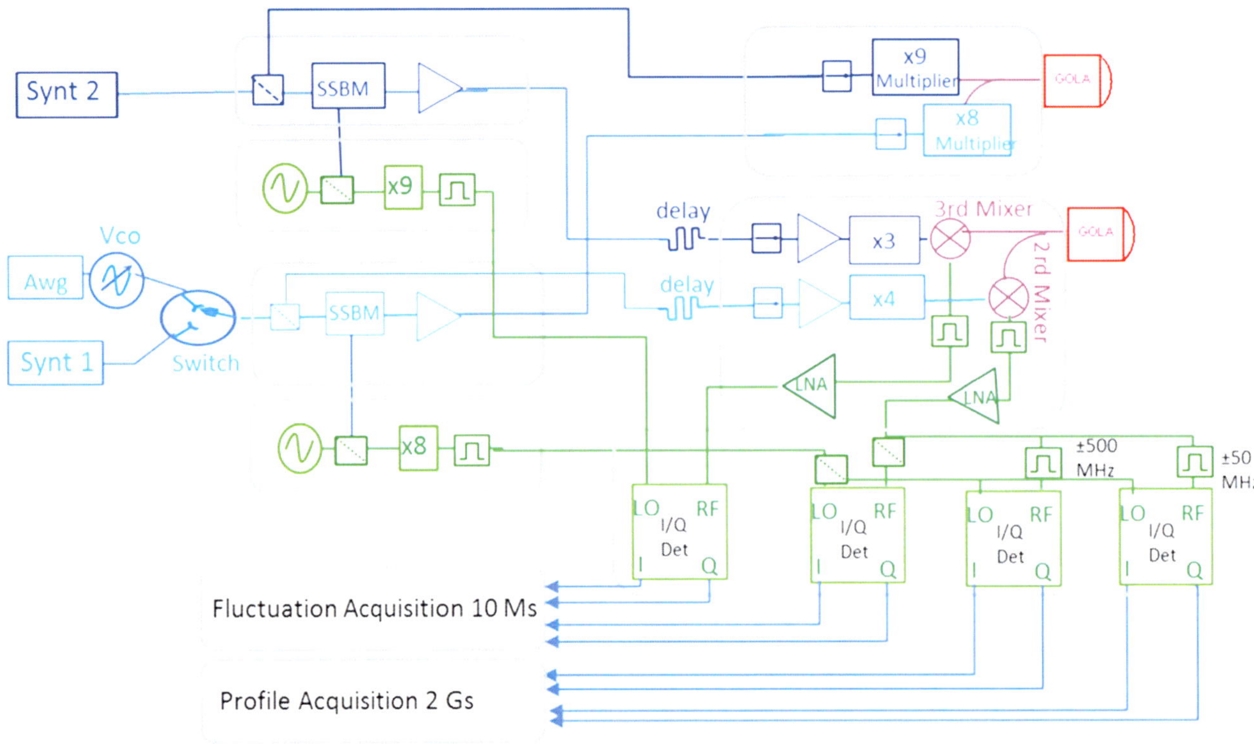

Fig. 2 Microwave set-up of the core reflectometer channels

correlation channel is built with a 9 order multiplier and a third harmonic mixer. The two channels are combined into a 3 dB directional coupler and then launched into the plasma though a high gain Gaussian antenna. The reflected wave is received by a second antenna and then split by another direction coupler. A sub-harmonic mixer down-converts the reflected wave frequency before filtering and amplifying. The phase φ and the amplitude are measured with an I/Q detection. Most multiplier and mixer components were provided by the Virginia Diodes Company.

Decreasing the sweeping time T_{sweep} by an order of magnitude increases the beat frequency by the same ratio:

$$\Delta F = \frac{1}{2\pi}\frac{\partial\varphi}{\partial t}\frac{\Delta F}{T_{sweep}} \tag{4}$$

where $\tau = \frac{1}{2\pi}\frac{\partial\varphi}{\partial t}$ is the time of flight and ΔF the frequency bandwidth.

To cope with this higher beat frequency, the detection part along with the acquisition system had to be replaced following the pioneer upgrade of the Tore Supra edge reflectometer system several years ago. Microsecond sweeps were performed on Tore Supa [25] and Asdex Upgrade [26].

The VME (Virtual Machine Environment) acquisition system was replaced by Linux diskless PC [27]. The acquisition system for the profiles, a 4 channel module ACQIRIS DC282, is similar to the edge reflectometer one [25]. The maximum sampling frequency is 2 GHz with 10 bit vertical resolution. The 32 Mbytes memory depth per channel allows several thousand profiles per discharge. The arbitrary waveform generator was also replaced by a 1.2 GS/s device to cope with much shorter sweeping times. The acquisition card is housed in a compactPCI board. A general purpose interface bus (GPIB) card controls the arbitrary waveform generator and the frequency synthesizers.

To prevent the beat frequency from overlapping, the SSBM modulation frequency of the profile channel was increased from 100 to 300 MHz. The filters and demodulation line were thus changed to accommodate the higher intermediate frequency (2.4 GHz) and much larger bandwidth (± 500 MHz). As only 2 acquisition channels are needed, a second demodulation line with narrower filter (± 50 MHz) was installed. A narrower IF bandwidth and a longer sweeping time (30–60 μs) might be needed to improve the signal to noise ratio in lower hybrid (LH) heated discharges since the background ECE noise level can increase by several orders of magnitude with LH.

Fluctuation Measurements

The acquisition system for the fluctuations measurements was also upgraded. It is based on a 4 channel module NI PXI 6115, with 12 bit vertical resolution and up to 10 MS/s sample rate. The memory depth allows several seconds of measurements per shot at a sampling rate of 1 MS/s.

The fluctuation detection part was also modified to widen the fluctuations bandwidth up to ± 5 MHz and to benefit for the differential input of the acquisition card. In the hopping mode, the diagnostic performs typically 20 frequency steps of 10 ms several times per shots. The diagnostic can also perform longer acquisition up to 1 s at a fixed frequency to follow the time evolution of the density fluctuations.

The diagnostic implantation was moved to the Q5B port and the diagnostic sight angle is now in the middle of the port i.e. where the ripple variations are the lowest. The diagnostic, the control system and the acquisition are housed in a HF shielded cubicle. 7 to 10 m cables connect the cubicles to the millimetric components (multipliers and mixers). The emission and detection boxes are mounted on the antenna supporting frame which is bolted on the vacuum window flange. The Gaussian optics lens antenna (GOLA) are roughly at 1.8 m to the plasma center and their line of sight crosses near the plasma center.

The Core Reflectometer Accessibility on WEST

Although the reflectometer covers the same frequency range, its accessibility in WEST is modified. At the nominal magnetic field ($B_0 = 3.6$ T), the lowest frequency ($\sim$ 105 GHz) cut-off layer is expected to be close to the pedestal with a local density around $n_e \sim 3.10^{19}\ \mathrm{m}^{-3}$. The plasma center can be probed up to a central density as high as $n_e(0) = 10^{20}\ \mathrm{m}^{-3}$. The HFS accessibility is smaller than on Tore Supra because of the smaller minor radius (0.5 now vs. 0.72 m on Tore Supra) and an expected lower density peaking in H-mode. The cut-off layer profile exhibits often a maximum at mid-radius on the high field side which limits the HFS accessibility (see Fig. 3). At a lower magnetic field ($B_0 \sim 3$T), the accessibility is limited.

First Measurements

The diagnostic was installed during the summer 2018 and was in operation for the WEST C3b experimental campaign. Profile measurements were performed in few microseconds with 0.5 µs dead time between profiles. Few thousands profiles were measured in each discharge. Data are transferred to WEST databased at the end of the discharges.

The burst mode will be in operation for the next campaign. Thousands of sweeps with dead time as low as 0.5 µs could be measured in sequence to follow fast density profile evolution or MHD instabilities. As the size of the acquisition card can recorded only around 5000 sweeps, there is a trade-off between the time resolution and the duration of the burst mode.

WEST ECEI Characteristics

ECEI Principle

Charged particles in a magnetic confined devices, gyrate around the magnetic field lines and, as a consequence, emit electromagnetic radiation at the cyclotron frequency f_c and its harmonic [16]:

$$f_c = \frac{1}{2\pi}\frac{qB}{m} \tag{5}$$

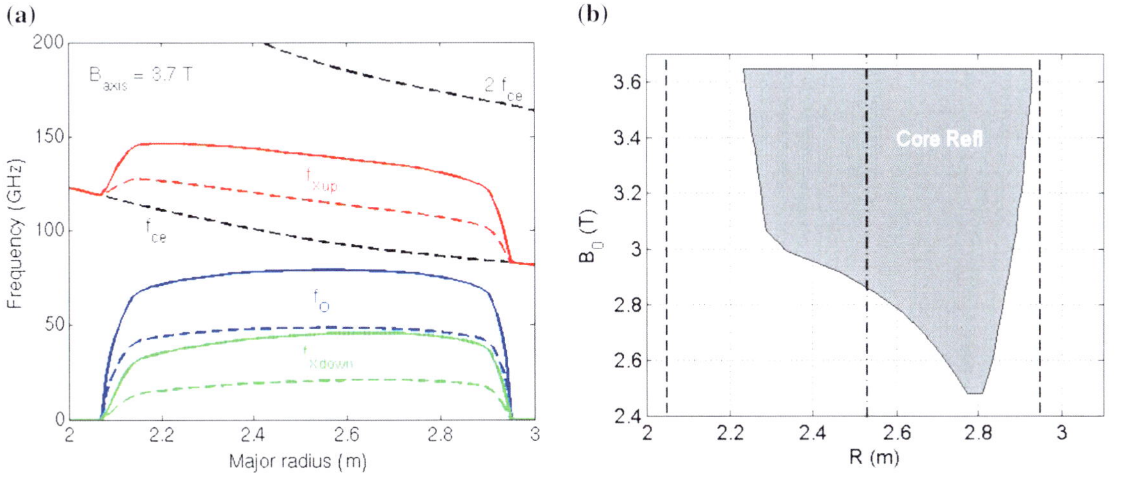

Fig. 3 **a** Profiles of X and O-mode cut-off frequencies at B = 3.6 T for high and low central density ($n_e(0) = 8.10^{19}\mathrm{m}^{-3}$ solid line, $n_e(0) = 3.10^{19}\mathrm{m}^{-3}$ dots line) for typical WEST plasma configuration. **b** Core reflectometer accessibility with magnetic field for a high central density case ($n_e(0) = 7.10^{19}\mathrm{m}^{-3}$, top pedestal density $n_{e,\ ped}$=-2.5 $10^{19}\mathrm{m}^{-3}$)

where m and q are the particle mass and electric charge, B is the magnetic field.

For electrons, these frequencies are in the microwave range ($f_c = 28\,\text{GHz}$ for $B = 1\text{T}$) and the optical depth of the 1st or the 2nd harmonic is usually short in the core of magnetic fusion devices. A local thermodynamic equilibrium is then reached and the ECE intensity approaches the blackbody level. As the low frequency limit (Rayleigh–Jeans law) applies, the ECE intensity is proportional to the local electron temperature $T_e(R)$:

$$\mathrm{I}(\omega) = \frac{\omega^2 T_e(R)}{8\pi^3 c^2} \quad (6)$$

In a tokamak, the magnetic field in toroidal direction decreases monotonically along the major radius R as 1/R, resulting in an ECE spectrum which is a broad continuum. A radiometer measuring the emission intensity along a line of sight can thus recover the electron temperature profiles. The radial resolution depends on the radiometer filter banks. It is limited by ECE relativistic broadening.

An Electron Cyclotron Emission Imaging diagnostic can be seen as a set of 1D ECE radiometers stacked vertically (Fig. 4). The single antenna of a classic 1D ECE is replaced by an array of antennas [7]. The ECEI principle was first validated on the Text-U, RTP and TEXTOR tokamaks [28–30]. Due to the strong divergence of microwave beams, a large vacuum window, large optics and refocusing elements are required to reach a beam waist of the order of few centimetres inside the plasma. The development of mini-lenses at the University of California at Davis team has simplified optical designs and has reduced optical aberrations, leading to improved focal plane patterns [31, 32] with spot diameters as low as 2–3 cm. The localisation is determined by the local oscillator (LO) frequency, but the field depth is usually smaller than in a 1D ECE diagnostic. The LO frequency must thus be tuned together with the focus position of the optics. ECEI diagnostics give 2-D images of the temperature fluctuations with typically 8 (radial) pixels by 16–24 (vertical) pixels. Images are usually monitored with a time resolution around 1–2 microseconds.

Implementation in the Man Access Port

On WEST, the maintenance access was the only port to implement the large vacuum window for the ECEI beam. The flange and the port are large enough for a person to crawl inside the torus. This access was available because the tangential view angle, a flange at several meters from the plasma edge and the requirement to remove all equipment for maintenance operation made this port poor for diagnostics. The strong reduction of the plasma radiation flux due to the oblique small view angle would however ease the thermal constraint on the vacuum window. Because of this unusual implantation, innovative designs had to be developed.

In-Vessel-Mirrors

The large composite lens which usually makes the primary focusing optic could not be implemented on WEST because of the tangential view and the heat radiation during long pulses. Y. Nam et al. proposed a design with two in-vessel mirrors implemented at roughly midway [33]. The ECEI diagnostics on the GAMMA 10 tandem mirror and the large helical device (LHD) were the first to use in vessel focusing mirrors [34]. The WEST mirrors are installed in a Y duct at roughly 2.1 m from the plasma centre and also 2.1 m to the port flange (Fig. 5). The large oblong vacuum window (772 × 272 × 24 mm) is mounted with all-metal (Helicoflex®) seals. The design sought to maximize the mirror height (H ~ 59 cm) to provide good beam properties at all vertical position. The width was restricted to 20 cm (external width 27 cm) to avoid self-shadowing. Both mirrors have cylindrical curved surfaces: the curvature of the first mirror from the plasma is in the toroidal direction (4.0 m radius) and in the poloidal direction (3.5 m radius) for the second.

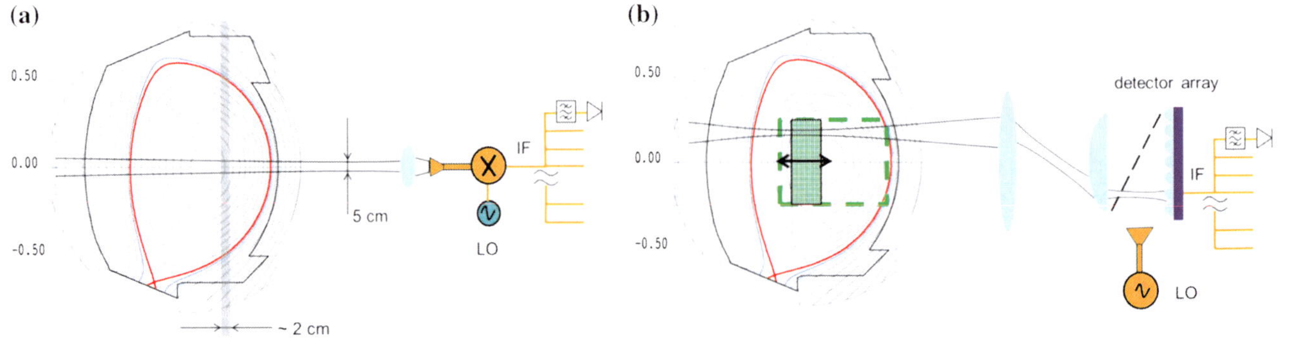

Fig. 4 Schematic diagram illustrating the principles of WEST **a** 1D ECE radiometer and **b** ECEI diagnostic

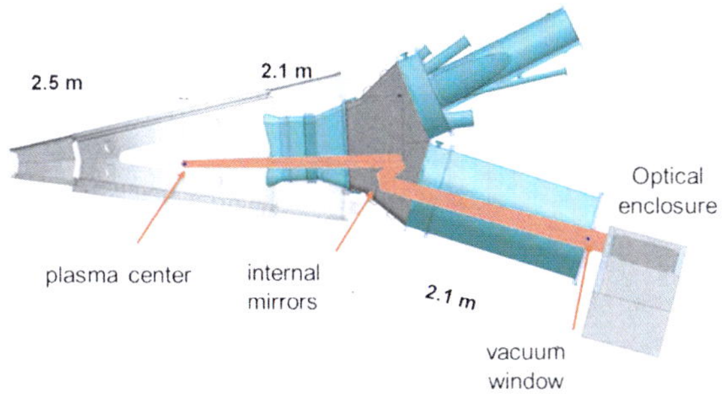

Fig. 5 ECEI beam path (purple color) in the Q5A port. The in-vessel mirrors are installed in the Y duct. The maintenance access is thought the bottom cylinder duct. The pellet injector and the 2D X spectrometer diagnostic view the plasma though the other side of the Y duct (top) (Color figure online)

During the maintenances, the mirrors will remain inside the vacuum vessel to avoid safety controls imposed on elements exiting from the vacuum vessel. To facilitate the mirror manipulation, the mirrors are suspended on a rail track. The rails are similar to those installed in the WEST Articulated Inspection Arm (AIA) [35]. They can withstand machine baking and they have shown reliable operation under vacuum. Although the mirrors will not be manipulated under vacuum, reliability is a key safety concern. The mirrors will be moved on the right to free the path for personnel access to the vessel (see Fig. 6b). The mirrors are mounted on a shaft suspended on the rail. Two pivot points enables to preserve the alignment after a manipulation: the top pivot is used to free the rotation while fine angle tuning is done with a second pivot behind the mirror.

The mirrors are attached to the middle of the shaft to ensure free and symmetric dilation during long pulses. Thermomechanical analysis predicts that the temperature in the centre of the first mirror would reach 430 °C at the end of the reference WEST long pulse (1000 s, 9 MW of RF heating). The thermal expansion remains small (few millimetres) and the optical quality of the microwave beam is not affected (Fig. 7). Water cooling of the first mirror is thus not required.

Optical Enclosure

The integration of the diagnostic outside the vacuum vessel was also challenging since the space for the ex-vessel optics was very tight. Roughly 1 m^2 was available between the flange and the access lobby to house the focusing lens and the detection array. Taking advantage from the height of the WEST equatorial plane ($\sim$ 2.5 m above the ground floor), the imaging optical components are vertically aligned enabling a considerable reduction of the footprint (0.8 m $\times$ 1.2 m) compared to typical ECEI diagnostics. This housing includes a provision space for the installation of a second channel at higher frequencies.

The lens at the optical enclosure entrance will first refocus the beam to the detectors (see Fig. 8). The 45° mirror will reflect the beam downward. The polarization is then rotated to match with the antenna pattern of the detector array [36]. The focus position of the beam in the plasma will be set by the position-controllable lens. The last lens focuses the elongated beam to the detector array made of two rows of 12 mini-lenses and mixers [32]. A dichroic plate set above the detector array rejects the image frequencies. Extra effort was invested in design to make the imaging system and the local oscillator as compact as possible. In particular, a couple of aspheric lenses is used to shorten the optical path length of the LO optics to $\sim$ 0.5 m [37]. The design includes a provision for a beam splitter for high field side (HFS) view. A notch rejection filter [38] will be inserted if the ECRH gyrotrons at 118 GHz resume operation. It will be mounted behind the movable lens.

To recover the optical enclosure position after each operation without any re-alignment effort, the optical enclosure is mounted on a supporting frame. This frame enables to adjust the lateral position, the orientation, the height and the tilt of the optical enclosure (see Fig. 9). Due to its height, the optical enclosure will be lifted by the torus hall crane. The supporting frame will be removed with the optical enclosure.

The optical enclosure will rest on dedicated plates fixed to the metallic beams below the ground floor. Socket ball

Fig. 6 **a** mirror positions during plasma operation with the beam axis in yellow. **b** Mirrors during maintenance (Color figure online)

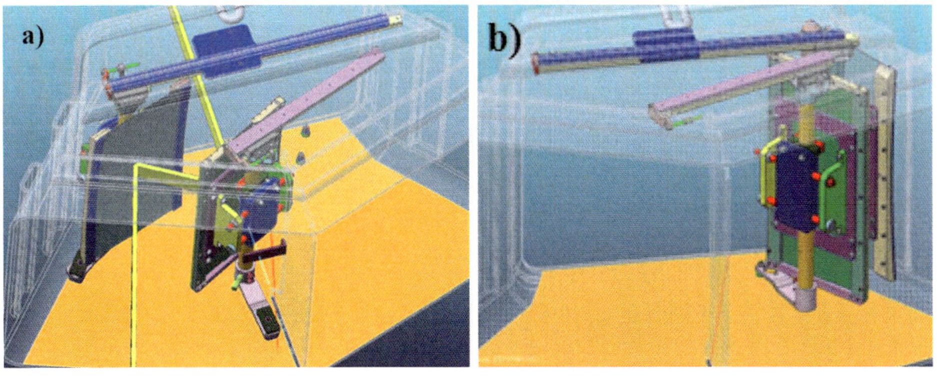

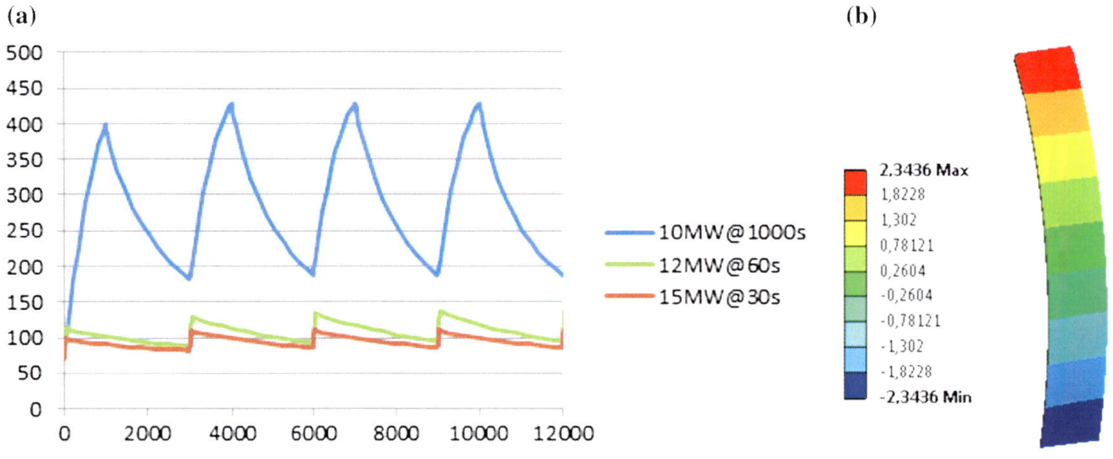

Fig. 7 **a** Rise of the first mirror temperature after the 3 WEST reference shots [1000 s discharges with 10 MW addition heating (blue), 60 s discharges with 12 MW (green), 30 s discharges with 15 MW (red)]. **b** Mirrors maximum vertical elongation (millimeters) after 1000 s shots (Color figure online)

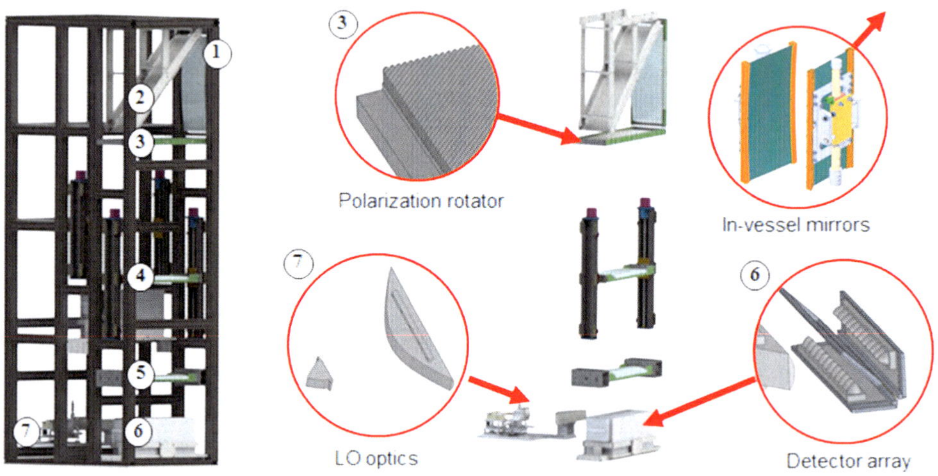

Fig. 8 Optical enclosure: ① entrance lens, ② 45° mirror, ③ polarization rotator, ④ focusing lens (moveable), ⑤ collimation lens, ⑥ detector array box ⑦ LO optics

Fig. 9 The optical enclosure supporting frame. The white squares show the position of the dedicated plates that have latter been installed on the metallic floor. The vacuum vessel is on the left, and the access lobby on the right, the lobby rails can be seen

like feet will guarantee a good reproducibility of the supporting frame positioning after each manipulation. During the maintenance periods, the optical enclosure will be placed in a dedicated area inside the torus hall. This dedicated area is close to the entrance port so the optical enclosure can be connected, microwave cables included, to the basement cubicles for diagnostic testing.

Control and Acquisition Cubicles

The IF electronics, the power supplies, the data acquisition system and an industrial personal computer are housed in cubicles located in the torus hall basement below the optical enclosure. Three electromagnetic and radiofrequency noise shielded cubicles are assembled side by side. The first cubicle houses the LFS channel. The central one is dedicated to the power supplies. The last cubicle is empty and will house the inner channel. The connection cables between the cubicles and the optical enclosure comprise 24 low loss 20 m long cables, several optical fibers for two ethernet networks and a power supply cable.

The intermediate frequency (IF) electronics and the associated power supply have been provided by the UNIST team. The diagnostic probes 8 radial positions with 900 MHz separation. The IF amplitude are tuned with controlled voltage attenuators to fully exploit the acquisition card dynamic range.

The 192 channels are recorded by 6 acquisition units with 32 channels each at a sampling rate up to 2 Ms/s (ACQ132CPCI from D-tAcq), similar to the KSTAR ECEI acquisition system [39]. Up to 10 s at 500 kS/s can be recorded. The acquisition can be split in several triggers to get synchronized measurements with other fast acquisition diagnostics. The first trigger starts before the plasma initiation to measure the background noise. Data are transferred at the end of the discharge. The WEST timing network provides the time-stamping and the event triggers for the data acquisition clock and for starting the acquisition.

The industrial PC running LINUX OS controls also several devices in the cubicles and in the optical enclosure such as the LO frequency synthesizer, the lens step motors etc.

Frequency Range and Plasma Accessibility

The first channel of the diagnostic will operate in the W-band covering the frequency range 85–113 GHz. At the WEST nominal magnetic field (B = 3.7 T), the diagnostic will fully cover the low field side (LFS). Figure 10 shows the evolution of the focus position and of the poloidal size for the central frenquency when moving the focus lens from the top to the bottom of the cubicle enclosure under geometrical optics assumption. The dotted points shows the extended coverage on the inner side (and edge) by the channels above (below) the middle one when the lens is at its extremal positions. Measurements on the inner side (from R = 2.3) are even possible without degrading the beam focusing properties. The second channel will cover the range 100–130 GHz for the inner side access (R = 2 to R = 2.6 m) with some overlapping with the LFS channel.

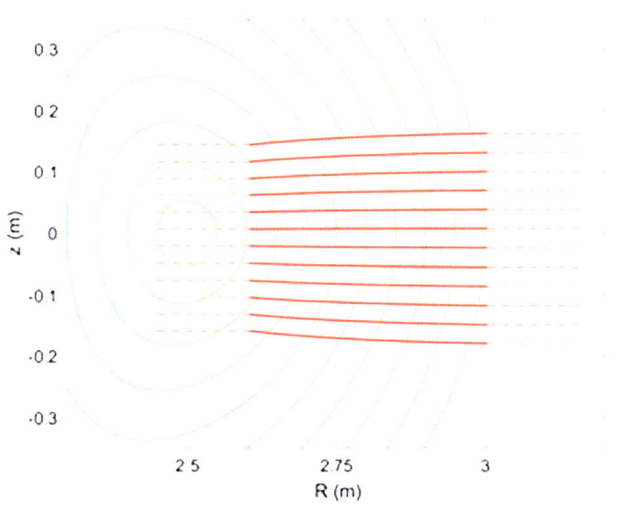

Fig. 10 The moveable lens enables a good coverage of the LFS by the focusing plane. The plain line shows the evolution of the focus position and the vertical elongation at the middle frequency when the focusing lens is moved from the top to the bottom of the optical enclosure. The dotted area show the coverage of the plasma core by the channels above the middle one when the lens is at the top position. Same on the edge coverage by channel below the middle one for the lens at the bottom position

The 1st harmonic O-mode was chosen instead of the 2nd harmonic X-mode because of much easier and cheaper component aviability in W-band than in G-band (140–220 GHz). Moreover, the 2nd harmonic X-mode would have been more strongly perturbed by the downshifted 3rd harmonic emission from suprathermal electron accelerated by LH waves. Although the diagnostic runs in O-mode, strong refraction are not expected as long as the density is not too high. Based on the criteria established for the Tore Supra 1D ECE [40], the threshold density for a strong beam refraction is $n_e(0) \sim 8.510^{19}\,\mathrm{m}^{-3}$ at the center and $n_e \sim 5.510^{20}\,\mathrm{m}^{-3}$ at the top of the pedestal.

The diagnostic probes 8 radial positions with 900 MHz separation. The 7.2 GHz IF bandwidth provides a radial coverage around 18 cm in the plasma centre and 24 cm at the edge.

Installation Timeline

The metallic mirrors were send to the UNIST laboratory in 2017 for an integration test of the entire imaging optics. The tests were performed at 90 GHz. The focusing capability was checked. It was verified that imaging optics can focus on any radial position on the LFS by controlling the vertical focusing lens. The size at the focus is around 1.6 cm and the divergence is small enough for the beam size to be almost constant over the extension of the probing zone. The beam pattern along the vertical direction were also measured for the focus position at the plasma edge, mid-radius and the plasma centre. The measured and simulated design beam patterns well match to each other. This demonstrated that the optical components are fabricated consistent with the design. More details on the test can be found in [41].

Korean elements were delivered at Cadarache in April 2018. The detection and IF units were tested to check that they were not damaged during the shipping. The optical tower was then assembled. The control command system and data acquisition card are being integrated in the WEST plasma control system.

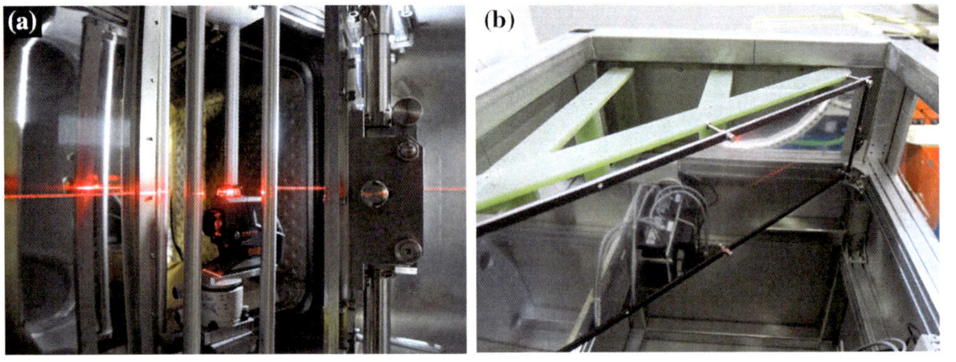

Fig. 11 **a** The cross line laser and the in-vessel mirrors, **b** the 45° mirror at the optical enclosure top after alignment: the laser beam is in the middle of the mirror

The in-vessel mirrors were installed on WEST during the 2019 winter shutdown. A cross laser (deviation less than 2 mm at 10 m, see Fig. 11a) was suspended on the innermost rail to get both the equatorial and vertical planes aligned with reference markers on the inner wall and the back of the Y duct [42]. The positions and orientations of the outer then inner mirrors were tuned to get the laser beam at the centre of the outer flange. The mirror were moved back and forth to check that the beam position at the outer flange was not displaced within the laser beam width uncertainty ($\sim$ 1 mm).

The optical tower was then installed and aligned. Its position and orientation were adjusted to get the laser beam at the center of the entrance lens and the 45° mirror (see Fig. 11b). The orientation of this mirror was then corrected to get the beam on the detection array marks. The optical tower was then lifted up and down, no displacement (within the laser beam width) of the laser beam was observed.

Taking advantage of the proximity of the depository area during maintenance, the entire acquisition from the optical enclosure to the acquisition board will be tested before the end of the shutdown. First measurements on plasma are expected late May or early June during the restart phase of the WEST C4 campaign.

Conclusion

The design of two microwave imaging diagnostic have been adapted to deal with the constraints of long pulse operation and the limited access, The core reflectometer was upgraded to measures density profiles in few microseconds which will enable the reconstruction of core density maps with very good time resolution. The issues associated to the ECEI large vacuum window and large optics were overcome with in-vessel mirrors inside the personnel access port. An innovative design ensures an easy manipulation of the mirrors to free the tokamak access, as well as a good reproducibility of the optic enclosure positioning. The ECEI diagnostic is now in the last installation phase. Qualification measurements are foreseen before the 2019 campaign. MHD and turbulence studies during the 2019 campaign will thus benefit from the detailed and simultaneous images provided by these diagnostics.

Acknowledgements The ECEI diagnostic development was supported by a Grant from the Région Sud Provence-Alpes-Côte d'Azur of France under Contract DEB 15-1376 and by the NRF of Korea under Contract Nos. NRF-2014M1A7A1A03029881 and NRF-2014M1A7A1A03029865.

References

1. H. Park, C.C. Chang, B.H. Deng, C.W. Domier, A.J.H. Donné, K. Kawahata, C. Liang, X.P. Liang, H.J. Lu, N.C. Luhmann, A. Mase, H. Matsuura, E. Mazzucato, A. Miura, K. Mizuno, T. Munsat, Y. Nagayama, M.J. van de Pol, J. Wang, Z.G. Xia, W.-K. Zhang, Recent advancements in microwave imaging plasma diagnostics. Rev. Sci. Instrum. **74**, 4239–4262 (2003). https://doi.org/10.1063/1.1610781
2. Y. Wang, B. Tobias, Y.-T. Chang, J.-H. Yu, M. Li, F. Hu, M. Chen, M. Mamidanna, T. Phan, A.-V. Pham, J. Gu, X. Liu, Y. Zhu, C.W. Domier, L. Shi, E. Valeo, G.J. Kramer, D. Kuwahara, Y. Nagayama, A. Mase, N.C. Luhmann, Millimeter-wave imaging of magnetic fusion plasmas: technology innovations advancing physics understanding. Nucl. Fusion **57**, 072007 (2017). https://doi.org/10.1088/1741-4326/aa5e30
3. P. Smeulders, Evolution of the m = 1 tearing mode during internal sawtooth disruption. Nucl. Fusion **23**, 529 (1983). https://doi.org/10.1088/0029-5515/23/4/010
4. E. Westerhof, P. Smeulders, N.L. Cardozo, Observations of sawtooth postcursor oscillations in JET and their bearing on the nature of the sawtooth collapse. Nucl. Fusion **29**, 1056 (1989). https://doi.org/10.1088/0029-5515/29/6/016
5. Y. Nagayama, R. Büchse, A. Cavallo, E.D. Fredrickson, A. Janos, K.M. McGuire, G.K. Petravic, C. Sule, G. Taylor, Image reconstructions of ECE and X-ray signals for high β plasmas on TFTR. Rev. Sci. Instrum. **61**, 3265–3267 (1990). https://doi.org/10.1063/1.1141609
6. Y. Nagayama, K.M. McGuire, M. Bitter, A. Cavallo, E.D. Fredrickson, K.W. Hill, H. Hsuan, A. Janos, W. Park, G. Taylor, M. Yamada, Analysis of sawtooth oscillations using simultaneous measurement of electron cyclotron emission imaging and X-ray tomography on TFTR. Phys. Rev. Lett. **67**, 3527–3530 (1991). https://doi.org/10.1103/PhysRevLett.67.3527

7. R.P. Hsia, W.R. Geck, S. Cheng, W.-M. Zhang, C.W. Domier, N.C. Luhmann, ECE imaging array diagnostic development of TEXT-U. Rev. Sci. Instrum. **66**, 834–836 (1995). https://doi.org/10.1063/1.1146239
8. TEXTOR team, H.K. Park, N.C. Luhmann, A.J.H. Donné, I.G.J. Classen, C.W. Domier, E. Mazzucato, T. Munsat, M.J. van de Pol, Z. Xia, Observation of high-field-side crash and heat transfer during sawtooth oscillation in magnetically confined plasmas. Phys. Rev. Lett. **96**, 195003 (2006). https://doi.org/10.1103/physrevlett.96.195003
9. T.E.X.T.O.R. Team, H.K. Park, A.J.H. Donné, N.C. Luhmann, I.G.J. Classen, C.W. Domier, E. Mazzucato, T. Munsat, M.J. van de Pol, Z. Xia, Comparison study of 2D images of temperature fluctuations during sawtooth oscillation with theoretical models. Phys. Rev. Lett. 96 **195004**, 96.195004 (2006). https://doi.org/10.1103/PhysRevLett
10. G.S. Yun, W. Lee, M.J. Choi, J. Lee, H.K. Park, B. Tobias, C.W. Domier, N.C. Luhmann, A.J.H. Donné, J.H. Lee, KSTAR Team, Two-dimensional visualization of growth and burst of the edge-localized filaments in KSTAR H-mode plasmas. Phys. Rev. Lett. **107**, 045004 (2011). https://doi.org/10.1103/PhysRevLett.107.045004
11. B.J. Tobias, I.G.J. Classen, C.W. Domier, W.W. Heidbrink, N.C. Luhmann, R. Nazikian, H.K. Park, D.A. Spong, M.A. Van Zeeland, Fast ion induced shearing of 2D Alfvén eigenmodes measured by electron cyclotron emission imaging. Phys. Rev. Lett. **106**, 075003 (2011). https://doi.org/10.1103/PhysRevLett.106.075003
12. T. Nicolas, H. Lütjens, J.-F. Luciani, X. Garbet, R. Sabot, Impurity behavior during sawtooth activity in tokamak plasmas. Phys. Plasmas **21**, 012507 (2014). https://doi.org/10.1063/1.4861859
13. J. Bucalossi, A. Argouarch, V. Basiuk, O. Baulaigue, P. Bayetti, M. Bécoulet, B. Bertrand, S. Brémond, P. Cara, M. Chantant, Y. Corre, X. Courtois, L. Doceul, A. Ekedahl, F. Faisse, M. Firdaouss, J. Garcia, L. Gargiulo, C. Gil, C. Grisolia, J. Gunn, S. Hacquin, P. Hertout, G. Huysmans, F. Imbeaux, G. Jiolat, M. Joanny, L. Jourd'heuil, M. Jouve, A. Kukushkin, M. Lipa, S. Lisgo, T. Loarer, P. Maget, R. Magne, Y. Marandet, A. Martinez, D. Mazon, O. Meyer, M. Missirlian, P. Monier-Garbet, P. Moreau, E. Nardon, S. Panayotis, B. Pégourié, R.A. Pitts, C. Portafaix, M. Richou, R. Sabot, A. Saille, F. Saint-Laurent, F. Samaille, A. Simonin, E. Tsitrone, Feasibility study of an actively cooled tungsten divertor in Tore Supra for ITER technology testing. Fusion Eng. Des. **86**, 684–688 (2011). https://doi.org/10.1016/j.fusengdes.2011.01.114
14. R. Sabot, P. Hennequin, L. Colas, Fluctuation measurements and their link with transport on tore supra. Fusion Sci. Technol. **56**, 1253–1272 (2009). https://doi.org/10.13182/FST09-A9176
15. P. Moreau, A. Le-Luyer, P. Spuig, P. Malard, F. Saint-Laurent, J.F. Artaud, J. Morales, B. Faugeras, H. Heumann, B. Cantone, M. Moreau, C. Brun, R. Nouailletas, E. Nardon, B. Santraine, A. Berne, P. Kumari, S. Belsare, The new magnetic diagnostics in the WEST tokamak. Rev. Sci. Instrum. **89**, 10J109 (2018). https://doi.org/10.1063/1.5036537
16. I.H. Hutchinson, Principles of plasma diagnostics. Cambridge Core (2002). https://doi.org/10.1017/CBO9780511613630
17. H.J. Hartfuss, T. Geist, M. Hirsch, Heterodyne methods in millimetre wave plasma diagnostics with applications to ECE, interferometry and reflectometry. Plasma Phys. Controlled Fusion **39**, 1693–1769 (1997). https://doi.org/10.1088/0741-3335/39/11/001
18. R. Sabot, D. Elbèze, W. Lee, Y. Nam, H. Park, J. Shen, G. Yun, M. Choi, J.-C. Giacalone, T. Nicolas, C. Bottereau, F. Clairet, P. Lotte, D. Molina, Microwave imaging of magnetohydrodynamic instabilities in fusion plasma. C. R. Phys. **17**, 1018–1026 (2016). https://doi.org/10.1016/j.crhy.2016.07.021
19. H. Lütjens, J.-F. Luciani, XTOR-2F: a fully implicit Newton–Krylov solver applied to nonlinear 3D extended MHD in tokamaks. J. Comput. Phys. **229**, 8130–8143 (2010). https://doi.org/10.1016/j.jcp.2010.07.013
20. T. Nicolas, R. Sabot, X. Garbet, H. Lütjens, J.-F. Luciani, Z. Guimaraes-Filho, J. Decker, A. Merle, Non-linear magnetohydrodynamic simulations of density evolution in Tore Supra sawtoothing plasmas. Phys. Plasmas **19**, 112305 (2012). https://doi.org/10.1063/1.4766893
21. F. Clairet, R. Sabot, C. Bottereau, J.M. Chareau, M. Paume, S. Heuraux, M. Colin, S. Hacquin, G. Leclert, *X*-mode heterodyne reflectometer for edge density profile measurements on Tore Supra. Rev. Sci. Instrum. **72**, 340–343 (2001). https://doi.org/10.1063/1.1319865
22. R. Sabot, F. Clairet, C. Honoré, C. Bottereau, J.-M. Chareau, F. Gabillet, P. Hennequin, S. Heuraux, G. Leclert, A. Sirinelli, A. Truc, L. Vermare, Advances of reflectometry on Tore-Supra: from edge density profile to core density fluctuations. Int. J. Infrared Millim. Waves **25**, 229–246 (2004). https://doi.org/10.1023/B:IJIM.0000017896.97839.24
23. R. Sabot, A. Sirinelli, J.-M. Chareau, J.-C. Giaccalone, A dual source D-band reflectometer for density profile and fluctuations measurements in Tore-Supra. Nucl. Fusion **46**, S685–S692 (2006). https://doi.org/10.1088/0029-5515/46/9/S04
24. R. Sabot, C. Bottereau, J.-M. Chareau, F. Clairet, M. Paume, Single sideband modulator, a key component of Tore-Supra heterodyne reflectometers. Rev. Sci. Instrum. **75**, 2656–2659 (2004). https://doi.org/10.1063/1.1777379
25. F. Clairet, S. Heuraux, C. Bottereau, D. Molina, L. Ducobu, F. Leroux, A. Barbuti, Fast sweeping reflectometry upgrade on Tore Supra. Rev. Sci. Instrum. **81**, 10D903 (2010). https://doi.org/10.1063/1.3460624
26. F. Clairet, C. Bottereau, A. Medvedeva, D. Molina, G.D. Conway, A. Silva, U. Stroth, 1 μs broadband frequency sweeping reflectometry for plasma density and fluctuation profile measurements. Rev. Sci. Instrum. **88**, 113506 (2017). https://doi.org/10.1063/1.4991789
27. F. Leroux, T. Saint-Laurent, Linux diskless system for the Tore Supra data acquisition system, in: 2010 17th IEEE-NPSS Real Time Conference (2010), pp. 1–4. https://doi.org/10.1109/rtc.2010.5750392
28. B.H. Deng, S.R. Burns, C.W. Domier, T.R. Hillyer, R.P. Hsia, N.C. Luhmann, D.L. Brower, G. Cima, A.J.H. Donné, T. Oyevaar, ECE imaging of plasma Te profiles and fluctuations. Fusion Eng. Des. **53**, 77–85 (2001). https://doi.org/10.1016/S0920-3796(00)00480-4
29. B.H. Deng, C.W. Domier, N.C. Luhmann, A.J.H. Donné, M.J. van de Pol, Electron cyclotron emission imaging diagnostic on TEXTOR. Rev. Sci. Instrum. **72**, 368–370 (2001). https://doi.org/10.1063/1.1316748
30. C.W. Domier, Z.G. Xia, P. Zhang, N.C. Luhmann, H.K. Park, E. Mazzucato, M.J. van de Pol, I.G.J. Classen, A.J.H. Donné, R. Jaspers, Upgrades to the TEXTOR electron cyclotron emission imaging diagnostic. Rev. Sci. Instrum. **77**, 10E924 (2006). https://doi.org/10.1063/1.2352727
31. P. Zhang, C.W. Domier, T. Liang, X. Kong, B. Tobias, Z. Shen, N.C. Luhmann, H. Park, I.G.J. Classen, M.J. van de Pol, A.J.H. Donné, R. Jaspers, The next generation of electron cyclotron emission imaging diagnostics (invited). Rev. Sci. Instrum. **79**, 10F103 (2008). https://doi.org/10.1063/1.2967342
32. B. Tobias, X. Kong, T. Liang, A. Spear, C.W. Domier, N.C. Luhmann, I.G.J. Classen, J.E. Boom, M.J. van de Pol, R. Jaspers, A.J.H. Donné, H.K. Park, T. Munsat, Advancements in electron cyclotron emission imaging demonstrated by the TEXTOR ECEI

diagnostic upgrade. Rev. Sci. Instrum. **80**, 093502 (2009). https://doi.org/10.1063/1.3233913
33. Y. Nam, W. Lee, G.S. Yun, H.K. Park, D. Elbeze, J.L. Segui, R. Sabot, J.P. Chenevoix, Design of the reflective optics for Tore Supra ECEI system. Rev. Sci. Instrum. **83**, 10E318 (2012). https://doi.org/10.1063/1.4732852
34. A. Mase, H. Negishi, N. Oyama, K. Watabe, K. Mizuno, Y. Nagayama, K. Kawahata, H. Matsuura, K. Uchida, A. Miura, T. Tamano, K. Yatsu, ECE-imaging work on GAMMA 10 and LHD. Fusion Eng. Des. **53**, 87–95 (2001). https://doi.org/10.1016/S0920-3796(00)00481-6
35. J.-C. Hatchressian, V. Bruno, L. Gargiulo, D. Keller, Y. Perrot, P. Bayetti, J.-J. Cordier, J.-P. Friconneau, J.D. Palmer, F. Samaille, Development of an inspection robot under iter relevant vacuum and temperature conditions. J. Phys: Conf. Ser. **100**, 062031 (2008). https://doi.org/10.1088/1742-6596/100/6/062031
36. J. Lee, G.S. Yun, M. Kim, W. Lee, H.K. Park, Large-aperture broadband polarization rotator for the KSTAR ECE imaging system. J. Inst. **7**, C01037–C01037 (2012). https://doi.org/10.1088/1748-0221/7/01/C01037
37. Y.B. Nam, D.J. Lee, J. Lee, C. Kim, G.S. Yun, W. Lee, H.K. Park, New compact and efficient local oscillator optic system for the KSTAR electron cyclotron emission imaging system. Rev. Sci. Instrum. **87**, 11E130 (2016). https://doi.org/10.1063/1.4961290
38. T. Liang, C.W. Domier, Z. Shen, N.C. Luhmann, Quasi-optical notch filters for plasma imaging applications, in: 2008 33rd International Conference on Infrared, Millimeter and Terahertz Waves (2008), pp. 1–2. https://doi.org/10.1109/icimw.2008.4665804
39. J.B. Kim, W. Lee, G.S. Yun, H.K. Park, C.W. Domier, N.C. Luhmann, Data acquisition and processing system of the electron cyclotron emission imaging system of the KSTAR tokamak. Rev. Sci. Instrum. **81**, 10D931 (2010). https://doi.org/10.1063/1.3479023
40. J.L. Ségui, D. Molina, G. Giruzzi, M. Goniche, G. Huysmans, P. Maget, M. Ottaviani, The Tore Supra Team, An upgraded 32-channel heterodyne electron cyclotron emission radiometer on Tore Supra. Rev. Sci. Instrum. **76**, 123501 (2005). https://doi.org/10.1063/1.2140225
41. Y. Nam, in: 45th European Physical Society Conference on Plasma Physics (Prague, 2018)
42. R. Sabot, Integration of an Electron Cyclotron Imaging diagnostic system on the WEST tokamak, Accepted for Publication in EPJ Web of Conferences (2019)

Journal of Fusion Energy (2019) 38:406–436
https://doi.org/10.1007/s10894-018-0162-5

ORIGINAL RESEARCH

Plasma Magnetic Control in Tokamak Devices

Gianmaria De Tommasi[1]

Published online: 21 May 2018

Abstract

In tokamak experimental reactors, the magnetic control system is one of the main plasma control systems that is required, together with the density control, since the very beginning, even before first operations. Indeed, the magnetic control drives the current in the external poloidal circuits in order to first achieve the breakdown conditions and, after plasma formation, to track the desired plasma current, shape and position. Furthermore, when the plasma poloidal cross-section is vertically elongated, the magnetic control takes also care of the vertical stabilization of the plasma column, and therefore it is an essential system for operation. This chapter introduces a reference architecture for plasma magnetic control in tokamaks. Given the proposed architecture, the techniques to design all the required control algorithms is also presented. Experimental results obtained on the JET and EAST tokamaks and simulations for machines currently under construction are shown to prove the effectiveness of the proposed architecture and control algorithms.

Keywords Systems and control · Plasma magnetic control · Tokamaks

Introduction

Confinement of the hot plasma in toroidal fusion devices is achieved by means of several magnetic fields. In particular, a set of coils wrapped around the vacuum vessel produce the toroidal magnetic field, as shown in the simplified tokamak schematic reported in Fig. 1. An additional external field is produced by a set of toroidal coils, called Poloidal Field (PF) coils. The produced poloidal magnetic field is needed first to achieve the conditions for plasma formation inside the vacuum chamber. Soon after plasma formation, the currents flowing in the PF coils (and hence the produced field) need to be controlled in order to induce current into the plasma and to control its value during each phase of the tokamak discharge (ramp-up, flattop and ramp-down).

Other than controlling the plasma current, also the plasma boundary and position need to be carefully controlled, in order to achieve the desired experimental objectives. Indeed, the time evolution of PF coil currents, along with the plasma geometrical and physical parameters that define the so called *scenario*, is obtained as a sequence of plasma equilibria. Feedforward PF coil currents, and sometimes voltages, are then determined to obtain the desired magnetic field [25]. Being these nominal currents (and voltages) computed with numerical codes, they are affected by model uncertainty; moreover, unexpected disturbances always occur in a tokamak discharge. It follows that a plasma position and shape control system is necessary for tokamak operations. Furthermore, in the case of plasmas with a vertically elongated poloidal cross-section (see the examples reported in Fig. 2), the active control of the current in some of the PF coils is mandatory in order to generate the radial field needed to vertically stabilize the plasma column (see some examples in [7, 39, 72]).

It turns out that plasma magnetic control is one of the crucial issues to be addressed since the very beginning during the design of a tokamak. Indeed a plasma magnetic control system, although not necessarily at its full capacity, is needed since day one (examples of existing architecture can be found in [71, 83]). Furthermore, a robust and reliable plasma magnetic control system is required to successfully control high performance plasmas, such as the ones envisaged for ITER and DEMO.

✉ Gianmaria De Tommasi
detommas@unina.it

1 Dipartimento di Ingegneria Elettrica e delle Tecnologie dell'Informazione, Università degli Studi di Napoli "Federico II" – Consorzio CREATE, Naples, Italy

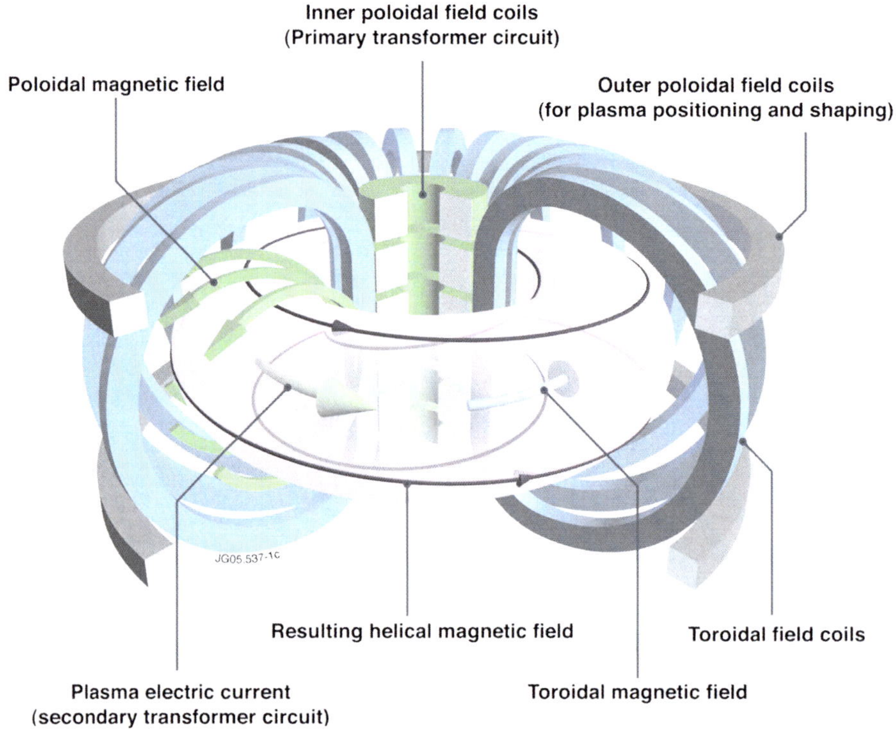

Fig. 1 Simplified scheme of a tokamak

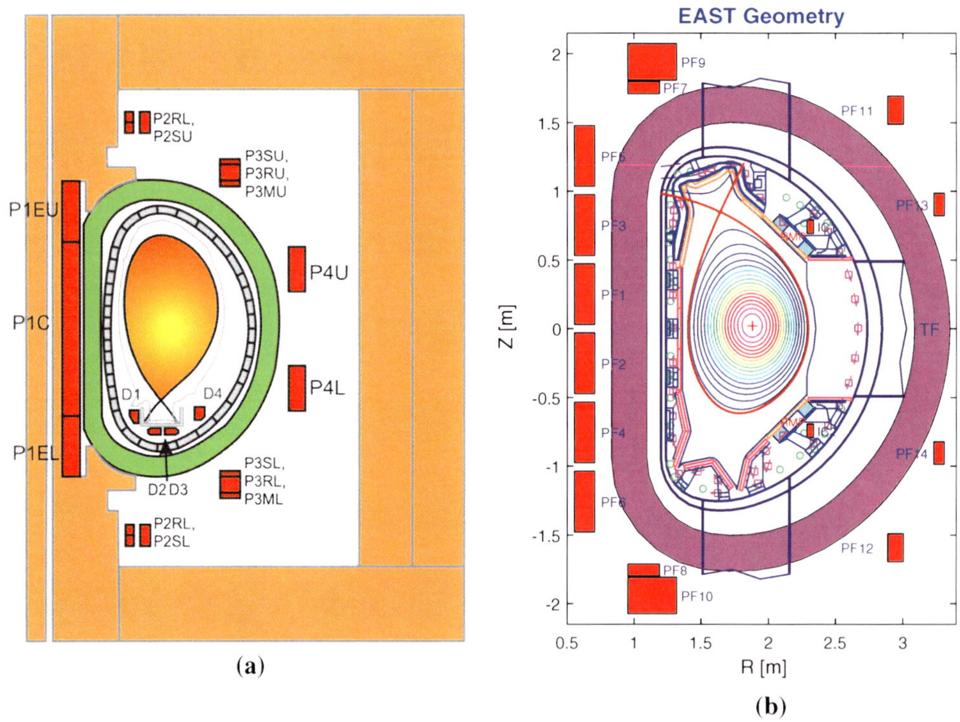

Fig. 2 Poloidal cross-sections of two tokamaks with vertically elongated plasmas. **a** JET poloidal cross-section, **b** EAST poloidal cross section

The design of magnetic control systems in tokamaks relies on different design approaches, from the *classic* single-input single-output (SISO) ones, to more advanced multi-input multi-output (MIMO) techniques.

In early tokamaks with circular cross-section, the plasma current and position control was separated in the following two *independent* problems:

- horizontal position and current control by means of up-down symmetric control currents flowing into sets of

poloidal field coils symmetrically placed with respect to the tokamak equatorial plane;
- vertical position control by means of updown anti-symmetric currents flowing into the poloidal field coils.

In this case a decoupled system of poloidal windings is used. This system consists of an ohmic heating winding, the so-called *central solenoid*, that controls the ohmic magnetic flux and thus the plasma current, as well as a vertical field circuit that controls the plasma major radius. For these tokamaks, the simplest controller structure consists of two separate SISO controllers. As an example, see the control system of the FTU tokamak [20, Ch. 6] and of ISTTOK [27]. A SISO approach is adopted also at JET to control the shape of vertically elongated plasmas; however in this case more than two plasma shape descriptors can be controlled (see [60]).

For more complex systems, or when high performance is required, MIMO decoupling controllers have been envisaged since the early eighties [42, 65], and used to solve the plasma current position and shape control problem. In general, in order to achieve better performance, model-based design approaches have been adopted.

As far as plasma boundary shape control is concerned, different solutions have been proposed either based on isoflux [41, 51, 83], or on plasma-wall gaps control [17, 19, 47, 60], whose design has been based using a wide range of techniques, including $\mathscr{H}_\infty$ and model-predictive control. Moreover, in tokamaks with vertically elongated and unstable plasmas different solutions have been proposed to solve the vertical stabilization problem: simple SISO controllers [53, 59], optimal linear-quadratic control [64], predictive control [48], nonlinear adaptive control [73], and robust control [1, 78]. In [79] a full multivariable model of the vertical instability is used to perform a matrix pencil analysis in order to provide for a rigorous demonstration of necessary conditions for stabilization of the plasma by proportional–derivative feedback of vertical displacement. In [72] an anti-windup synthesis is proposed to allow operation of the vertical controller in the presence of saturation.

This chapter introduces a reference architecture for plasma magnetic control in tokamaks, and describes the control algorithms required in each component of such architecture. In particular, the chapter is made by four main sections. "Systems and Control Basics" section introduces the systems and control jargon and backgrounds, since the reader is assumed to have some knowledge about control in order to understand the details of the proposed algorithms. Those reader who are already familiar with control theory can skip this section. "Linear Model for the Plasma and the Surrounding Conductive Structures" section presents the linearized model used to simulate the behaviour of the plasma, of PF circuits, and of the surrounding passive structures, under the axisymmetric assumption. The algorithms for plasma magnetic control, together with the overall control architecture are presented in details in "Plasma Magnetic Control" section, at the beginning of which the control magnetic problem is also stated. The effectiveness of the overall proposed magnetic control system is shown in "Experiments and Simulations" section by means of experimental results, as well as by simulations performed for machines that are currently under construction. Eventually, some conclusive remarks are given.

Those readers who want to deepen the topics discussed in this chapter, may refer to the survey [8] and to the monograph [20] for a complete overview of magnetic control in tokamaks. As far as the systems and control theory is concerned, a lot of textbooks are available for those who wants to go deeper in the subject; among the various, the author would suggest [22, 26, 43, 55, 76, 84].

Systems and Control Basics

In this section some fundamentals on systems and control theory are introduced, together with the notation the will be used in the next sections. It should be noticed. Note that, for the sake of brevity, the discussion will necessarily skip many details. The reader already familiar with this subject can move directly to "Linear Model for the Plasma and the Surrounding Conductive Structures" section.

State-Space Dynamical Systems

In the systems and control context, given a physical system, we will refer to the *dynamical model* in the so-called *state-space* form, as the following set of differential equations

$$\dot{x}(t) = f(x(t), u(t), t), \quad x(t_0) = x_0 \tag{1a}$$

$$y(t) = h(x(t), u(t), t) \tag{1b}$$

where

- $x(t) \in \mathbb{R}^n$ is the system *state*, where n is called the *order* of the system
- $x(t_0) \in \mathbb{R}^n$ is the *initial condition*
- $u(t) \in \mathbb{R}^m$ is the *input* vector
- $y(t) \in \mathbb{R}^p$ is the *output* vector

and where (1a) and (1b) are called state and output equation, respectively. Although it is a mathematical representation of the real physical system (see also Fig. 3), the model (1) is commonly referred to as the *dynamical system* in the state-space form.

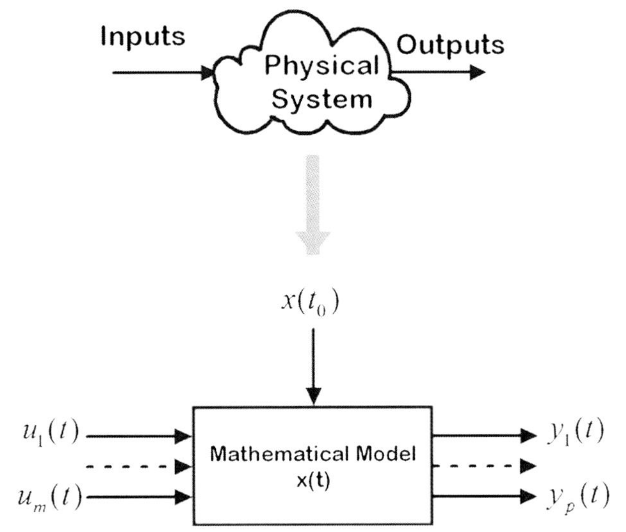

Fig. 3 The physical system under consideration and the mathematical model used to described its behaviour

When the behaviour of a physical process can be assumed linear and time-invariant (LTI), the state space model (1) can be rewritten as

$$\dot{x}(t) = Ax(t) + Bu(t), \quad x(0) = x_0, \tag{2a}$$

$$y(t) = Cx(t) + Du(t), \tag{2b}$$

where $A \in \mathbb{R}^{n\times n}$, $B \in \mathbb{R}^{n\times m}$, $C \in \mathbb{R}^{p\times n}$ and $D \in \mathbb{R}^{p\times m}$. A dynamical system is called SISO when $m = 1$ and $p = 1$, while is called MIMO otherwise. Note that, for a time-invariant system the initial time t_0 can be assumed to be equal to 0 without loss of generality.

Given the nonlinear and time-invariant system

$$\dot{x}(t) = f(x(t), u(t)), \quad x(0) = x_0, \tag{3a}$$

$$y(t) = h(x(t), u(t)) \tag{3b}$$

it is possible to describe its behaviour around equilibrium state using a LTI system. Indeed, if the input is constant, i.e. if $u(t) = \bar{u}$, let consider the solutions $x_{e_1}, x_{e_2}, \ldots, x_{e_q}$ of the homogeneous equation

$$f(x_e, \bar{u}) = 0. \tag{4}$$

The solutions of (4) represent the equilibrium states of (3). Given the equilibrium state x_{e_i}, the correspondent output is given by

$$y_{e_i} = h(x_{e_i}, \bar{u}).$$

Given an asymptotically stable equilibrium x_e, if $x_0 = x_e + \delta x_0$ and $u(t) = \bar{u} + \delta u(t)$, with δx_0 and $\delta u(t)$ *sufficiently small*, then the behaviour of (3) around the considered equilibrium state is described by the LTI system

$$\delta\dot{x}(t) = \left.\frac{\partial f}{\partial x}\right|_{\substack{x = x_e \\ u = \bar{u}}} \delta x(t) + \left.\frac{\partial f}{\partial u}\right|_{\substack{x = x_e \\ u = \bar{u}}} \delta u(t), \quad \delta x(0) = \delta x_0, \tag{5a}$$

$$\delta y(t) = \left.\frac{\partial h}{\partial x}\right|_{\substack{x = x_e \\ u = \bar{u}}} \delta x(t) + \left.\frac{\partial h}{\partial u}\right|_{\substack{x = x_e \\ u = \bar{u}}} \delta u(t). \tag{5b}$$

The total output can be then computed as

$$y(t) = h(x_e, \bar{u}) + \delta y(t).$$

When dealing with control system design it is desirable that the process to be controlled is modeled using LTI systems. This will be the case also for plasma magnetic control. In particular, the behaviour of the plasma, of the currents in the active coils and in the surrounding passive structures will be described by the LTI system that will be presented in "Linear Model for the Plasma and the Surrounding Conductive Structures" section. Although this *engineering-oriented* LTI models may appear too simplified with respect to the codes used to run detailed *physics-oriented* simulations [46, 68], it is sufficiently detailed to capture the main features of the process to be controlled, as far as magnetic control is considered. It will be the robustness of the control system that will assure both performance and stability, also in the presence of uncertainty and disturbances.

Stability of Dynamical Systems

Let $x_e \in \mathbb{R}^n$ be an equilibrium state for system (1), i.e. the solution of (4) for a given constant input $\bar{u}$. Given this input $\bar{u}$, according to the definition of Lyapunov stability (see [56, Ch. 4]), x_e is said to be *stable* if, for each $\varepsilon > 0$, there exists a positive scalar δ (possibly depending on t_0 and ε) such that, for all $t_0 \in \mathbb{R}^+$, if $\|x_0 - x_e\| < \delta(\varepsilon, t_0)$, then[1]

$$\|x(t) - x_e\| < \varepsilon, \quad t \geq t_0.$$

On the other hand, the equilibrium state x_e is said to be *unstable* if it is not stable, while is said to be *asymptotically stable* if it is stable and δ can be chosen such that

$$\|x_0 - x_e\| < \delta \Rightarrow \lim_{t\to+\infty} \|x(t) - x_e\| = 0.$$

In the case of LTI systems (2), it can be shown that the stability property is related to the model rather than to a

[1] The symbol $\|\cdot\|$ denotes any p-norm, i.e.

$$\begin{cases} \|x\|_p = (|x_1|^p + \cdots + |x_n|^p)^{\frac{1}{p}}, & 1 \leq p < +\infty, \\ \|x\|_\infty = \max_i |x_i|. \end{cases}$$

specific equilibrium point, as in the case of nonlinear dynamical systems. In particular, asymptotic stability of LTI systems roughly asserts that, when the input $u(t) = 0$, it is

$$\lim_{t \to +\infty} x(t) = 0,$$

for any initial state x_0. The following well known results hold for LTI systems (see also [55, Sec. 2.6.1]).

Theorem 1 *System (2) is* ***asymptotically stable if and only if*** *A matrix is Hurwitz, that is if every eigenvalue λ_i of A has strictly negative real part ($\Re(\lambda_i) <, \forall \lambda_i$).*

Theorem 2 *System (2) is* ***unstable if*** *A has at least one eigenvalue $\bar{\lambda}$ with strictly positive real part, that is*

$\exists\, \bar{\lambda}$ s.t. $\Re(\bar{\lambda}) >$.

As we have already seen, for nonlinear system the stability property is related to a specific equilibrium x_e. However, the stability of x_e can be assessed analyzing the stability property of the corresponding linearized model (5). Indeed, the following results hold (for more details about the stability of equilbria in nonlinear systems, the interested reader may refer to [56, Ch. 4]).

Theorem 3 *The equilibrium state x_e of the nonlinear system (1) is* ***asymptotically stable if*** *all the eigenvalues of the correspondent linearized system (5) have strictly negative real part.*

Theorem 4 *The equilibrium state x_e of the nonlinear system (1) is* ***unstable if*** *there exists at least one eigenvalue of the correspondent linearized model (5) with strictly positive real part.*

Transfer Functions, Zeros and Poles of LTI Systems

Given a LTI system (2) the corresponding *transfer matrix* from the input u to the output y is defined as

$$G(s) = C\big(sI - A\big)^{-1}B + D, \tag{6}$$

where $s \in \mathbb{C}$ and I denotes $n \times n$ identity matrix. If we denote with $U(s)$ and $Y(s)$ the Laplace transforms of $u(t)$ and $y(t)$, then it holds that

$$Y(s) = G(s)U(s),$$

when the initial condition of system (2) is $x(0) = 0$. In the case of SISO systems, (6) is called *transfer function* and it can be shown that it is equal to the Laplace transform of the impulsive response of system (2) with zero initial condition. Moreover, given the transfer function $G(s)$ and the Laplace transform of the input $U(s)$, the time response of system (2) can be computed as inverse transform of $G(s)U(s)$, without solving differential equations. As an example, the *step response* of a system can be computed as

$$y(t) = \mathscr{L}^{-1}\left[G(s)\frac{1}{s}\right].$$

Given a SISO LTI system, its transfer function is a rational function of s, i.e.

$$G(s) = \frac{N(s)}{D(s)} = \rho\frac{\Pi_i(s - z_i)}{\Pi_j(s - p_j)},$$

where $N(s)$ and $D(s)$ are polynomials in s with real coefficients and such that $deg(N(s)) \leq deg(D(s))$, while p_j and z_i are the *poles* and *zeros* of $G(s)$, respectively.

From the definition of transfer function and from the definition of eigenvalue of a matrix, it readily follows that each pole of $G(s)$ is an eigenvalue of the A matrix, while the converse is not necessary true. If all the poles of $G(s)$ have strictly negative real part—i.e. they are located in the left half of the s-plane (LHP)—then the SISO system is said to be *Bounded–Input Bounded–Output (BIBO) stable*. In particular, a system is said to be BIBO stable if a bounded input to the system results in a bounded output over the time interval $[0, +\infty)$.

A transfer function can be also specified in terms of the following parameters

- time constants (τ,T)
- natural frequencies (ω_n,α_n)
- damping factors (ξ,ζ)
- gain (μ)
- system type (i.e., the number g of poles/zeros in 0)

$$G(s) = \mu\frac{\Pi_i(1 + T_i s)\Pi_j\left(1 + 2\frac{\zeta_j}{\alpha_{n_j}}s + \frac{s^2}{\alpha_{n_j}}\right)}{s^g\Pi_k(1 + \tau_k s)\Pi_l\left(1 + 2\frac{\xi_l}{\omega_{n_l}}s + \frac{s^2}{\omega_{n_l}}\right)}.$$

Moreover, when dealing with transfer functions, it is usual to resort to *block diagrams* which permits a graphical representation of the interconnections between systems in a convenient way. The basic interconnections between two transfer functions are shown in Fig. 4. In particular, given two asymptotically stable LTI systems whose transfer functions are $G_1(s)$ and $G_2(s)$ the following conclusions can be drawn

- the *series* connection $G_2(s)G_1(s)$ is asymptotically stable
- the *parallel* connection $G_1(s) + G_2(s)$ is asymptotically stable
- the *feedback* connection $\frac{G_1(s)}{1 \pm G_1(s)G_2(s)}$ *is not necessarily stable*

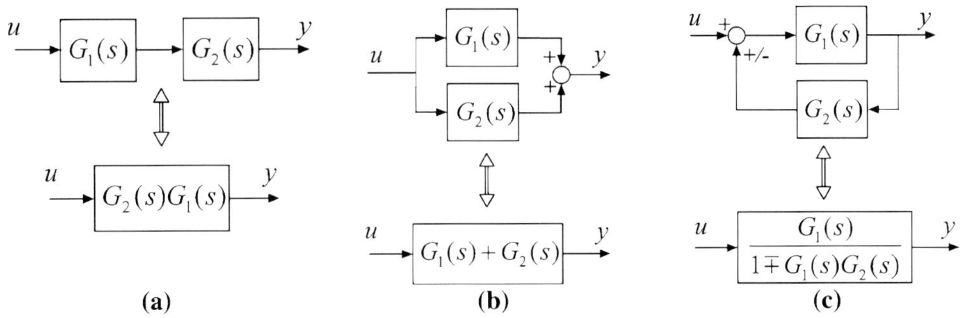

Fig. 4 Basic interconnections in block diagrams and correspondent equivalent block. **a** Series connection, **b** parallel connection, **c** feedback connection. In this case it is assumed that $G_1(s)$ and $G_2(s)$ are SISO transfer functions

Indeed, since all the poles are also eigenvalues of a given LTI system, in the case of series and parallel connections, the poles remain the same as in the interconnected systems, and therefore the stability property is preserved. On the other hand, the poles of the feedback connection are not the ones of the interconnected systems, hence stability of the closed-loop system cannot be guaranteed *a priori*.

From this fact, it follows that the design of a feedback control loop should be carried out with care, in order not to lose stability. On the other hand, it is also true that feedback control can be used to *stabilize* unstable plant.

Frequency Response of LTI Systems

Given the LTI system (2), the complex function

$$G(j\omega) = C\left(j\omega I - A\right)^{-1}B + D,$$

with $\omega \in \mathbb{R}^+$ is called *frequency response* of (2), and it permits to evaluate the system steady-state response to a sinusoidal input when the system is asymptotically stable. In particular if

$$u(t) = A\sin(\bar{\omega}t + \varphi),$$

then the steady-state response of a LTI system is given by

$$y(t) = |G(j\bar{\omega})|A\sin(\bar{\omega}t + \varphi + \angle G(j\bar{\omega})).$$

Given a LTI system $G(s)$ the *Bode diagrams* are a graphical representation of the magnitude of $G(j\omega)$ (specified in dB, that is $|G(j\omega)|_{dB} = 20\log_{10}|G(j\omega)|$) and of the phase of $G(j\omega)$ (specified in degree), as a function of the frequency ω (expressed in rad/s) on a base 10 semi-log scale. Figure 5a shows an example of Bode diagram for a third order system.

The *Nyquist plot* is a polar plot of the frequency response $G(j\omega)$ on the complex plane. This plot combines the two Bode plots—magnitude and phase—on a single graph, with the frequency ω as a parameter along the curve, ranging from $-\infty$ to $+\infty$. This alternative graphical representation of the frequency response is typically useful to check stability of closed loop systems (see "Feedback Control Systems" section). Figure 5b shows the Nyquist for the same third order system considered in Fig. 5a.

Another possible graphical representation for the frequency response of a LTI system is the *Nichols plot*. This plot combines the two Bode diagrams on a single plot; indeed in a Nichols plot both the magnitude and the phase of $G(j\omega)$ are plotted on a single chart, with frequency ω as a parameter along the curve.[2] Figure 5c shows the Nichols plot for the third order system considered in Fig. 5a. Nichols plots are useful for the design of control systems, in particular for the design of lead, lag, and lead-lag compensators; furthermore they permit to easily estimate the stability margins of a SISO closed loop system.

Feedback Control Systems

In this section we will mainly focus on the basics for the design of SISO (negative) feedback control systems. More details on the design of MIMO control systems can be found in [76, 84].

The objective of a control system is to make the output of a plant $y(t)$ behave in a desired way by manipulating the plant input $u(t)$. Referring to the scheme reported in Fig. 6, a controller should keep the output $y(t)$ close to the reference $r(t)$, by counteracting the effect of the disturbance $d(t)$ and by minimizing the effect of the measurement noise $n(t)$, even in the presence of plant unmodeled dynamics.

More in details, a controller should in principle guarantee:

- *Nominal stability*—The closed loop system is stable when the nominal (without uncertainty) model is considered.
- *Nominal Performance*—The closed loop system satisfies the performance specifications when the nominal model is considered.
- *Robust stability*—The closed loop system is stable for all perturbed plants (i.e., taking account uncertainty)

[2] Differently from Nyquist plots, in Nichols plots the frequency ranges from 0 to $+\infty$.

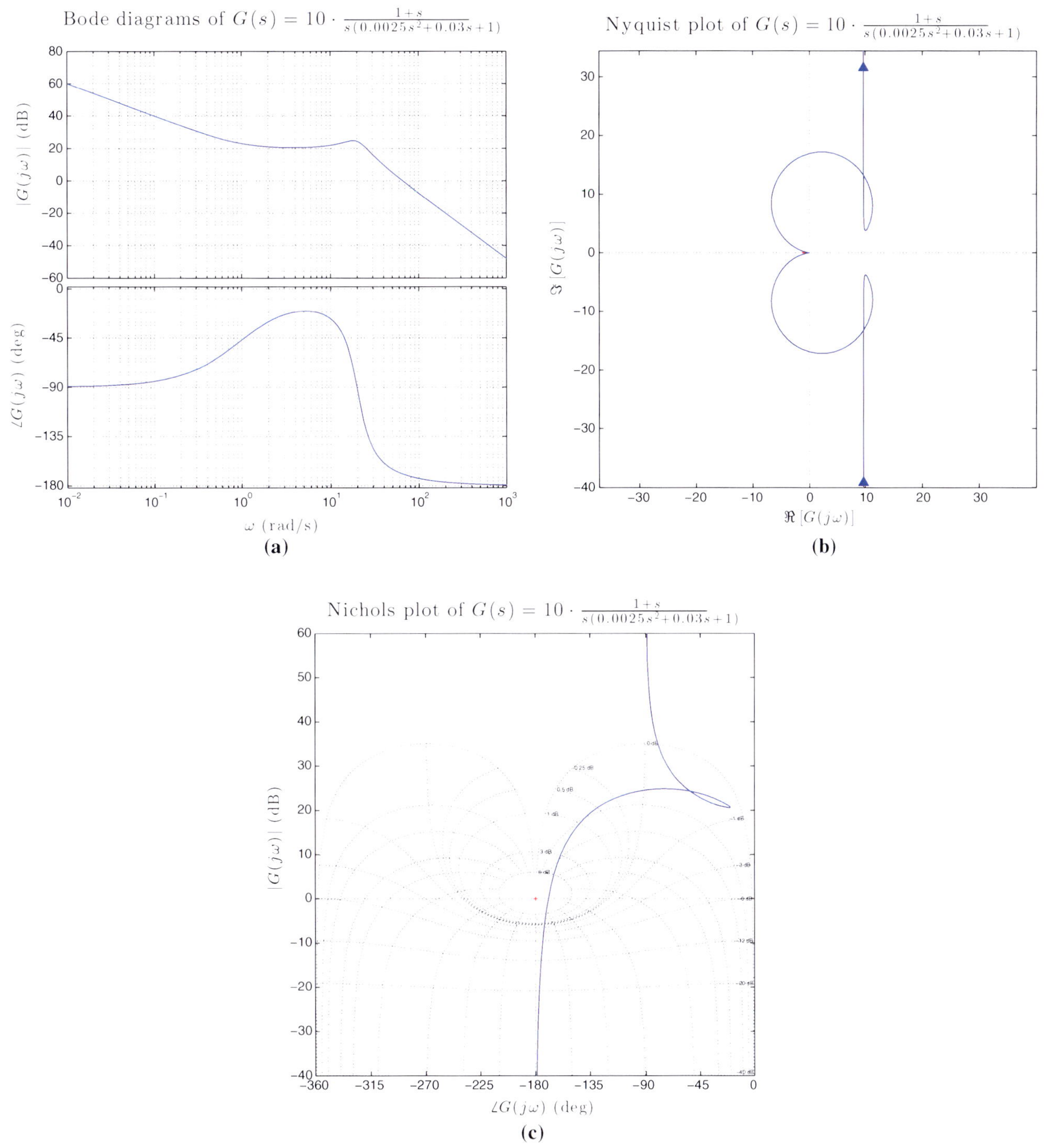

Fig. 5 The different graphical representation of the frequency response of the LTI system described by the transfer function $G(s) = 10\frac{1+s}{s\left(\frac{s^2}{400}+2\frac{0.3}{20}s+1\right)} = 10\frac{1+s}{s(0.0025s^2+0.03s+1)}$. **a** Bode diagrams, **b** Nyquist plot, **c** Nichols plot

- *Robust performance*—The closed loop system satisfies the performance specifications for all perturbed plants

The main sources of difficulty in achieving the above mentioned objectives are that

- the plant model $G(s)$ and the disturbance model $G_d(s)$ may be affected by uncertainty and/or may change with time
- the disturbance is usually not measurable
- the plant can be unstable

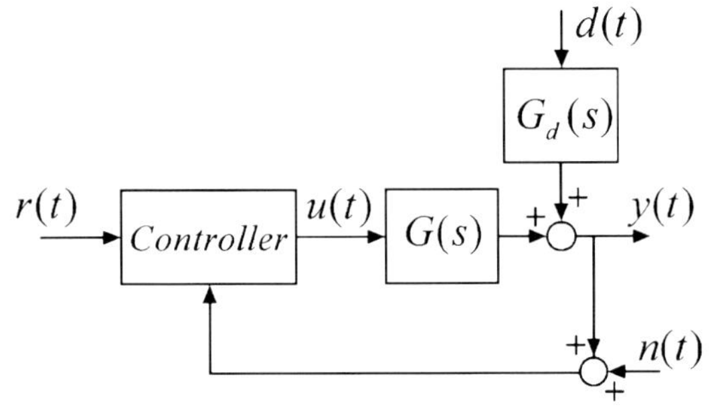

Fig. 6 Reference block diagram for a SISO (negative) feedback control system. $r(t)$ is the reference signal, $u(t)$ is the control input, while $y(t)$ is the controlled output. The two exogenous signals $d(t)$ and $n(t)$ represent the external disturbance and the measurement noise, respectively

Feedback control represents a reliable solution to robustly guarantee the desired performance. However, the benefits of feedback come at a price: designing a feedback control system is not straightforward and instability is always around the corner.

With reference to the one-degree-of-freedom SISO control system shown in Fig. 7, let us introduce the following transfer functions

- $K(s)$ is the *controller transfer function*
- $L(s) = G(s)K(s)$, which is the (open) *loop transfer function*
- $S(s) = \left(I + L(s)\right)^{-1}$, which is the *sensitivity function*
- $T(s) = \left(I + L(s)\right)^{-1}L(s)$, which is the *complementary sensitivity function*.

Note that the above definition (and many of the following introduced in this section) are written using the notation for transfer matrices, hence they hold also for MIMO control systems. For example, in the case of SISO systems, the two sensitivity functions are

$$S(s) = \frac{1}{1 + L(s)},$$

and

$$T(s) = \frac{L(s)}{1 + L(s)}.$$

From definition of the two sensitivity transfer functions it follows that

$$T(s) + S(s) = I. \tag{7}$$

Moreover, exploiting the linearity and applying the composition rules for block diagrams, it is possible to write the following relationships in the Laplace domain

$$Y(s) = T(s) \cdot R(s) + S(s) \cdot G_d(s) \cdot D(s) - T(s) \cdot N(s), \tag{8a}$$

$$E(s) = -S(s) \cdot R(s) + S(s) \cdot G_d(s) \cdot D(s) - T(s) \cdot N(s), \tag{8b}$$

$$U(s) = K(s) \cdot S(s) \cdot R(s) - K(s) \cdot S(s) \cdot G_d(s) \cdot D(s) - K(s) \cdot S(s) \cdot N(s), \tag{8c}$$

where $R(s)$, $D(s)$, $N(s)$ and $E(s)$ are the Laplace transforms of the reference signal $r(t)$, of the disturbance $d(t)$, of the measurement noise $n(t)$, and of the control error $e(t)$. it should be noticed that, from (8a) it follows that the complementary sensitivity function $T(s)$ correspond to the *closed loop* transfer function between the reference signal and the plant output.

If we now consider Eq. (8a), it can be noticed that:

- in order to reduce the effect of the disturbance $d(t)$ on the output $y(t)$, the sensitivity function $S(s)$ should be kept *small* (particularly in the *low frequency* range, where the disturbance is assumed to *act*)
- in order to reduce the effect of the measurement noise $n(t)$ on the output $y(t)$, the complementary sensitivity function $T(s)$ should be kept *small* (particularly in the *high frequency* range, where the noise should assume significant values)

However, for all frequencies the relationship (7) holds. Given this *control dilemma*, designing a controller $K(s)$ is about finding a good trade-off solution between the minimization of the disturbance effect, and the minimization of the effect of the measurement noise. Moreover, since the sensitivity function describes also the relative sensitivity of the closed-loop transfer function $T(s)$ to the relative plant model error due to model uncertainty, i.e. it is also

$$\frac{\frac{dT}{T}}{\frac{dG}{G}} = S(s),$$

it follows that keeping $S(s)$ (especially at *high frequency*, where there are usually more uncertainties in the model) permits to increase the robustness of the closed-loop system.

Furthermore, when designing a feedback control system, stability of the closed-loop system may become an issue. As a simple example, Fig. 8 shows the behaviour of a SISO control system for different values of the overall gain of a proportional-integral (PI) controller.

Usually the frequency response $|L(j\omega)|$ of the loop transfer function has a *low-pass* behaviour, and the *crossover frequency* ω_c is defined as the frequency such that $|L(j\omega_c)| = 1$ (see Fig. 9). In most of the cases the crossover frequency is a good estimation of the closed-loop

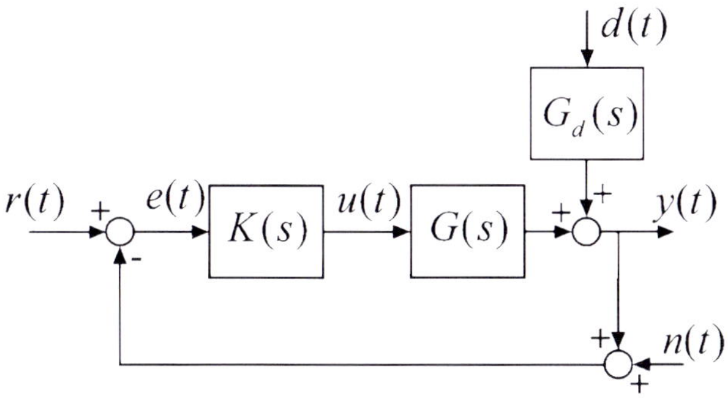

Fig. 7 Block diagram of a one-degree-of-freedom SISO controller

bandwidth. The frequency response $L(j\omega)$ of the loop transfer function is used to define the stability margins, which are used as quantitative descriptors of the robust stability in SISO control systems.

With reference to the Nyquist plot of a typical $L(j\omega)$ (which is assumed to have a low-pass behaviour) shown in Fig. 10, the *gain margin* (*GM*) is defined as

$$GM = 1/|L(j\omega)_{180}|,$$

where ω_{180} is the *phase crossover frequency*, that is the frequency such that $\angle L(j\omega)_{180} = 180°$.

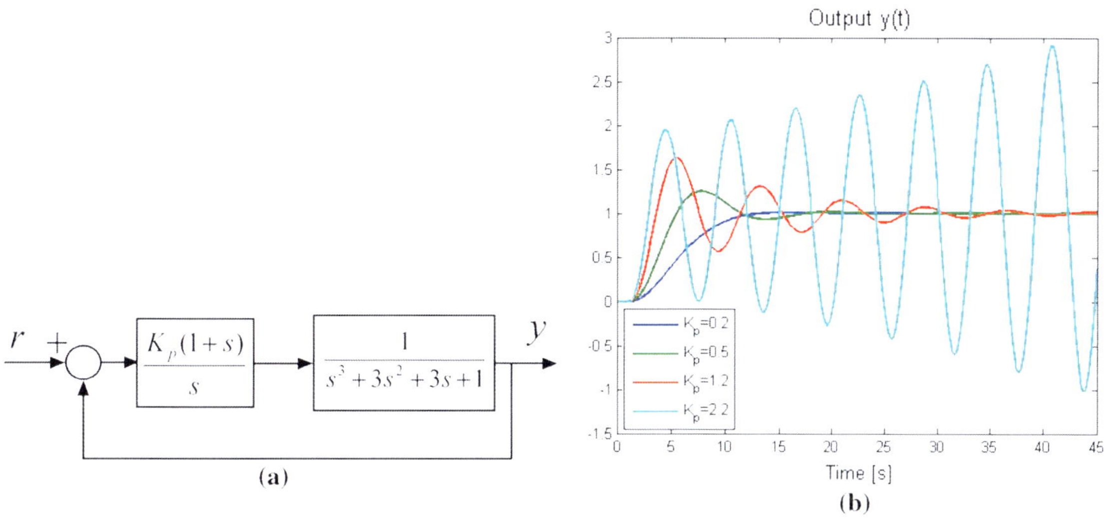

Fig. 8 Example of controller overreaction that may cause instability of the closed-loop system. **a** SISO plant controlled by a proportional-integral controller, **b** step response of the closed-loop system for different values of the proportional gain K_p

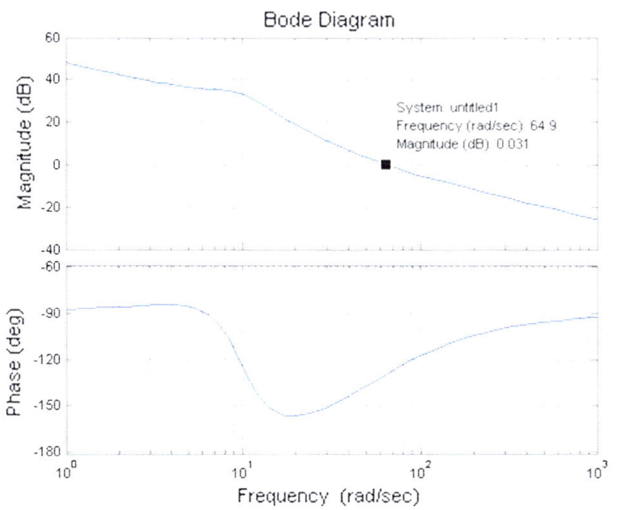

Fig. 9 Typical frequency response $L(j\omega)$ of a loop transfer function with low-pass behaviour. The crossover frequency ω_c is also shown

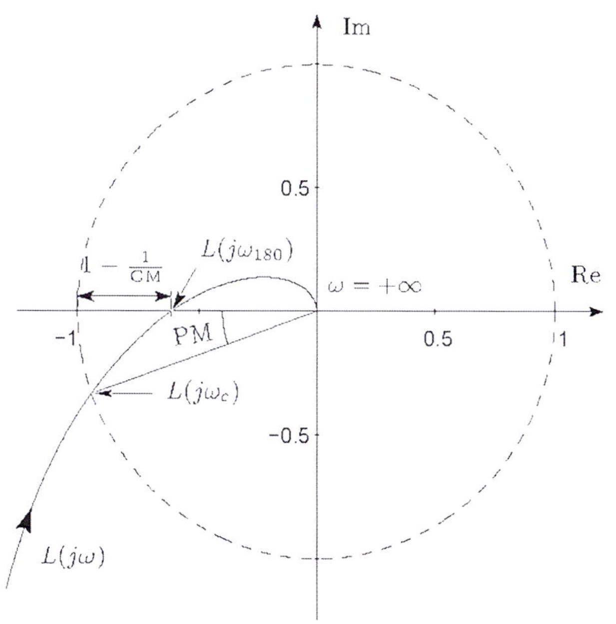

Fig. 10 Nyquist plot of a typical $L(j\omega)$ with a low-pass behaviour and graphical representation of the correspondent stability margins

The *phase margin* (*PM*) is defined

$$\angle L(j\omega_c) + 180^\circ.$$

The *Nyquist Criterion* makes use of the Nyquist plot of $L(j\omega)$ (i.e., of the open loop transfer function) in order to check the stability of a closed-loop system. The readers interested to the proof of this result can refer to [76, Sec. 4.9.2]

Nyquist Stability Criterion Consider a loop frequency response $L(j\omega)$ and let

- P be the number of poles of $L(s)$ with strictly positive real part
- Z be the number of zeros of $L(s)$ with strictly positive real part

The Nyquist plot of $L(j\omega)$ makes a number of encirclements N (assumed positive if *clockwise*) about the point $(-1, j0)$ equal to

$$N = Z - P.$$

It follows that, the closed-loop system is asymptotically stable *if and only if* the Nyquist plot of $L(j\omega)$ encircles (*counter clockwise*) the point $(-1, j0)$ a number of times equal to P.

The following remarks apply to the Nyquist Stability Criterion

- If $L(s)$ has a pole in zero whose multiplicity is equal to l, then the Nyquist plot has a discontinuity at $\omega = 0$. Further analysis indicates that the poles in zero should be neglected, hence if there are no unstable poles, then the loop transfer function $L(s)$ should be considered stable, i.e. $P = 0$.
- If $L(s)$ is stable, then the closed-loop system is unstable for any encirclement (clockwise) of the point -1.
- If $L(s)$ is unstable, then there must be one counter clockwise encirclement of $(-1, j0)$ for each unstable pole of $L(s)$.
- If the Nyquist plot of $L(j\omega)$ intersect the point $(-1, j0)$, then deciding upon even the marginal stability of the system becomes difficult and the only conclusion that can be drawn from the plot is that there exist poles on the imaginary axis.

From the Nyquist Stability Criterion and from the definition of the stability margins, it turns out that, given an asymptotically stable $L(s)$

- the GM is the factor by which the loop gain $|L(j\omega)|$ may be increased before the closed-loop system becomes unstable, and thus it is a direct safeguard against uncertainties on steady-state gain;
- the PM tells how much phase lag can added to $L(j\omega)$ at the crossover frequency ω_c before the phase at this frequency becomes 180°, which corresponds to closed-loop instability. Hence, the PM is a direct safeguard against uncertainty on the time delays in the closed-loop.

As conclusion of this section, we briefly introduce the *root locus* of the loop transfer function $L(s)$. This alternative graphical representation of $L(s)$ permits to evaluate how changes in $L(s)$ affect the location of the closed-loop poles.

Indeed, the time behaviour of a closed-loop system is strictly related to the position of its poles on the complex plane. For example, for a second order closed-loop system it is possible to relate the features of the step response such as the rise and settling times and the overshoot, to the location of its poles.

The closed-loop poles are given by the roots of

$$1 + L(s). \tag{9}$$

Assuming that $L(s) = \rho L'(s)$, the root locus plots the locus of all possible roots of (9) as ρ varies in the range $[0, \infty)$. Moreover, the root locus can be used to study the effect of additional poles and zeros in $L'(s)$, i.e. in the controller $K(s)$. Hence, the root locus can be effectively used to design SISO controllers, as it will be shown in "Algorithms for Plasma Magnetic Control" section for the vertical stabilization system.

Figure 11 shows and example of root locus, when the loop transfer function is

$$L'(s) = K(s) \cdot G(s) = \frac{1+s}{s} \cdot \frac{1}{s^3 + 3s^2 + 3s + 1} = \frac{1}{s(s+1)^2}.$$

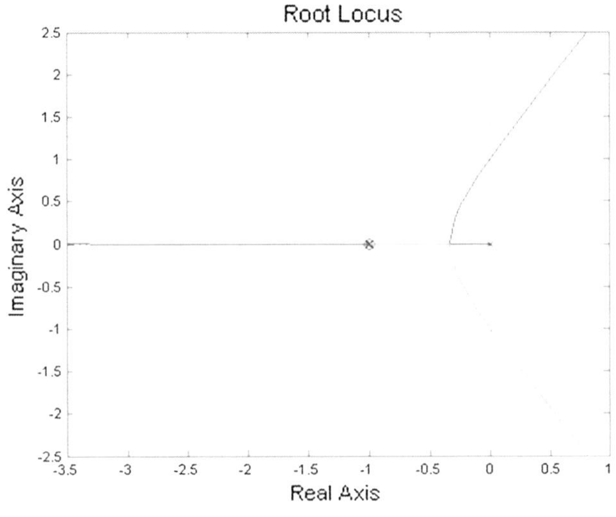

Fig. 11 Example of root locus for the loop transfer function $L'(s) = \frac{1}{s(s+1)^2}$

Linear Model for the Plasma and the Surrounding Conductive Structures

This section presents the linear model used to simulate the behaviour of the plasma, of the PF circuits, and of the surrounding passive structures.

In toroidal and magnetically confined fusion devices, the Grad–Shafranov equation permits to determine two-dimensions magnetic equilibria under axisymmetric assumption [74]. The Grad–Shafranov partial differential equation is solved by using numerical codes [3, 50, 54, 57], and returns the poloidal flux ψ, and in particular its value in the conductive structures that surrounds the plasma, both passive and active (i.e., the PF coils). This information can be used to compute the parameters of the following nonlinear lumped parameters circuital model, that can be obtained under axisymmetric assumption and describes the behaviour of the plasma and of the currents that flow in the surrounding conductive structures (more details can be found in [4, Sec. 2.2] and in [20, Ch. 2]

$$\frac{\mathrm{d}}{\mathrm{d}t}\left[\mathcal{M}\big(y(t),\beta_p(t),l_i(t)\big)I(t)\right]+RI(t)=U(t), \qquad (10a)$$

$$y(t)=\mathcal{Y}\big(I(t),\beta_p(t),l_i(t)\big). \qquad (10b)$$

where

- $y(t)$ are the outputs to be controlled; other than the plasma current and the current in the active and passive structures, this vector may contain plasma shape and position descriptors, e.g. the fluxes at fixed point inside the vacuum chamber, the position of the plasma current centroid, the descriptors of the plasma boundary, such as the plasma-wall gaps (see also the examples reported in Fig. 12, where the ITER poloidal cross-section is shown).
- $I(t)=\big(I_{PF}^T(t)\ I_e^T(t)\ I_p(t)\big)^T$ is the vector that includes the currents in the active coils $I_{PF}(t)$, the eddy currents in the passive structures $I_e(t)$, and the plasma current $I_p(t)$;
- $U(t)=\big(U_{PF}^T(t)\ 0^T\ 0\big)^T$ is the input voltages vector;
- $\mathcal{M}(\cdot)$ is the mutual inductance nonlinear function; this function depends on the plasma internal profiles[3] (this dependency is taken into account using the poloidal beta β_p and the internal inductance l_i), and on the plasma shape and position, whose descriptors are included in the output vector $y(t)$.
- R is the resistance matrix;

[3] For a given *shape* of the plasma internal profiles, e.g., *bell* shaped profiles, the two parameters β_p and l_i measure the plasma internal distributions of pressure and current, respectively. Moreover, their variations can be considered as disturbances as far as the magnetic control is concerned.

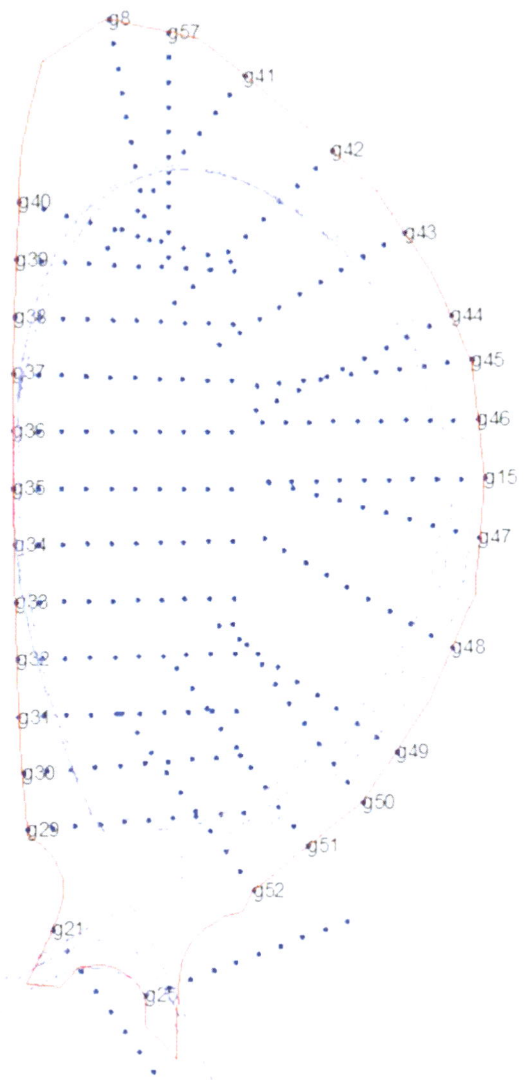

Fig. 12 Example of plasma-wall segments used to define *gaps* between the plasma boundary and the first wall for the ITER tokamak (figure taken from [14]). Along the same segments it is possible to define also points where virtual flux probes can be placed. Gaps and fluxes on these virtual probes can be added as outputs of the model presented in "Linear Model for the Plasma and the Surrounding Conductive Structures" section

- $\mathcal{Y}(\cdot)$ is the output nonlinear function.

It is worth to notice that, other than on axisymmetry, the nonlinear lumped parameters model (10) relies also on the two further main assumptions, namely:

- the inertial effects are neglected at the time scale of interest, since plasma mass density is low;
- the magnetic permeability is assumed homogeneous and equal to 0 everywhere.

Mass Versus Massless Plasma Although a massless plasma assumption is usually made when dealing with magnetic modeling, in [49, 79] the authors proved that there are certain cases when neglecting plasma mass may lead to erroneous conclusion on plasma vertical stability.

From the magnetic control point of view, a plasma equilibrium is specified in terms of nominal values of the plasma current $I_{p_{eq}}$, of the currents in the PF circuits $I_{PF_{eq}}$, and of the disturbances, namely $\beta_{p_{eq}}$ and $l_{i_{eq}}$. Following the linearization procedure briefly described in "State-Space Dynamical Systems" section, it is possible to specify the behaviour of the plasma and of the surrounding coils around a given equilibrium by means of the following state space model

$$\dot{x}(t) = Ax(t) + Bu(t) + E\dot{w}(t) \tag{11a}$$

$$y(t) = Cx(t) + Fw(t), \tag{11b}$$

where

- $x = \begin{pmatrix} \delta I_{PF}^T & \delta I_e & \delta I_p \end{pmatrix}^T$ is the vector the current variations in the PF circuits, in the passive structures, and of the plasma current;
- u is the vector of the voltage variations applied to the PF circuits;
- $w = \begin{pmatrix} \delta\beta_p & \delta l_i \end{pmatrix}^T$ is the disturbances vector, i.e. the variations of β_p and l_i.

It is worth to notice that the linearization of (10) can be computed either analytically (as an example, see [4, Sec. 2.3]), or numerically as in [2].

As already mentioned in "State-Space Dynamical Systems" section, although simpler than the physics-oriented simulation codes, the availability the engineering-oriented linear model (11) is essential not only to enable model-based design of control systems. Indeed, it permits also to automate both the validation and deployment of the plasma axisymmetric magnetic control (relevant examples on existing tokamak devices can be found in [7, 24, 28, 32, 40, 52]. Furthermore, reliable linear plasma models are also used to support the design and commissioning of the magnetic diagnostic [69], as well as to run inter-shot simulations aimed at optimising the controller parameters [38].

Linear models (11) can be generated starting from nonlinear equilibrium codes, by numerically computing the derivatives in (5). For example, plasma/circuits models in the form (11) are generated by the CREATE 2D nonlinear equilibrium codes [3, 4], which are available in the Matlab/Simulink® environment, which in turn represents the de facto standard tool for control systems design and validation. The CREATE equilibrium codes, together with the correspondent linearized models, have been extensively validated against different magnetically confined fusion devices, among which there are JET [5], TCV [17], and more recently the EAST tokamak [7, 31]. The same modeling tools have been also used to perform preliminary studies and code benchmarking for ITER [67], DEMO [81], JT-60SA [28] and DTT [6]. It shall be remarked that the CREATE-NL+ equilibrium code [70] can be integrated with transport solvers [70], in order to receive as input the plasma internal profiles, hence running coupled nonlinear simulations.

Plasma Magnetic Control

The Plasma Axisymmetric Control Problems

The plasma *axisymmetric* magnetic control in tokamaks includes the following control problems

- the *Vertical Stabilization* problem;
- the *Plasma Current Control* problem.
- the *Plasma Shape Control* problem;
- the *PF Currents Control* problem.

Although it may apparently seem that the *PF Current Control* is not directly related to the plasma magnetic control, it is important to remark that the regulation of the currents into the external active coils is usually included in the magnetic control system. Indeed, as anticipated at the beginning of "Introduction" section, plasma scenarios are defined in terms of nominal currents into the PF coils. Furthermore, in the architecture proposed in "A Flexible Architecture for Plasma Magnetic Control in Tokamaks" section the plasma current and shape controller will be designed so to compute additional references for the PF current controller. In the author's opinion, the solution of the PF Current Control problem is a precondition to implement advanced plasma current and shape control system.

It should be also remarked that the axisymmetric control problems are not the only ones that arise during tokamak operations and which are related to the magnetic configuration. Non axisymmetric control problems, such as the control of *Resistive Wall Modes* or of the *Neoclassical Tearing Modes* are also related to magnetic instabilities. In order to tackle these problems, usually more advanced control systems are required, which integrate both plasma magnetic and kinetic control (see [16, 18, 21, 63, 80]).

A brief description of the main objectives for each of the control problems listed above is given in the next sections.

The Vertical Stabilization Problem

High performances plasmas have diverted shape with an elongated poloidal cross-section. One X-point (for *single-null* configurations) or even two X-points (for *double-null* configurations) can be active into the vacuum chamber. Such vertically elongated plasmas turn out to be vertically unstable.

In order to illustrate the plasma vertical instability, let us consider the simplified filamentary model depicted in Fig. 13, where two rings, modeling two active coils, are kept fixed and in symmetric position with respect to the r axis, while the third, which models the plasma, can freely move along the vertical direction. If the currents in the two fixed rings are equal, then the vertical position $z = 0$ is an equilibrium point for the system. Given the plasma current I_p and the current in the active coils I, the following two cases can be considered:

$\mathrm{sgn}(I_p) \neq \mathrm{sgn}(I) \rightarrow$ the plasma poloidal cross-section is circular and the equilibrium is vertically stable (see Fig. 14);

$\mathrm{sgn}(I_p) = \mathrm{sgn}(I) \rightarrow$ the plasma poloidal cross-section is elongated and the equilibrium is vertically unstable (see Fig. 15).

It is important to note that the current in the two active coils of this simplified model represent differential control currents. This additional currents act on top of the equilibrium ones that counteract plasma radial expansion. Indeed, the plasma toroidal geometry generates a force that tends to expand the plasma itself outward along the major radius. An externally applied vertical field is then required to balance this force. Such a vertical field can be generated by currents in the two active coils shown in Fig. 13 that flow in the opposite direction with respect to I_p (see also [44, Sec. 11.7]).

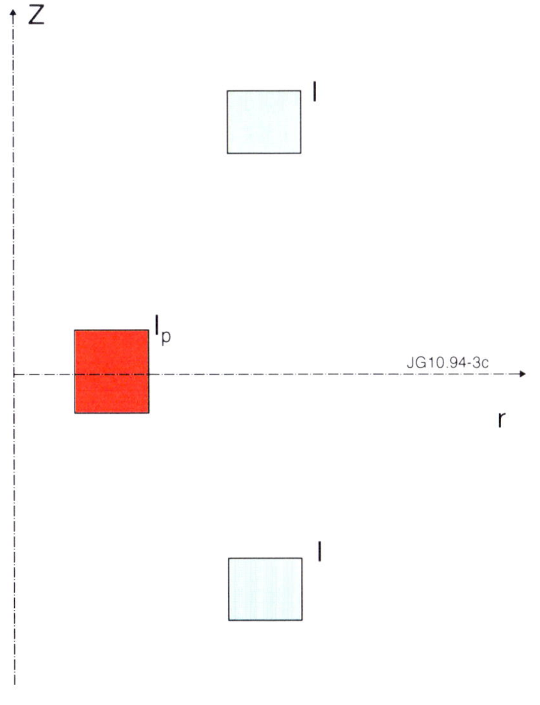

Fig. 13 Simplified electromechanical model with three conductive rings (figure taken from [33]). Two rings are kept fixed and in symmetric position with respect to the r axis and model the active surrounding coils. The third ring can freely move vertically and is used to model the plasma

The plasma vertical instability reveals itself in the linearized model (11) by the presence of an unstable eigenvalue (i.e., real and positive) in the matrix A. Hence, it follows that an active control system is needed to vertically stabilize the plasma column. It should be recalled that, thanks to the presence of the conducting structures that surround the plasma, which slow down the growth time of the vertical instability,[4] it is possible to use a feedback control system for the stabilization. it should be remarked that, without this passive stabilizing action, it would not be possible to actively stabilize the plasma.

Hence, it turns out that the Vertical Stabilization system is essential, since, in the case of elongated plasmas, without this system the tokamak discharge cannot be run. Other that vertically stabilize elongated plasmas, the objective of any VS system includes also the counteraction of the disturbances, such as Edge Localized Modes (ELMs), fast disturbances modeled as Vertical displacement Events (VDEs, [9]), etc.

As conclusion of this section, it is important to remark that in order to achieve plasma vertical stabilization it is not necessarily required to control the vertical position of the centroid. At least in principle, to vertically stabilize the plasma column it should be sufficient to *stop* the plasma, that is controlling to zero its vertical speed, as it will be shown in "The VS System" section.

The Plasma Current Control Problem

During a tokamak discharge, tracking the desired value for the plasma current I_p is another mandatory task that must be performed by the magnetic control system (in Fig. 16 a typical time behaviour for I_p is shown). To accomplish this task the current in the external PF coils should be driven by a controller so to regulate I_p to the desired value. Indeed, from the I_p control point of view, the plasma can be regarded as the secondary coil of a transformer. Robustness is a key requirement for the plasma current control, since it is usually required to have a single control algorithm that

[4] Given a plasma equilibrium and the corresponding linear model (11), an estimation of the growth time for the vertical instability is given by the inverse of the positive eigenvalue of the A matrix.

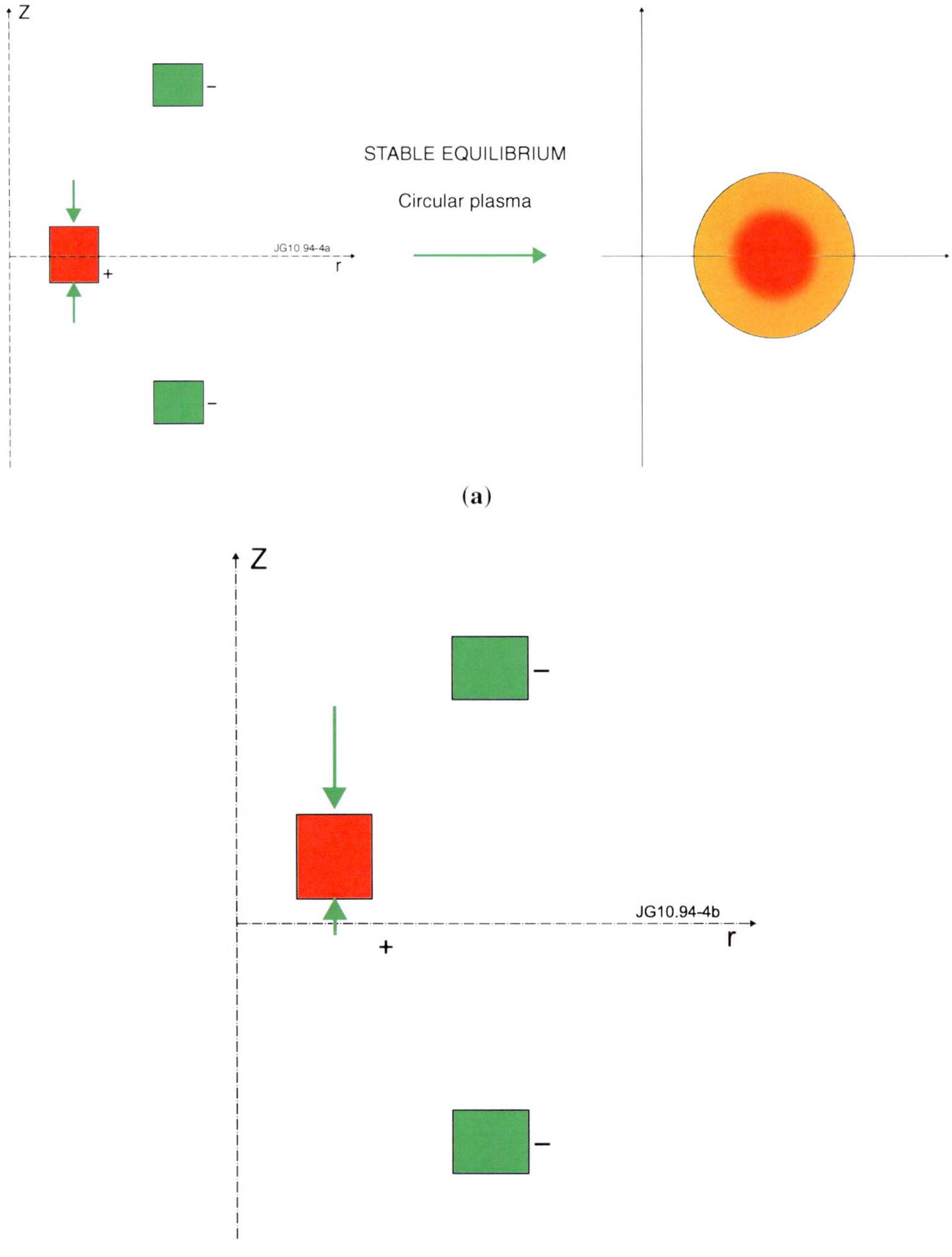

Fig. 14 Stable equilibrium for the simplified electromechanical model depicted in Fig. 13 (figures taken from [33]). **a** Force balance at the equilibrium for a circular plasma, **b** stable equilibrium

works for all the possible operative scenarios, independently of the desired plasma shape.

It should be noticed that both plasma current and shape control may use the current in the PF coils as *control variables* (see also "The Plasma Shape Control Problem" section). Therefore there is also a problem of *actuator sharing* to be taken into account when using this approach for the design of the plasma current and shape control algorithms. However, different solutions can be considered to tackle this issue, from the design of a single integrated controller (see [11, 17]), to the design of two decoupled controllers, by exploiting the possibility of generating a flux that is spatially uniform across the plasma (but with a desired temporal behavior). Such a flux will drive I_p without affecting the plasma shape too much, as it will be shown in "Algorithms for Plasma Magnetic Control" section.

The Plasma Shape Control Problem

As it was already mentioned in "Introduction" section, the control of the plasma boundary is essential to achieve high performance in tokamak operations and to deal with the problem of the power exhaust. Usually, during the first

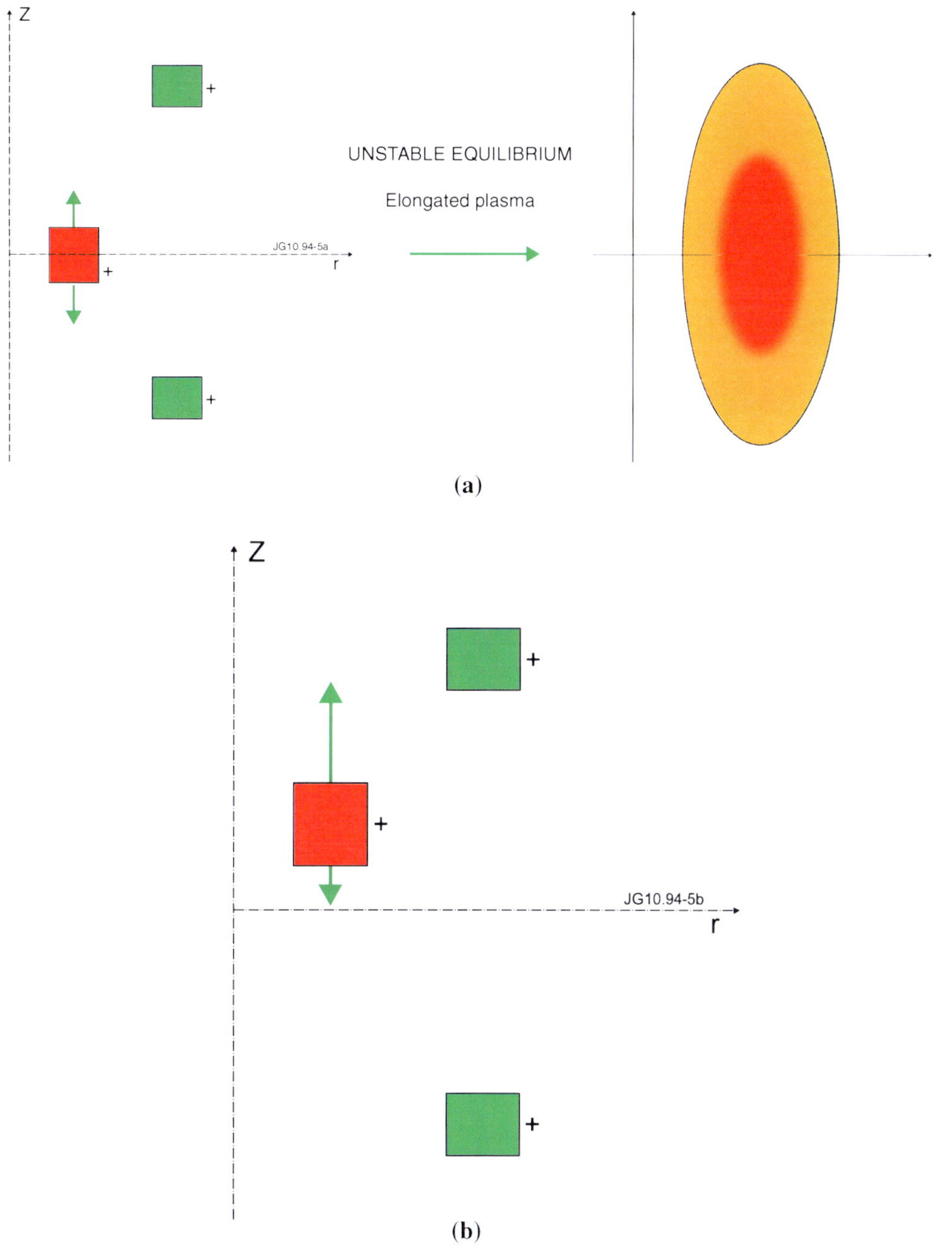

Fig. 15 Unstable equilibrium for the simplified electromechanical model depicted in Fig. 13 (figures taken from [33]). **a** Force balance at the equilibrium for an elongated plasma, **b** unstable equilibrium

phases of a discharge, when the plasma is still in the *limiter* configuration,[5] the control of the plasma current centroid position (both horizontal and vertical), together with the feedforward currents in the PF coils, is typically sufficient to obtain the desired performance. Example of of the plasma centroid controllers can be found in [27, 83]. However, such a solution becomes insufficient when the plasma shape changed from limiter to divertor configuration, and when high confinement configurations are achieved, since the effect of the disturbances on the plasma shape becomes not negligible. In this phases of the discharge the control of the whole plasma boundary is required. This objective can be achieved by simultaneously regulating the position of a set of points along the plasma boundary, as it will shown in "Algorithms for Plasma Magnetic Control" section.

The PF Currents Control Problem

A tokamak scenario is usually defined as a sequence of nonlinear equilibria to be achieved during a discharge. Among the various parameters that define a single snapshot there are also the currents in the PF coils. Hence, there is

[5] For the definition of *limited* and *diverted* plasmas, the reader can refer to Tutorial 7 in [23].

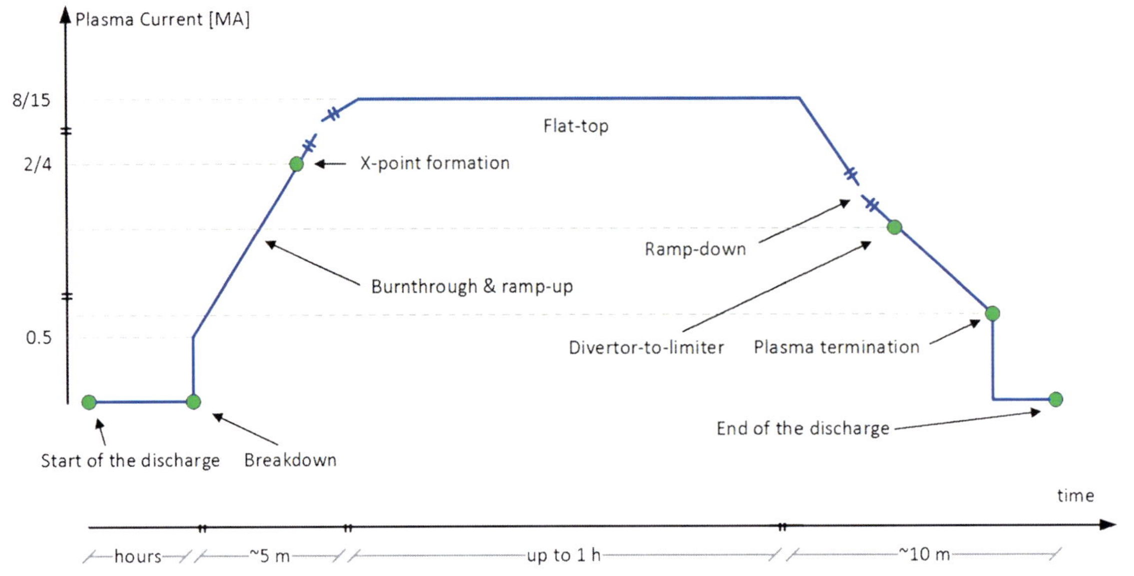

Fig. 16 The typical time behaviour of the plasma current during a tokamak discharge. The I_p values and time scale shown in this figure are the ones expected for an ITER discharge. The different phases of the discharge are also reported

the need to control in closed loop the currents that flow in the PF coils, in order to track nominal scenario currents, as well as the variations requested by the plasma current and shape controllers. In order to increase the reliability of the overall system, the PF current control problem should include also the management of the current saturation in the coils. If the nominal currents are within the plant limits, this latter problem is usually tackled when designing the plasma current and shape control system (for example see [13] and [29]).

A Flexible Architecture for Plasma Magnetic Control in Tokamaks

In this section an overview of the proposed architecture for the magnetic control system is given.

A block diagram of the overall architecture is reported in Fig. 17, where the main components are shown. In particular, the proposed architecture envisages a dedicated block for each of the control problems introduced in "The Plasma Axisymmetric Control Problems" section, namely

- the *Vertical Stabilization (VS) System*, which takes care of the stabilization of the elongated and hence unstable plasmas. This system relies on a set of dedicated actuators, i.e. power supplies that are usually faster than the one used for plasma current and shape control. In the more general case, these dedicated power supplies may feed both ex-vessel and in-vessel coils. In the proposed architecture, other than the current in the actuators, the VS system receives as input the plasma vertical speed, but not the vertical position; indeed this system will only stop the plasma, leaving to the shape controller the task of bringing the plasma back to the desired position;
- the *PF Current (PFC) Decoupling Controller*; this block represents the inner control loop of a nested architecture that includes also the plasma current and shape controllers. This block aims at solving the PF current control problem presented in "Plasma Shape Controller" section; hence it guarantees that the currents in the PF coils track the scenario references, as well as the corrections received from the two outer loops;
- the *Plasma Current Controller*, which tracks the plasma current reference by generating additional requests for the PFC Decoupling Controller;
- the *Plasma Shape Controller*, which controls shape of the last closed flux surface within the vacuum chamber by tracking a set of plasma shape descriptors; in the proposed architecture also this block generates additional requests for the PFC Decoupling Controller.

The next sections will introduce the proposed control algorithms for each of the main blocks shown in Fig. 17.

Algorithms for Plasma Magnetic Control

We will now describe a set of possible control algorithms to be deployed in each of the main blocks reported in the architecture described in the previous section. It should be noticed that the proposed architecture can be used the implement control different control approaches with respect to the ones presented in this section.

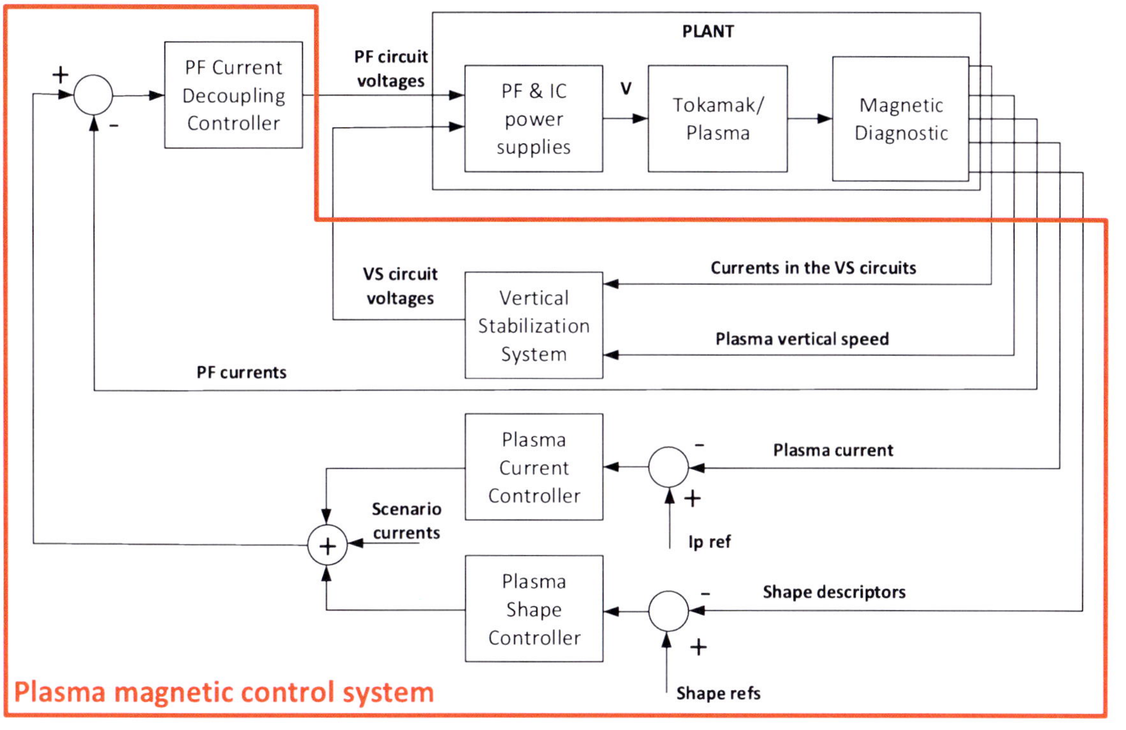

Fig. 17 The proposed architecture for the magnetic control system. The main blocks are shown. The scenario currents represent the PF nominal currents

The VS System

In all the existing machines with elongated plasmas, the vertical stabilization problem (see "The Vertical Stabilization Problem" section) is solved by designing a controller that drives the current in dedicated PF circuits that mainly generate a radial magnetic field, which in turns is needed to apply the vertical force used to stop the plasma column. One example of such dedicated circuit is the JET tokamak *Radial Field Circuit* [66]. At least one pair coils fed in *anti-series*[6] are installed in other existing machines, and the same solution is envisaged for the tokamaks currently under construction, in order to deal with the vertical stabilization problem. A generic setup, that fits with the one existing on different machines (e.g. JET, EAST, ITER) includes both a pair of in-vessel and ex-vessel coils. For this reason in the architecture shown in Fig. 17 the VS system receives as input the vertical speed of the plasma centroid and the current flowing in the dedicated circuits, the latter possibly including both the current in dedicated in-vessel and ex-vessel circuits.

Referring to the this general case, where the VS system can use both an in-vessel and a ex-vessel circuits, the following MIMO controller can be used to vertically stabilize the plasma column

$$U_{IC}(s) = F_{VS}(s) \cdot \left(K_v \cdot \bar{I}_{p_{ref}} \cdot V_p(s) + K_{ic} \cdot I_{IC}(s)\right), \quad (12a)$$

$$U_{EC}(s) = K_{ec} \cdot I_{IC}(s), \quad (12b)$$

where U_{IC} and U_{EC} are the voltages to be applied to the in-vessel and ex-vessel circuits, respectively, while I_{IC} is the current flowing in the in-vessel coil, and V_p is the plasma vertical speed. The controller parameters are

- K_{ic} and K_{ec} are the current gains;
- K_v is the speed gain which, at each time sample, must be scaled by the nominal value of the plasma current $\bar{I}_{p_{ref}}$ (it is assumed that the plasma current controller is tracking the reference with a small steady-state error);
- $F_{VS}(s)$ is the transfer function of a dynamic compensator, which is usually a lead compensator (see [76, Section 2.6.5]).

Being not shielded by the passive structures, if available, the in-vessel circuit loop (12a) is used to achieve stabilization, while, the ex-vessel (12b) is used to reduce the current in the in-vessel circuit. Indeed, the ex-vessel circuit is used to "drain" current from the in-vessel circuit, hence reducing the root mean square value of I_{IC}, in order to meet the specific constraints on the power dissipation for the copper coils (see the ITER example reported in [9, 10]). It is worth to notice that, if the in-vessel coils are not installed (as in the JET case), a control law similar to (12a) can be

[6] Two coils are said to be connected in anti-series if they are connected together with windings in the opposite direction.

used to determine the voltage to be applied to the ex-vessel coil in order to vertically stabilize the plasma column; i.e. the voltage U_{EC} is given by

$$U_{EC}(s) = F_{VS}(s) \cdot \left(K_v \cdot \bar{I}_{p_{ref}} \cdot V_p(s) + K_{ec} \cdot I_{EC}(s)\right),$$

where I_{EC} is the current flowing in the ex-vessel circuit.

Moreover, it should be noticed once again that the proposed VS controller achieves stabilization of the plasma column without the need of controlling the vertical position. Indeed, as it was remarked in "The Vertical Stabilization Problem" section, the main objective of the VS system should be to stop the plasma, rather than to move it back to the original position. This latter task can be performed by the plasma shape controller. Moreover, the proposed controller structure is rather simple, since there are few parameters to be tuned against the operational scenario. Such a simple structure permits to envisage effective *adaptive* algorithms, as it is usually required during operation.

In order to better understand how the in-vessel loop achieves vertical stabilization let us consider the case of the EAST tokamak, whose block diagram is reported in Fig. 18 (more details can be found in [7, 30]). In the EAST case, two in-vessel coils (shown as red rectangles labeled as "IC" in Fig. 2b) are connected in anti-series and represent the actuator dedicated to the vertical stabilization; hence the VS system shown in Fig. 18 includes only the in-vessel loop (12a).

In order to design the controller parameters in (12a) the state space model (11) is used to derive the following input–output relationship

$$Y(s) = \begin{pmatrix} Y_1(s) \\ Y_2(s) \end{pmatrix} = W_p(s) \cdot \widetilde{U}_{IC}(s) = \begin{pmatrix} W_{p_1}(s) \\ W_{p_2}(s) \end{pmatrix} \cdot \widetilde{U}_{IC}(s), \qquad (13)$$

where $\widetilde{U}_{IC}(s)$ is the actual voltage applied to the IC circuit

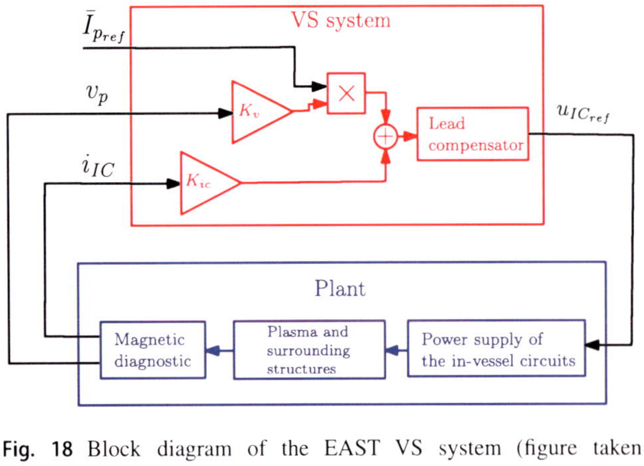

Fig. 18 Block diagram of the EAST VS system (figure taken from [31]). The plant block includes the power supply, the plasma and surrounding conductive structures, and the magnetic diagnostic

by the power supply, while $Y_1(s) = Z_p(s)$ is the plasma vertical position, and $Y_2(s) = I_{IC}(s)$ is the current flowing in the IC circuit. The transfer matrix (13) models the behaviour of the plasma and of the surrounding conductive structures; however, when designing the VS system, the models of both the power supply and of the relevant magnetic diagnostic should be included in the plant. Indeed, these two components usually have a major impact on the performance of the closed loop system. To this aim, the power supply is modeled as a first order filter with a delay, i.e.

$$\widetilde{U}_{IC}(s) = \frac{e^{-\delta_{ps}s}}{1 + s\tau_{ps}} \cdot U_{IC}(s), \qquad (14)$$

where $\tau_{ps} = 100\,\mu s$ and $\delta_{ps} = 550\,\mu s$ are the estimated values for the power supply time constant and delay, respectively. Note that the presence of a time delay is modeled by the exponential term $e^{-\delta_{ps}s}$ in (14). Moreover, at EAST the plasma vertical speed $V_p(s)$ is estimated by means of a derivative filter applied on the measured vertical position of the centroid $Z_p(s)$, i.e.

$$V_p(s) = \frac{s}{1 + s\tau_v} \cdot Z_p(s), \qquad (15)$$

where the time derivative of the derivative filter is set equal to $\tau_v = 1$ ms. As far as the current in the in-vessel circuit is concerned, it is assumed that there is no delay on its measure. The overall plant model can be obtained by computing the series of (13)–(15), which is equal to

$$W_{plant}(s) = \begin{pmatrix} 1 \\ \dfrac{s}{1 + s\tau_v} \end{pmatrix} \cdot W_p(s) \cdot \frac{e^{-\delta_{ps}s}}{1 + s\tau_{ps}}. \qquad (16)$$

In order to use a design approach for (12a) which is based on the root locus, the 550 µs time delay of the IC power supply can be replaced by its third order Padé approximation (see [76, Section 4.1.6]), that is

$$e^{-\delta_{ps}s} \simeq \frac{-(s - 8444)(s^2 - 1.34 \cdot 10^4 s + 8.54 \cdot 10^7)}{(s + 8444)(s^2 + 1.34 \cdot 10^4 s + 8.54 \cdot 10^7)}. \qquad (17)$$

By exploiting the *parity interlacing property* (PIP, [82]), in [30] it was shown that, given the plant model (16), it is not possible to stabilize the plasma by feeding back only the vertical speed V_p. Indeed, given the two positive zeros introduced by the power supply delay (see (17)), the PIP requires the presence of another non negative pole in the open loop transfer function of the $U_{IC} - \dot{V}_p$ channel, otherwise it is not possible to move the unstable pole on the left half of the complex plane using a stable controller. In order to add such a non negative pole, either an integral action on V_p or a feedback on I_{IC} is needed. However, the former solution is equivalent to the feedback of the vertical position Z_p, hence it would add a coupling between the VS system and the plasma shape control, which is not desirable. For this reason, in (12a) it is proposed to add the feedback term on I_{IC}. By closing the loop on I_{IC} another unstable pole is added in the $U_{IC} - V_p$ channel, as it is shown in Fig. 19a; then, by closing the proportional loop on V_p, it is possible to achieve closed loop stabilization, as it is shown in Fig. 19b. Note that the root locus diagrams shown in Fig. 19 are obtained considering a reduced model of order 10 for (16) (see [84, Ch. 7]).

Before concluding this section, it is worth to mention that two different techniques have been proposed to robustly design the controller parameters: in [9, 10] the gains are given by the solution of an optimization problem with Bilinear Matrix Inequality constraints (BMI [77]); multi-objective optimization is used in [31] to tune the controller gains and the parameters of a lead compensator $F_{VS}(s)$.

PFC Decoupling Controller

In the proposed architecture, the Plasma Current and the Plasma Shape Controllers compute their control action as deviations with respect to the PF currents of the nominal scenario. The PFC Decoupling Controller block implements the algorithm that tracks these deviations, as well as the nominal values (see Fig. 17). In particular, it receives as input the currents in the PF circuits together with their desired values, and it generates the voltage references for the PF power supplies. When designing the control algorithm for this block, other than achieving PF decoupling, it must be taken into account that the PF currents should exhibit almost the same dynamic response, since the plasma shape controller will rely on such a behaviour (see "Plasma Shape Controller" section).

The design of the controller can be carried out starting from a plasmaless model. Such a model can be written in the following form

$$L\dot{x}(t) + Rx(t) = u(t), \tag{18}$$

where L and R are the inductance and resistance matrices, $x(t)$ is the vector of the current in the circuits (both active and passive), and $u(t)$ is the vector of the voltages applied to the circuits (which are equal to zero for the passive structures, see also "Linear Model for the Plasma and the Surrounding Conductive Structures" section).

The design of the PFC Decoupling Controller is based on a modified version of the inductance matrix $\widetilde{L}_{PF} \in \mathbb{R}^{n_{PF}} \times \mathbb{R}^{n_{PF}}$, where n_{PF} is the number of PF circuits. $\widetilde{L}_{PF}$ is calculated by neglecting the effect of the passive structures. Furthermore, in order to minimize the control effort, in each row of $\widetilde{L}_{PF}$, all the mutual

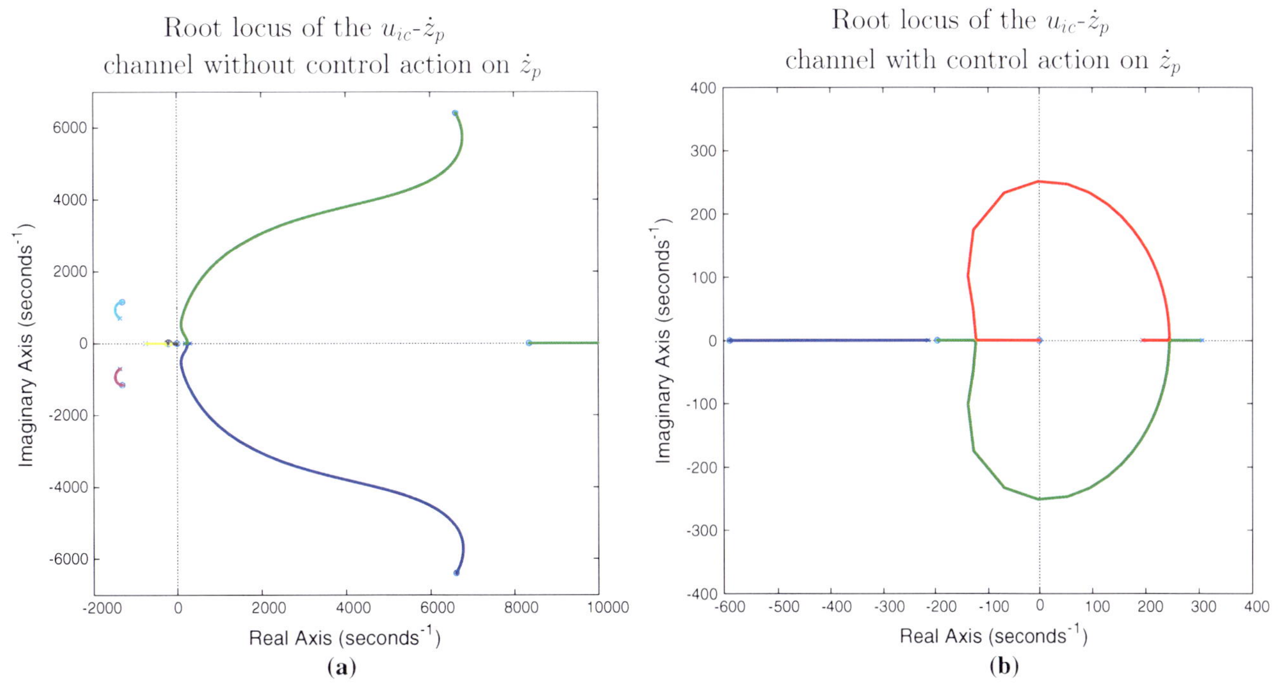

Fig. 19 Root locus of the $U_{IC} - \dot{V}_p$ channel when the controller (12a) is used and the plasma model for the EAST pulse #60938 at $t = 6$ s is considered (figures taken from [30]). **a** Root locus of the $U_{IC} - \dot{V}_p$ channel, when both the loops provided in (12a) are closed, **b** root locus of the $U_{IC} - \dot{V}_p$ channel, when the loop on the IC current provided in (12a) is closed

inductance terms which are less than the 10% of the circuit self-inductance are also neglected. Thanks to this choice, the current in each circuit is controlled only by the circuits that are more coupled with it, avoiding to easily saturate the voltages without gaining any practical improvement in the control of the PF currents.

The proposed design approach aims at obtaining a first order response for all the PF circuits when the PFC Decoupling Controller is closed. Given the time constants τ_{PF_i} for the response of the ith circuit, we define:

$$\Lambda = \begin{pmatrix} 1/\tau_{PF1} & 0 & \dots & 0 \\ 0 & 1/\tau_{PF2} & \dots & 0 \\ \dots & \dots & \dots & \dots \\ 0 & 0 & \dots & 1/\tau_{PF_n} \end{pmatrix}.$$

Note that, in order to have the same dynamic on all the PF circuits, the time constants τ_{PF_i} should be all equal.

The voltages to be applied to the PF circuits are then calculated as

$$U_{PF}(t) = K_{PF} \cdot \left(I_{PF_{ref}}(t) - I_{PF}(t)\right) + \widetilde{R}_{PF} I_{PF}(t), \qquad (19)$$

where $\widetilde{R}_{PF} \in \mathbb{R}^{n_{PF} \times n_{PF}}$ is an estimation of the resistance matrix, while the control gain matrix K_{PF} is given by

$$K_{PF} = \widetilde{L}_{PF} \cdot \Lambda.$$

Taking into account that the estimation of the PF circuit resistances can be usually performed with good accuracy, i.e. $\widetilde{R}_{PF} \cong R_{PF}$, by replacing the controller output (19) in (18), it follows that the closed loop behaviour for the currents in the PF circuits is well described by the differential equation

$$\dot{I}_{PF}(t) = \Lambda \cdot \left(I_{PF_{ref}}(t) - I_{PF}(t)\right).$$

Since Λ is diagonal the desired decoupling is assured by (19). Moreover, the feedback term $\widetilde{R}_{PF} I_{PF}(t)$ is equivalent to an integral action or, in other word, it allows to treat any coil as a supercondutive one.[7]

The Bode diagrams of the closed loop system designed for the JT-60SA tokamak are shown in Fig. 20. It can be noticed that the diagonal channels (from the ith voltage to the ith current, Fig. 20a) exhibit very similar dynamic responses, while the off-diagonal channels (whose Bode diagrams are shown in Fig. 20b) have a much lower magnitude. It follows that good decoupling is achieved.

The design approach described in this section was originally proposed for the JET tokamak in [45, 71]; the same approach was recently tested on the EAST tokamak [37].

Plasma Current Controller

The Plasma Current Controller is one of the two feedback loops that send requests to the PFC Decoupling controller. It receives as input the plasma current together with the corresponding reference waveform, and it computes the PF current deviations needed to obtain the desired I_p.

Let $k_{p_{curr}} \in \mathbb{R}^{n_{PF}}$ be the vector of PF currents that causes a constant flux variation along a closed line containing the envisaged plasma boundary; the field generated by the linear combination of PF currents in $k_{p_{curr}}$ is called the *transformer field*. The elements of the $k_{p_{curr}}$ vector can be obtained via an optimization procedure based on the plasmaless model introduced in "The PF Currents Control Problem" section. Moreover, by definition, the linear combination of PF currents specified by $k_{p_{curr}}$ is such that the effect on the plasma shape is minor.

It follows that the $k_{p_{curr}}$ vector can be used to design a plasma current control algorithm that has a loose coupling with the plasma shape controller. One way to achieve this objective is to consider the following single-input–multi-output (SIMO) control law

$$\delta I_{PF}(s) = k_{p_{curr}} \cdot F_{I_p}(s) \cdot I_{p_e}(s), \qquad (20)$$

where $I_{p_e} = I_{p_{ref}} - I_p$ is the error on I_p, and $F_{I_p}(s)$ is the dynamic response of the controller. This latter SISO controller is usually designed as a proportional–integral-derivative (PID) regulator, where the integral action assures zero steady-state error during the current flat-top. However, when special requirements need to be met, other structures for $F_{I_p}(s)$ can be considered. As an example, since for ITER it is important to track the reference with zero error also during the ramp-up and ramp-down phases, the transfer function $F_{I_p}(s)$ should be designed with a double integral action [14]. It should be remarked that, for tokamaks with an iron core, such as JET, the transformer field is usually obtained by driving the current in the central solenoid, hence the $k_{p_{curr}}$ becomes a scalar and the control law (20) becomes a SISO one (see [71]).

Plasma Shape Controller

The Plasma Shape Controller is the second block that generates additional reference for the PFC Decoupling Controller. In modern tokamaks, the plasma shape control problem is solved using different approaches during the different phases of the plasma discharge reported in Fig. 16. The various solutions typically deal with a different choice of the shape descriptors to be controlled, from the *simple* one that controls only the position of the current centroid (both horizontal and vertical, [83, Section 3.2]), to the full boundary control solution [19].

[7] Superconductive coils provides an integral action by themselves.

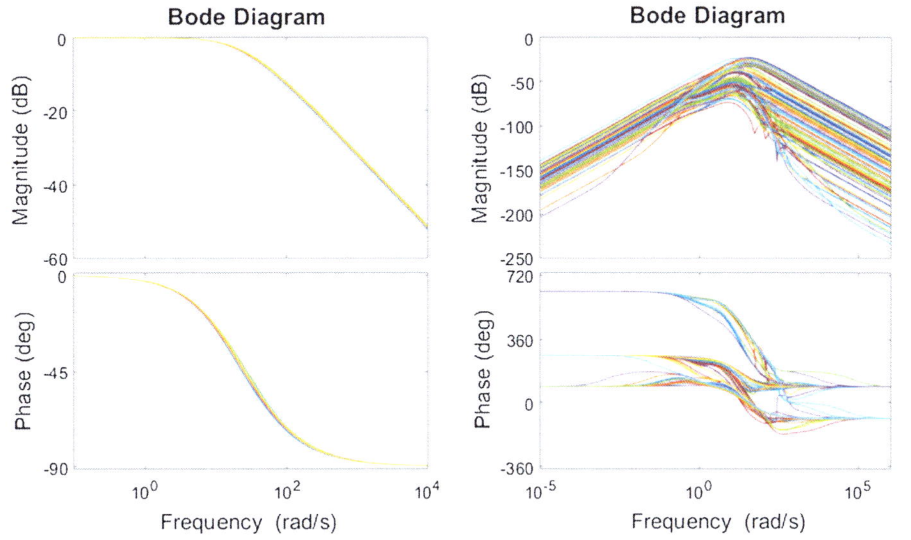

Fig. 20 Bode diagrams for the closed loop PFC decoupling control system designed for the JT-60SA tokamak (figures taken from [28]). **a** Bode diagrams for the diagonal channels, **b** Bode diagrams for the off-diagonal channels

Taking into account that at low values of I_p, due to measurement noise, the reconstruction of the plasma-wall gaps is not reliable, the following controlled variables are typically used during the different phases of the discharge

- during the *limiter* phase, the controlled parameters are the *limiter point position* and a set of *flux differences*;
- during the *limiter-to-divertor transition* the *X-point position* and a set of *flux differences* are controlled;
- during the *diverted* phase the controlled variables are the *plasma-wall* gaps;
- after the *divertor-to-limiter transition*, it may be convenient to move back to the control of the *limiter point position* and of a set of *flux differences*.

It follows that the proposed architecture must be flexible enough to be able to implement both the *isoflux boundary control* and the *gap boundary control*. As an example, let us consider the ITER poloidal cross-section and the control segments shown in Fig. 12.

Let g_i be the abscissa along i-th control segment, being $g_i = 0$ at the first wall. *Gap boundary control* is achieved by imposing

$$g_{i_{ref}} - g_i = 0$$

on a sufficiently large number of control segments.

Moreover, if the plasma shape intersects the i-th control segment at g_i, the following condition is satisfied

$$\psi(g_i) = \psi_B,$$

where ψ_B is the flux at the plasma boundary; hence *isoflux boundary control* is achieved also by controlling to 0 the difference

$$\psi(g_{i_{ref}}) - \psi_B,$$

where ψ_B can be taken equal either to the flux at the limiter point ψ_L or at the X-point ψ_X; the limiter-to-divertor and divetor-to-limiter transitions can be achieved by properly select the reference for ψ_B.

The proposed control algorithm for the Plasma Shape Controller block is based on the *eXtreme Shape Controller* (XSC, [19]), since this approach can be easily adapted both to gap and isoflux control. Moreover, a *current limit avoidance systems* that tries to avoid the saturation of the PF currents can be easily integrated within the XSC.

The main advantage of the XSC approach is the possibility of tracking a number of shape parameters larger than the number of PF circuits, by minimizing a weighted steady-state quadratic tracking error, when the references are constant signals.

The XSC control relies on the PFC Decoupling Controller presented in "The PF Currents Control Problem" section. Indeed, for the XSC design, it is assumed that each PF circuits can be treated as an independent SISO channel whose dynamic response is modeled by

$$I_{PF_i}(s) = \frac{I_{PF_{ref,i}}(s)}{1 + s\tau_{PF}}.$$

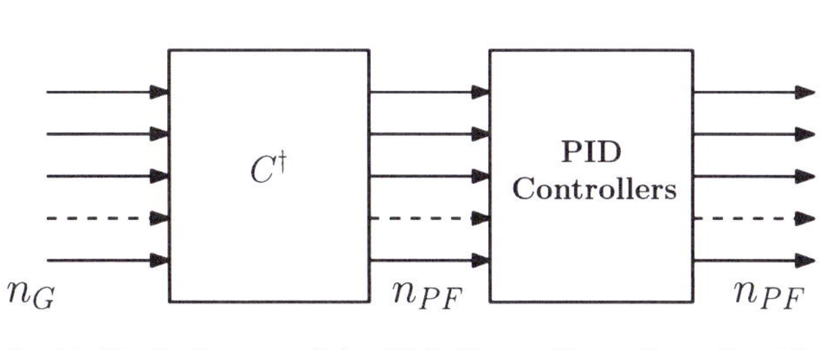

Fig. 21 Block diagram of the XSC Plasma Shape Controller. The pseudo-inverse $C^\dagger$ is usually computed using the largest singular values that result from the SVD of the C matrix

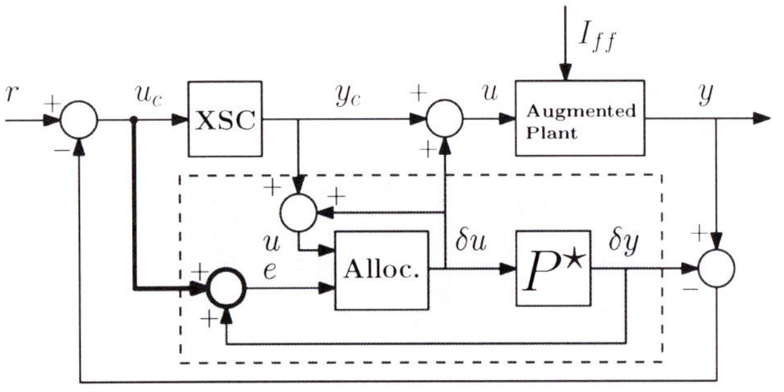

Fig. 22 Block diagram of the XSC Plasma Shape Controller with the Current Limit Avoidance (CLA) system (figure taken from [35]). The CLA made by the blocks in the gray shaded box, which includes the *Allocator*, and takes both the XSC input and output signals as inputs. In this figure the *Augmented Plant* represents the plant with the PFC Decoupling Controller

It is also assumed that all the PF circuits exhibit the same bandwidth (i.e., they have the same time constant τ_{PF}).

Denoting as $\delta Y(s)$ are the variations of the n_G shape descriptors—which can include flux differences, position of the X-point, gaps, etc.—from the plasma linear model (11) it follows that

$$\delta Y(s) = C\frac{I_{PF_{ref}}(s)}{1+s\tau_{PF}},$$

which, at steady-state, implies

$$\delta Y(s) = CI_{PF_{ref}}(s).$$

If the number of controlled plasma shape descriptors n_G is such that $n_G > n_{PF}$, the XSC computes the additional current references as

$$\delta I_{PF_{ref}} = C^{\dagger}\delta Y. \tag{21}$$

which are then sent to the PFC Decoupling Controller, in order to track the desired shape in a *least-mean-square* sense. In (21) the matrix $C^{\dagger}$ denotes the pseudo-inverse of C that can be computed via the singular value decomposition (SVD). Two diagonal matrices $Q \in \mathbb{R}^{n_G \times n_G}$ and $N \in \mathbb{R}^{n_{PF} \times n_{PF}}$ can be used to weight the shape descriptors and the currents in the PF circuits, respectively.

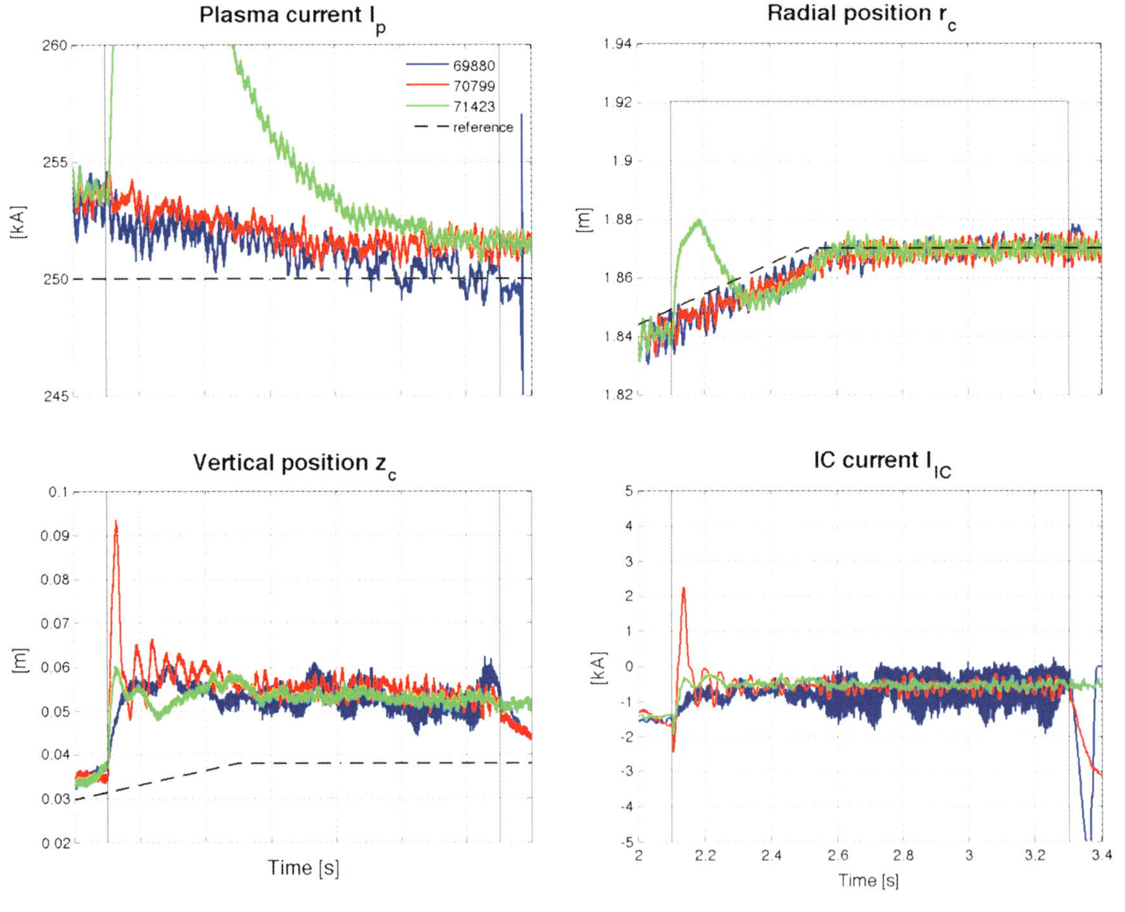

Fig. 23 Comparison between the EAST pulse #69880, which was the first one where the *ITER-like* VS was enabled for more than 1 s (from $t = 2.1$ s to $t = 3.3$ s), and the pulses #70799 and #74123, during which the VS controller gains were finely tuned exploiting the model. Note that during the considered time window the vertical position of the plasma current centroid was left uncontrolled in order to prove the effectiveness of the control law (12a)

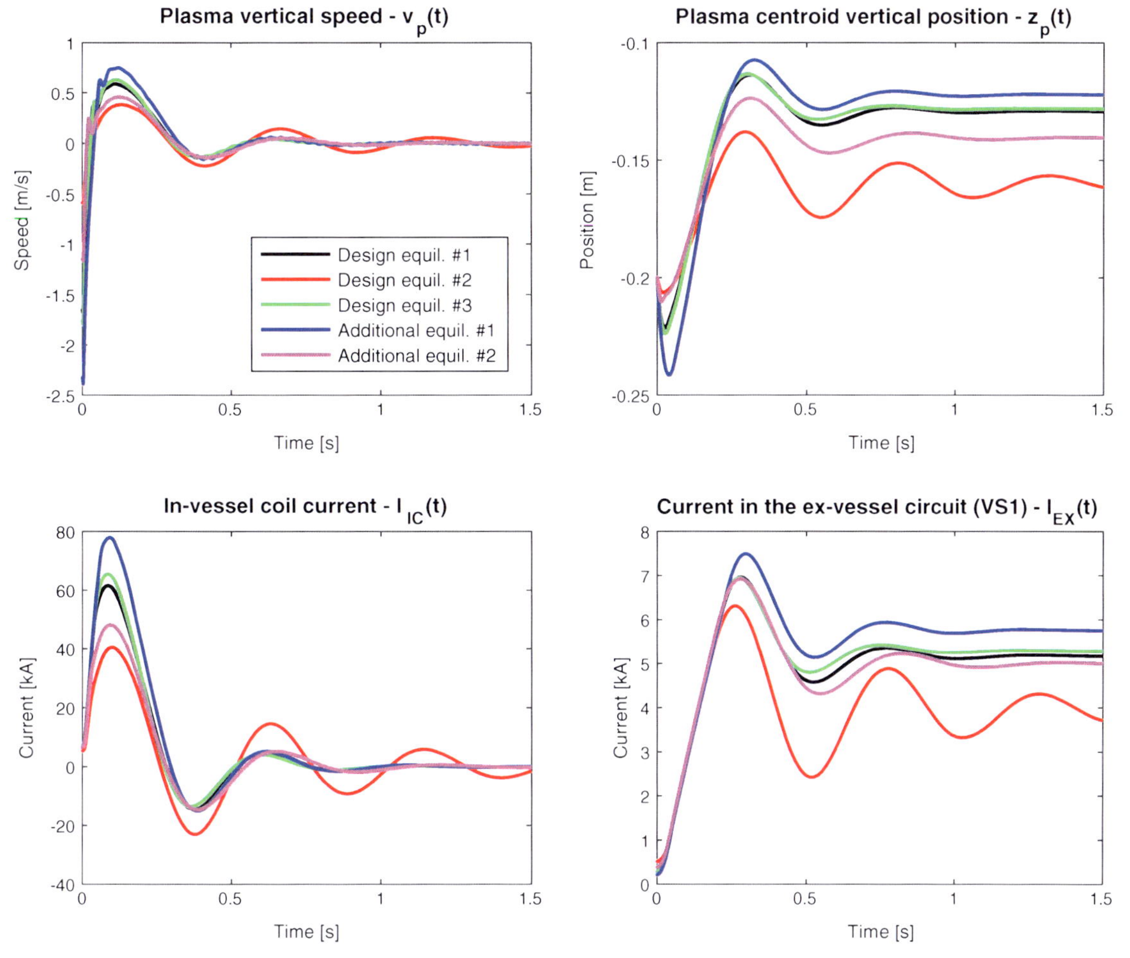

Fig. 24 Closed loop response to a 20 cm VDE (figure taken from [10]). This figures reports the time traces of the in-vessel coils I_{IC}, of the current I_{EX} in the ITER ex-vessel circuit (which is called *VS1* in the ITER jargon), of the plasma vertical speed v_p, and of the vertical position of the plasma centroid z_p

In this case the current to be sent to the PFC Decoupling Controller can be computed using the SVD of the weighted output matrix

$$\widetilde{C} = QCN = USV^T.$$

It follows that the XSC algorithm controls to zero the error on n_{PF} linear combinations of geometrical descriptors. Controlling to zero such an error is equivalent to minimizing the following steady-state performance index

$$J_{XSC} = \lim_{t \to +\infty} (\delta Y_{ref} - \delta Y(t))^T Q^T Q (\delta Y_{ref} - \delta Y(t)). \quad (22)$$

where δY_{ref} are constant references for the geometrical descriptors. When SVD of the $\widetilde{C}$ matrix is used to minimize (22), it may happen that some singular values (depending on the plasma configuration) are one order of magnitude smaller than the others. This fact implies that minimizing the performance index (22) retaining all the singular values results in a large control effort at the steady-state, that is a large request on some PF currents which have only a minor effect on the plasma shape. For this reason, the XSC achieves a trade-off condition, by minimizing a modified quadratic cost function that penalizes both the error on the controlled shape descriptors, and the control effort. In order to do that, only the $\bar{n} < n_{PF}$ linear combinations of PF currents which are related to the largest singular values are controlled. This is achieved by using only the $\bar{n}$ singular values when computing the pseudo-inverse $C^\dagger$.

The block diagram of the XSC Plasma Shape Controller is shown in Fig. 21; a set of PIDs is added in order to improve the dynamic response of the controller.

For more details on the XSC the reader is referred to [12, 19]. It worth to remark that SVD-*based* control approaches similar to the XSC have been used not only for magnetic control [15], but also for kinetic control (see [58, 63, 75] among the various).

Although in the XSC design the weighting matrix N can be used to reduce the use of the coils whose currents are close to saturation, the design procedure presented so far

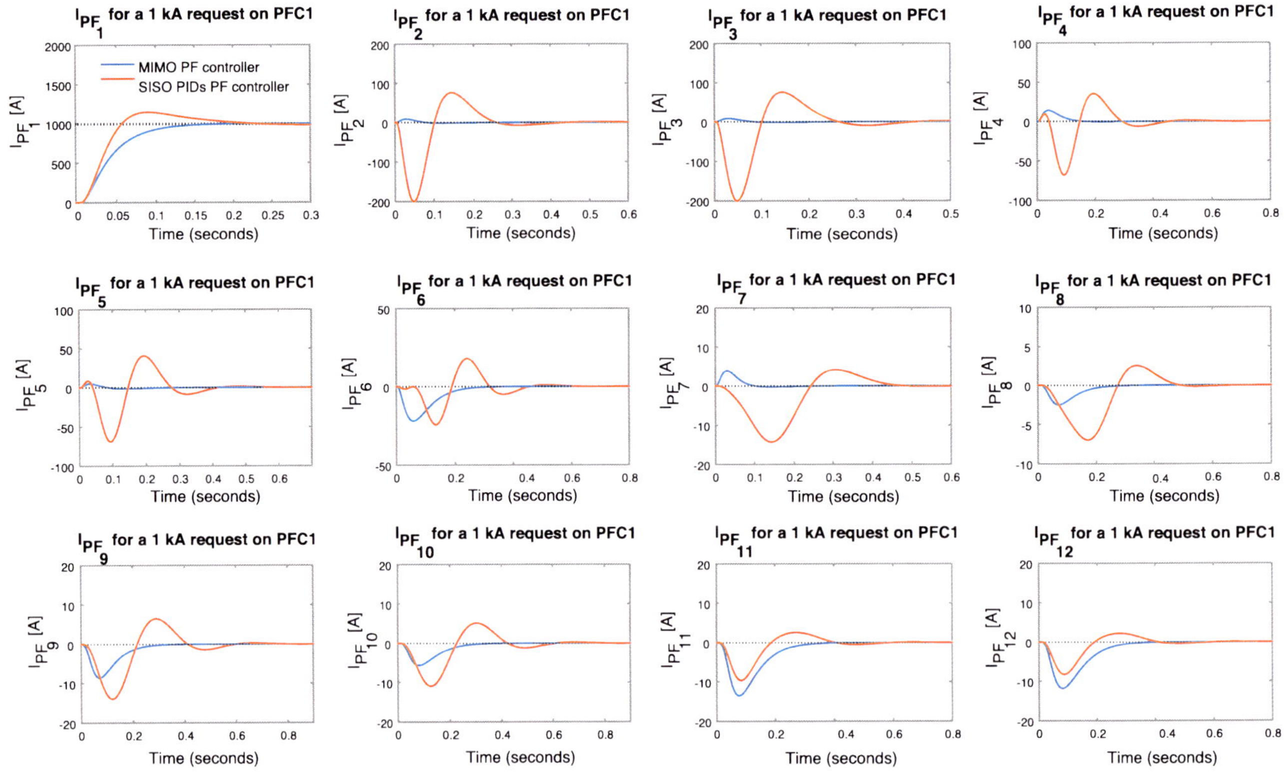

Fig. 25 Simulation showing the comparison between the PFC Decoupling Controller (19) and the EAST SISO PIDs

does not take explicitly into account the saturation constraints. It turns out that the PF current may saturate during the experiment, which may trigger the stop of the plasma discharge. In order to avoid current saturations it is possible to rely on a *Current Limit Avoidance (CLA)* system.The CLA algorithm exploits the following idea: keep the PF currents within their limits without degrading too much the plasma shape, by finding an optimal trade-off between these two objectives.

In particular, the CLA aims at keeping the value of the plant inputs u, i.e. the PF currents, inside a desirable region, meanwhile ensuring a small tracking error $e = r - y$, i.e. a small error on the plasma shape. In order to quantify this trade-off, a continuously differentiable cost function $J_{CLA}(u^\star, e^\star)$ is introduced, where the superscript $\star$ on a signal denotes its steady-state value.

The CLA corresponds to the gray shaded box in Fig. 22, which receives inputs from the XSC and modifies the request to the plant, i.e., the plasma controlled by the PFC Decoupling Controller (represented by the *Augmented Plant* block in Fig. 22). If we denote by $x_a \in \mathbb{R}^{n_a}$ the *Allocator* internal state, and by $B_0 \in \mathbb{R}^{n_{PF} \times n_a}$ a suitable full column rank matrix, then the two allocator outputs are equal to

$$\delta u = B_0 x_a, \tag{23}$$

and

$$\delta y = P^\star B_0 x_a, \tag{24}$$

where $P^\star$ is the steady-state input/output gain of the plant. The output (23) modifies the PF current requests generated by the XSC, while (24) *hides* the resulting steady-state change in the plasma shape to the XSC. Hiding the plasma shape change to the XSC is required in order to prevent the controller to react to these changes.

The key property of the current allocator algorithm is that, under suitable assumption on the cost J_{CLA}, for each constant current request of the XSC, it has a unique globally asymptotically stable equilibrium $x_a^\star$ that coincides with the unique global minimizer of J_{CLA}.

The CLA system has been successfully deployed at JET and has made the operation with the XSC safer. More details on the CLA system and on its implementation at JET can be found in [29, 34, 36].

Experiments and Simulations

In this section both experimental and simulation results are presented in order to show the effectiveness of the control algorithms presented in "Algorithms for Plasma Magnetic Control" section.

Let us first consider the VS system described in "The VS System" section. As was already mentioned, the

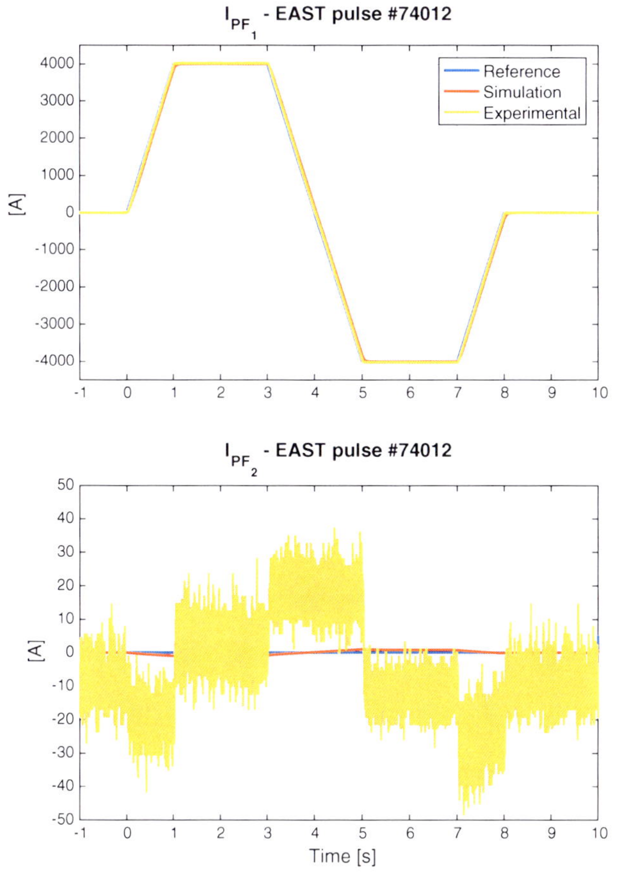

Fig. 26 Comparison between the simulated and the experimental values for the currents in the EAST $PF1$ and $PF2$ circuits during the EAST pulse #74012. Note that, taking into account the measurement noise, the experimental current in PF2 is practically zero

control law (12a) has been successfully deployed at EAST. Figure 23 compares the time traces of I_p, r_p, z_p, and I_{IC} for the EAST pulses #69880, #70799, and #71423. During these pulses only I_p and r_p were controlled with the standard EAST controller described in [83], while z_p was left uncontrolled. This was made on purpose in order to confirm that the proposed VS stabilizes the plasma by controlling v_p and I_{IC}, without the need of a feedback on z_p.

The EAST pulse #69880 was the first one where the proposed VS was enabled for more than 1 s, from $t = 2.1$ s to $t = 3.3$ s. In pulse #70799 the VS controller gains were tuned with the objective of reducing the maximum absolute value of the current requested in IC. Indeed, if the absolute value of the current in IC exceeds a given threshold, the power supply is tripped and the discharge is shut down. A further fine tuning was made for pulse #71423, in order to increase the stability margins, and hence to reduce the oscillatory behaviour. Note that, during the EAST pulse #71423 a fictitious disturbance was induced on the plasma current by a bump on the correspondent control loop. The overall behaviour is equivalent to a step disturbance to be rejected by both the VS and r_p control loops.

The VS system (12) has been proposed also for the ITER tokamak. In this case both the in-vessel and the ex-vessel circuits are available. A BMI optimization problem was solved to design a static and robust version of the VS system (12). Robustness was imposed by considering three different plasma equilibria when solving the optimization BMI problem. Figure 24 shows the simulated response of the closed loop system for a 20 cm VDE, for the three equilibria considered for the design, as well as for two additional ones (more details can be found in [10]).

The PFC Decoupling Controller introduced in "PFC Decoupling Control" section is routinely used at JET [62, 71]. More recently, the same algorithm has been tested at EAST, in order to improve the decoupling of the PF circuits. Indeed, according to the simulation results shown in Fig. 25, the MIMO controller (19) achieves a better decoupling when compared with the SISO PIDs used at EAST. The simulation results have been confirmed by the preliminary tests run during plasmaless discharges. In particular, Fig. 26 shows the currents in the two circuits $PF1$ and $PF2$; it is possible to note that the simulated and experimental results are in good agreement, and that the desired decoupling is achieved. Indeed, taking into account the measurement noise, the experimental current in $PF2$ is practically zero, when a current request is made on the $PF1$ circuit; the same behaviour was observed also on the other circuits.

The XSC plasma shape controller was successfully deployed at JET in 2003, and since then has been used during hundreds of pulses. Figure 27 shows the time traces of some of the controlled shape descriptors for the JET pulse #68953, while Fig. 28 shows how the XSC was used during the JET pulse #83202 to linearly change the plasma elongation while I_p was ramping down.

As it was claimed in "Plasma Shape Controller" section, the XSC approach can be also adopted when isoflux control is used to track the desired shape. As an example, Fig. 29 shows a simulation of the plasma current ramp-up for JT-60SA. The simulation results reported in this figure have been obtained simulating all the controllers of the proposed architecture.

In order to made the operation with the XSC safer, the CLA system was also deployed at JET. Figures 30 and 31 shows the results of an experiment aimed at producing a severe limitation for the XSC, and hence to prove the effectiveness of the CLA system. In order to do that, up to four, out of the eight available PF circuits, have been limited, and the following strategy has been adopted to carry out the experiment. First a reference pulse was run (pulse #81710), during which the XSC without CLA

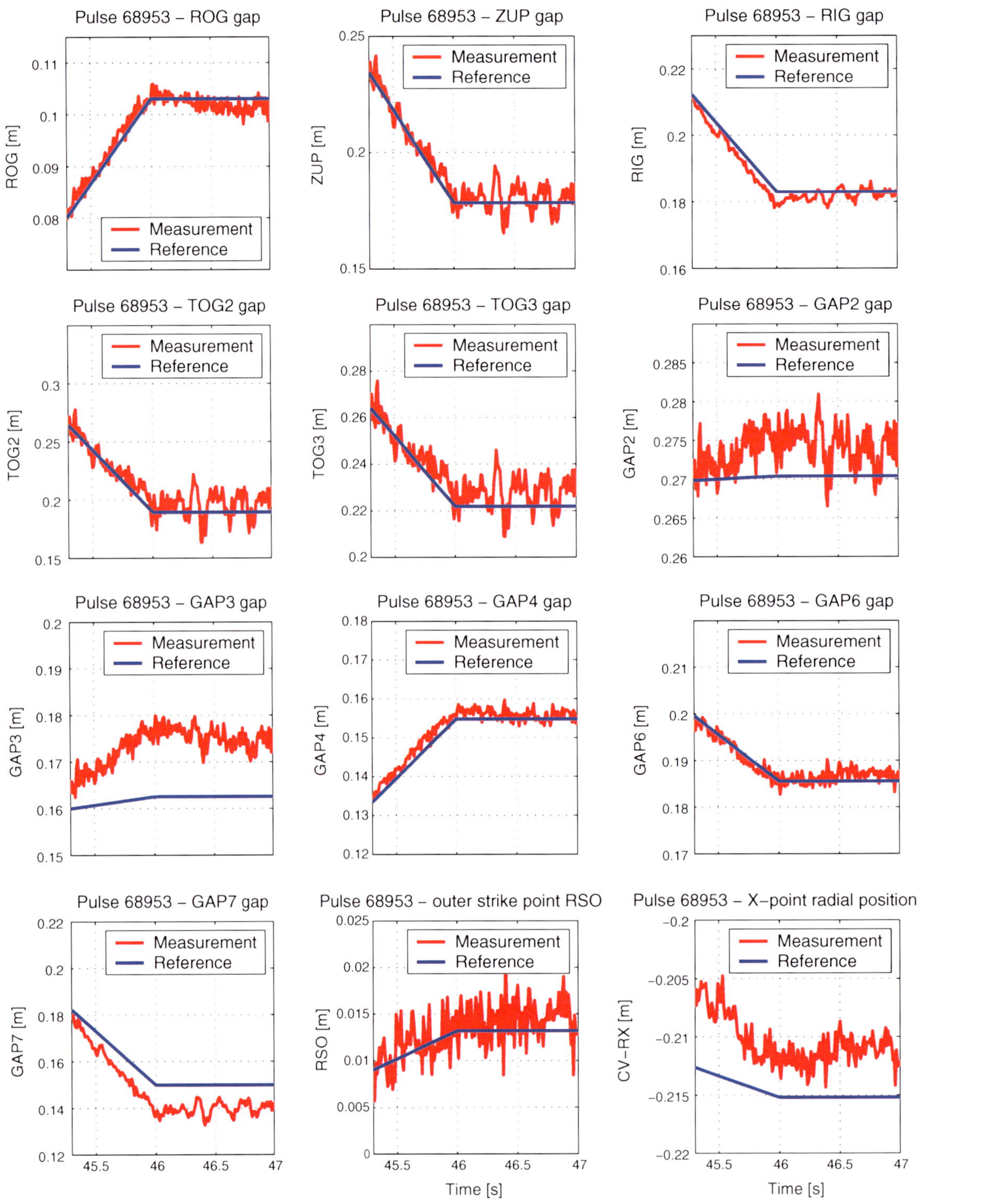

Fig. 27 JET pulse #68953. During this pulse the XSC has been used to control the plasma shape using the gaps (figure taken from [33]). This figure shows the time evolution of some of the controlled plasma shape descriptors

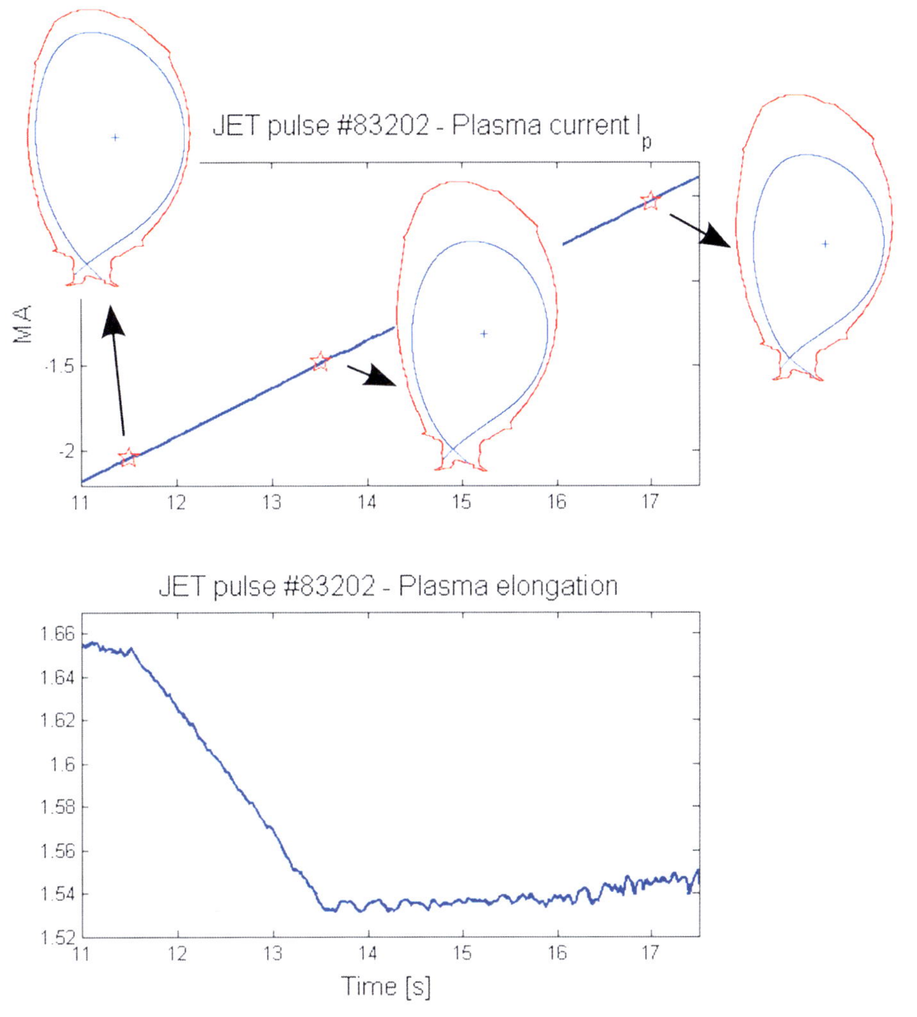

Fig. 28 Change of the plasma elongation during the ramp-down of the JET pulse #83202) (figure taken from [36]). The elongation changes from ~ 1.66 to ~ 1.54, while I_p is ramping down

successfully controlled the plasma shape between 20 s and 23 s. Afterwards, during the JET pulse #81715, the CLA system was enabled from 21 s, in order to limit the currents in the four divertor coils $D1$–$D4$ within a range smaller than the one actually available.

As expected, when a PF currents went outside their saturation limits, the CLA tried to bring them back to the permitted range while obtaining almost the same plasma shape as shown in Fig. 30. Moreover, Fig. 31 shows a comparison between the currents in the JET divertor coils for pulses #81710 and #81715. Taking into account that the limitation of more than two control currents represents already a challenging scenario at JET, the performance obtained during pulse #81715 is fully satisfactory. Furthermore, it is important to note that the CLA parameters used during this experiment included a hard constraint on the X-point position. That means that the CLA *prefers* to increase the shape error on the top-outer region of the

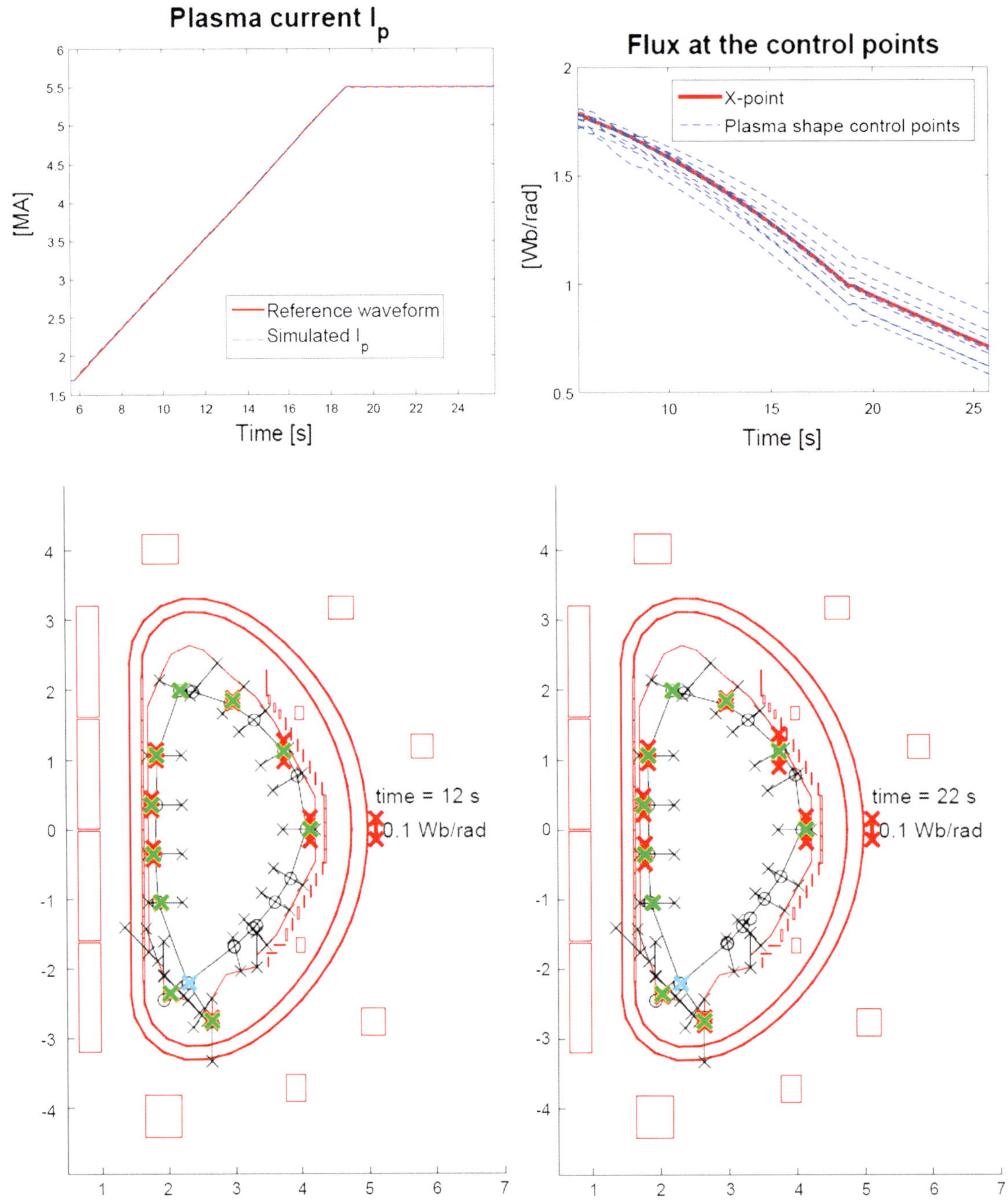

Fig. 29 Closed loop simulation of the plasma current ramp-up (from 1.7 MA to 5.5 MA) of JT60-SA Scenario 2 (figures taken from [28]). Plasma shape control done using an isoflux version of the XSC. In particular, the XSC controls to zero the difference between the flux at the control points (shown as green crosses) and the flux at the X-point. Top left: plasma current versus time during the ramp-up phase. Top right: magnetic flux at the X-point and at the control points. Bottom: plasma poloidal cross-section during the plasma current ramp-up ($t = 12$ s, left) and at the flat-top ($t = 22$ s, right). The blue cross shows the desired position for the X-point (Color figure online)

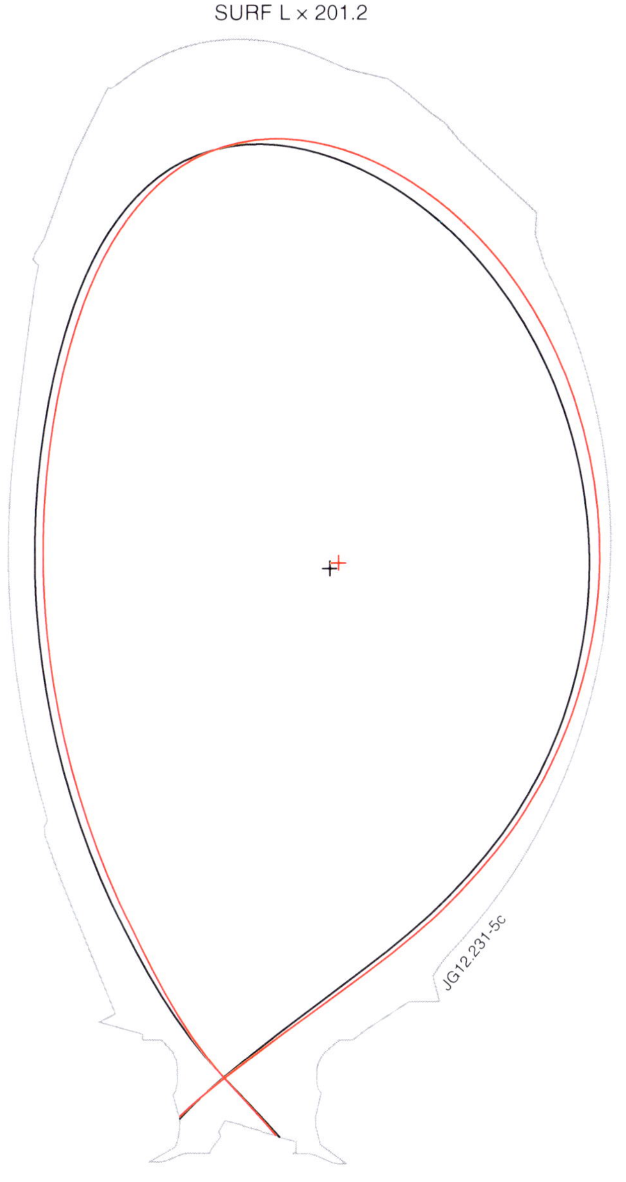

Fig. 30 JET pulse #81715. Shape comparison at 22.5 s (figure taken from [35]). The black shape is the one obtained in pulse #81710 without CLA, while the red shape is the one obtained when the CLA is enabled (Color figure online)

plasma, rather than to change the position of the X-point, as it is shown in Fig. 30.

Conclusive Remarks

In this chapter the magnetic control problems in tokamak devices have been presented together with a generic architecture for the deployment of the related controllers. The proposed architecture and control algorithms result from the experience made by the members of the CREATE Consortium in the last two decades. Despite their relative simplicity if compared with other control approaches (such as model predictive control [47, 61]), the proposed control algorithms can be easily adapted to different machines, and they have been deployed, partially or completely, on various tokamaks. The same algorithms have been also proposed for ITER [14], JT-60SA [28], DEMO and DTT [6]. Finally, it is worth to remark that the design of all the proposed control algorithms are based on dynamic model of the plasma and of the surrounding coils; hence the availability of reliable plant models is a key factor for the design and deployment of plasma magnetic controllers.

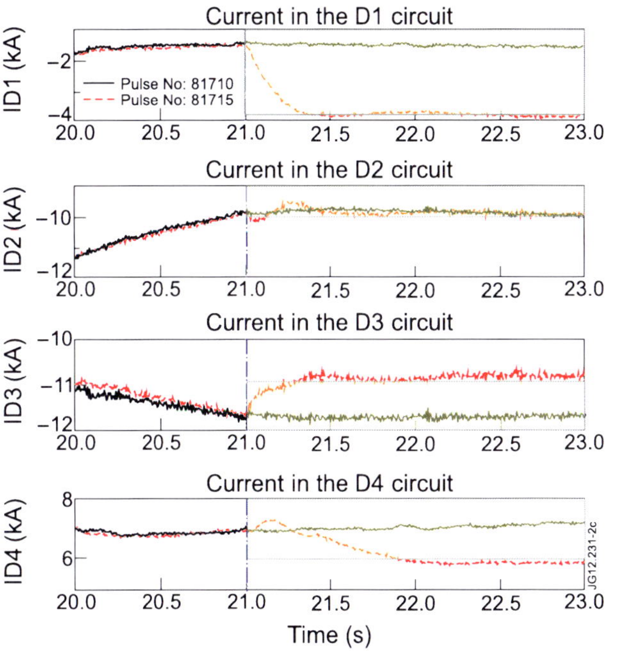

Fig. 31 Currents in the divertor circuits (figure taken from [35]). Comparison between pulse #81710 (reference pulse without CLA) and pulse #81715 (with CLA). The shared areas correspond to regions beyond the current limits enforced by the CLA parameters

Acknowledgements This work has been partially carried out within the framework of the EUROfusion Consortium and has received funding from the Euratom research and training programme 2014–2018 under Grant Agreement No. 633053. The views and opinions expressed herein do not necessarily reflect those of the European Commission. The author would like to thank all the colleagues in the CREATE Consortium for having shared with him their experience on plasma magnetic modeling and control during the last twenty years. A special thanks goes to Prof. Ariola for carefully proofreading of "Systems and Control Basics" section, and to all the colleagues from all around the world in the fusion community, together with whom the author had the honor and pleasure to work.

References

1. M. Al-Husari et al., Vertical stabilization of tokamak plasmas, in *Proceedings of 30th IEEE Conference on Decision and Control, Brighton, UK* (1991)
2. R. Albanese, R. Ambrosino, M. Mattei, CREATE-NL+: a robust control-oriented free boundary dynamic plasma equilibrium solver. Fus. Eng. Des. **96–97**, 664–667 (2015)
3. R. Albanese, G. Calabrò, M. Mattei, F. Villone, Plasma response models for current, shape and position control at JET. Fus. Eng. Des. **66–68**, 715–718 (2003)
4. R. Albanese, F. Villone, The linearized CREATE-L plasma response model for the control of current, position and shape in tokamaks. Nucl. Fus. **38**(5), 723–738 (1998)
5. R. Albanese et al., Design, implementation and test of the XSC extreme shape controller in JET. Fus. Eng. Design **74**(1–4), 627–632 (2005)
6. R. Albanese et al., Diagnostics, data acquisition and control of the divertor test tokamak experiment. Fus. Eng. Des. **122**, 365–374 (2017)
7. R. Albanese et al., ITER-like vertical stabilization system for the EAST tokamak. Nucl. Fus. **57**(8), 086,039 (2017)
8. G. Ambrosino, R. Albanese, A survey on modeling and control of current, position and shape of axisymmetric plasmas. IEEE Control Syst. Mag. **25**(5), 76–91 (2005)
9. G. Ambrosino, M. Ariola, G. De Tommasi, A. Pironti, Plasma vertical stabilization in the ITER tokamak via constrained static output feedback. IEEE Trans. Control Syst. Technol. **19**(2), 376–381 (2011)
10. G. Ambrosino, M. Ariola, G. De Tommasi, A. Pironti, Robust vertical control of ITER plasmas via static output feedback, in *Proceedings of the 2011 IEEE Multi-Conference on System Control Applications, Denver, Colorado* (2011), pp. 276–281
11. G. Ambrosino, M. Ariola, G. De Tommasi, A. Pironti, A. Portone, Design of the plasma position and shape control in the ITER tokamak using in-vessel coils. IEEE Trans. Plasma Sci. **37**(7), 1324–1331 (2009)
12. G. Ambrosino, M. Ariola, A. Pironti, Optimal steady-state control for linear non-right-invertible systems. IET Control Theory Appl. **1**(3), 604–610 (2007)
13. G. Ambrosino, M. Ariola, A. Pironti, A. Portone, M. Walker, A control scheme to deal with coil current saturation in a tokamak. IEEE Trans. Control Syst. Technol. **9**(6), 831–838 (2001)
14. R. Ambrosino et al., Design and nonlinear validation of the ITER magnetic control system, in *Proceedings of the 2015 IEEE Conference on Control Applications, Sydney, Australia* (2015), pp. 1290–1295
15. H. Anand, S. Coda, F. Felici, C. Galperti, J. Moret, A novel plasma position and shape controller for advanced configuration development on the TCV tokamak. Nucl. Fus. **57**(12), 126,026 (2017)
16. F. Argomedo, E. Witrant, C. Prieur, *Safety Factor Profile Control in a Tokamak* (Springer, Berlin, 2014)
17. M. Ariola, G. Ambrosino, A. Pironti, J. Lister, P. Vyas, Design and experimental testing of a robust multivariable controller on a tokamak. IEEE Trans. Control Syst. Technol. **10**(5), 646–653 (2002)
18. M. Ariola, G. De Tommasi, A. Pironti, F. Villone, Control of resistive wall modes in tokamak plasmas. Control Eng. Pract. **24**, 15–24 (2014)
19. M. Ariola, A. Pironti, The design of the eXtreme Shape Controller for the JET tokamak. IEEE Control Syst. Mag. **25**(5), 65–75 (2005)
20. M. Ariola, A. Pironti, *Magnetic Control of Tokamak Plasmas*, 2nd edn. (Springer, Berlin, 2016)
21. M. Ariola et al., Integrated plasma shape and boundary flux control on JET tokamak. Fus. Sci. Technol. **53**(3), 789–805 (2008)
22. K. Åström, B. Wittenmark, *Computer-Controlled Systems: Theory and Design* (Prentice-Hall, Upper Saddle River, 1997)
23. A. Beghi, A. Cenedese, Advances in real-time plasma boundary reconstruction. IEEE Control Syst. Mag. **25**(5), 44–64 (2005)
24. T. Bellizio et al., A MARTe based simulator for the JET Vertical Stabilization system. Fus. Eng. Des. **86**, 1026–1029 (2011)
25. J. Blum, *Numerical Simulation and Optimal Control in Plasma Physics* (1989)
26. E. Camacho, C. Alba, *Model Predictive Control* (Springer, Berlin, 2013)
27. I. Carvalho et al., ISTTOK real-time architecture. Fus. Eng. Des. **89**(3), 195–203 (2014)
28. N. Cruz et al., Control-oriented tools for the design and validation of the JT-60SA magnetic control system. Control Eng. Pract. **63**, 81–90 (2017)
29. G. De Tommasi, S. Galeani, A. Pironti, G. Varano, L. Zaccarian, Nonlinear dynamic allocator for optimal input/output performance trade-off: application to the JET tokamak shape controller. Automatica **47**(5), 981–987 (2011)
30. G. De Tommasi, A. Mele, Z. Luo, A. Pironti, B. Xiao, On plasma vertical stabilization at EAST tokamak, in *2017 IEEE Conference on Control Technology and Applications (CCTA), Kohala Coast, Hawaii* (2017), pp. 511–516
31. G. De Tommasi, A. Mele, A. Pironti, Robust plasma vertical stabilization in tokamak devices via multi-objective optimization, in *Optimization and Decision Science: Methodologies and Applications, Springer Proceedings in Mathematics & Statistics*, vol. 217 (2017), pp. 305–314
32. G. De Tommasi et al., XSC tools: a software suite for tokamak plasma shape control design and validation. IEEE Trans. Plasma Sci. **35**(3), 709–723 (2007)
33. G. De Tommasi et al., Current, position, and shape control in tokamaks. Fus. Sci. Technol. **59**(3), 486–498 (2011)
34. G. De Tommasi et al., A software tool for the design of the current limit avoidance system at the JET tokamak. IEEE Trans. Plasma Sci. **40**(8), 2056–2064 (2012)
35. G. De Tommasi et al., First experimental results with the current limit avoidance system at the JET tokamak. Fus. Eng. Des. **88**(5), 400–407 (2013)
36. G. De Tommasi et al., Shape control with the eXtreme Shape Controller during plasma current ramp-up and ramp-down at JET tokamak. J. Fusion Energ. **33**(2), 233–242 (2014)
37. G. De Tommasi et al., Model-based plasma vertical stabilization and position control at EAST. Fus. Eng. Des. **129**, 152–157 (2018)
38. J. Farthing et al., Status of ITER remote experimentation centre. Fus. Eng. Des. **128**, 158–162 (2018)
39. J.Y. Favez, J.B. Lister, P. Müllhaupt, B. Srinivasan, Improving tokamak vertical position control in the presence of power supply voltage saturation. Plasma Phys. Control Fus. **47**(10), 1709 (2005)
40. F. Felici et al., Development of real-time plasma analysis and control algorithms for the TCV tokamak using Simulink. Fus. Eng. Des. **89**, 165–176 (2014)
41. J. Ferron et al., Real time equilibrium reconstruction for tokamak discharge control. Nucl. Fus. **38**(7), 1055–1066 (1998)

42. M. Firestone, Analysis of modern optimal control theory applied to plasma position and current control in TFTR. IEEE Trans. Plasma Sci. **10**, 105–115 (1982)
43. G. Franklin, J. Powell, M. Workman, *Digital Control of Dynamic Systems*, vol. 3 (Addison-Wesley, Menlo Park, 1998)
44. J. Freidberg, *Plasma Physics and Fusion Energy* (Cambridge University Press, Cambridge, 2007)
45. M. Garibba et al., First experimental experience with the new plasma position and current control system of jet, in *Proceedings of 18th Symposium of Fusion Technology*, vol. 1 (1994), pp. 747–750
46. G. Gennacchi, A. Taroni, JETTO: a free boundary plasma transport code (basic version). Tech. rep., ENEA, RT/TIB 1988(5), Rome (Italy) (1988)
47. S. Gerkšič, G. De Tommasi, ITER plasma current and shape control using MPC, in *Proceedings of 2016 IEEE Conference on Control Applications (CCA), Buenos Aires, Argentina* (2016), pp. 599–604
48. J. Gossner et al., Application of cautious stable predictive control to vertical positioning in COMPASS-D tokamak. IEEE Trans. Control Syst. Technol. **7**, 580–587 (1999)
49. J. Helton, K. McGown, M. Walker, Conditions for stabilization of the tokamak plasma vertical instability using only a massless plasma analysis. Automatica **46**(11), 1762–1772 (2010)
50. H. Heumann et al., Quasi-static free-boundary equilibrium of toroidal plasma with CEDRES++: computational methods and applications. J. Plasma Phys. **81**(3), 35 (2015)
51. D. Humphreys, M. Walker, J. Leuer, J. Ferron, Initial implementation of a multivariable plasma shape and position controller on the DIII-D tokamak, in *Proceedings of 2000 IEEE International Conference on Control Applications (CCA), Anchorage, Alaska* (2000), pp. 412–418
52. D. Humphreys et al., DIII-D Integrated plasma control solutions for ITER and next-generation tokamaks. Fus. Eng. Des. **83**(2–3), 193–197 (2008)
53. S. Jardin, D. Larrabee, Feedback stabilization of rigid axisymmetric modes in tokamaks. Nucl. Fus. **22**, 1095–1098 (1982)
54. S. Jardin, N. Pomphrey, J. DeLucia, Dynamic modeling of transport and positional control of tokamaks. J. Comput. Phys. **66**(2), 481–507 (1986)
55. T. Kailath, *Linear Systems* (Prentice-Hall, Englewood Cliffs, 1980)
56. H. Khalil, *Nonlinear Systems* (MacMillan Publishing Company, Basingstoke, 1992)
57. R. Khayrutdinov, V. Lukash, Studies of plasma equilibrium and transport in a tokamak fusion device with the inverse-variable technique. J. Comput. Phys. **109**(2), 193–201 (1993)
58. S. Kim, J. Lister, A new approach to plasma profile control in ITER. Fus. Eng. Des. **87**(12), 1921–1925 (2012)
59. M. Lennholm et al., Plasma vertical stabilisation at JET using adaptive gain control, in *Proceedings of 17th SOFE conference, San Diego, CA*, vol. 1 (1997), pp. 539–542
60. M. Lennholm et al., Plasma control at JET. Fus. Eng. Des. **48**(1), 37–45 (2000)
61. M. Mattei, C. Labate, D. Famularo, A constrained control strategy for the shape control in thermonuclear fusion tokamaks. Automatica **49**(1), 169–177 (2013)
62. F. Maviglia et al., Improving the performance of the JET Shape Controller. Fus. Eng. Des. **96**, 668–671 (2015)
63. D. Moreau et al., A two-time-scale dynamic-model approach for magnetic and kinetic profile control in advanced tokamak scenarios on JET. Nucl. Fus. **48**(10), 106,001 (2008)
64. S. Moriyama et al., Analysis of optimal feedaback control of vertical plasma position in a tokamak system. Jpn. J. Appl. Phys. **24**, 849–855 (1985)
65. G. Neilson, G. Dyer, P. Edmonds, A model for coupled plasma current and position feedback control in the ISX-B tokamak. Nucl. Fus. **24**, 1291–1302 (1984)
66. A. Neto et al., Exploitation of modularity in the JET tokamak vertical stabilization system. Control Eng. Pract. **20**(9), 846–856 (2012)
67. A. Neto et al., Conceptual architecture of the plant system controller for the magnetics diagnostic of the ITER tokamak. Fus. Eng. Des. **96–97**, 887–890 (2015)
68. G. Pereverzev, P. Yushmanov, *ASTRA automated system for transport analysis in a tokamak* Tech. rep, Max-Planck-Institut fuer Plasmaphysik, Garching (Germany) (2002)
69. S. Peruzzo et al., Installation and commissioning of the JET-EP magnetic diagnostic system. Fus. Eng. Des. **84**, 1495–1498 (2009)
70. M. Romanelli et al., JINTRAC: a system of codes for integrated simulation of tokamak scenarios. Plasma Fus. Res. **9**(Special issue 2), 3403023-1–3403023-4 (2014)
71. F. Sartori, G. De Tommasi, F. Piccolo, The joint European torus. IEEE Control Syst. Mag. **26**(2), 64–78 (2006)
72. E. Schuster, M. Walker, D. Humphreys, M. Krstić, Plasma vertical stabilization with actuation constraints in the DIII-D tokamak. Automatica **41**(7), 1173–1179 (2005)
73. L. Scibile, B. Kouvaritakis, A discrete adaptive near-time optimum control for the plasma vertical position in a Tokamak. IEEE Trans. Control Syst. Technol. **9**, 148–162 (2001)
74. V. Shafranov, Plasma equilibrium in a magnetic field. Rev. Plasma Phys. **2**, 103 (1966)
75. W. Shi et al., A two-time-scale model-based combined magnetic and kinetic control system for advanced tokamak scenarios on DIII-D, in *Proceedings of IEEE Conference on Decision and Control, Maui, Hawaii* (2012), pp. 4347–4352
76. S. Skogestad, I. Postlethwaite, *Multivariable Feedback Control: Analysis and Design* (Wiley, New York, 2007)
77. J. VanAntwerp, R. Braatz, A tutorial on linear and bilinear matrix inequalities. J. Process Control **10**, 363–385 (2000)
78. P. Vyas, A. Morris, D. Mustafa, Vertical position control on COMPASS-D. Fusion Tech. **33**, 97–105 (1998)
79. M. Walker, D. Humphreys, On feedback stabilization of the tokamak plasma vertical instability. Automatica **45**(3), 665–674 (2009)
80. M. Walker et al., Emerging applications in tokamak plasma control. IEEE Control Syst. Mag. **26**(2), 35–63 (2006)
81. R. Wenninger et al., Advances in the physics basis for the European DEMO design. Nucl. Fus. **55**(6), 063,003–7 (2015)
82. D. Youla, J. Bongiorno Jr., C. Lu, Single-loop feedback-stabilization of linear multivariable dynamical plants. Automatica **10**(2), 159–173 (1974)
83. Q. Yuan et al., Plasma current, position and shape feedback control on EAST. Nucl. Fus. **53**(4), 043,009 (2013)
84. K. Zhou, J. Doyle, *Essentials of Robust Control*, vol. 104 (Prentice Hall, Upper Saddle River, 1998)

Journal of Fusion Energy (2019) 38:437–444
https://doi.org/10.1007/s10894-018-0190-1

ORIGINAL RESEARCH

Advancements in Plasma Control Supervision

Kevin Meyer[1] · Jayanethie Padayachee[1]

Published online: 11 September 2018

Abstract

Early fusion projects were small enough with so few control variables that it was possible to define and execute experiments without sophisticated software tools. The session leaders/engineers-in-charge knew the key settings well enough that they could tell that the experiment was "safe". If something went wrong, then they learned to check for it in the future. As projects grew in size and experiments grew in complexity, the number of control variables grew significantly, and it became standard to use software tools to support the definition, set-up, validation and execution of experiment programs (i.e. the pre-planned set of tasks that defines what the machine must do). However, in all cases, the tools typically still only applied to the experiment. One of these tools is the familiar "plasma control system" (PCS) which is the real-time software that typically has responsibility for operating the fusion machine during a pulse. Other tools handle system configuration definition and experiment program editing. Future fusion projects like ITER and DEMO will be even more complex than current projects. The high infrastructure and human resource running costs means that these fusion projects will need to be efficient, and not waste resources or time. Such fusion projects will therefore benefit from a suite of integrated supervision and control software tools that facilitates the operation of the machine over its entire lifetime. These software tools would include a more capable PCS together with expanded system configuration definition and experiment program editing tools, and additional tools to support operations outside of the experimental phase.

Keywords Fusion · Supervision · Operation · Plasma control system

Introduction

Fusion projects have always been designed on the basis of specific physics principles to support the overall field of fusion research. Contributions from this research have brought us to the current state of knowledge where for the first time in history we can realistically expect to achieve $Q > 1$.

On the road towards fusion as a viable energy source, fusion research projects of the future will inherently be more complex from construction and operational standpoints. In a climate of reduced science funding, this means that these projects must be designed to be highly economic (cost) and efficient (time and resources). Apart from the high initial investment to construct the facility, these projects will also have high operating costs and so every experiment must be designed with contingencies and alternatives to always yield a useful result.

Early fusion experiments were short (a few seconds) and were done mostly on a trial-and-error basis. The systems controlling these experiments were simple, executing sequential commands that occurred at specified times. Future fusion experiments are expected to run for significantly longer ($\sim$ 30 min and more). This means that the systems that control these experiments, and in a larger context, the entire machine, must be equivalently smarter to adapt to the changing environment and requirements.

ITER [1] and DEMO [2], the next generation of fusion projects, should therefore not be expected to be operated using the same philosophies as their predecessors and so their designs must incorporate the next-generation concepts in supervision and control.

✉ Kevin Meyer
kevin.meyer@cosylab.com

1 Cosylab d.d., Gerbičeva ulica 64, Ljubljana, Slovenia

Operational Phases

Every fusion project has many phases that apply at different times in its operational life. Some of the key phases are:

- Experiment
- Shift
- Day
- Campaign
- Operational lifetime

Experiment

Figure 1 shows the three phases of a typical experiment: pre-experiment setup, the real-time phase and post-experiment processing. The key element of the experiment phase is the experiment program which is a pre-planned set of tasks that defines what the machine must do.

The aim of the pre-experiment setup phase is to prepare the machine for the experiment. This starts with retrieving the experiment program and all related configuration information necessary to setup the various physical systems for executing the experiment program. Once the information has been retrieved, it is checked for consistency, limits, known problems, etc. Thereafter, the information is loaded onto the various systems to prepare them for executing the experiment.

When all systems are ready, the engineer-in-charge can start the experiment. From this point, the plasma control system (real-time phase) runs until the experiment program ends. Typically experiments can end in one of three ways:

- Completes successfully
- A protection system detects an issue and ends the experiment (ideally a "soft landing")
- An interlock or safety system detects an issue and hardware systems perform a shut-down

Post-experiment processing is traditionally when data acquisition systems, including diagnostics, write out buffered data to archives. This is also when scientific data analysis tools extract and archive high-level experiment results. If any protection events have occurred, additional analyses are performed to check for damage. It is during this phase that a summary of the experiment is captured in the appropriate information system for later use in data-mining.

Shift

Figure 2 shows that a shift is mostly a collection of experiments, but there are pre-shift setup and end-of-shift process phases where operators hand-over operation of the machine.

Day

Figure 3 shows the phases that make up a day for a machine that does not operate for 24 h. In a continuously running machine, there will not be regular shut-downs and start-ups, but the conditioning may be done regularly.

For a machine that does not operate for 24 h, a day starts with a start-up phase that brings the machine up out of standby. This is typically followed by a conditioning experiment, which is a special experiment designed to check that the machine functions and behaves as expected. If all is well, then the regular shifts continue.

At the end of the day there is a daily shutdown that puts the machine into standby.

Campaign

A campaign (Fig. 4) is characterized by the addition of new infrastructure that extends the capabilities of the machine.

The new infrastructure must be commissioned and tested before it can be used. The new capabilities must be added to configuration definition databases and experiment programs must be updated to reflect these changes.

Once this has been completed, the normal daily cycle can continue.

At the end of the campaign, there is a maintenance shutdown to put the machine into a state that allows for maintenance and upgrades.

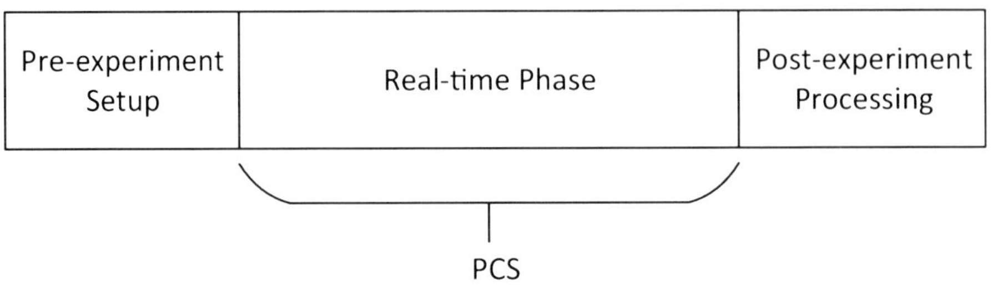

Fig. 1 The three phases in a typical experiment. The plasma control system (PCS) traditionally operates in the real-time phase

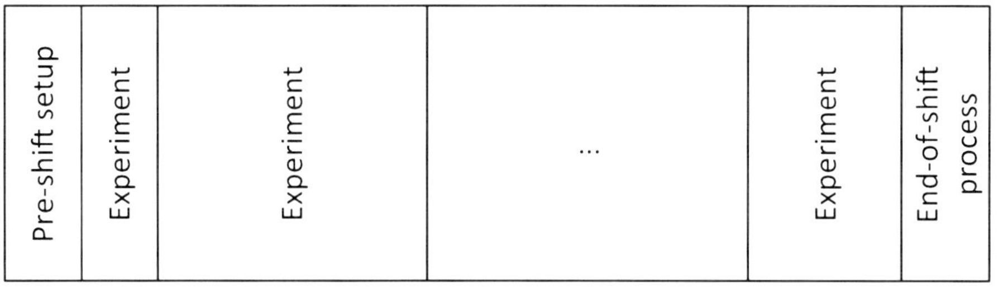

Fig. 2 The three phases in a typical shift

Fig. 3 A day is simply a series of two or three shifts

Fig. 4 A campaign is characterized by new capabilities added to the machine and consists of four phases

Operational Lifetime

Figure 5 shows the full operational lifetime of a project. This is characterized by construction, operation and decommissioning. The operational phase consists of a series of campaigns.

Current Software Support Tools

The earliest fusion machines and experiments had so few control variables that it was possible for the experiment program to be constructed manually. Then, as the machines grew in size and experiments grew in scope, software tools were used but typically these only supported editing and executing the experiment program.

These software tools covered the following major areas:

- System configuration definition
- Experiment program editing
- Plasma control system

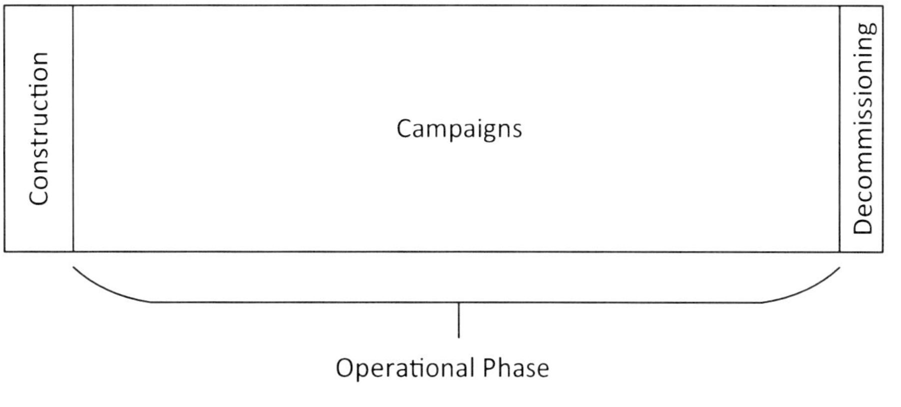

Fig. 5 The operational lifetime of a project

System Configuration Definition and Experiment Program Editing

Each system that makes up a fusion machine must be appropriately configured before it can be incorporated into an experiment. "System configuration definition" involves the identification of the different inputs and outputs of the system and the association of these inputs and outputs with control variables. Software tools like W-7X's ConfiX [3, 4] and ITER's Self-Description Data [5] Editor support the definition of system configuration.

Current experiment program editing software consists of suites of applications that assist with all stages of developing an experiment program. These suites of tools allow experiment developers to specify their sequence of experiment tasks and also include validation as part of the standard workflow to minimise errors and inconsistencies during the development. Even though these software tools, e.g. JET's Level 1 [6, 7] and W-7X's Xedit [3, 4], have grown in functionality, they still essentially only support the experiment phase.

Plasma Control System

The plasma control system (PCS) is considered to be the heart of the running fusion experiment. The PCS hosts a variety of control algorithms that are designed to achieve research goals (what the physicist wants) by executing the experiment program (what the engineer-in-charge can make the machine do).

Real-Time Strategies for Experiment Execution

Real-time control systems are categorised by the event distribution system used to signal events. Event distribution systems can be either time-based or condition-based. In early plasma control systems, the experiment program was driven by time-based events distributed on the timing system. By definition, this was a purely feed-forward control system.

PCSs evolved to implement feed-back algorithms where the experiment program is driven by condition-based events that use inputs from diagnostics to signal when conditions are satisfied. However, for legacy reasons, these events are often distributed on the timing system.

Real-Time Tasks

Independently of the experiment execution control, there are other lower-level control processes that continually run within certain phases of the experiment that fulfil critical control needs, e.g. vertical stabilisation. These algorithms are initiated and terminated by appropriate events.

Other Software Tools

Diagnostics data and real-time processed data are stored in the data archives. However, certain parameters can typically only be computed after the experiment is completed. There are separate software tools to support these analyses.

There are also high-level management applications that are used to coordinate experiments to be conducted within a campaign. These tools range from accepting submissions for experiments (based on the campaign capabilities), the evaluation and selection of experiments and then the day-to-day planning for which experiments are to be run.

Future of Fusion Supervision and Control

Future fusion control will benefit from broader supervision support than just in the experiment phase. This support could take the form of an integrated suite of software tools inspired by JET's Level 1 and W-7X's Xedit among many others to provide enhanced orchestration and monitoring support to all operational phases. Such a suite of supervision and control software tools would also include a plasma control system with enhancements that provide greater flexibility.

New Supervision and Control Philosophies

In order to support the next generation of fusion projects, a new mind-set is needed for the operation of the machine. There are three main philosophies that need to be incorporated:

- Flexible task execution
  - Any task, including but not limited to experiments, must be able to reach a goal without having the exact path specified.
  - Control systems must select from available goals based on current operating conditions.
  - The ability to achieve multiple goals in a single experiment or operational task.
- Automation of repetitive tasks
  - Routine and repetitive tasks must be managed with minimal operator intervention.
- Expansion of scope
  - Supervision and control applies to more than just the experimental phase, e.g. also during commissioning.

- Supervision and control applies to more than just actuators and diagnostics, e.g. protection systems can be included in experiment development.

Supervision Outside the Plasma Control System

The supervision software tools can be used during the various operational phases of the project to support activities that happen continuously at a machine, campaign and experiment level.

Operational Lifetime: Engineering Limits

Every machine has engineering limits defined by the materials used in its construction. For fusion machines, some of the engineering limits that are applicable are the neutron budget (the material degradation caused by neutron irradiation), number of times the magnetic coils can be energised/de-energised (due to material fatigue caused by the change in magnetic fields) and the number of times that the superconducting magnetic coils can be warmed up (due to material fatigue caused by thermal stresses). These limits must be constantly updated and monitored through the lifetime of the machine to reduce the probability of premature/unexpected failure.

Support for managing all engineering limits, over the long-term, i.e. the operational lifetime of the machine, as well as short-term, i.e. for an individual experiment program, should be inherently included in the supervision software tools. This makes sure that all engineering limits are consistently updated and monitored.

The software tools can also be used to assist with engineering limit planning. Budget plans are constructed for each limit on a per campaign basis, and then per experiment within each campaign. For each experiment program, before it is executed, a simulation is done to predict a value of a specific engineering limit, e.g. compute the predicted neutron flux. Then, after each experiment is completed, data from all the diagnostics is processed automatically to determine the actual values and the specific engineering limit is updated accordingly.

Operational Lifetime: Localised Version of Functional Interfaces During Construction

To support construction, localised versions of the functional interfaces provided by required site services, e.g. ITER's mini-CODAC [8, 9], should be available. The supervision software tools must be able to function within this environment.

Campaign: Parallel Activities During Commissioning

The supervision software tools must be flexible to support the parallel activities that can take place during campaign commissioning and shut-down. Parallel activities usually require a resource to be share-locked or exclusive-locked, in order to prevent activities from interfering with each other. An example of an exclusive lock would be on the power supplies of the magnetic coils that grant control to one activity, e.g. to perform a controlled magnetic sensor calibration. An example of a shared lock would be on the power supplies of the magnetic coils to prevent power to the magnetic coils.

Many commissioning activities use the standard experiment program control software. Therefore, the experiment program control software must be able to operate in a "commissioning mode" where parallel experiment programs can be executed simultaneously.

Campaign: Collaborative Experiment Program Development

The supervision and control software tools should support collaborative development of the experiment programs that are to be executed within the campaign. The experiment designers must be able to use the software from locations other than the fusion project site, e.g. their home institutions.

Campaign: Management of Engineering Limits

Part of the experiment program design includes simulations to model plasma behaviour to verify that the experiment will do what is intended. Additional results of the simulation include estimates of engineering limits of e.g. neutron fluxes and forces on components, which are used to validate the experiment will not exceed any of the planned limits. The information is integrated with the supervision and control tools to automatically facilitate the engineering limit management processes.

Campaign: Configuration of Safety and Interlock Systems

The supervision and control software tools should allow for experiment programs to include expected configurations for the safety and interlock systems. Configuration in this sense means, e.g. which functions are expected to be enabled/disable and what thresholds are expected. This allows for a comprehensive validation of the actual and expected configurations before an experiment is executed.

Campaign: Support in Localised Version of Functional Interfaces During Site Acceptance Testing

In addition, site acceptance tests are performed on new equipment in isolation of all existing hardware. This requires localised versions of the functional interfaces provided by all site services, e.g. ITER's mini-CODAC [8, 9]. The supervision software tools must be able to function within this environment.

Day

For machines that do not operate for 24 h, a normal operational day (Fig. 3) starts with bringing the machine out of stand-by. To do this, operators will carry out the same set of actions. Therefore, the supervision and control software tools must support the encoding and execution of these repeatable actions. This improves efficiency, reduces operator error and ensures a level of consistency. However, these should be goal-based actions and not simply writing a set of macros and scripts and having them blindly executed. The repeatable actions must take into account the daily conditions, i.e. the actual resource availability on the day. This includes material resources, such as available power from the grid, and human resources for who is on-duty.

Once the machine is operational, facilities typically perform a conditioning experiment. The supervision software tools must be able to process the diagnostics data from the conditioning experiment to evaluate system health. In addition, the overall long-term state of the machine can be checked by a detailed analysis of the system behaviour. This data can be used to teach machine-learning systems about the machine state which can then be used to inform pre-emptive maintenance, thereby reducing failures.

At the end of the day, shutdown tasks, like glow-discharge cleaning, may need to be performed and the supervision software tools should facilitate this, ideally without continuous/direct human oversight.

Shift: Commissioning

During commissioning, the supervision software tools must support multiple teams commissioning their equipment in parallel. This is where shared and exclusive locks are used to manage resources to prevent conflicts. In such a situation, a high-level planning view allows for the consequences of delays to be communicated. The supervision software tools must additionally support and perform checks during commissioning to ensure that what one commissioning team is about to do will not affect any of the other commissioning teams.

Shift: Normal Operation

Classical shift planning involves preselecting a set of candidate experiments with additional contingency experiments. Experiments are executed based on how their planned resource requirements compare to the actual resource availability during that shift. If a planned experiment cannot be executed, then one of the contingency experiments should possibly be performed instead.

Supervision software tools in future systems could automatically analyse current resource availability and select the experiments to be executed in a shift from a pool of experiments planned for a campaign. Other information that would automatically be considered during the experiment selection would be the authorisation level of the key personnel on duty during the shift (e.g. operator licencing). In addition, the supervision software tools can assist with ordering the experiments so that similar experiments are executed together. This kind of flexible machine-assisted planning makes best use of actual resources.

Experiment: Resource Checks

The experiment program would encode expected values of key parameters like neutron dose, maximum plasma current and, maximum heating power. After the experiment program is loaded, the supervision software tools automatically check the expected values against the available systems to confirm whether these demands can be satisfied. Engineering budgets would also be checked but these depend on the history of the day/shift. Additionally, all the required infrastructure (e.g. communication and archiving networks) would also be checked to confirm that they are functional. The operators/engineer-in-charge would be checked that they have the authorisation to execute the experiment. These "passive" checks are identical to those performed for the experiment selection process in the shift but are repeated to confirm that the current situation is still suitable for the experiment to be executed.

Experiment: Configuration Checks

If these checks pass, then the next step looks at the expected configuration for the experiment compared to the actual configuration of the machine. While existing software tools update the configuration of the various subsystems to execute the planned experiment, a significant opportunity for advancement is checking and updating the protection system configuration. Generally, protection system configuration refers to the limits and protection functions that determine the behaviour of the protection system. The actual protection system configuration must be available for comparison with the expected configuration encoded into the experiment program. Any changes can be flagged for approval before updating the protection system. This automated process ensures that all necessary changes are identified, that the responsible personnel review and approve the changes and the correct changes are applied by avoiding transcription errors.

Experiment: Post-experiment Checks

When the experiment is running, it is under the complete control of the plasma control system, however the supervision software tools still have a high-level overview of the process. When the experiment concludes, the supervision software tools can check that all systems have operated as expected, that resulting data has been processed and archived and that related experiment summary information is computed and written. Engineering budgets are also automatically updated with actual information.

Plasma Control Supervision

Older plasma control systems were designed to run experiment programs that drove the plasma through a single series of states (i.e. a trajectory) to achieve a single goal. Any single goal could be achieved in multiple ways, based on the different system configuration options, for example, using the heating systems in different combinations to achieve the same total power, but each configuration would have to be encoded into a specific experiment program that was manually selected based on the current situation.

However, for the next-generation of fusion projects, a single experiment program must capture a range of options of multiple or different series of states (multiple trajectories), with the plasma control system deciding at runtime which to follow [10, 11]. The real-time plasma control system would be a part of the suite of integrated supervision and control software tools. These tools should also include plasma control supervision that provides flexible real-time strategies for experiment execution.

Under the new paradigm, the ideal situation is a single experiment program that can achieve multiple goals in multiple ways. After the experiment program is loaded, plasma control supervision determines which is the most appropriate goal and the most appropriate trajectory to achieve that goal and instructs the plasma control system to execute it. If conditions change as the experiment progresses so that the original trajectory is no longer feasible, then the supervision system automatically selects a new trajectory to achieve the same goal and communicates this to the plasma control system. If, however, it is no longer possible to achieve the selected goal, then a new goal is selected. Once a goal has been achieved, a new goal is selected out of those remaining, including any that were previously abandoned, if conditions are now suitable.

This level of flexibility in task execution guarantees that each experiment produces a useful result.

Diagnostics as a Support to Supervision

All fusion research projects have diagnostics that produce data for research purposes and so when people think of diagnostics, the natural tendency is to associate it only with the few systems used for research purposes. Future projects are more focused towards power production and continuous operation so the definition of "diagnostics" for these systems need to broader to take into account whole-system performance and behaviour.

When one considers all the inputs that needs to be taken into account, it is clear that diagnostics extends far beyond the realm of a conventional data acquisition systems. Even such elements as the protection system configuration becomes a diagnostic input.

Conclusions

The high-cost of future fusion projects demands a high level of efficiency and utilisation. An integrated suite of supervision and control software tools that operate throughout the lifetime of the project supports this aim. Such tools make it possible to design and execute fusion experiments encompassing multiple goals with multiple ways of achieving each goal. In addition, these tools provide support during all operational phases of the machine including commissioning.

Supervision and control software tools include the more familiar plasma control system, but also include a variety of other supervisory tools, experiment program editing tools and planning tools to manage aspects outside of the times when the plasma control system is functional.

Acknowledgements The authors thank Anett Spring (IPP - Greifswald), Wolfgang Treutterer (IPP-Garching), Remy Nouailletas and Philippe Moreau (WEST), Adam Stephen (JET), Axel Winter (ITER/IPP - Greifswald) and Bertrand Bauvir (ITER) for many insightful discussions.

References

1. K. Tomabechi et al., ITER conceptual design. Nucl. Fusion **31**(6), 1135 (1991)
2. G. Federici et al., European DEMO design strategy and consequences for materials. Nucl. Fusion **57**(9), 092002 (2017). https://doi.org/10.1088/1741-4326/57/9/092002

3. A. Spring et al., Establishing the Wendelstein 7-X steady state plasma control and data acquisition system during the first operation phase. Fusion Eng. Des. **123**, 579–583 (2017)
4. H. Lacqua et al., Experiences with the segment control system at Wendelstein 7-X operation. Fusion Eng. Des. **123**, 588–592 (2017)
5. D. Stepanov et al, Integrated approach to the development of the ITER control system configuration data, in *Proceedings of 13th International Conference on Accelerator and Large Experimental Physics Control Systems (ICALEPCS 2011)*, MOMAU005 (2011), pp. 52–55. https://accelconf.web.cern.ch/accelconf/icalepcs2011/papers/momau005.pdf
6. H. van der Beken et al, Level 1 software at JET: a global tool for physics operation, in *Proceedings of IEEE 13th Symposium on Fusion Engineering*, vol. 1, no. 2–6 (1989), pp. 201–204
7. J.W. Farthing et al, Data management at JET with a look forward to ITER, in *Proceedings of ICALEPCS07 (2007)*, pp. 74–78. https://accelconf.web.cern.ch/accelconf/ica07/PAPERS/TOPA01.PDF
8. F. Di Maio et al., Development of the ITER CODAC core system, in *Proceedings of 12th International Conference on Accelerator and Large Experimental Physics Control Systems (ICALEPCS 2009) (CD-ROM)*, MOC005 (2009), pp. 23–24. https://accelconf.web.cern.ch/accelconf/icalepcs2009/papers/moc005.pdf
9. F. Di Maio, *CODAC Core System Overview*. http://static.iter.org/codac/cs/CODAC_Core_System_Overview_34SDZ5_v6_1.pdf. Accessed 25 July 2018
10. N. Ravenal et al., Architecture of WEST plasma control system. Fusion Eng. Des. **89**, 548–552 (2014)
11. R. Nouailletas et al., From the conceptual design to the first simulation of the new WEST plasma control system. Fusion Eng. Des. **96–97**, 680–684 (2015)

Journal of Fusion Energy (2019) 38:445–457
https://doi.org/10.1007/s10894-018-0205-y

ORIGINAL RESEARCH

Bayesian Data Analysis for Gaussian Process Tomography

T. Wang[1,2,3] · D. Mazon[2] · J. Svensson[5] · A. Liu[6] · C. Zhou[6] · L. Xu[7] · L. Hu[7] · Y. Duan[7] · G. Verdoolaege[3,4]

Published online: 20 November 2018

Abstract
Bayesian inference is used in many scientific areas as a conceptually well-founded data analysis framework. In this paper, we give a brief introduction to Bayesian probability theory and its application to the tomography problem in fusion research by means of a Gaussian process prior. This Gaussian process tomography (GPT) method is used for reconstruction of the local soft X-ray (SXR) emissivity in WEST and EAST based on line-integrated data. By modeling the SXR emissivity field in a poloidal cross-section as a Gaussian process, Bayesian SXR tomography can be carried out in a robust and extremely fast way. Owing to the short execution time of the algorithm, GPT is an important candidate for providing real-time feedback information on impurity transport and for fast MHD control. In addition, the Bayesian formulism allows for uncertainty analysis of the inferred emissivity.

Keywords Bayesian inference · Data analysis · Plasma physics · Tomography · Soft X-ray · Gaussian process · Nuclear fusion · Tokamak

Data analysis in Fusion Experiments

Introduction

Magnetic confinement fusion experiments generate large quantities of complex data. At a basic level, the data reflects the state of the machine and plasma, enabling safe and reliable operation of the device, i.e. well within the design limits of the machine and compatible with the scientific goals of the experiment. In addition, the data is an essential ingredient to interpreting the experiment at a higher level, deepening the physical understanding of the various plasma processes and their interaction with the device.

In a fusion reactor, most of the data will be used for machine operation. The main criteria will be safety and plasma performance, maximizing the fusion output. Modelling codes produce even more data, which needs to be compared to the experimental results. Depending on the requirements, different analysis techniques are needed to extract as much useful information as possible from the raw data. This ranges from very basic tools for signal resampling or motion correction in images to more advanced statistical analysis or pattern recognition methods. As the available computing power increases, real-time high-level data analysis also starts to become feasible for plasma control.

The throughput volume and complexity of fusion experimental data demand special attention for the analysis. As an illustration, the JET experiment presently generates ~ 55 GB of data per pulse and ITER will produce

✉ T. Wang
tianbo.wang@ugent.be

1 Southwestern Institute for Physics, CNNC, Chengdu 610200, People's Republic of China

2 CEA, IRFM, 13108 St Paul lez Durance, France

3 Department of Applied Physics, Ghent University, 9000 Ghent, Belgium

4 Laboratory for Plasma Physics, Royal Military Academy (LPP-ERM/KMS), 1000 Brussels, Belgium

5 Max-Planck-Institut für Plasmaphysik, 17491 Greifswald, Germany

6 CAS Key Laboratory of Basic Plasma Physics, Department of Modern Physics, University of Science and Technology of China, Hefei 230026, People's Republic of China

7 Institute of Plasma Physics, Chinese Academy of Sciences, Hefei 230031, People's Republic of China

up to 1 TB in each discharge. The complexity of the data is essentially due to the complexity of the physics of a fusion plasma and to uncertainty introduced by the diagnostic techniques. Indeed, the variables through which the plasma is characterized usually interact in an intricate, nonlinear way and their measurements can be affected by considerable uncertainty. Furthermore, different diagnostics may provide information on the same plasma quantity, resulting in a certain degree of redundancy that can be exploited for the purpose of joint analysis of multiple diagnostics, but may also entail issues of data consistency (through measurement error and incorrect assumptions regarding the measurement model) and data heterogeneity. Modern information science can play an important role in the improvement of data accuracy and robustness and the unravelling of useful physical quantities or relations from the data.

Though modern tools from probability theory and machine learning have been applied to fusion experiments, they are still relatively uncommon in fusion science. The situation is quite different in other data-intensive disciplines of the physical sciences (e.g. high-energy physics, astronomy and ecology), where advanced techniques have been part of the common practice for many years. Therefore, development and application of state-of-the-art techniques based on probability theory and machine learning need to be stimulated in fusion science, in order to address complex data analysis problems. The goal should be to raise the part of the scientific process related to data analysis to the same high level as the experimentation, modelling and theory building.

We start by introducing some basic concepts from Bayesian probability theory, with a view to the application to Gaussian process tomography presented in "Gaussian Process Tomography" section. We do not strive for comprehensiveness; rather mean to give a flavour of typical method and references to further readings, which we hope will prove to be useful to the beginning practitioner.

A Note About Probability

We begin our discussion with a short note about the view we choose to adopt on probability. Over time, various interpretations of probability have been proposed, the most popular by far being the *frequentist* and the *Bayesian* approaches. Both interpretations come with their associated set of methods and tools, although practices vary across disciplines (e.g. social sciences vs. physical sciences). In addition, there exist various flavours of Bayesian thinking, some of which are half frequentist and use frequentist terminology. However, in the physical sciences and fusion in particular, the point of view of [3, 11] has come to prevail, wherein probability theory is considered as an extension of logic to uncertain propositions. The theory is based on a set of common-sense axioms (or desiderata), including the basic requirement of consistency. In other words, different observers should arrive at the same conclusions when having been provided with the same information. From this point of view, a probability always depends on two pieces of information: the proposition or hypothesis A, the probability of which we seek, and the information B, conditioning the probability. The standard notation used to denote such a probability is $P(A|B)$ and corresponds to a real number in the interval $[0, 1]$ expressing the plausibility of A provided B is true. Put differently, it is a measure of the extent to which B *implies* A. As such, a probability denoted by $P(.|.)$ always has two slots that need to be filled to produce a meaningful numerical output. The conditioning information B can be a problem statement, some measured data or any other available background information. The whole of information on which a probability statement is conditioned is often summarized by the generic symbol I, e.g. $P(.|I)$. Specific additional information conditioning the probability may be stated explicitly, e.g. $P(.|B, I)$.

The classical frequentist definition of probability is based on the frequency of an event in the long run and necessitates identical repeats of experiments or ensembles of the system under study. As such, it contrasts with the Bayesian interpretation, which enables a direct evaluation of the probability of any statement or single event. Examples include the probability of the plasma density in a fusion machine being within certain bounds or the probability that a plasma is in the H-mode. In the Bayesian view, probabilities are useful whenever there is a general lack of certainty, e.g. details of plasma particle orbits causing fluctuations in the macroscopic quantities or the electron thermal motion in a detector introducing noise in the measurement. No appeal is made to any notion of 'randomness' and although the term 'random variable' occurs in some Bayesian literature, we will avoid it. Neither will we follow the habit of using capital letters to denote random variables since, from our point of view, the information carried by any physical variable is subject to uncertainty. This approach is adopted by several excellent textbooks on Bayesian probability theory geared to physicists; see [9, 11, 15, 18].

Bayesian Diagnostic Modelling

Introduction

An important application of Bayesian probability theory that has gained considerable use within the fusion community can be found in fusion diagnostics modelling [5–7]. This practice is motivated by the frequent and substantial uncertainties entering the derivation of physical quantities from raw (measured) data, often a voltage at a sensor. Without proper measures, these uncertainties can enhance and complicate the total uncertainty on the quantity of interest, deteriorating the quality and reliability of the result. In extreme cases the result may become meaningless, which, for instance, may occur in the calculation of profiles from line-integrated measurements through an inversion technique. Not only is this problem heavily underdetermined (multiple solutions fit the data), but the noise on the line-integrated measurements also propagates through the inversion and generally worsens the result. Limiting the number and smoothness of the solutions by regularization techniques is the standard solution, but further improvements are often possible. Bayesian probability theory offers a solid and coherent framework for better profile reconstruction. It consistently handles any kind of uncertainty by assigning probabilities (or probability distributions) and manipulating them using a set of well-defined rules. One of these is *Bayes' rule*—hence the designation 'Bayesian probability theory'—which we will introduce below. The result of a Bayesian calculation is a probability (distribution) for the quantity of interest or a hypothesis, depending on the available information.

In addition to modelling of individual diagnostics, Bayesian theory enables combination of measurements of individual diagnostics, resulting in an increase in accuracy of common physics parameters of interest [14]. By modelling each diagnostic in the Bayesian framework, a probability for a specific plasma quantity can be derived that is consistent with the data from all diagnostics involved in the analysis. Therefore, when multiple diagnostic measurements contain information regarding a certain plasma quantity (e.g. interferometry and Thomson scattering for the electron density of a plasma), it is possible to exploit the advantages of each diagnostic, hence improving the quality of the estimate of the desired plasma quantity. This quality can be reflected in smaller error bars, higher resolution or enhanced robustness, since the analysis involves information from diagnostics that are based on different measurement principles or different hardware components.

Concepts of Bayesian probability Theory

In this section, we introduce some basic concepts from Bayesian probability theory. We do this in the context of a simple example involving estimation of the parameters of a normal distribution based on suitable prior information and repeated measurements of a physical quantity.

In fusion, as in many scientific activities, we do not have a direct access to the physical properties of the plasma. The experimentalist must instead devise a measurement technique for a diagnostic, which returns numbers (usually voltages) that are related to the quantity of interest. If the properties of a physical system were known precisely, together with the full details of the measurement process, the corresponding measurement values could be computed straightforwardly. Calculating the measurements would require the measurement process to be encoded in a mathematical model: the *forward model*. However, since the details of the physical system and the measurement process are unknown, the raw measurement itself is never a truly precise number.[1] We may nevertheless be able to model this uncertainty by assigning a probability for the measured value to lie within certain bounds. Representing the measurement by a real number x, it is convenient to introduce the *probability density function* (PDF) of x.[2] The PDF is usually denoted $p(x|I)$, where $p(x|I)\mathrm{d}x$ can be interpreted as the probability to find the measurement value between x and $x + \mathrm{d}x$, given the background information I.[3] Strictly speaking, for continuous variables, the term *distribution function* refers to the cumulative distribution function; it is nevertheless used colloquially to denote the PDF. To indicate that a variable x is assumed to be distributed according to (or sampled from) the model with PDF $p(x|I)$, one writes $x \sim p$. One of the most common models is the *normal* or *Gaussian distribution*, the density function of which is the well-known function:

$$p(x|\mu, \sigma, I) = \frac{1}{\sqrt{2\pi}\sigma} \exp\left[-\frac{(x-\mu)^2}{2\sigma^2}\right]. \tag{1}$$

This is often written as $x \sim \mathcal{N}(\mu, \sigma^2)$, where μ is the mean (expectation value) of the distribution and σ the standard deviation. Note that, in the Bayesian spirit, we have explicitly written the PDF as a two-slot function $p(x|\mu, \sigma, I)$, where the second slot emphasizes that the probability density can only be calculated for known values of μ and σ. Any additional information (e.g. the fact that it is a Gaussian distribution) has been summarized in the symbol I. With the same knowledge, the probability that x is found to lie within two given bounds can be calculated by integration (involving the error function erf). The measurement uncertainties that can be described by probability models such as (1) are often referred to as *stochastic* or *statistical uncertainties*. This is opposed to *systematic*

[1] At a certain point, one also hits the rather philosophical question of whether there is such a thing as an 'exact value' of a (microscopic) physical quantity.

[2] We will treat only continuous variables here. This represents the most common situations in physical sciences.

[3] We will follow the common slight abuse of notation of using the symbol p to denote a specific model (e.g. Gaussian) of the probability density of a specific variable or, in general, the probability density of a variable, possibly unknown or unspecified.

uncertainties, which result in a 'deterministic' deviation of the result from the 'correct' measurement.

An entirely different question concerns the assignment of probabilities to the parameters μ and σ, given a set of measurements for the variable x, denoted by $\vec{x}$. The solution, known as *Bayes' rule* or *Bayes' theorem*, is generally credited to Thomas Bayes (1763) and Pierre-Simon Laplace (1812). This is a direct consequence of the well-known product rule of probability theory, which in this context reads

$$p(\vec{x}, \mu, \sigma|I) = p(\vec{x}|\mu, \sigma, I)p(\mu, \sigma|I). \tag{2}$$

Of course, the same rule also yields

$$p(\vec{x}, \mu, \sigma|I) = p(\mu, \sigma|\vec{x}, I)p(\vec{x}|I),$$

which, together with (2), results in Bayes' rule (applied to the current example):

$$\overbrace{p(\mu, \sigma|\vec{x}, I)}^{\text{posterior}} = \frac{\overbrace{p(\vec{x}|\mu, \sigma, I)}^{\text{likelihood}}\,\overbrace{p(\mu, \sigma|I)}^{\text{prior}}}{\underbrace{p(\vec{x}|I)}_{\text{evidence}}}. \tag{3}$$

Thus, the rules of probability theory naturally enable solving the inverse problem by using a forward model. Determining appropriate values for the quantities of interest—μ and σ—from (3) is referred to as the task of *parameter estimation*. We will discuss this task in more detail below, still in the context of a normal distribution.

Bayes' rule can also be interpreted in terms of a learning process. In this sense, information regarding the parameters μ and σ is considered before the data is acquired, using the so-called *prior distribution*, indicated in (3). The prior can be obtained from background knowledge about the problem or other experiments. Alternatively, we may choose an *uninformative prior*, which, as the term implies, allows us to adopt a maximally ignorant point of view before performing the experiment. However, one should be careful in keeping the problem well-posed (or *identifiable*), i.e. there should be sufficient information in the data and the prior to estimate the parameters with reasonable accuracy (e.g. the discussion in [17]). Various criteria and methods exist to assign uninformative prior distributions, including invariance under basic transformations (e.g. translations and rotations), *maximum entropy distributions* in the presence of specific pieces of *testable* information (e.g. knowledge of the mean of a distribution), smoothness-ensuring priors for regularizing ill-posed problems and others (see [15, 18] for practical advice and [11] for more in-depth discussions). Whether one chooses an uninformative or informative prior, in the objective Bayesian view the prior distribution is not a subjective judgment by an individual scientist, but rather the result of quantifying the available prior information. In this sense, two individuals, which have the same prior information to the experiment, should arrive at the same prior distribution, hence should obtain the same conclusions. Many practitioners use a very pragmatic rule to assign prior distributions, which is related to the analytical tractability of the posterior. This gives rise to the so-called *conjugate priors*, described for instance in [8]. However, one could object that more objective prior information is to be preferred over a criterion based on computational ease, especially as the likelihood and the experimental setup also depend, to some extent, on 'subjective' choices.

Continuing with our example of inferring the parameters of a normal distribution, we could next perform a series of n measurements, here described by $\vec{x}$. We will assume, given a common μ and σ, that the measurements are performed independently from each other. Fixed model parameters μ and σ imply stationary conditions in the system of interest and the measurement process. Under these assumptions, the distribution of $\vec{x}$, conditioned on μ and σ, can be factorized into a product of marginal distributions for each of the x_i components of $\vec{x}$ ($i = 1\ldots n$):

$$\begin{aligned} p(\vec{x}|\mu, \sigma, I) &= \prod_{i=1}^{n} \frac{1}{\sqrt{2\pi}\sigma} \exp\left[-\frac{(x_i - \mu)^2}{2\sigma^2}\right] \\ &= \frac{1}{(2\pi)^{n/2}\sigma^n} \exp\left[-\sum_{i=1}^{n} \frac{(x_i - \mu)^2}{2\sigma^2}\right]. \end{aligned} \tag{4}$$

This is the *likelihood distribution*. According to (3), the product of the likelihood and the prior is proportional to the *posterior distribution* for the quantities of interest. The posterior quantifies our knowledge regarding the parameters of interest μ and σ, based on the prior information and the data. In this respect, an important observation is that, in estimating μ and σ, we do not need to care about the normalization of the posterior. Only the posterior's dependence on μ and σ matters, since it determines the shape of their joint distribution. For this reason, the denominator in (3) is irrelevant in the context of parameter estimation and may be absorbed in a proportionality constant. The denominator is referred to as the *evidence* (or marginal likelihood) since it gives the probability of the data in the light of the background information, which includes knowledge about the model that we use for the physical and measurement systems. This term provides evidence for our model and becomes important in comparing different models. In the Bayesian approach, the task of *model selection* is also performed according to Bayes' rule; we will not go deeper into this topic here, but the reader may refer to [18].

It is possible to show that the uniform distribution is an adequate uninformative prior for μ, while Jeffrey's scale prior ($\propto 1/\sigma$) is suitable for σ [18]. Under these assumptions, the joint posterior distribution for μ and σ obeys the following proportionality:

$$p(\mu, \sigma|\vec{x}, I) \propto \frac{1}{(2\pi)^{n/2}\sigma^{n+1}} \exp\left[-\sum_{i=1}^{n} \frac{(x_i - \mu)^2}{2\sigma^2}\right]. \quad (5)$$

From this, the marginal distribution for μ can be obtained by integrating over σ—an operation that is ubiquitous in Bayesian probability theory and that is referred to as *marginalization*. For μ, this leads to Student's t-distribution, while for σ one obtains a χ^2 marginal distribution [18].

Gaussian Process Tomography

In this section, we give an example application of Bayesian inference in fusion research, using the relatively new technique of Gaussian process tomography.

Tomography in Fusion Devices

Reconstruction of a space-resolved image or profile of a plasma quantity using tomography has been considered on many fusion devices [10, 12, 13]. The tomographic problem essentially involves prediction of high-dimensional physics parameters by an inversion of a limited number of measurements. This is an ill-posed problem, as the number of measurements is always lower than the number of unknowns.

In our present example we consider tomography for a soft X-ray (SXR) diagnostic at WEST (Tungsten Environment in Steady-State Tokamak) using synthetic data. Then, we provide another example using real data from EAST (Experimental Advanced Superconducting Tokamak). SXR spectroscopy can give valuable information about the plasma in a two-dimensional poloidal cross-section, such as magnetohydrodynamic activity and impurity transport. In the past, various tomographic reconstruction techniques have been applied to SXR, such as the Cormack method [2], the maximum entropy method, the minimum Fisher information method [4], etc. Particularly the minimum Fisher information technique has been widely adopted in the fusion community [1, 10, 12, 13]. This reconstruction method involves χ^2 optimization, regularized by the Fisher information. Intuitively, the goal is to find the least complex solution that is compatible with the data. The method is often implemented, e.g. on Tore–Supra and WEST, using additional information concerning the location of the equilibrium magnetic flux surfaces, obtained from magnetic measurements. It provides a good trade-off between the initial magnetic flux surface information and the SXR data.

Bayesian Tomographic Model

The purpose of SXR tomography in magnetic fusion devices is to reveal the spatial distribution of SXR emissivity in a poloidal cross-section, by the inversion of a number of noisy line-integrated emissivity measurements. A common and simple approach to discretize the emissivity field in a poloidal cross-section uses a square grid. We here impose a 100 × 100 grid comprising square cells with a dimension of 16 mm × 16 mm. The SXR emissivity within each pixel can reasonably be assumed to be constant, so the SXR line-integrated emissivity $\bar{d}_m$ along m viewing chords can be written in the following matrix form:

$$\bar{d}_m = \bar{\bar{R}}_{m\times n} \cdot \bar{E}_n + \bar{\varepsilon}. \quad (6)$$

Here, $\bar{E}_n$ is the unknown vector of local emissivities in $n = 10^4$ cells, while $\bar{\bar{R}}$ is the geometry matrix, whose elements R_{ij} represent the physical length of chord i through cell j. $\bar{\varepsilon}$ denotes an error term to account for measurement uncertainty, which is usually limited to statistical errors only.

From the Bayesian perspective, (6) is the basic forward model of the tomography problem. The classic optimization approach uses a criterion that minimizes the difference between the measured line integrals and the prediction by the model. Because of the ill-posedness, the optimization has to be combined with some regularization technique, e.g. assuming a spline model for the local emissivity field, or by optimizing at the same time some information measure like the Shannon entropy or the Fisher information. In this paper we choose a probabilistic methodology, which provides a probability distribution $p(\bar{E}_n)$ of the emissivity in all cells rather than a single solution. In Bayesian inference, one starts from the prior probability distribution of the emissivity field, which can be used to encode the regularization. This is then updated through Bayes' theorem as data become available (the background information I is implicit):

$$p(\bar{E}_n|\bar{d}_m) = \frac{p(\bar{d}_m|\bar{E}_n)p(\bar{E}_n)}{p(\bar{d}_m)} \sim p(\bar{d}_m|\bar{E}_n)p(\bar{E}_n) \quad (7)$$

$$p(\bar{d}_m) = \int p(\bar{d}_m, \bar{E}_n)d\bar{E}_n = \int p(\bar{d}_m|\bar{E}_n)p(\bar{E}_n)d\bar{E}_n. \quad (8)$$

$\bar{E}_n$ Vector of emissivity values in all n pixels at a particular time t. $\bar{d}_m$ Vector of m line-integrated GEM array measurements at time t

In Eq. (7), the likelihood term $p(\bar{d}_m|\bar{E}_n)$ assesses the mismatch between the measured line integrals $\bar{d}_m$ and their predictions by the forward model, under the assumption of some emissivity field $\bar{E}_n$. In Eq. (8), we have marginalized over the emissivity field. The resulting evidence (marginal likelihood) $p(\bar{d}_m)$ depends on the particular forward measurement model, which we will assume to be fixed. Therefore, again it can be considered as a normalization factor, independent of the emissivity. The posterior probability distribution $p(\bar{E}_n|\bar{d}_m)$ quantifies our uncertainty on the estimated emissivity field, given our model, prior knowledge and the measured data. Thus, Bayesian inference yields probabilities for all possible results consistent with our model. In principle, systematic uncertainties can also be estimated, provided some knowledge is available about them from other sources of information, such as other experiments.

Gaussian Process Prior and Posterior

Gaussian process tomography (GPT) is a new technique whereby the prior distribution regularizes the tomographic reconstruction process [16], by imposing a smoothness level dictated by the correlation between pixels. Briefly, a Gaussian process (GP) is a generalization of the multivariate normal (Gaussian) distribution to a function space. It is described by a mean function $\bar{\mu}\left(\bar{r}\right)$ and a covariance function $\bar{\bar{\Sigma}}\left(\bar{r},\bar{r}'\right)$, where $GP\sim\mathcal{N}\left(\bar{\mu},\bar{\bar{\Sigma}}\right)$. The distribution of a Gaussian process is the joint distribution of infinitely many normally distributed random variables, and as such, it is a distribution over functions with a continuous domain, e.g. time or space. GPT is related to Gaussian process regression (or "*kriging*" in geoscience), a nonparametric regression technique widely used in machine learning. Being nonparametric, Gaussian process regression does not assume any functional form for the regression function, hence leaving a lot of flexibility. Instead, the regression surface is regularized through the covariance matrix of the Gaussian process. Likewise, GPT assumes that the prior joint distribution of the emissivity in the n cells with coordinates $\bar{r}_i$ is multivariate Gaussian (Fig. 1) with covariance matrix $\bar{\bar{\Sigma}}_E$ given by:

$$\bar{\bar{\Sigma}}_E=\begin{pmatrix} k\left(\bar{r}_1,\bar{r}_1\right) & \cdots & k\left(\bar{r}_1,\bar{r}_n\right)\\ \vdots & \ddots & \vdots\\ k\left(\bar{r}_n,\bar{r}_1\right) & \cdots & k\left(\bar{r}_n,\bar{r}_n\right)\end{pmatrix}. \tag{9}$$

Here, $k\left(\bar{r}_i,\bar{r}_j\right)=cov\left[E\left(\bar{r}_i\right),E\left(\bar{r}_j\right)\right]$, with $E\left(\bar{r}_i\right)=E_i$ the emissivity in pixel i, is the *covariance kernel function*, for which we choose the common squared-exponential form:

$$k_{SE}=\sigma_f^2\exp\left(-\frac{d^2}{2\sigma_l^2}\right),\quad d=\left\|\bar{r}_i-\bar{r}_j\right\|. \tag{10}$$

In turn, the kernel function depends on two parameters σ_f and σ_l, referred to as the signal standard deviation and length scale, respectively. In Bayesian terminology, any parameters of the prior distribution are called *hyperparameters*. Here, the key hyperparameter is σ_l, as it determines the smoothness of the prior emissivity field. On the other hand, σ_f sets the overall scale of the prior covariance matrix, and as such it determines the weight of the smoothing prior in the posterior covariance. Summarizing σ_f and σ_l by $\bar{\theta}$, the total inference problem can be written as

$$p\left(\bar{E}_n|\bar{d}_m,\bar{\theta}\right)=\frac{p\left(\bar{d}_m|\bar{E}_n,\bar{\theta}\right)\cdot p\left(\bar{E}_n|\bar{\theta}\right)}{p\left(\bar{d}_m|\bar{\theta}\right)}\sim p\left(\bar{d}_m|\bar{E}_n,\bar{\theta}\right)\cdot p\left(\bar{E}_n|\bar{\theta}\right), \tag{11}$$

where the prior is given by

$$p\left(\bar{E}_n|\bar{\theta}\right)=\frac{1}{(2\pi)^{\frac{n}{2}}\left|\bar{\bar{\Sigma}}_E\right|^{\frac{1}{2}}}exp\left[-\frac{1}{2}\left(\bar{E}_n-\bar{\mu}_E\right)^T\bar{\bar{\Sigma}}_E^{-1}\left(\bar{E}_n-\bar{\mu}_E\right)\right]. \tag{12}$$

Here, $\bar{\mu}_E$ is the prior mean, which will be fixed at 0, or it may be chosen on the basis of earlier experiments or prior knowledge of the involved domain experts. In principle, the hyperparameters can be marginalized from the problem (i.e. integrated out), but this would greatly increase the computational complexity of the method. Instead, we will employ a common approximation wherein a fixed set of hyperparameters is determined by maximizing the evidence $p\left(\bar{d}_m|\bar{\theta}\right)$, and plugging those estimates into Eq. (11). This procedure is motivated in the "Appendix".

The next step in the inference process consists of choosing a likelihood function $p\left(\bar{d}_m|\bar{E}_n,\bar{\theta}\right)$, containing the forward model. Under the reasonable assumption of a normal distribution of the measurement uncertainty on the emissivity line integrals, described by the variable $\bar{\epsilon}$ in Eq. (6), the likelihood can be written as

$$p\left(\bar{d}_m|\bar{E}_n,\bar{\theta}\right)=\frac{1}{(2\pi)^{\frac{m}{2}}\left|\bar{\bar{\Sigma}}_d\right|^{\frac{1}{2}}}exp\left[-\frac{1}{2}\left(\bar{\bar{R}}\cdot\bar{E}_n-\bar{d}_m\right)^T\bar{\bar{\Sigma}}_d^{-1}\left(\bar{\bar{R}}\cdot\bar{E}_n-\bar{d}_m\right)\right]. \tag{13}$$

Here, $\bar{\bar{\Sigma}}_d$ is the covariance of the measured data, describing measurement uncertainty and correlation pertaining to the vector $\bar{d}_m$ of measured line integrals. We will assume that the various line-integrated measurements are uncorrelated and choose a 5% noise level, based on previous experience at Tore–Supra. Therefore,

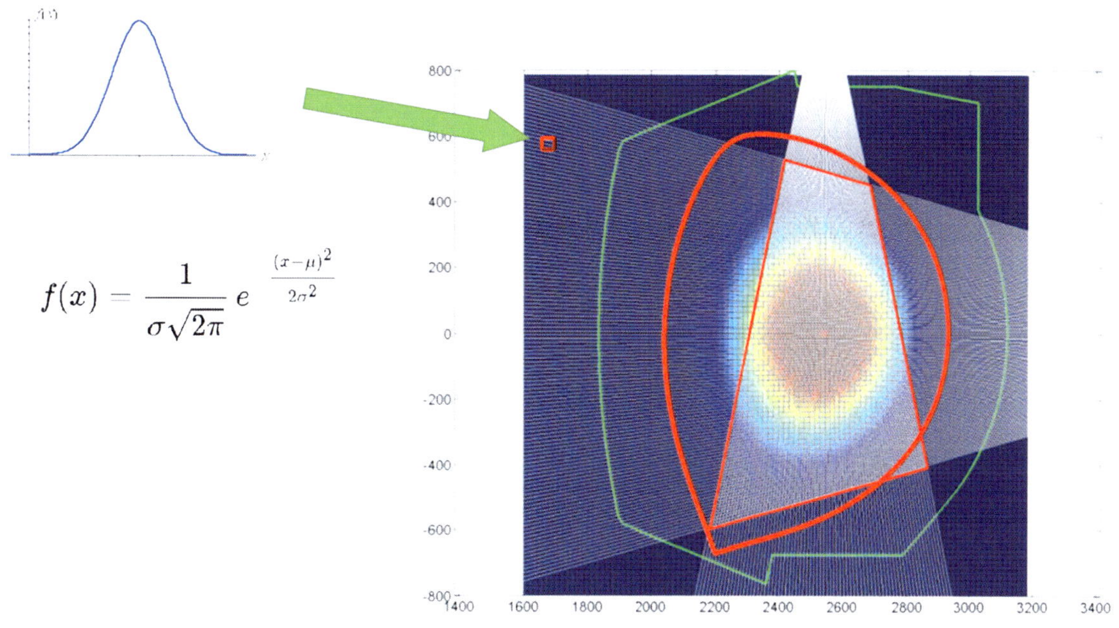

Fig. 1 Tomographic capabilities of the WEST SXR system based on GEM detectors. The horizontal camera views along 128 lines-of-sight. The vertical camera is inside the vertical port and is coupled to 75 sight lines. In the Gaussian process framework, the emissivity in each cell follows a Gaussian distribution, while the joint distribution of every subset of pixels is multivariate normal. This imposes structure on the emissivity field, avoiding wildly fluctuating emissivity in neighboring cells

$$\bar{\bar{\Sigma}}_d = \begin{pmatrix} 0.05 \cdot d_1 & \cdots & 0 \\ \vdots & \ddots & \vdots \\ 0 & \cdots & 0.05 \cdot d_m \end{pmatrix}. \qquad (14)$$

Finally, the posterior distribution reads up to a constant factor,

$$\begin{aligned} p(\bar{E}_n|\bar{d}_m, \bar{\theta}) &\sim p(\bar{d}_m|\bar{E}_n, \bar{\theta}) \cdot p(\bar{E}_n|\bar{\theta}) \\ &\sim exp\left[-\frac{1}{2}\left(\bar{\bar{R}} \cdot \bar{E}_n - \bar{d}_m\right)^T \bar{\bar{\Sigma}}_d^{-1}\left(\bar{\bar{R}} \cdot \bar{E}_n - \bar{d}_m\right)\right] \\ &\cdot exp\left[-\frac{1}{2}\left(\bar{E}_n - \bar{\mu}_E\right)^T \bar{\bar{\Sigma}}_E^{-1}\left(\bar{E}_n - \bar{\mu}_E\right)\right]. \end{aligned} \qquad (15)$$

The major advantage of normal distributions and a linear forward model now becomes clear. Indeed, it follows from standard probability calculus that the product of two normal distributions is also Gaussian, with mean vector and covariance matrix given by

$$\bar{\mu}_E^{post} = \bar{\mu}_E^{prior} + \left(\bar{\bar{R}}^T \bar{\bar{\Sigma}}_d \bar{\bar{R}} + \bar{\bar{\Sigma}}_E^{-1}\right)^{-1} \bar{\bar{R}}^T \bar{\bar{\Sigma}}_d^{-1}\left(\bar{d}_m - \bar{\bar{R}} \cdot \bar{\mu}_E\right). \qquad (16)$$

$$\bar{\bar{\Sigma}}_E^{post} = \left(\bar{\bar{R}}^T \bar{\bar{\Sigma}}_d \bar{\bar{R}} + \bar{\bar{\Sigma}}_E^{-1}\right)^{-1}. \qquad (17)$$

The posterior mean is thus available in a closed form, representing a computationally light-weight estimate of the emissivity field. In addition, the diagonal elements of the posterior covariance matrix $\bar{\bar{\Sigma}}_E^{post}$ quantify the uncertainty on the inference result. It should be noted that the GPT method implemented here only uses the SXR line integral measurements and no assumptions whatsoever are made regarding the magnetic equilibrium. This renders the method very flexible, potentially allowing detection of structures in the emissivity field (e.g. local impurity concentrations) that do not show up in the equilibrium reconstruction. It also prevents misguided information to enter the SXR reconstruction process in case the equilibrium information is incorrect, e.g. during sawtooth activity.

GPT Phantom Test on WEST

WEST Phantom Data

Our implementation of the GPT algorithm has been validated using phantom data, i.e. a set of synthetic SXR emissivity fields characterized by various emissivity patterns, some of which create a challenging tomography problem. Line integrals with added noise were calculated, followed by reconstruction of the emissivity field by GPT and comparison with the original emissivity phantom.

Four different shapes were used for the phantom tests, corresponding to various situations that are expected to be relevant for WEST SXR emission: Gaussian shape, hollow shape, left–right kidney shape and up-down kidney shape, as shown in Fig. 2. The kidney shapes correspond to asymmetric impurity distributions, e.g. due to toroidal plasma rotation.

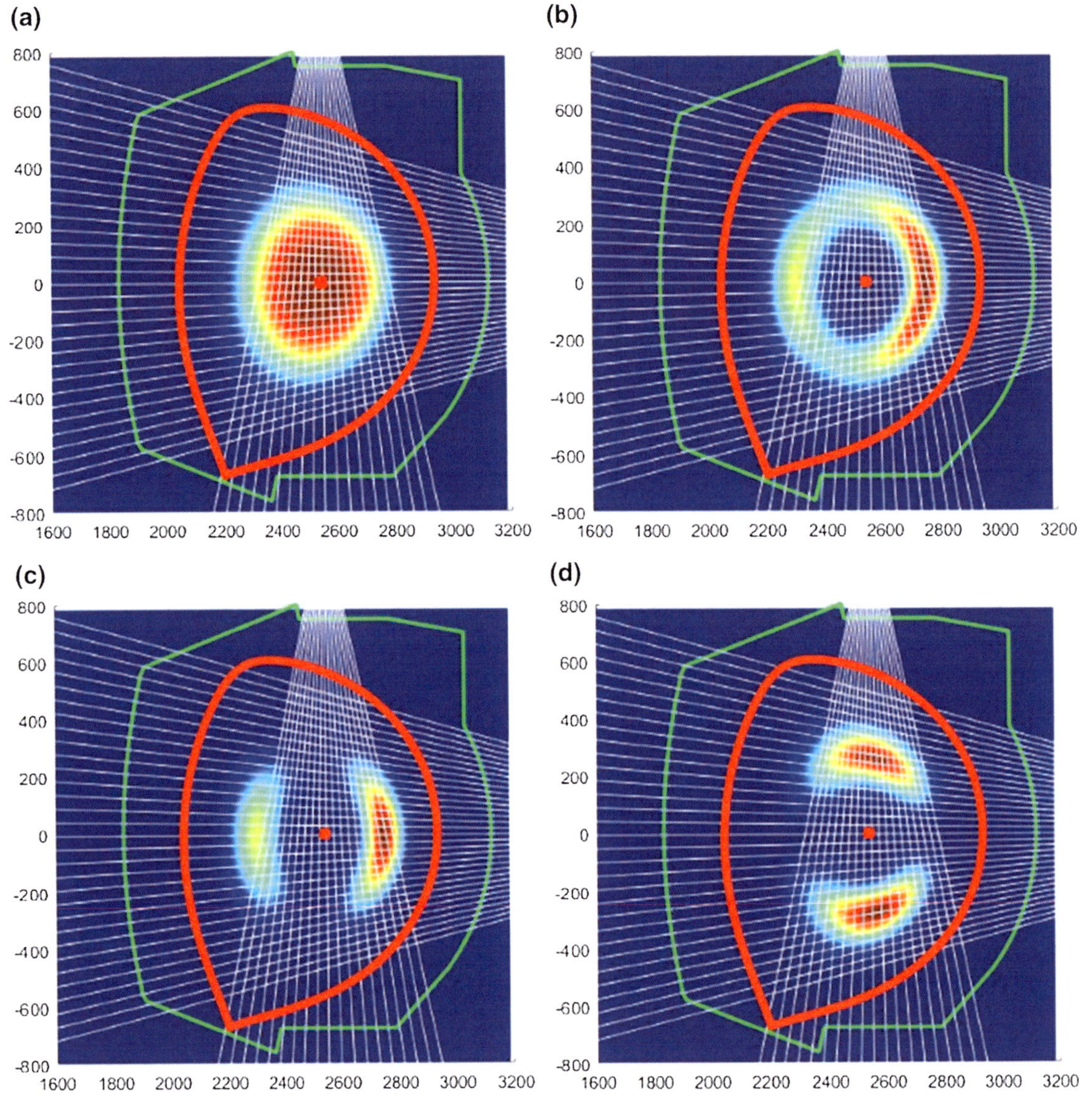

Fig. 2 Four phantom emissivity fields are used in our test: (**a**) Gaussian shape, (**b**) hollow shape, (**c**) left–right kidney shape, (**d**) up-down kidney shape. The green curve in the figure represents the vacuum vessel and the red curve is the last-closed flux surface (Color figure online)

Tomography Results

The reconstructed emissivity fields based on line integrals with a 5% noise level are shown in Fig. 3. The quality of the reconstructions can be quantified through a relative error map, showing the difference between the phantom and reconstructed field, normalized by the maximum phantom emissivity:

$$\xi_i = \frac{\left|E_{n,i}^{(rec)} - E_{n,i}\right|}{max\{\bar{E}_n\}}. \quad (18)$$

In case of the Gaussian shape, the maximum relative error is around 6.9%, 15% for the hollow shape, 12% for the left–right kidney shape and 15% for the up-down kidney shape. In general, more asymmetric emissivity fields are more difficult to reconstruct, the uncertainty level depending greatly on the coverage and field of view of the optical system. Nevertheless, in all cases the characteristic shape of the phantom is recovered relatively well by GPT. In addition, one can compare the line integrals obtained from the original phantom, with those calculated from the reconstructed emissivity field. As shown in Fig. 3, good agreement is achieved in all cases. The calculation is also fast: on a typical PC environment with Matlab, each time slice takes about 100 ms computing time.

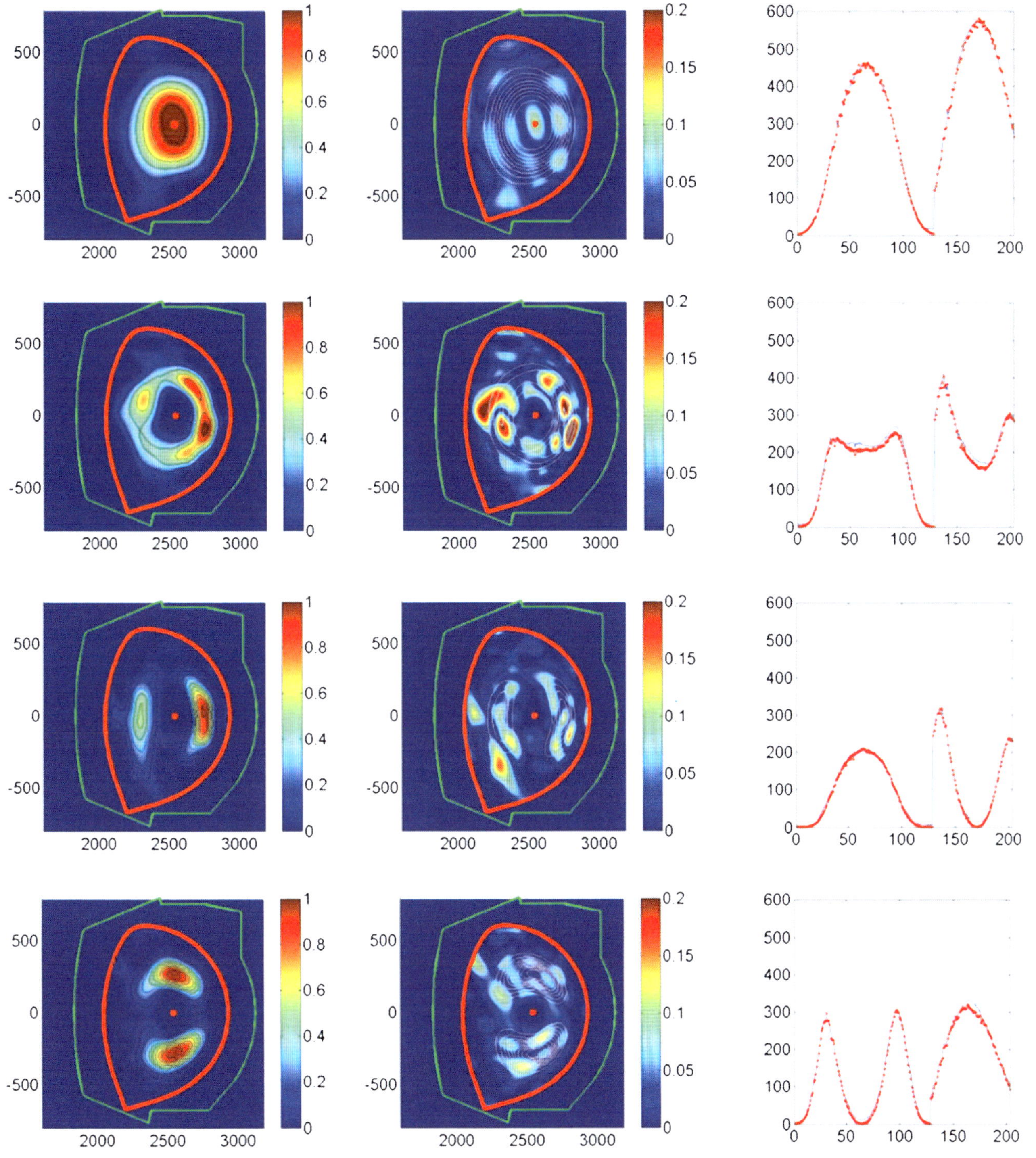

Fig. 3 GPT phantom test with 5% noise level. From left to right, the first column contains the reconstructions, the second column shows the relative error maps according to Eq. (18) (the white contours represent the original phantom), and the third column gives the comparison between the line integrals obtained from the phantom (red dots) and from the reconstructed emissivity fields (blue curves). Note that the phantom emissivity has been normalized for the benefit of numerical computation (Color figure online)

Gaussian Process Tomography Uncertainty Analysis

A valuable advantage of GPT is that it provides uncertainty estimates on the reconstructed emissivity field through the posterior covariance matrix; Eq. (17). Indeed, one expects the deviation of the reconstruction from the original phantom to be largest in those pixels where the posterior variance is large. This is confirmed by comparing the posterior variance map with the relative error field, as shown in Fig. 4. Naturally, the relative error field will not be available when performing tomography on real WEST data, but the posterior variance can still be calculated.

GPT Application to EAST SXR

A quantitative comparison between the minimum Fisher information (MFI) tomographic technique and GPT has been demonstrated for SXR data from Tore–Supra in [19]. Here, we present a first application of GPT to SXR data obtained at EAST. A sawtooth-triggered kink mode has been observed, as shown in the reconstructed emissivity fields in Fig. 5. Note that this reconstruction incorporates information regarding the magnetic equilibrium, as described in [20]. The redistribution of plasma due to the sawtooth and the kink mode is clearly visible in the reconstructions.

Fig. 5 Examples of GPT using data from SXR spectroscopy at EAST ▶ during a sawtooth-triggered kink mode. The time trace in the center panel corresponds to one of the SXR channels (units a.u.). The panels below present reconstructed emissivity fields during the sawtooth evolution, while the upper panels show the evolution of the kink mode, which is triggered by the preceding sawtooth

Summary

In this article, some elementary notions from Bayesian data analysis have been presented and applications of SXR tomography on WEST and EAST have been shown. Bayesian analysis is well suited for solving tomographic problems on the basis of line-integrated data from fusion diagnostics. Using a flexible non-parametric Gaussian process as a prior distribution, Gaussian process tomography regularizes the reconstruction without the need for additional information regarding the magnetic equilibrium. Moreover, the Bayesian formalism provides uncertainty estimates on the reconstructed SXR emissivity field, through the posterior distribution. This brings into reach the possibility of an automated self-check of the reconstruction performance, which could be very useful in future devices. Owing to its reliance on Gaussian distributions, a further important advantage of the GPT method is its computational efficiency, rendering the tomography

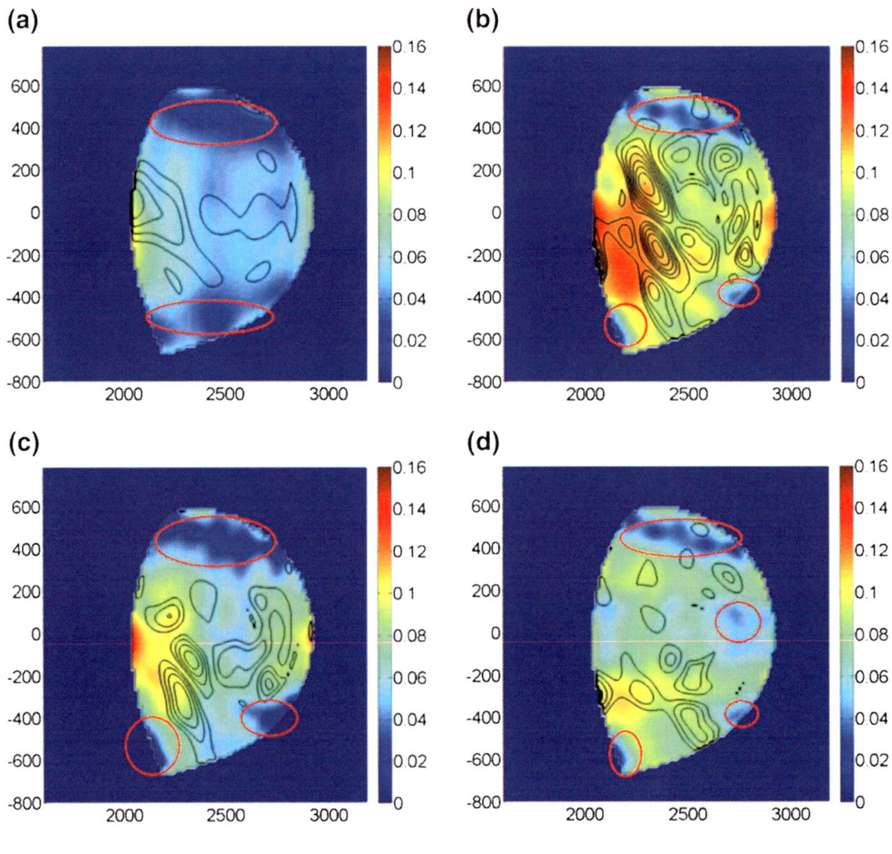

Fig. 4 Examples of a comparison between the posterior variance map (color map) and relative error map (black contours) on a 5% noise level: (**a**) Gaussian shape, (**b**) hollow shape, (**c**) left–right kidney shape, (**d**) up-down kidney shape. In the areas marked by red ellipses, both the posterior variance and relative error are low (the units are arbitrary) (Color figure online)

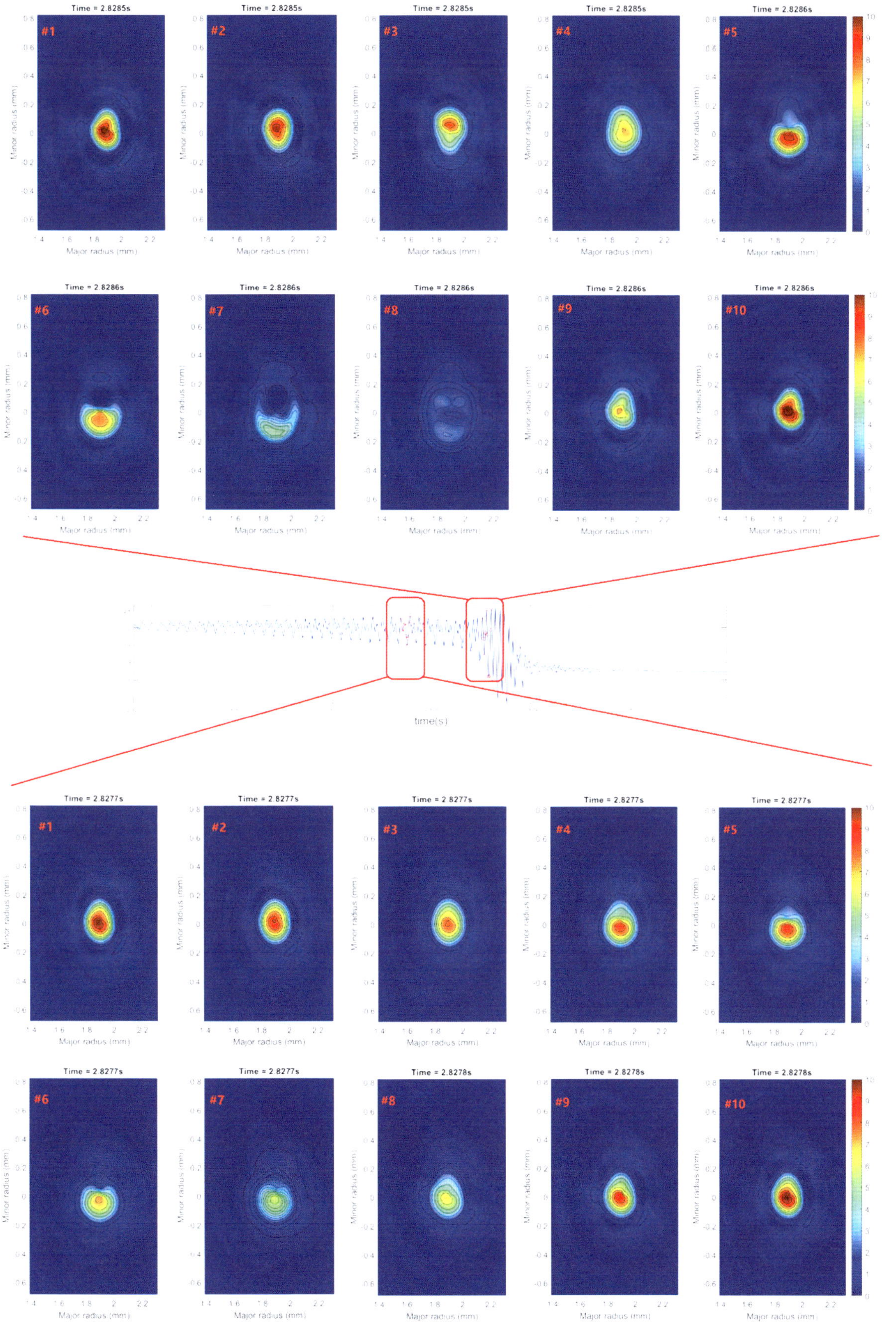
Time = 2.8285s
#1
#2
#3
#4
Time = 2.8286s
#5
#6
#7
#8
#9
#10
Major radius (mm)
Minor radius (mm)
time(s)
Time = 2.8277s
Time = 2.8278s

sufficiently fast for real-time applications. Thus, in view of its many advantages over traditional tomographic reconstruction techniques, Gaussian process tomography in the Bayesian framework has excellent potential in fusion science.

Funding Funding was provided by China Scholarship Council (Grant No. 201508510100).

Appendix

The choice of suitable hyperparameters is a key issue for the GPT method, as they determine the degree of smoothness of the reconstructed emissivity field. A full Bayesian analysis would determine the hyperparameters together with the emissivity parameters, but this would not be feasible in real time. As a workaround, we determine the hyperparameters $\bar{\theta}$ from the data by maximizing the evidence. The rationale is that the marginal posterior for the hyperparameters (i.e. with the parameters $\bar{E}_n$ marginalized) can be written as

$$p(\bar{\theta}|\bar{d}_m) \sim p(\bar{d}_m|\bar{\theta}) \cdot p(\bar{\theta}). \quad (19)$$

$\bar{\theta}$ *hyperparameters*
$\bar{d}_m$ *GEM measurements*

Now, assuming a non-informative uniform hyperprior distribution $p(\bar{\theta})$ we see that the posterior for the hyperparameters is proportional to the evidence $p(\bar{d}_m|\bar{\theta})$, which also occurs in Eq. (11). Hence, by maximizing the evidence w.r.t. $\bar{\theta}$, we find the maximum a posteriori estimates of the hyperparameters. Using these values in the posterior Eq. (15) is usually a good approximation to the full Bayesian solution. From Eq. (8), the marginal posterior for the hyperparameters is given by

$$p(\bar{\theta}|\bar{d}_m) \sim p(\bar{d}_m|\bar{\theta}) = \int p(\bar{d}_m|\bar{E}_n, \bar{\theta}) p(\bar{E}_n|\bar{\theta}) d\bar{E}_n. \quad (20)$$

This results in the following expression:

$$\log(p(\bar{\theta}|\bar{d}_m)) = -\frac{1}{2} \left\{ m\log(2\pi) + \log \bar{\bar{\Sigma}}_d + \bar{\bar{R}}^T \bar{\bar{\Sigma}}_E \bar{\bar{R}} + \bar{d}_m^T (\bar{\bar{\Sigma}}_d + \bar{\bar{R}}^T \bar{\bar{\Sigma}}_E \bar{\bar{R}})^{-1} \bar{d}_m \right\}. \quad (21)$$

The hyperparameters are contained in $\bar{\bar{\Sigma}}_E$, see Eq. (9). An example of the evidence as a function of the two hyperparameters is given in Fig. 6. The data were obtained from a hollow shape phantom test and the optimization results in a length scale $\sigma_l = 144$ mm and signal standard deviation value $\sigma_f = 0.2427$.

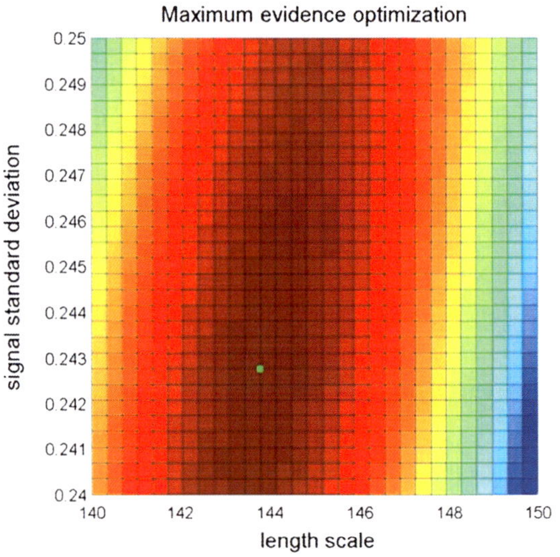

Fig. 6 An example of log(evidence) maximization for a 5% noise level using a hollow shape phantom test. The maximum is obtained at the green point (Color figure online)

References

1. M. Anton et al., X-ray tomography on the TCV tokamak. Plasma Phys. Control. Fusion **38**, 1849 (1996)
2. A. Cormack, Representation of a function by its line integrals, with some radiological applications. II. J. Appl. Phys. **35**, 2908 (1964)
3. R.T. Cox, *The Algebra of Probable Inference* (Johns Hopkins Press, Baltimore, MD, 1961)
4. K. Ertl, W. von der Linden, V. Dose, A. Weller, Maximum entropy based reconstruction of soft X-ray emissivity profiles in W7-AS. Nucl. Fusion **36**, 1477 (1998)
5. R. Fischer et al., Bayesian modelling of fusion diagnostics. Plasma Phys. Control. Fusion **45**, 1095–1111 (2003)
6. R. Fischer et al., Integrated data analysis of fusion diagnostics by means of the Bayesian probability theory. Rev. Sci. Instrum. **75**, 4237–4239 (2004)
7. R. Fischer et al., Integrated data analysis of profile diagnostics at ASDEX upgrade. Fusion Sci. Technol. **58**, 675–684 (2010)
8. A. Gelman, J.B. Carlin, H.S. Stern, D.B. Dunson, A. Vehtari, D.B. Rubin, *Bayesian Data Analysis*, 3rd edn. (Chapman & Hall/CRC, Boca Raton, 2013)
9. P. Gregory, *Bayesian Logical Data Analysis for the Physical Sciences: A Comparative Approach with Mathematica Support* (Cambridge University Press, Cambridge, 2005)
10. A. Jardin et al., Tomographic capabilities of the new GEM based SXR diagnostic of WEST. J. Instrum. **11**, C07006 (2016)
11. E.T. Jaynes, *Probability Theory. The Logic of Science* (Cambridge University Press, Cambridge, 2004)
12. D. Mazon et al., Soft x-ray tomography for real-time applications: present status at Tore Supra and possible future developments. Rev. Sci. Instrum. **83**, 063505 (2012)
13. J. Mlynar et al., Soft X-ray tomographic reconstruction of JET ILW plasmas with tungsten impurity and different spectral response of detectors. Fusion Eng. Des. **96–97**, 869–872 (2015)
14. S. Rathgeber et al., Estimation of profiles of the effective ion charge at ASDEX upgrade with integrated data analysis. Plasma Phys. Control. Fusion **52**, art. no. 095008 (2010)
15. D.S. Sivia, J. Skilling, *Data Analysis: A Bayesian Tutorial*, 2nd edn. (Oxford University Press, Cambridge, 2006)

16. J. Svensson, Non-parametric tomography using Gaussian processes, in *JET Internal Report EFDA-JET-PR(11)24* (2011)
17. G. Verdoolaege et al., Potential of a Bayesian integrated determination of the ion effective charge via bremsstrahlung and charge exchange spectroscopy in tokamak plasmas. IEEE Trans. Plasma Phys. **38**, 3168–3196 (2010)
18. W. von der Linden, V. Dose, U. von Toussaint, *Bayesian Probability Theory. Applications in the Physical Sciences* (Cambridge University Press, Cambridge, 2014)
19. T. Wang et al., Gaussian process tomography for soft-X ray spectroscopy at WEST without equilibrium information. Rev. Sci. Instrum. **89**, 063505 (2018)
20. T. Wang et al., Incorporating magnetic equilibrium information in Gaussian process tomography for soft X-ray spectroscopy at WEST. Rev. Sci. Instrum. **89**, 10F103 (2018)

Journal of Fusion Energy (2019) 38:458–466
https://doi.org/10.1007/s10894-018-0178-x

ORIGINAL RESEARCH

Current Research into Applications of Tomography for Fusion Diagnostics

Jan Mlynar[1] · Teddy Craciunescu[3] · Diogo R. Ferreira[4] · Pedro Carvalho[4] · Ondrej Ficker[1,2] · Ondrej Grover[1,2] · Martin Imrisek[1] · Jakub Svoboda[2] · JET contributors

Published online: 23 August 2018

Abstract
Retrieving spatial distribution of plasma emissivity from line integrated measurements on tokamaks presents a challenging task due to ill-posedness of the tomography problem and limited number of the lines of sight. Modern methods of plasma tomography therefore implement a-priori information as well as constraints, in particular some form of penalisation of complexity. In this contribution, the current tomography methods under development (Tikhonov regularisation, Bayesian methods and neural networks) are briefly explained taking into account their potential for integration into the fusion reactor diagnostics. In particular, current development of the Minimum Fisher Regularisation method is exemplified with respect to real-time reconstruction capability, combination with spectral unfolding and other prospective tasks.

Keywords Tokamaks · Plasma tomography · Plasma diagnostics

Introduction

Sufficient spatial resolution is required in experimental studies of high-temperature plasma physics including the fusion research and development. However, determining the spatial behaviour is not straightforward due to the very poor accessibility of fusion relevant plasmas. Active diagnostic systems which determine plasma properties from its interaction with a physical probe (e.g. laser light, particle beam) offer an excellent spatial resolution in a very specific region of plasma only. Majority of the passive diagnostic systems, which can diagnose plasma from outside its border, allow for spatial resolution of the integral plasma emission, i.e. of the plasma projections. The aim of plasma tomography is to determine local plasma properties from the measured projections.

Spatial resolution of the reconstructed plasma image is, however, rather poor due to the sparse data caused by the limited accessibility, with detector positions typically confined to ports of the vacuum vessel. Implementation of additional constraints (e.g. non-negativity of the emission, zero border emission) and a-priori information (e.g. increased smoothess along field lines) can improve the resolution considerably. On the other hand, temporal resolution of the plasma tomography can be sufficiently high for studies of rapid emissivity evolution [1–3].

Plasma tomography has been applied on several different diagnostic systems. It is rather widespread with Soft X-ray (SXR) diagnostics, often using linear pinhole cameras, see e.g. [2, 4, 5]. Due to the high temporal resolution, the SXR tomography can contribute to the MHD analyses [2] and to the impurity transport studies [6]. At several facilities, data from bolometric diagnostic systems are regularly analysed using tomography, e.g. at the Joint European Torus JET [1]. In the perspective of fusion reactors the key application of tomography is linked to analyses of data from the neutron cameras, see [7, 8],

JET contributors: See the author list of X. Litaudon et al 2017 Nucl. Fusion 57 102001.

✉ Jan Mlynar
mlynar@ipp.cas.cz

1 Institute of Plasma Physics of the CAS, 18200 Prague, Czech Republic

2 Faculty of Nuclear Sciences and Physical Engineering, Czech Technical University in Prague, 11519 Prague, Czech Republic

3 National Institute for Laser, Plasma and Radiation Physics, 077125 Bucharest, Romania

4 Instituto de Plasmas e Fusão Nuclear, Instituto Superior Técnico, 1049-001 Lisbon, Portugal

where it is expected to contribute substantially to regular monitoring of the fusion power.

In this paper, our recent experience with the plasma tomography development for fusion research (in particular at JET) is summarised. In "Current Methods of Plasma Tomography" section the main methods currently applied in plasma tomography are outlined, with a focus on the Tikhonov regularisation. In "Progress in the Minimum Fisher Regularisation" section, the Minimum Fisher Regularisation method is briefly explained and its further development into current applications, fostered by our team, is presented. In particular, a novel approach to the combined neutron tomography with spectral unfolding is proposed. The key recommendations for tomography applications and development in fusion reseach are recapitulated in the conclusions.

Current Methods of Plasma Tomography

The aim of a typical two-dimensional tomographic reconstruction is to retrieve spatial distribution of a scalar field in a plane (e.g. emissivity cross-section image) from its line integrated values (e.g. projections of the emissivity, provided the object is optically thin). Tomographic reconstruction is a special case of an inversion task. Even with the complete knowledge of the line integrals (for all angles and distances) the inversion task is **ill-posed**. As a consequence, a minor error in the data may result in substantial errors in the reconstructed image. In order to mitigate this issue, different **regularisation** methods are usually applied. The regularisation process typically introduces some a-priori knowledge, e.g. some restriction on the result complexity (commonly enforcing smoothness of the result).

In plasma tomography, however, the inversion task is further complicated by the fact that the experimental constraints usually prevent measurements in arbitrary directions. Therefore, many plasma projections are missed, i.e. data are sparse, making the tomographic inversion **underdetermined** for any practical spatial resolution, see Fig. 1. Semi-analytical inversion methods, e.g. based on filtered backprojection (FBP) [1], encounter diverse difficulties under these circumstances. Therefore, dedicated methods with reliable and robust regularisation principles had to be developed for plasma tomography.

In order to process digitised projection data, FBP or Fourier transform based approaches to tomography inversion implement digitisation of the analytical inversion, i.e. of the inverse Radon transform. However, given sparse data, digitisation of the unknown image prior to the tomography inversion proves to be better adapted to implementation of a-priori knowledge [1]. Therefore, in a large majority of current computer codes for plasma tomography, the unknown emissivity distribution $g(\mathbf{r})$ is decomposed into basis functions $b_j(\mathbf{r})$. The basis functions can be global, each covering the whole size of the reconstructed image (e.g. special functions) or local, decomposing the image into regions (e.g. square pixels), see [1]. Consequently, the plasma tomography task is reduced to finding the amplitudes g_j:

$$g(\mathbf{r}) = \sum_{j}^{\infty} g_j b_j(\mathbf{r})\,, \qquad (1)$$

While with infinite series the image can be (in principle) reconstructed to full perfection, in practice the sum is truncated. In other words, the choice of a limited amount of basis functions $g_j\,,\, j = 1, \ldots, N$ contribute to the regularisation of the task. In this respect, implementing any other additional information, in particular the magnetic field geometry directly into the shape of the basis functions is discouraged. Indeed, knowledge of the magnetic field geometry is burdened with errors and, consequently, any adapted shape of the basis functions enhances artefacts. Notice that this holds in particular for the notorius Abel inversion (see e.g. [9]), where the unfolding procedure stems from the unprecise knowledge of the magnetic axis position. In the current plasma tomography, it is recommended to apply standard 2D tomography with preferential smoothness along magnetic flux surfaces (see "Progress in the Minimum Fisher Regularisation" section, Eq. 11) instead of the 1D Abel inversion.

In the vast majority of diagnostic methods, the projections are determined for a finite number of chords (lines of sight, views) $f_i\,,\, i = 1, \ldots, P$ which will be further referred to as the line integrated measurements, although in general the chords need not be strictly linear in the projection plane. Actually, the real width of the chords is often evaluated within the tomographic algorithms. A simple relation between the finite number of amplitudes g_j and the finite number of projection measurements f_i can then be proposed:

$$f_i = \sum_{j}^{N} T_{ij} g_j + \xi_i, \quad i \in 1, \ldots, P, \qquad (2)$$

where T_{ij} stands for elements of the contribution matrix (sometimes referred to as the geometric matrix) and ξ_i for misfits which compensate, in a real experimental setup, both systematic errors and noise.

Different regularisation methods are applied in algorithms of tomographic codes in order to find unknown g_j from sparse data f_i. A reliable performance of the algorithm, with sufficient precision, robustness against artefacts caused by ξ_i, low demand on CPU etc. is usually requested from any applicable tomography method. Some of the methods will actually find a reconstruction matrix $\mathbf{M}$:

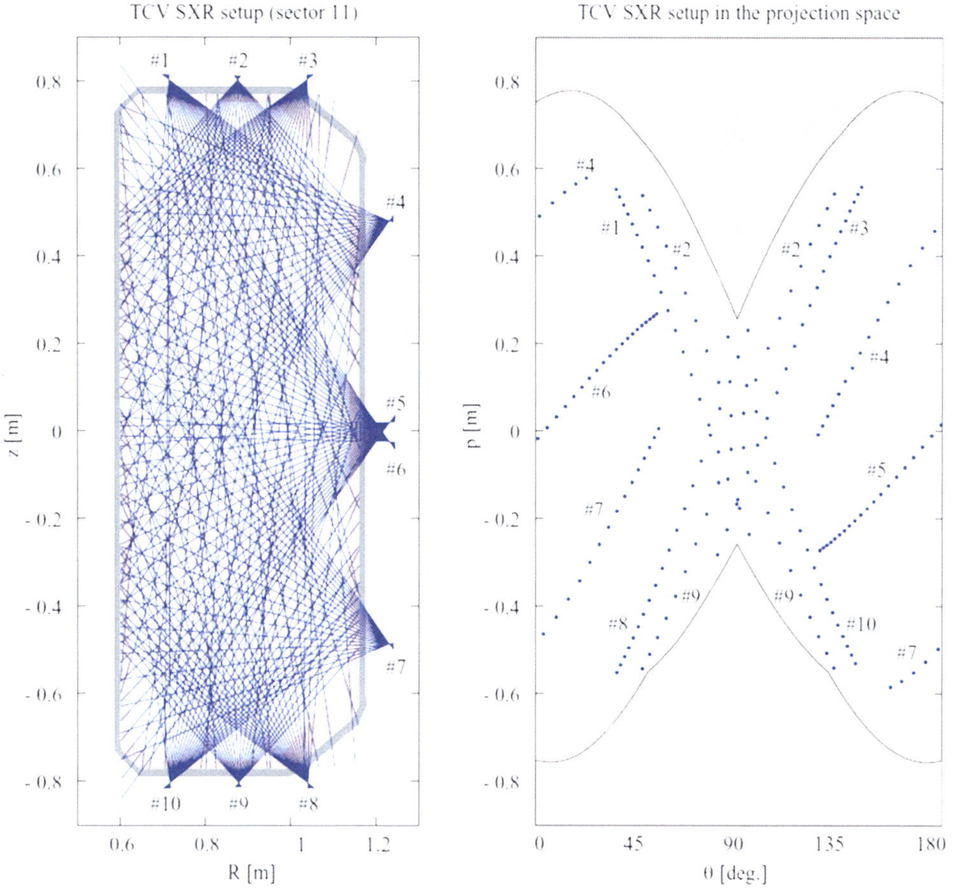

Fig. 1 The Soft X-ray tomography setup at tokamak TCV [3]. While number of the measured line integrated projections is rather high (left, with R the radius and z the vertical coordinate) the actual coverage in the projection space is still quite sparse (right, with θ the angle and and p distance of lines from the vessel centre at $R = 0.88\,\mathrm{m}, z = 0\,\mathrm{m}$)

$$g_j = \sum_i^P M_{ji} f_i \,, \tag{3}$$

Tikhonov regularisation is probably the most widespread method where the reconstruction matrix **M** is derived. In the Tikhonov regularisation

$$M_{ji} = \sum_k^N \left(\sum_l^P T_{kl}^T T_{lj} + \lambda H_{kj} \right)^{-1} T_{ki}^T \tag{4}$$

where λ is the regularisation factor which sets the relative strength of the regularisation, and the regularisation matrix **H** is given, in the simplest case, by

$$H_{kj} = \sum_m^N B_{km}^T B_{mj} \,, \tag{5}$$

where the smoothing matrix **B** corresponds to numerical differentiation, for details see e.g. [2]. The regularisation factor λ can be determined iteratively with the Pearson's χ^2 test:

$$\chi^2 = \frac{1}{P} \sum_i^P \frac{\xi_i^2}{\sigma_i^2} \rightarrow 1 \,, \tag{6}$$

where ξ_i is given by Eq. 2 and σ_i represents expected data errors (standard deviations of the projection data f_i), see e.g. [2]. Figure 2 gives a flowchart of the Tikhonov regularisation as implemented in practice. Another efficient and reliable implementation of the Tikhonov regularisation can be made using the Generalised Singular Value Decomposition (GSVD), see e.g. [10].

Besides the Tikhonov regularisation, considerable efforts in plasma tomography have been invested into the development and application of reconstruction methods based on Bayesian approach, see e.g. [5, 11]. At present, the Maximum likelihood (ML) tomography represents application of a probabilistic method where systematic expertise in experimental data analyses has been collected, see [12]. It is a nonlinear iterative algorithm that attempts to find the estimate of the emissivity distribution that is most consistent with the measured tomographic projections in the sense of maximizing the likelihood (conditional probability of the data given the parameters). The emission is considered to be a Poisson process and therefore f_i presents a sample from a Poisson distribution, whose expected value is f. Consequently, the probability of obtaining the measurement $f = f_i | i = 1, \ldots, P$ from the emissivity $g = g_j | j = 1, \ldots, N$ is given by the likelihood function:

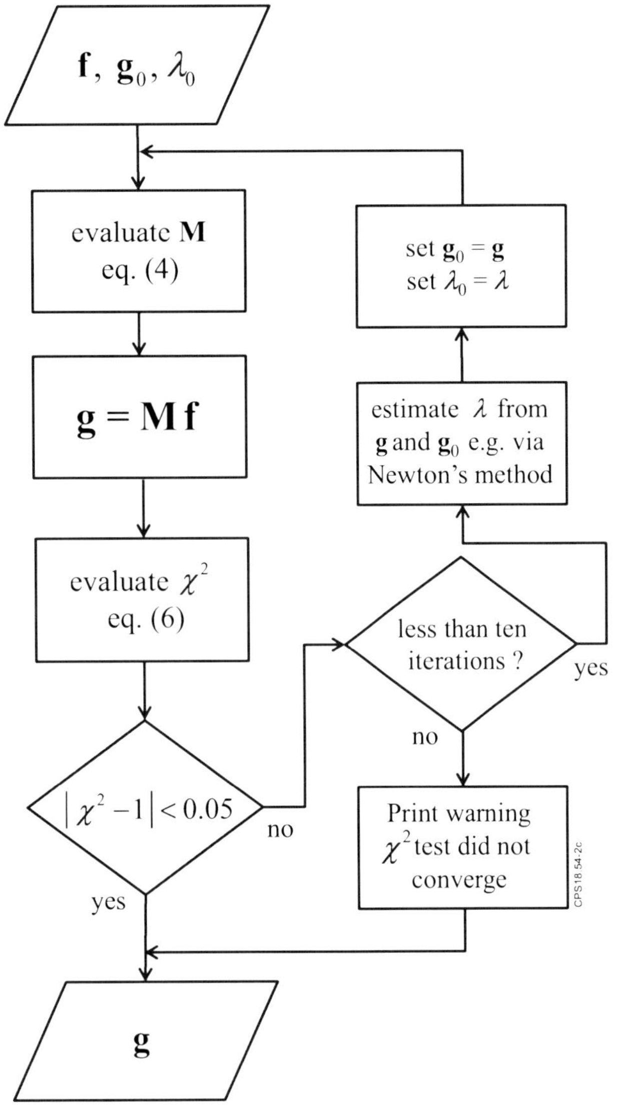

Fig. 2 Flowchart of algorithm solving the Tikhonov regularisation, including the implemented value of the convergence threshold of the Pearson's test and the typical maximum number of iteration loops

$$L(f/g) = \prod_i \frac{1}{f_i!} (f)^{f_i} \times \exp(f) \quad (7)$$

The ML estimate is obtained by maximizing the above expression:

$$g_{\mathrm{ML}} = \mathrm{argmax}_g L(f/g) \quad (8)$$

The mathematical basis for a broadly applicable algorithm has been first applied to images by Richardson [11] and Lucy [13] but the method has been started to be used extensively in tomography only after the introduction of an iterative solution for finding the ML estimate by Shepp and Vardi [14] and, independently, by Lange and Carson [15]:

$$g_j^{(k+1)} = \frac{g_j^{(k)}}{s_j} \sum_i \frac{f_i}{\sum_m T_{im} g_m^{(k)}} T_{ij} \quad (9)$$

where k indexes the iterations and $s_j = \sum_i T_{ij}$ is the probability that emission originating in pixel n will be recorded in a projection bin.

As already mentioned the tomography problem is a highly undetermined inversion leading to an ill-posed mathematical problem. In order to obtain realistic and robust solutions, it is therefore strongly recommended also for the ML method to introduce a regularisation procedure which consists of imposing smoothness along the magnetic surfaces, given by the plasma equilibrium. The ML method uses 1-D average filtering on a sliding window, which moves along the magnetic contour lines. The iterative reconstruction formula 9 has a particular form, which is advantageous with respect to modelling of the projection noise propagation. In other words, it is possible to retrieve a variance image which accompanies the reconstructions. Accurate modelling of the projection noise propagation is important for both qualitative interpretation and quantitative analysis of the reconstructed images. Following the ideas first introduced by Barret [16] two approximations can be introduced for obtaining the variance image: to consider that the noise is small compared to the mean value of the reconstruction and to assume that the convergence of the ML algorithm is fast enough, so that the projection of the current estimate is close to the noise-free projection. Details of the implementation for JET tomography are given in [17].

In JET, the ML method has been applied for gamma and neutron tomography (see e.g. [12, 18] for representative examples). More recently it has been implemented also for bolometry [19], where the evaluation of the uncertainties accompanying the reconstruction is particularly important, allowing the estimation of the confidence intervals for the radiated power (Fig. 3).

The tomographic inversion process can also be performed with neural networks. A first attempt at doing this involved the use of a neural network to find the parameters of a two-dimensional gaussian distribution that would best fit the measurements of a horizontal and a vertical soft X-ray camera [20]. This assumed that the plasma profile could be approximated by a 2D gaussian shape, and the neural network would learn to predict the amplitude, the horizontal/vertical position, and the horizontal/vertical width of the gaussian distribution.

Later, the assumption that the plasma profile had a particular shape was relaxed to allow for models with more degrees of freedom. Namely, a model was developed to take into account—among other parameters—ellipticity, triangularity and the Shafranov shift [21]. A neural network

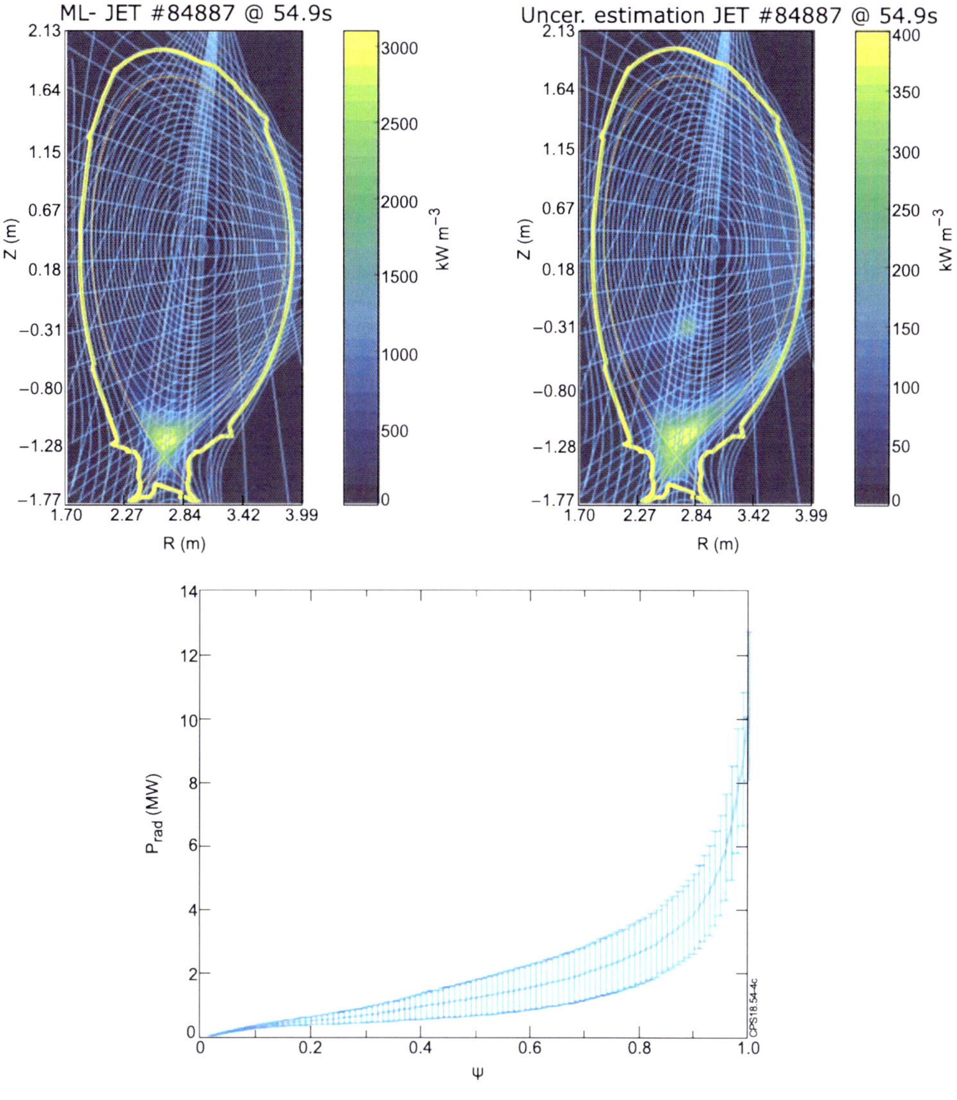

Fig. 3 Illustration of the ML bolometric tomography at JET: reconstruction (top-left), image variance (top-right) and radiation profile versus the normalised Ψ coordinate with the estimate of the uncertainties in the emitted power (bottom)

was used to learn the parameters of this model from measurements of a horizontal and a vertical neutron camera [22]. In this case, the neural network could learn up to 16 model parameters.

With the advent of deep learning [23], it became possible to train neural networks with many more layers and with a much larger number of parameters. One of the most successful applications of deep learning was image classification using convolutional neural networks (CNNs), where a 2D input image is transformed into a 1D output vector of class probabilities. By reversing the architecture of a CNN, it is possible to devise a deconvolutional neural network for plasma tomography, where a 1D input vector of bolometer measurements is transformed into a 2D output reconstruction of the plasma profile [24].

In this case, the neural network was trained to reproduce each single pixel of the tomographic reconstruction. For reconstructions with a resolution of 200 × 120 pixels, the network had 24,000 outputs, and was able to achieve a similarity score above 90% on previously unseen data. The main advantage is that, once trained, such network can compute hundreds or even thousands of reconstructions per second, making it possible to visualise the plasma profile over the course of an entire discharge [25] and, potentially, in real-time applications.

To conclude, the neural network tomography is the fastest among the three methods presented above, however, it is critically dependent on the quality and range of the training data. The Tikhonov regularisation is highly versatile, robust and independent on previous knowledge. Although it requires more computation time than the network tomography, it is considerably faster than any proper implementation of the Bayesian approach, including ML. The ML tomography represents the most sophisticated method which in various tests results in higher accuracy of tomographic reconstructions than the Tikhonov regularisation [26] and allows for semi-analytic calculation of the projection noise propagation. In this context it should be reminded that any tomographic method solves an ill-posed task, so that its accuracy can be significantly damaged by minor systematic errors including detector misalignments or varying sensitivities of individual detectors.

Progress in the Minimum Fisher Regularisation

The Minimum Fisher Regularisation (MFR) presents a widespread method of tomography in current tokamak research. It relies on Tikhonov regularisation with an iterative optimisation of the results so that the Fisher information of the reconstructed image is minimised:

$$I_F = \int \sum_{i,j=1}^{2} \left| \frac{\partial g(x_1, x_2)}{\partial x_i} \frac{\partial g(x_1, x_2)}{\partial x_j} \frac{1}{g(x_1, x_2)} \right| \mathrm{d}x_1 \, \mathrm{d}x_2 \,. \quad (10)$$

In plasma tomography on tokamaks, x_1 and x_2 correspond to the R and z coordinates (Figs. 1, 3). Optimisation according to Eq. 10 allows—in simple terms—for better spatial resolution (less smoothness) in regions with high levels of emissivity. The MFR method was first developed and applied at TCV [2] and at present it is implemented and updated, among others, at JET [9], COMPASS [4] and Tore Supra [27].

The first major extension of the MFR aimed at a possibility of rapid analyses of large amounts of data via temporal averaging of the smoothess matrix, see [3]. Nowadays the temporal averaging is hardly ever applied in practice due to the substantial increase in the CPU performance of the computers. With robust and rapid performance of MFR, further efforts have been invested into the development of real-time relevant version of MFR. As a result the idea of a rolling iteration was introduced, see [28]. In the rolling iteration, the time index of analysed data is increased by one with every new round of the iterative process. This allows for sufficient precision in the case of smooth, slowly evolving data from line integrated measurements. Importantly, in [28] it is demonstrated that the artefacts linked to sudden changes in data have also rather low and rapidly decreasing amplitude in MFR, i.e. the rolling iteration is stable. It can be concluded that a real-time version of the MFR is foreseeable.

Furthermore, preferential smoothness of the reconstructed image along the flux surfaces was introduced at JET due to the low number of the lines of view. As a result, the reconstructed image features slowly changing emissivity along the flux surfaces, while the emissivity gradient in plasma radius may be steep. This corresponds to the expected plasma emissivity behaviour. In oder to enforce this smoothness anisotropy, a new recipe for the regularisation matrix was implemented instead of Eq. 5:

$$H_{kj} = S(\eta) \sum_{m}^{N} B^T_{\parallel km} w_m B_{\parallel mj} + S(-\eta) \sum_{m}^{N} B^T_{\perp km} w_m B_{\perp mj} \,, \quad (11)$$

In this set of equations, $\mathbf{B}_{\parallel}$ is the smoothing matrix in the direction parallel to the magnetic field, $\mathbf{B}_{\perp}$ the smoothing matrix in the direction perpendicular to the magnetic field, the weights w_m allow e.g. for implementation of the Minimum Fisher Information according to Eq. 10 (see [2] for details), and the function $S(\eta)$ controls the anisotropy amplitude. In practice, the logistic sigmoid function $S(\eta) = (1 + e^{-\eta})^{-1}$ is applied. This amendment of the matrix $\mathbf{H}$ proved so reliable and beneficial [9] that it is nowadays applied as a routine feature in MFR.

Due to the non-linear character of the Minimum Fisher Information and to the sophisticated smoothing procedure it is not possible to determine analytically the error transmision from line integrated data to the emissivity distribution like in the ML method, see "Current Methods of Plasma Tomography" section. Instead, the Monte Carlo method was introduced, which tests statistically the MFR reconstruction response to random errors in data. The method was first used in extensive studies of the MFR performance at JET [29]. In this work it was shown that the MFR is stable against artefacts and that a Gaussian noise in data transmits with a good precision to a Gaussian noise in the reconstruction. Recently MFR contributed to detailed evaluation of accuracy and precision of the ITER neutron profile reconstruction from the simulated Radial Neutron Camera (RNC) data. The studies proved high robustness of the MFR method and acceptable level of precision of the neutron profiles with a sufficient temporal resolution, in particular in the high performance discharges [8].

Besides plasma tomography, MFR was successfully applied in unfolding of neutron spectra Φ_j from the pulse height data Ψ_i measured by neutron scintillation detector with detailed knowledge of its response function with matrix elements R_{ij}, for details see [30] :

$$\Psi_i = \sum_j R_{ij} \Phi_j, \quad (12)$$

Unlike in plasma tomography, in the case of unfolding the task need not be underdetermined so that the L-curve principle based on data statistics can be employed in search of the value of the regularisation factor λ instead of the χ^2 Pearson's method, see [30, 31]. Notice that the unfolding is still an ill-posed problem with a tendency to create artefacts, therefore a reliable calibration of the response matrix $\mathbf{R}$ was indispensable.

Based on the observed robustness of the MFR method as well as on experience with spectral unfolding, a new approach to analyses of spatially resolved pulse height data from the RNC scintillation detectors at ITER can be proposed. In contrast to the step-by-step approach presented in [32], where tomography analyses is run for separate energy bins, it is recommended to directly combine the contribution (i.e. geometric) matrix and the response matrix of the energy calibrated detectors into a single inversion problem as follows:

- Denote f_{ik} elements of a matrix of data where the rows correspond to $i = 1, \ldots, P$ line integrated measurements from the P scintillation spectrometers and the columns to $k = 1, \ldots, C$ pulse height measurements from each spectrometer.
- Seek elements of a matrix of spectrally resolved emissivities g_{jl}, where the rows $j = 1, \ldots, N$ correspond to the pixel index in the spatially resolved mesh of pixels, and the columns $l = 1, \ldots, B$ to discrete bins of the unfolded neutron energies.
- The double inversion problem is described—according to Eqs. 2 and 12—by the following set of equations

$$f_{ik} = \sum_{l}^{B} R_{kl} \sum_{j}^{N} T_{ij} g_{jl} + \xi_{ik}, \qquad (13)$$

- This set of equations can be re-indexed so that a standard regularisation procedure (e.g. the MFR code) can be applied

$$f_{\alpha} = \sum_{\beta}^{B\times N} S_{\alpha\beta} g_{\beta} + \xi_{\beta} \quad \rightarrow \quad g_{\beta} = \sum_{\alpha}^{C\times P} G_{\beta\alpha} f_{\alpha} \qquad (14)$$

The main advantage of the proposed procedure is that the combined tomography and spectral unfolding problem is treated as a single ill-posed task, so that the amplification of artefacts is prevented.

The method was tested on phantom functions with random noise. The contribution matrix **T** was based on geometry of the JET neutron emissivity profile monitor [12], and the response matrix **R** had values of the well calibrated liquid scintillation detector that was successfully tested at JET as a neutron spectrometer [33]. One of the test functions and the resulting reconstruction is exemplified in Fig. 4. In this example, a simple gaussian profile of neutron emissivity was used as a phantom function on the grid of 10×15 pixels, with 100 energy bins in each. The phantom neutron spectrum was based on the DT neutron emission (14.1 MeV) with Doppler broadening corresponding to a parabolic plasma temperature profile. As a result, the combined procedure resulted in considerably improved precision of reconstruction (by approx. 15%) compared to the step-by-step unfolding and tomography. It can be concluded that the preliminary results are promising.

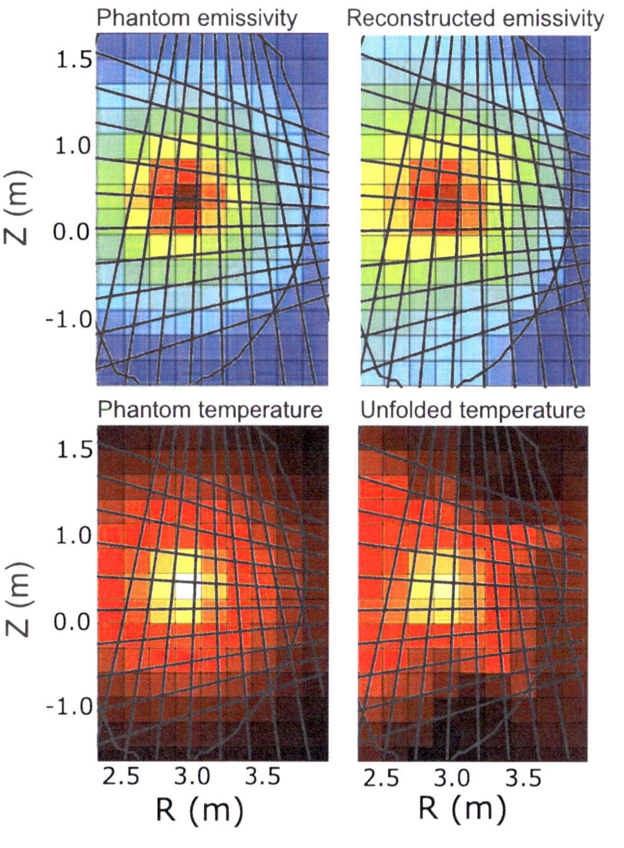

Fig. 4 Preliminary results of the combined tomography and unfolding MFR code, with phantom functions in left column and combined 2D tomography and unfolding results in the right column. In the top row, the spatial distribution of the DT neutron emissivity is shown, while the bottom row presents the model temperature profile and the unfolded temperature distribution, based on spectral width of the DT neutron energy. The simulation was run for the JET neutron emissivity profile monitor geometry [12]

Since its introduction more than 20 years ago, the MFR also went through several code remakes and numerical optimisations. The most important extension happened in 2012 when the original MatLab MFR package was ported to the Python platform with new numerical options included [9]. The Python version was significantly refactored again in 2017, with a new streamlined hierarchic module structure and Mercurial version control. In the version currently maintained for the JET tomography, a new implementation of the anisotropic smoothness matrix based on gradient of the poloidal magnetic flux (instead of the flux surface interpolation) was introduced, see Fig. 5. Besides, the same mesh of pixels was pre-defined for the three JET tomography diagnostic systems: the neutron, the soft X-ray and the bolometry cameras, and a new simple access to JET data is under development. A new graphical user interface (GUI) was introduced, which allows, among

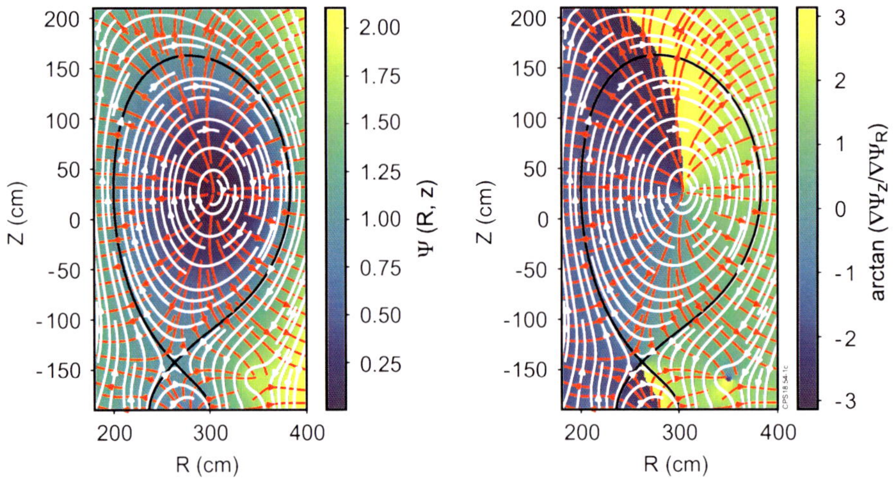

Fig. 5 Left: Two-dimensional plot of the normalised magnetic flux function $\Psi_N(R, z)$ as reconstructed by EFIT at JET, including Last Closed Flux Surface in black (the value of the flux is normalised to the value at this contour). The red arrows mark the gradient of the Ψ function and the white arrows the perpendicular direction. Right: A similar plot of inverse tangent of the direction perpendicular to the gradient of Ψ. This function is then used in order to find proper weights for the derivative (smoothing) matrix with preferential smoothing along the magnetic flux contours

others, to evaluate evolution of radiation in between regions of interest pre-defined by the user, profiting from data of all the three diagnostic systems.

The MFR method was also applied in reconstruction of data from fast 2D matrix cameras with tangential view of the plasma in the visible region (based on the plasma axial symmetry) on the COMPASS tokamak, see e.g. [34], and tentatively to unfolding in analyses of the data from the activation probes at JET, see [35]. Its potential to reconstruct the plasma current density from the magnetic data is under discussion. Last, but not least a possible merger of the MFR with the neural networks can be considered in future, as proposed in [36]. According to this scheme, MFR regularisation parameters in Eqs. 4 and 11 would be determined by a trained neural network, combining robustness of MFR with the real-time relevance of the neural networks.

Conclusions and Outlook

In this contribution, research and development efforts in plasma tomography were presented. Particular focus was given to the large number of applications that a reliable method resolving the inversion problem may offer in fusion data analyses. Indeed, the potential of the tomography development for fusion is still to be exploited. At present, several methods with their advantages and disadvantages compete, in which a rich set of constraints may (but need not) be applied. Importantly, the JET contributors involved in tomography analyses have been currently working on quantitative comparison of performance of the three methods presented in the second part of this contribution. The conclusions are yet to be drawn, however, the preliminary results demonstrate that (1) with sparse data, the room for improvement is limited so that in this respect, augmented data precision is rather called for, and (2) the correct attitude is to maintain a few different inversion methods in order to be able to critically compare their results in analyses of important data events.

For future fusion reactors, it will be instrumental to develop a real-time tomography algorithm with low susceptibility of developing major artefacts. This task is of a particular importance in determination of the fusion neutron emissivity distribution. In this respect, several works and conceptual studies based on Minimum Fisher Regularisation method were also presented in this contribution. Obviously, similar ideas can be also anticipated in other plasma tomography methods.

Acknowledgements This work has been carried out within the framework of the EUROfusion Consortium and has received funding from the Euratom research and training programme 2014–2018 under Grant Agreement No 633053. The views and opinions expressed herein do not necessarily reflect those of the European Commission.

References

1. L.C. Ingesson, B. Alper, B.J. Peterson, J.-C. Vallet, Tomography diagnostics: bolometry and soft X-ray detection. Fusion Sci. Technol. **53**, 528 (2008)

2. M. Anton et al., X-ray tomography on the TCV tokamak. Plasma Phys. Control. Fusion **38**, 1849 (1996)
3. J. Mlynar et al., Investigation of the consistency of magnetic and soft X-ray plasma position measurements on TCV by means of a rapid tomographic inversion algorithm. Plasma Phys. Control. Fusion **45**, 169 (2003)
4. J. Mlynar et al., Introducing minimum Fisher regularisation tomography to AXUV and soft X-ray diagnostic systems of the COMPASS tokamak. Rev. Sci. Instrum. **83**, 10E531 (2012)
5. T. Wang, D. Mazon, J. Svensson, D. Li, A. Jardin, G. Verdoolaege, Gaussian process tomography for soft X-ray spectroscopy at WEST without equilibrium information. Rev. Sci. Instrum. **89**, 63505 (2018). https://doi.org/10.1063/1.5023162
6. C. Angioni et al., Tungsten transport in JET H-mode plasmas in hybrid scenario, experimental observations and modelling. Nucl. Fusion **54**, 083028 (2014)
7. G. Bonheure et al., A novel method for trace tritium transport studies. Nucl. Fusion **49**, 085025 (2009)
8. D. Marocco et al., System level design and performances of the ITER radial neutron camera, in Proceedings of 26th IAEA Fusion Energy Conference, Kyoto, FIP/P4-16 (2016)
9. M. Odstrcil et al., Modern numerical methods for tomography optimisation. Nucl. Instrum. Methods A **686**, 156 (2012)
10. A. Jardin, D. Mazon, J. Bielecki, Comparison of two regularization methods for soft X-ray tomography at Tore Supra. Phys. Scripta **91**, 044007 (2016)
11. W. Richardson, Bayesian-based iterative method of image restoration. J. Opt. Soc. Am **62**, 55 (1972)
12. T. Craciunescu et al., The maximum likelihood reconstruction method for JET neutron tomography. Nucl. Instrum. Methods A **595**, 623 (2008)
13. L. Lucy, An iterative technique for the rectification of observed distributions. Astron. J. **79**, 745 (1974)
14. L.A. Shepp, Y. Vardi, Maximum likelihood reconstruction for emission tomography. IEEE Tram. Med. Imaging MI4 **1**, 113 (1982)
15. K. Lange, R. Carson, EM reconstruction algorithms for emission and transmission tomography. J. Comput. Assist. Tomogr. **8**(2), 306 (1984)
16. H.H. Barrett, D.W. Wilson, B.M.W. Tsui, Noise properties of the EM algorithm: I. Theory. Phys. Med. Biol. **39**, 833 (1994)
17. T. Craciunescu et al., Evaluation of reconstruction errors and identification of artefacts for JET gamma and neutron tomography. Rev. Sci. Instrum. **87**(1), 013502 (2016)
18. Y.O. Kazakov et al., Efficient generation of energetic ions in multi-ion plasmas by radio-frequency heating. Nat. Phys. **13**(10), 973 (2017)
19. T. Craciunescu et al., Maximum likelihood bolometric tomography for the determination of the uncertainties in the radiation emission on JET. Rev. Sci. Instrum. **89**(5), 053504 (2018)
20. G. Demeter, Tomography using neural networks. Rev. Sci. Instrum. **68**, 1438 (1997)
21. E. Ronchi et al., A parametric model for fusion neutron emissivity tomography for the KN3 neutron camera at JET. Nucl. Fusion **50**, 035008 (2010)
22. E. Ronchi et al., Neural networks based neutron emissivity tomography at JET with real-time capabilities. Nucl. Instrum. Methods Phys. Res. Sect. A **613**, 295 (2010)
23. Y. Lecun, Y. Bengio, G. Hinton, Deep learning. Nature **521**, 436 (2015)
24. F.A. Matos, D.R. Ferreira, P.J. Carvalho, Deep learning for plasma tomography using the bolometer system at JET. Fus. Eng. Des. **114**, 18 (2017)
25. D.R. Ferreira, P.J. Carvalho, H. Fernandes, Full-pulse tomographic reconstruction with deep neural networks. Fusion Sci. Technol. **74**, 47 (2018)
26. T. Craciunescu et al., A comparison of four reconstruction methods for JET neutron and gamma tomography. Nucl. Instr. Methods Phys. Res. A **605**, 374 (2009)
27. D. Mazon et al., Soft X-ray tomography for real-time applications: present status at Tore Supra and possible future developments. Rev. Sci. Instrum. **83**, 063505 (2012)
28. V. Loffelmann et al., Minimum Fisher Tikhonov regularization adapted to real-time tomography. Fusion Sci. Technol. **69**, 505 (2016)
29. M. Imrisek, Studies of error transmission in tomography of fusion neutrons. B.Sc. Thesis (in Czech), Czech Technical University in Prague (2008)
30. J. Mlynar et al., Neutron spectra unfolding with minimum Fisher regularisation, International Workshop on Fast Neutron Detectors and Applications, SISSA Proceedings of Science PoS(FNDA2006)063 (2006)
31. P.C. Hansen, D.P. OLeary, The use of the L-curve in the regularization of discrete Ill-posed problems. SIAM J. Sci. Comput. **14**, 1487 (1993)
32. D. Marocco, B. Esposito, F. Moro, Combined unfolding and spatial inversion of neutron camera measurements for ion temperature profile determination in ITER. Nucl. Fusion **51**, 053011 (2011)
33. A. Zimbal et al., Compact NE213 neutron spectrometer with high energy resolution for fusion applications. Rev. Sci. Instrum. **75**, 3553 (2004)
34. M. Odstrcil et al., Plasma tomographic reconstruction from tangentially viewing camera with background subtraction. Rev. Sci. Instrum. **85**, 013509 (2014)
35. O. Ficker et al., Unfolding of energies of fusion products measured by the activation probe at JET, in Proceedings of 18th Conference of Czech and Slovak Physicists, ISBN 978-80-244-4726-1, p. 29 (2015)
36. J. Mlynar, V. Weinzettl, G. Bonheure, A. Murari, JET-EFDA contributors, Inversion techniques in the soft-X-ray tomography of fusion plasmas: toward real-time applications. Fusion Sci. Technol. **58**, 733 (2010)

Journal of Fusion Energy (2019) 38:467–479
https://doi.org/10.1007/s10894-018-0167-0

ORIGINAL RESEARCH

Multichannel Data Acquisition System for GEM Detectors

Grzegorz Kasprowicz[1] · Wojciech M. Zabołotny[1] · Krzysztof Poźniak[1] · Maryna Chernyshova[2] · Tomasz Czarski[2] · Michał Gąska[1] · Piotr Kolasiński[1] · Rafał Krawczyk[1] · Paweł Linczuk[1,2] · Andrzej Wojeński[1]

Published online: 21 June 2018

Abstract

This paper presents the data acquisition system for GEM-detector (Sauli in Nucl Instrum Methods Phys Res Sect A Accel Spectrom Detect Assoc Equip 386(2–3): 531–534, 1997. https://doi.org/10.1016/S0168-9002(96)01172-2) based cameras and spectrometers (Chernyshova et al. in Fusion Eng Des, 2017. https://doi.org/10.1016/j.fusengdes.2017.03.107). The system is modular (Czarski et al. in Rev Sci Instrum 87(11): 11E336, 2016. https://doi.org/10.1063/1.4961559) and supports 1D and 2D GEM arrays with software-defined readout modes. Channel count can vary from 8 up to hundreds. The readout electronics consists of two units—radiation tolerant Analog Front End and rack-mount data processing unit. Data processing is split into two parts—real-time hardware, based on FPGA and software, based on embedded multicore CPUs and hardware accelerators. The FPGA subsystem together with PCIe interface and multi-core Xeon processors forms low latency, high-performance processing chain suitable for applications requiring feedback. The detectors in tokamaks operate in the high magnetic field, so the system was equipped with multipoint magnetic field measurement synchronized with detector readout. The system also includes custom HV supply, developed for triple GEM detectors operating in the high-rate mode. Dedicated protection and diagnostic subsystems were developed as well to ensure safe and reliable operation of the detector in harsh conditions. To support the operation of the detector in the high-temperature surrounding, the liquid cooling subsystem was developed (Wojenski et al. in J Instrum 11(11): C11035, 2016).

Keywords GEM · FPGA · MTCA · DAQ

Abbreviations

ADC	Analog to digital converter
AFE	Analog Front End
AMC	Advanced Mezzanine Card
COTS	Commercial off the shelf
CPU	Central processing unit
DAQ	Data acquisition
DC	Direct current
DCB	Data concentrator board
DDR SDRAM	Double data rate synchronous dynamic random-access memory
EEPROM	Electrically erasable programmable read-only memory
EMC	Electromagnetic compatibility
FPC	Flexible polymer cable
FPGA	Field programmable gate array
GEM	Gas electron multiplier
HP	Horozintal pitch
HV	High voltage
IC	Integrated circuit
IPMI	Intelligent Platform Management Interface
ITER	International thermonuclear experimental reactor
I2C	Inter integration circuit interface
JTAG	Joint Test Action Group
LCD	Liquid crystal display
LV	Low voltage
LVDS	Low voltage differential signalling
MCH	MTCA Carrier Hub
MTCA	Micro Telecommunications Computing Architecture
PCB	Printed circuit board
PCIe	Peripheral Component Interconnect Express
PSU	Power supply unit

✉ Grzegorz Kasprowicz
G.Kasprowicz@elka.pw.edu.pl

[1] Institute of Electronic Systems, Warsaw University of Technology, Nowowiejska 15/19, 00-665 Warsaw, Poland

[2] Institute of Plasma Physics and Laser Microfusion, Hery 23, 01-497 Warsaw, Poland

RF	Radio frequency
RPM	Revolutions per minute
RTM	Rear transition module
R&D	Research and Design
SAS	Serial Attached SCSI
SPI	Serial Peripheral Interface Bus
SXR	Soft X ray
SW	Software
USB	Universal Serial Bus
VHDCI	Very-high-density cable interconnect
WEST	Tungsten (W) Environment in Steady-state Tokamak
WPIS	WEST Protection and Interlock System

Introduction

Precise estimation of soft X-ray distribution of plasma in future research tokamak reactors will be a key factor in plasma parameter optimization. Two-dimensional tomography with single photon energy estimation will deliver new quality diagnostics of the tokamak plasma. Large plasma structure dimensions (i.e., ITER) require complex imaging system that must fulfill contrary requirements: large detection area, good spatial resolution, good energy resolution for X-ray photons and resistance to strong neutron and ionised particle flux. Another issue is the high level of electromagnetic noise. Proper processing and discrimination of such events caused by radiation and EMI from other diagnostics and reactor itself, is one of the most important challenges. Several data acquisition systems and GEM detectors were constructed to cope with such requirements [7]. This article describes recent achievement in this field—complex data acquisitions system that was installed at WEST [5]. In the last chapter, the next generation system based on MTCA [1] platform is presented.

Data Acquisition System Architecture

The architecture of the readout system is based on full modularity. Such approach gives great flexibility in system construction where both small and large, 1D and 2D imaging systems can be relatively quickly constructed using standardized building blocks. The only difference between 64 channel 1D linear array and 2048 channel 2D/3D readout is detector pixel board with detector backplane. These are relatively simple designs that can be designed and manufactured quickly. Further processing is realized using repeatable blocks that consist of Analog Front End (AFE) channels, ADC modules, FPGA backplane, Data Concentrator Board and computer subsystem. Figure 1 presents the block diagram of the system for WEST linear camera. The aim of detector's signals processing is an estimation of energy distribution for each channel. The system consists of:

- Detector stripboard with the backplane. Its role is to receive the electron cloud charge and transfer it to the AFE boards.
- 16 channel AFE boards they convert the charge into the voltage that is transferred using a symmetrical cable do the ADC modules.
- 16 channel ADC boards they convert the voltage to the digital value and transfer the result using differential high-speed LVDS lanes to the FPGA.
- The FPGA backplane modules with 4 ADC slots. It performs synchronization and pre-processing of all the 64 channel data.
- Data Concentrator Board (DCB) with PCIe switch, management FPGA and serial ports for trigger and interlocks.
- Protection and Interlock System module.
- Optional 56-core Intel Xeon PHI numerical accelerator.
- Server-class main board where most of the data processing takes place.
- 800 W ATX power supply.
- High and low voltage power supply unit (PSU).
- Temperature, humidity and pressure sensors.
- Fast feedback PCIe communication board (i.e., reflective memory).
- Ethernet Power strip as system watchdog.
- Liquid cooling distribution box.

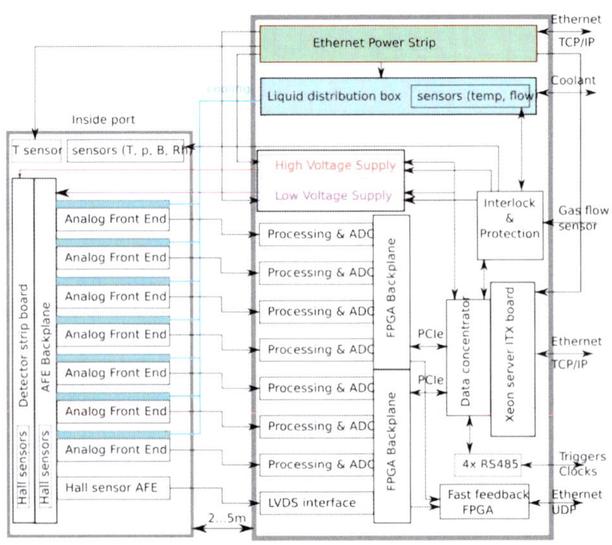

Fig. 1 Architecture of the system for WEST

All modules except for AFE and detector fit inside single 15U 19" computer rack called main processing unit. The main processing unit is not affected by radiation. Temperature is not an issue neither because it operates stand-alone with the independent cooling system.

GEM Detector Practical Realisation

GEM detector assembly for WEST vertical camera is installed inside of the vertical port. Mechanical construction (Figs. 2, 3) was optimised to fit within the constrained area. Due to the lack of the airflow, the liquid cooling was utilised.

The detector stripboard is connected with the AFE backplane using FPC jumpers and shield (Figs. 4, 5). That eliminates issues with the previous construction where both boards were connected using rigid Samtec connectors. It caused unwanted tension on the connectors caused by fixing of the detector assembly to the AFE cassette frame. Thanks to flex-rigid connectivity solution, such tension is also avoided during plugging and unplugging of the AFE boards. Moreover, it allows adjustment of the detector window relatively to the AFE cassette and pinhole. AFE boards are designed as Micro-TCA form-factor. Eight such boards can be installed in the backplane connectors (Fig. 4).

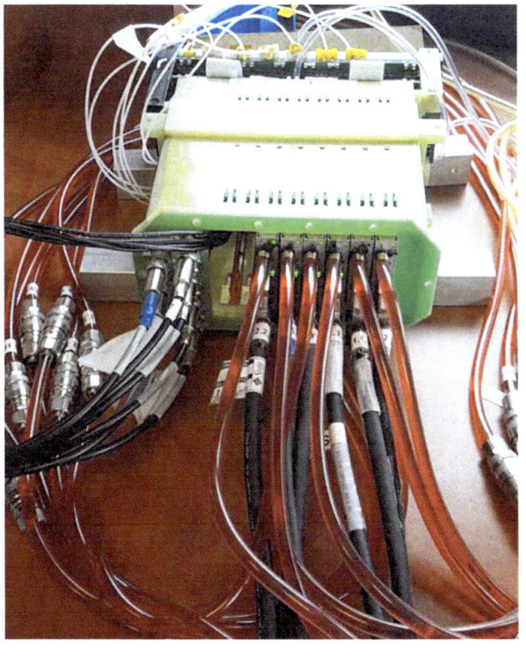

Fig. 2 GEM detector for WEST

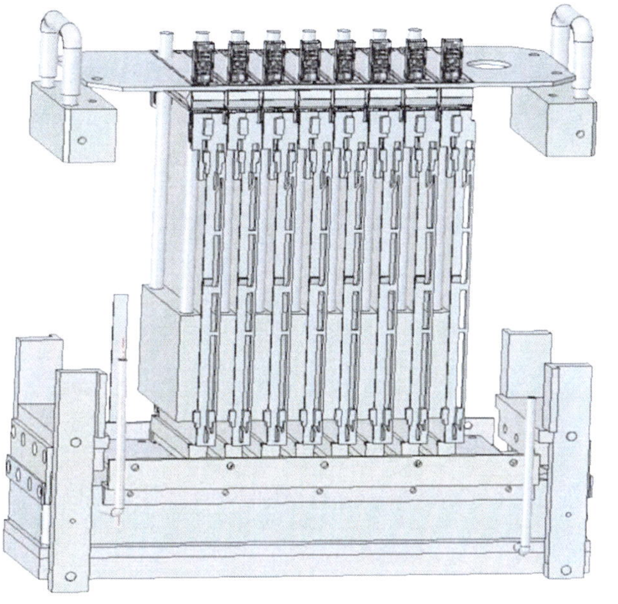

Fig. 3 Drawing of complete GEM detector (horizontal bottom part) with AFE electronics (vertical boards) for WEST. The dimensions are 230 × 88 × 414 mm

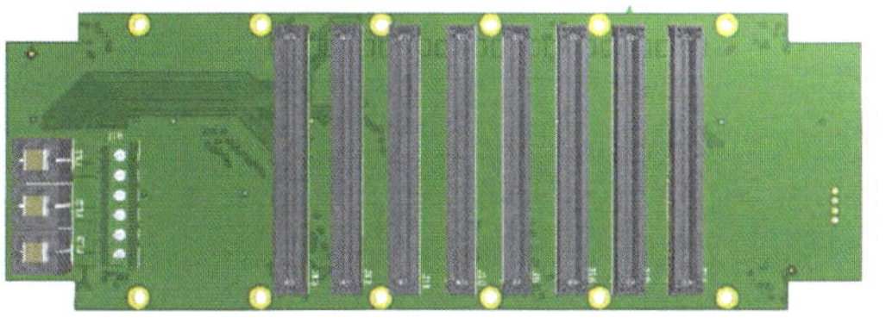

Fig. 4 Detector AFE backplane

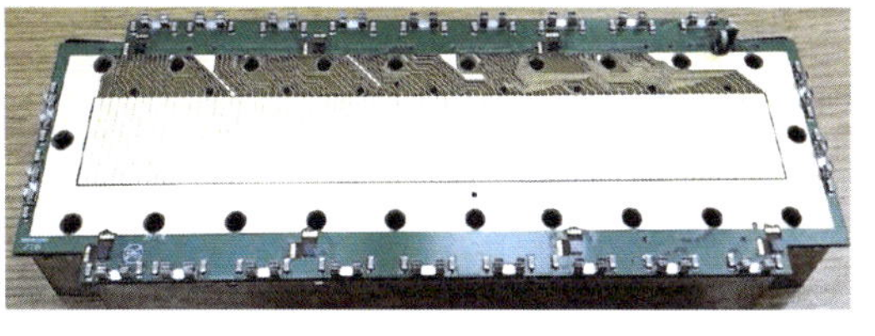

Fig. 5 Detector stripboard

AFE Modules for GEM Signals

MicroTCA form factor has several benefits:

- EMC compliance, fully shielded panel with EMC seal.
- Hotplug capability.
- The module can be plugged and unplugged using single hand. It enables replacement of the AFE module when the detector is installed deeply inside the port.
- The reliable locking mechanism that is released when the handle is pulled.
- The signalization of correct locking in the slot using miniature switch.

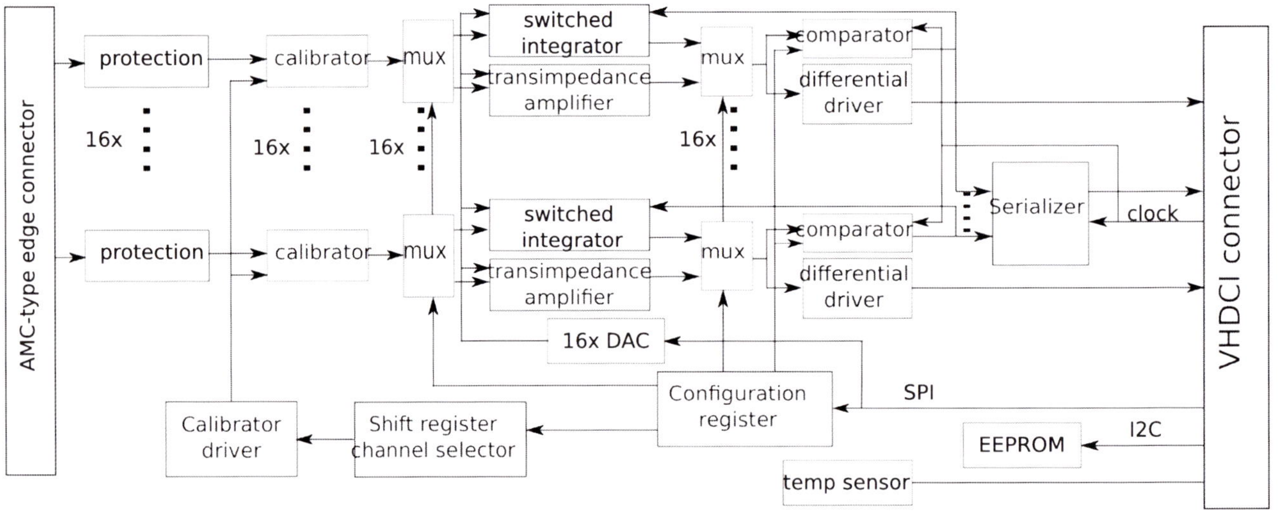

Fig. 6 AFE block schematic

- Rail guides with wide openings that help to position of the module inside the cassette.

Each AFE module (Figs. 6, 7) consists of 16 charge amplifiers followed by zero—deadtime switched integrator. Optionally, the signal can be routed directly to the AFE output. The charge amplifiers convert current signals from the strips and also shape the output voltage signal. High speed low noise current feedback amplifier is used. Analogue bandwidth is limited to 20 MHz. Each channel is equipped with dedicated calibration and protection circuit. The calibration circuit allows injecting known charge selectively to any combination of the AFE board inputs. This enables possibility of rapid diagnostic and calibration of the whole detector chain. Expected charge from the detector is 100 fC (it corresponds to 10 ns current pulse with 10 μA amplitude). Each channel has dedicated offset trimming circuit and fast discriminator with adjustable threshold. The AFE can work in one of 4 modes:

- Direct sampling of the detector signal. Bandwidth is limited to 20 MHz, ADC sampling is 125 MHz, 12 bits.
- Fast, switched hybrid integrator with zero dead time.
- Fast discriminator with adjustable threshold.
- Direct sampling and fast discrimination running in parallel.

Mode selection can be done during DAQ system runtime. To enable advanced diagnostics, each AFE board has dedicated temperature sensor and EEPROM memory. To limit the number of wires, outputs of the discriminators are routed to the serializer chip that informs FPGA that charge compensation took place or that there were multiple hits in a single sample, depending on selected AFE mode. All the signals to the ADC board are routed using single Very-high-density cable interconnect (VHDCI) connector and cable. Since more than 10 W of power is dissipated on every AFE board, the liquid cooling cold plate is attached to it with the bottom metal plate that receives the thermal energy from all integrated circuits. Due to radiation limitation, the p–n junction was used as temperature sensor. Active cooling is ensured using liquid cold plate. The prototype cold plates were made of copper, but due to radiation limitations, they were made of aluminium in the final version of the system. The detector and AFE are main components of the system that are affected by hazard conditions. In case of AFE, all the components were carefully selected to mitigate these threads. All amplifiers, multiplexers, serializers, diodes, transistors were selected from the list of components that are known to be radiation tolerant to the minimum of 50 kRad. That is why it is so important to mount correct components from specified vendors. The same type of, i.e., amplifier from two different vendors may have completely different behavior during irradiation.

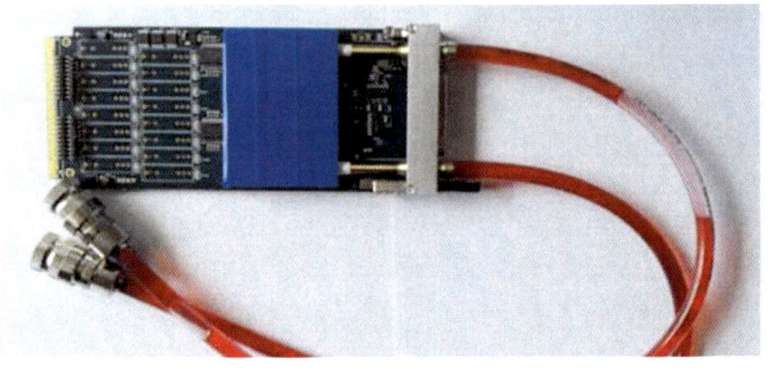

Fig. 7 AFE module with cold plate attached

GEM detector is very sensitive to the atmospheric pressure and temperature. Due to this reason, air-flow cooling cannot be used because it would create static pressure on the detector window foil. Liquid cooling does not cause such problems, and each AFE board is equipped with liquid cold plate connected to common distribution box (Detector & AFE Cooling section). External magnetic field seriously affects the operation of the GEM detector. It is possible to shield the detector against it using sheets of mu-metal. From the electronic system point of view, monitoring of magnetic field is assured using several magnetic field sensors distributed around the detector cassette volume.

AFE Modules for B Sensors

Due to the strong magnetic field which affects the GEM detector operation, it is important to know the distribution of magnetic field along the detector's active volume. For this purpose, each detector is equipped with 16 linear magnetic field sensors. Each sensor consist of 3 Hall ICs arranged perpendicularly to each other (Fig. 8). This setup enables 3D magnetic mapping of magnetic field distribution around the detector.

Eight sensors were installed at the GEM stripboard on the side of the GEM foils (Fig. 9), another eight sensors were installed on the backplane board (Fig. 10). That enables magnetic field estimation in the Z axis (along the AFE modules).

All Hall sensors installed around the GEM detector require 48-channel conditioning and digitization system. It is implemented as an additional board (Fig. 11) inserted to the backplane close to other AFE boards. Each channel consists of instrumentation amplifier with an offset correction potentiometer and 12 bit ADC. Measurement range is from − 0.5 to + 0.5 T. All the components were selected from ones known to be radiation tolerant. Each ADC has SPI output which after conversion to LVDS is transferred over VHDCI connector, identical as in case of AFE boards.

The LVDS signal is then transmitted to simple 3U interface board, installed in the same slots as ADC boards. Its role is LVDS signal buffering. Digital data are processed directly by FPGA circuit, in parallel with GEM detector data.

Fig. 8 3D magnetic field sensor

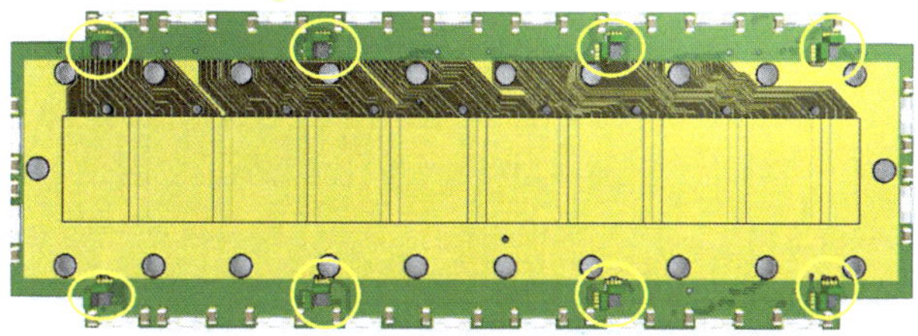

Fig. 9 AFE backplane with marked magnetic field sensors

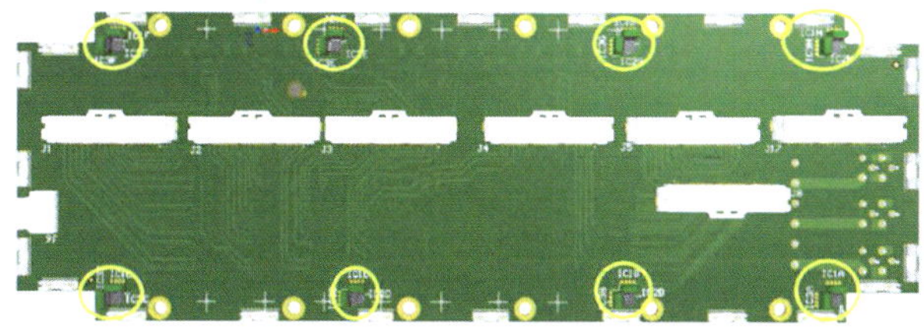

Fig. 10 GEM strip board with marked magnetic field sensors

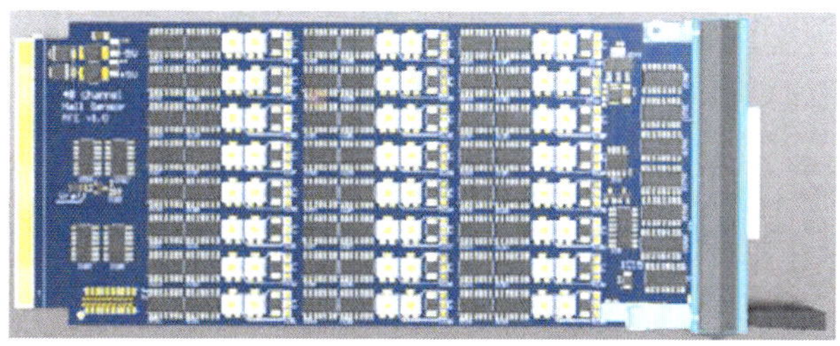

Fig. 11 AFE board for Hall sensors

ADC Boards (ADB)

The role of ADC boards (Fig. 12) is:

- Conversion of the analogue differential signal to digital form

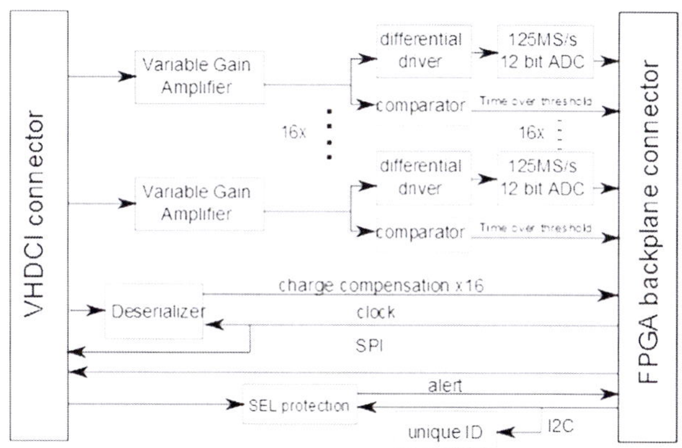

Fig. 12 ADC block schematic

- Readout of temperature sensor
- Configuration of AFE board using SPI protocol
- Measurement and limitation of DC power delivered to the AFE board
- Generation of calibration voltages for AFE board
- Adjustment of gain and offset in all measurement channels
- Distribution of clock to the ADC and AFE boards

The ADC boards are designed as 3U Euro module with the standard 3HP front panel (Fig. 13). The rear side is equipped with PCIe-type edge connector where ADCs have their serial outputs routed directly to the FPGA processing board slots. Linear Technology, 12 Bits, 125 MHz quad ADCs were used.

FPGA Processing Boards: ADC Backplane

Each FPGA backplane (Figs. 14, 15) supports up to 4 ADC boards using edge connectors. The FPGA backplane collects digital signals from 64 ADC channels and also provides control of 64 hybrid integrators. The FPGA also controls offset and calibration DACs and generates DAC level for the integrator. The reference clock signal is generated on the data concentrator board (Data Concentrator section) and distributed to the ADC boards via low jitter clock buffer. FPGA backplane is also equipped with 2 GB of fast DDR-3 SDRAM, four fast 25 Gbit/s links which utilize mini-SAS connectors. These links are used to transfer edge channels data to neighboring processing modules and provide PCIe x4 connectivity with the Data Concentrator Board.

Data Concentrator

The main role of the data concentrator (Fig. 16) is the collection of data through PCI Express Gen1 x4 links. DCB consists of following resources:

- 8 port, PCIe Gen3, switch with 16x upstream port and eight x4 downstream ports.
- Low jitter ADC and PCIe clock distribution circuits.

Fig. 13 Four ADC boards inserted into FPGA processing backplane

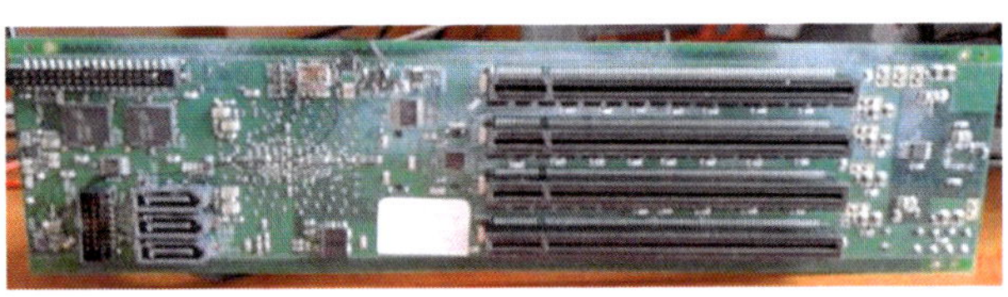

Fig. 14 FPGA processing backplane—front side

Fig. 15 FPGA processing backplane—rear side

Fig. 16 Data concentrator board

- DC/DC converters.
- Four RS485 ports for triggers and interlocks placed on separate module.
- 8 lane M-LVDS bus transceiver that connects all the FPGA. processing backplanes and provide synchronization, trigger and interlock signals.
- Sensor board placed on the separate module (temperature, humidity, magnetic field, pressure).
- USB to JTAG and I2C bridge that is used to configure all the FPGAs via JTAG interface. It also configures all ADCs, DACs, and registers using the I2C bus.
- FMC module socket for future extensions.

Thanks to the USB to JTAG bridge, all the FPGA bit streams are located in the file system of the server. With such approach, every single piece of code or configuration is loaded during booting process which ensures reliable reconfiguration and the same state of the system after every power up process.

Fig. 17 Dual x86 Xeon mainboard

Data Processing Xeon CPU Platform

Recent trends in High Energy Physics and other fields of science move towards software-defined data processing. Software Defined Radio is a popular example. Hardware components are used as front-end interfaces where amount of data is too high for computer systems to handle. The previous generation of systems [7] used FPGA-based data handling and processing which was characterized by low latency and high throughput. However, FPGA-driven approach has several disadvantages: the complexity of the algorithm, issues with maintenance, is inflexible, implementation of the new algorithm requires a highly skilled engineer who knows details of FPGA configuration.

Recent progress in multi-core computers caused a shift towards software-defined signal processing, and such approach was utilized in case of this architecture. The data processing subsystem was equipped with the embedded x86 platform. Server CPU Board (Fig. 17) provides GbE connectivity and 4 PCIe, Gen 3, x16 slots that are used to simultaneously transfer data from the Data Concentrator Boards to up to 24 Xeon CPU cores. Standard ETX form-factor main board is used. Up to two Xeon processors can be installed and optional 56-core Xeon PHI numerical accelerator can be used when additional processing power is needed.

Fig. 18 Protection system

Protection and Interlock System

GEM detectors are very fragile devices—they can be physically destroyed by overvoltage conditions, arching, too high temperature, wrong gas mixture or lack of gas flow. Moreover, the GEM detector is accompanied by power-hungry Analog Front End electronics that need effective cooling to evacuate the heat that otherwise would threaten the detector. The WEST Protection and Interlock System (WPIS) is designed to mitigate all hazard conditions.

The system is based on non-volatile FPGA technology to make it independent from the circuit that is being protected. Figures 18 and 19 present the block schematic and photo of the protection board. It performs following functions:

Gas Interlock

High voltage is enabled only when the gas is flowing and has correct parameters.

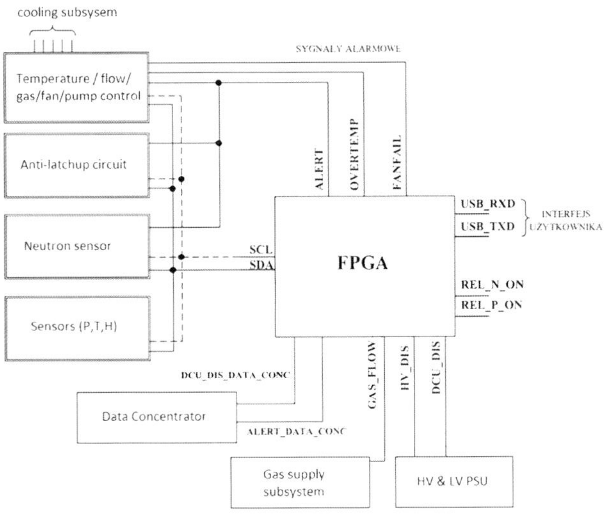

Fig. 19 Protection system schematic

ADC Cooling Interlock

Power supply for ADCs is delivered only when cooling fans are rotating, and temperature is below the threshold

AFE Cooling Interlock

Power supply for AFE modules is delivered only when the cooling liquid pumps are rotating, and the temperature is below the threshold. The power supply cannot be switched on without complete initialization of all the sensors.

Power Supply Interlock

AFE supply is enabled only when ADC board is supplied.

Temperature Alarm

All temperature sensors on AFE boards, ADC boards, in the detector proximity, in processing unit, on the FPGA backplane have programmable alarm output. All sensors outputs are wired together using the WIRED-AND scheme to make sure that power supply will be disabled when any of the sensors exceeds permitted temperature.

Latch-Up Detection

When high energy particle passes through the semiconductor device, it may cause the latch-up effect. It opens parasitic thyristor inside of the integrated circuit structure and causes the flow of excessive current from the positive to the negative terminal of the device. Such current, when not limited, causes quick overheating of the device and its breakdown. It may also cause an explosion of the device. So monitoring of supply current in such operating conditions is critical. The WPIS is connected to the sensitive current sensors installed on every ADC board, that directly measure the supply current on every AFE board. The current sensors outputs are sampled by specialised ADC with alarm function. Each channel of the ADC has dedicated threshold register and comparator that monitors measured value and when exceeded, in a few microseconds sets the alarm. The alarm causes immediate low voltage supply disconnect which protects the AFE. Afer programmable delay, the supply can be enabled.

He Buffer Humidity Monitor

Humidity of the helium buffer is critical because it indicates serious leaks of the cooling system or lack of helium in the buffer. SHT25 sensor is placed in the outlet of the helium buffer and continuously monitored by the embedded software.

Temperature Monitoring

The GEM detector is very sensitive to the temperature variations since it causes frame and window foil deflections and wrinkling. Moreover, the temperature exceeding 150 °C cause Mylar foil melting which damages the detector. The Analogue Front End Electronics (AFE) generates a large amount of heat it is over 100 W per detector. In case of AFE cooling failure it presents threat to the detector foils and the electronics itself due to lack of air circulation and limited heat exchange with walls. So the protection against over temperature has the highest priority. Following means to mitigate temperature hazards were foreseen:

- Each AFE has the independent temperature sensor. It is connected to the main processing unit and monitored by software. Moreover, each sensor has programmable threshold comparator which causes hardware action when the limit is exceeded. This can be programmed to physically disconnect the power supply from the detector.
- Since the solution above is based on some software and needs initialization to work, another redundant solution was foreseen. It is based on the external, independent Power Strip with temperature watchdog which physically disconnects the whole detector when the temperature exceeds the limit.
- The cooling unit has independent temperature and flow sensors that are connected directly to the main processing unit. In case of pump failure or heat exchanger problems, the detector will be shut down and errors will be reported.

Neutron Sensor

It is based on SRAM conception. 8 large area static memory chips are used. The supply of the RAM cells is programmable and can be lowered below value recommended by the manufacturer to increase device sensitivity to the neutrons. The software periodically writes memory with fixed pattern and then reads them in order to determine how many errors occurred in specified time.

Fan Speed Control

To limit the amount of noise generated by the processing unit, active fan speed control is implemented. The speed of fans in the ADC compartment is proportional to the temperature measured by two sensors.

Pump Wear Alarm

The rotation of the pump is constantly monitored. When significant loss of RPM is detected, warning condition is asserted to inform operator that pumps need replacement or something wrong happens in the installation

Fan Wear Alarm

The rotation of the fans is constantly monitored. When significant loss of RPM is detected, warning condition is asserted to inform operator that fans need replacement or something blocks the air flow

Redundant Supply for Fans and Pumps

The supply delivered to fans and pump can be sourced by HV & LV unit or processing unit. In case of failure of one of them, protection system will continue operation as well as pumps and cooling.

Watchdog

The software running on processing unit CPU periodically performs monitoring operations. In case of communication failure or over-temperature event, the external power strip (Fig. 20) will disable power supplies.

The WPIS is configured by data concentration block that bridges USB to several I2C interfaces. It is connected using several ruggedized M12 connectors with HV & LV unit, cooling distributor, and heat exchanger. Physically WPIS is installed inside the processing unit.

Fig. 20 Watchdog and protection unit

Detector and AFE Cooling

The liquid cooling system is a critical part of the data acquisition system. The cooling is equipped with 2 circuits primary and secondary. The primary is connected to the reactor cooling system that uses water. The secondary uses a special, not conductive, not flammable liquid. Special care was taken designing cooling system since it is crucial for detector operation. The decision was taken not to use water to cool the AFE boards because, in case of a leak, the electronics would be destroyed due to corrosion and electro-migration. To separate these two circuits, the heat exchanger was used. Additional pump, the fluid reservoir, and the flow sensor were foreseen to provide reliable cooling. The liquid cooling system is critical for the safety of the electronics. For this reason, long-term tests were conducted in simplified setup. The tests took 7 months and were performed in order to qualify the following components: pumps, flow meters, reservoir, couplings, heat exchanger, tubing, cold plates. The temperature of the electronics is not critical and does not need stabilisation. Calibration procedure performs periodic offset cancellation to mitigate temperature variation. One need to make sure that the circuits do not exceed 50 degrees at which point GEM detector window may get deformed.

During the test period we discovered several issues:

- One of the pumps failed due to leak that caused the inverter malfunction. The pump must be installed vertically to prevent such events.
- Both tested flow meters failed after a few weeks. It was caused by copper debris remaining of the cold plate machining process. Installation of particle filter fixed this issue.
- Plastic reservoir caused leaks. It was replaced with the solid aluminium one.
- Couplings were working reliably, no single leak was observed.
- Heat exchanger had some leak which was immediately removed.
- Tubing caused issues with the vapour collection. We had to replace them with cooling-agent compatible types.
- Cold plates didn not have any issues.

The final version of the cooling system was built (Fig. 21). The main difference between the test bench is the stronger pump and bigger reservoir with the transparent lid that protrudes through the front panel giving visual information about liquid level (Fig. 22).

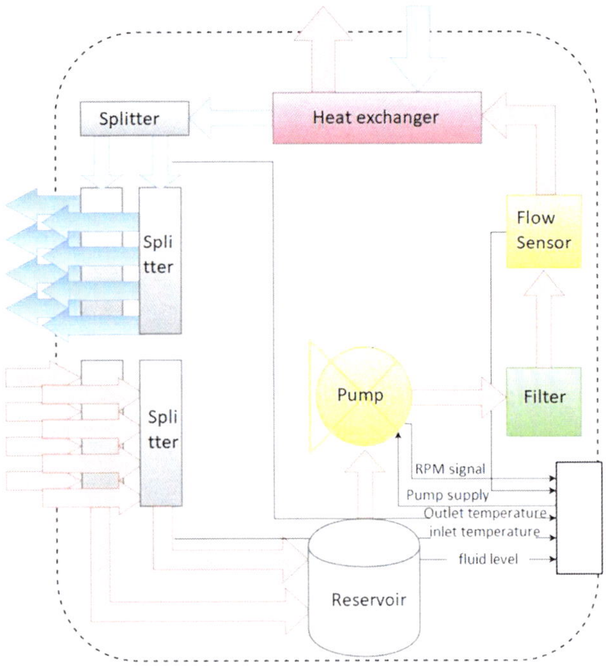

Fig. 21 Schematic of cooling subsystem

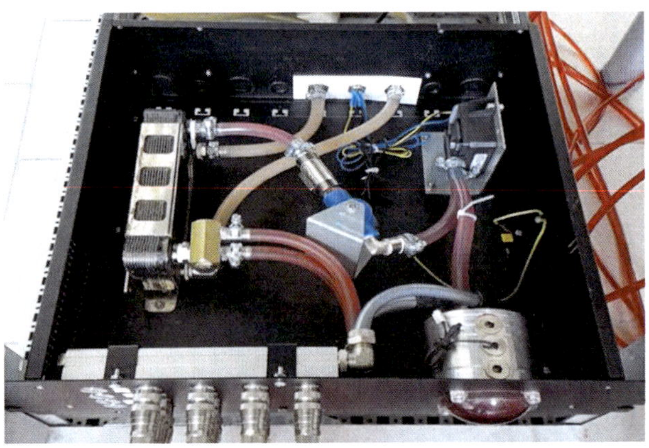

Fig. 22 Cooling subsystem—internal view

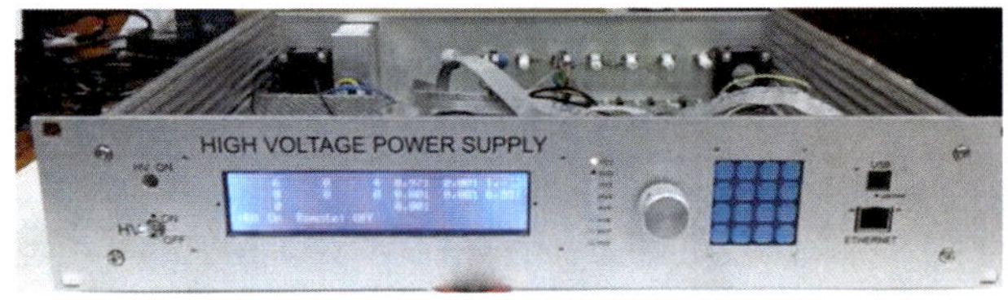

Fig. 23 High voltage power supply

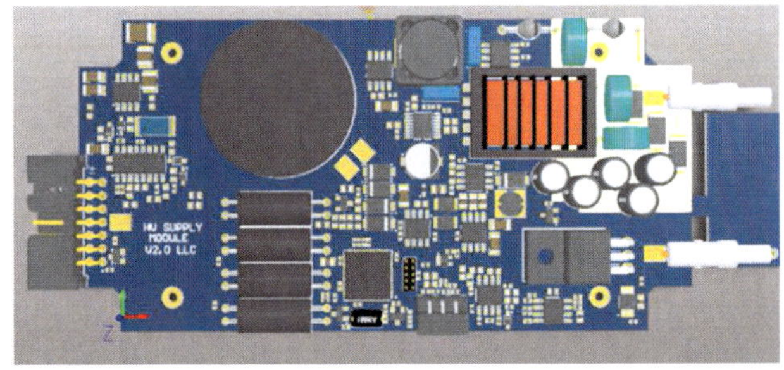

Fig. 24 High voltage power supply module

High Voltage Power Supply

To fulfill detector requirements (high rate and high linearity), dedicated High Voltage Power Supply was developed. It has a form of PCB modules within plastic enclosures that are mounted inside of dedicated 2U aluminium casing (Fig. 23). Unlike most of existing HV supplies it provides following features:

- Voltage range 0–1500 V, 0–10 μA.
- Modular construction (Fig. 24)—the number of channels can be customised. Resonant converter topology was used to provide high efficiency and low EMI level.
- Fully isolated inputs and outputs—the outputs can be connected in any manner in series or in a parallel way. Each module has 10 kV galvanic isolation from common control and power supply.
- A very low stored charge that is equivalent to 200 nF of output capacitance. It is 10x less than in the previous generation of HV supplies. It is critical in case of GEM foil breakdown.
- Very fast current limiter—critical to the operation of GEM detectors. It enables protection of detector in case of arching. It is critical in case of high gain detectors because can reduce the value of series protection resistor and significantly improve the linearity of the detector.
- Transient recorder on each channel—currents and voltages can be recorded thanks to embedded scope function. When arcing occurs, one can observe signals before and after the event.
- Voltage and current measurement in each channel. One can selectively measure current consumed by each GEM foil.
- Ethernet and USB interfaces.
- Local keyboard and LCD display.

Development of Next-Generation Systems

Continuous development of commercially available platforms (COTS) together with constantly changing design requirements inspired us to do the feasibility study on the application of standard measurement and control platforms for plasma physics diagnostic system. There are at least two approaches to the construction of equipment—build it yourself or adopt existing standards. Table 1 summarizes such approaches. In plasma diagnostics number of channels

Table 1 Comparison of custom and standard platforms

Feature	Custom platform	Standard platform
Production cost	Low or medium	High
Fitness for purpose	Optimal	Low or medium
R&D costs	High	Low or medium
Long term maintenance cost	High	Low or medium
COTS components compatibility	No	Yes

is usually low, rarely exceeding a few hundreds while experiments last several years. In such cases, R&D costs are often higher than the purchase of COTS equipment. The recent development of MictoTCA standard towards needs of High Energy Physics, especially detector readout and low-level RF application [3], makes this platform potentially suitable for plasma diagnostics as well.

MicroTCA for Detector Applications

MTCA is a modular system for data acquisition, control, management, and data processing [1]. It is derived from established telecommunication ATCA standard. Architecture is based on the high-speed backplane with the point to point connections up to 40 Gbit/s (Fig. 25). Recently developed version 4 of the standard provides extended synchronization, timing distribution and support for RF and analogue front-end electronics. Embedded computer, controller, and PCIe Gen 3 or 40G Ethernet hub make such platform compact and powerful solution. The standard defines several extensions Rear Transition Modules, ruggedized, convection cooled for military and aviation applications. The platform is managed using IPMI including cooling, power distribution, hot-plug and remote control of the chassis. The recent upgrade brings 128 Gbit link (Fig. 26) to external processing system so existing bottleneck between the hub, and the embedded computer was eliminated.

Feasibility of mTCA.4 platform for detector systems:

- Industrial standard with scientific community support.
- Plenty of board space, up to 12 AMC boards and Rear transition modules (6U height).
- Embedded data concentrator (PCIe Gen3.0 x4 / 40G Ethernet hub + 1Gbit Ethernet hub).
- Advanced IPMI including fully redundant architecture.
- High-speed links between the neighboring boards—perfectly suited for GEM detector data processing.
- FPGA oriented architecture (i.e., JTAG support on the backplane [10], PCIe clock distribution, remote firmware update provided by the crate controller).
- Centralized clock and trigger distribution.
- AMC board and RTM board plenty of space for components.

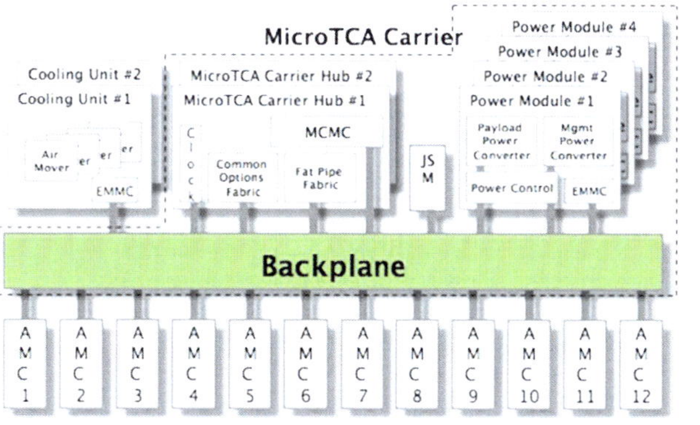

Fig. 25 Example MTCA architecture[1]

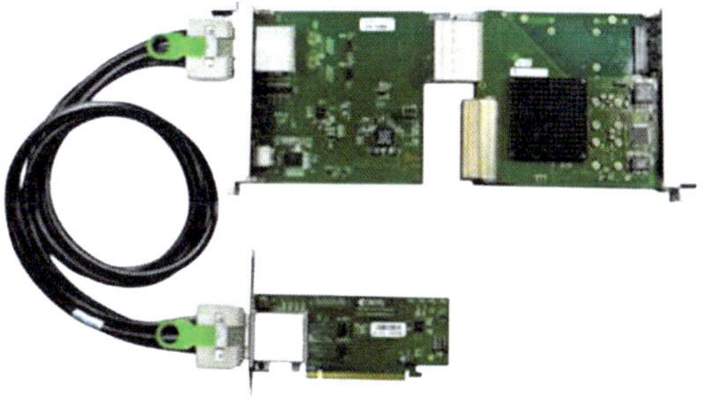

Fig. 26 128Gbit/s PCI Express interface

- Additional RF backplane [3] developed at WUT and DESY with low jitter clock, low noise analog connectivity, and dedicated low noise power supply.
- N.A.T. MCH with 128 Gbit/s uplink (by AIES). Enables SW processing in the powerful external server or inside of the MTCA crate [2].

MTCA.4 with recent developments (PCIe uplink, RTM, RF backplane) is the very attractive platform for detector processing systems. Intensive work with N.A.T on new Open Source White Rabbit timing module [8] takes place, so another limitation of the platform is eliminated. Next generation of DAQ system for GEM and other detectors will be based on the MicroTCA platform. Currently, the large scale processing system for GEM detectors based on MTCA for CBM experiment is under development [9]. Another implementation of GEM processing chain for tokamaks in MTCA.4 standard is under development.

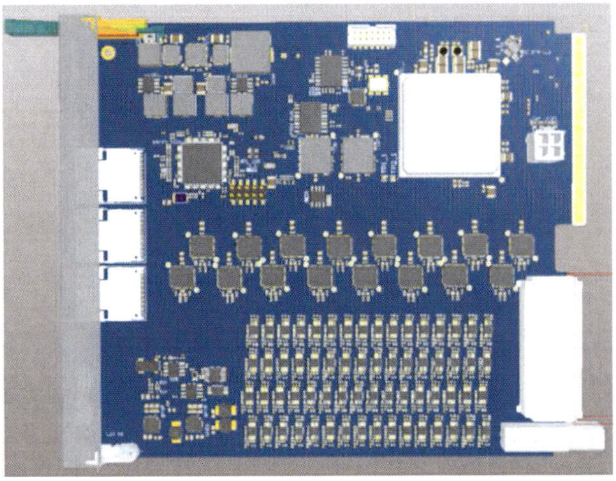

Fig. 27 GEM detector data processing board in AMC format

Fig. 28 GEM detector data processing board in RTM format

Figures 27 and 28 present 64 channel RTM and AMC solution which can be scaled up to 768 channels per MTCA crate. Each channel has adjustable gain and offset. To enable further extension beyond MTCA crate, 20 Gbit serial links (mSAS connectors) were installed on front panels. The RTM module has VHDCI AFE connectors compatible with existing electronics (WEST-type AFE) to make the migration to the next generation system straightforward.

Initial results

First laboratory experiments were performed with 1D GEM detector and Fe55 soft X-ray source. The measurement system used 64 analog input channels. One of the first results are presented in Figs. 29 and 30.

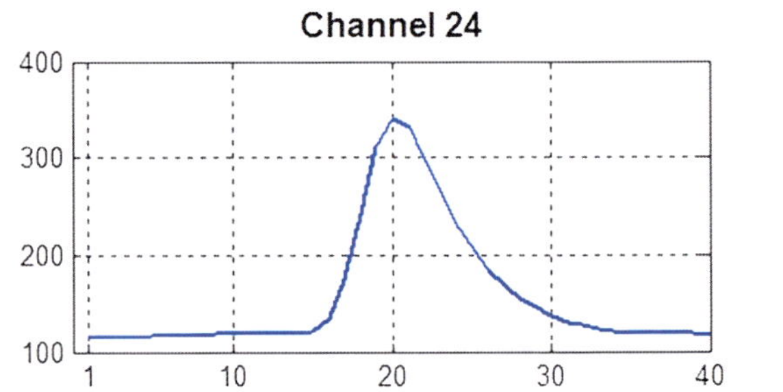

Fig. 29 Typical pulse acquired at channel 24 at sampling rate of 100 MHz, ADC units

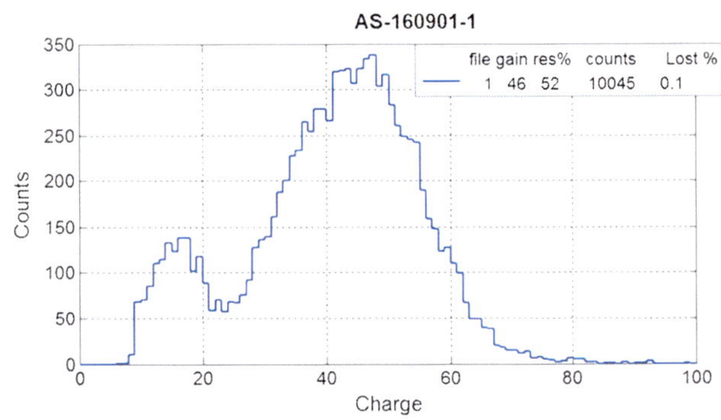

Fig. 30 Energy spectra obtained with Fe55 source

Conclusion

Complex real-time SXR measurement systems was built and installed at the WEST Tokamak. The system follows trends set by leading High Energy Physics laboratories where majority of the real time processing is performed in software, using high performance multi-core computer. Due to several threads and harsh conditions at the site, several measures had to be taken to provide safe and reliable operation of the system. Important role plays Protection and Interlock subsystem which after recent upgrade is able to physically remove whole detector subsystem from the tokamak port using custom made elevator mechanism. The system is currently tested with radioactive source and software integration with WEST control system is performed. Second system with curved GEM detector for horizontal plasma monitoring is in assembly phase. Both detection systems will form SXR tomography diagnostics.

Acknowledgements This scientific work was partly supported by Polish Ministry of Science and Higher Education within the framework of the scientific financial resources in the years 2016 and 2017 allocated for the realization of the international co-financed project—Agreement Nos. 332380/PnH/2016 and 3814/H2020-Euratom/2017/2.

References

1. MTCA specification. https://www.picmg.org/openstandards/microtca/. Accessed 2006
2. NAT PHYS. http://www.nateurope.com/products/NAT-MCH-PHYS80.html. Accessed 2018
3. RF backplane specification. http://www.nateurope.com/products/NAT-LLRF-Backplane.html. Accessed 2018
4. M. Chernyshova, T. Czarski, K. Malinowski, E. Kowalska-Strzęciwilk, J. Król, K.T. Poźniak, G. Kasprowicz, W. Zabołotny, A. Wojeński, R.D. Krawczyk, P. Kolasiński, I.N. Demchenko, Y. Melikhov, Development of gem detector for tokamak SXR tomography system: preliminary laboratory tests. Fusion Eng. Des. (2017). https://doi.org/10.1016/j.fusengdes.2017.03.107
5. Chernyshova, et al., Gem detector development for tokamak plasma radiation diagnostics: SXR poloidal tomography, in *Proceedings of SPIE*. SPIE (2015), pp. 966,231–966,231
6. T. Czarski, M. Chernyshova, K. Malinowski, K.T. Pozniak, G. Kasprowicz, P. Kolasinski, R. Krawczyk, A. Wojenski, W. Zabolotny, The cluster charge identification in the gem detector for fusion plasma imaging by soft X-ray diagnostics. Rev. Sci. Instrum. **87**(11), 11E336 (2016). https://doi.org/10.1063/1.4961559
7. G. Kasprowicz, et al., Readout electronics for the GEM detector. In: *Proceedings of SPIE*, pp. 80,080J–80,080J. SPIE (2011)
8. G. Kasprowicz, White rabbit timing receiver for micro tca-based data acquisition systems for applications in high energy physics. Acta Phys. Polon. Supp. **9**(2), 239–242 (2016). https://doi.org/10.5506/APhysPolBSupp.9.239
9. Guminski et al., Time and clock synchronization with AFCK for CBM, in *Proceedings of SPIE*. SPIE (2015), pp. 96,622V–96,622V
10. P. Kulik, Ethernet JTAG Switch Module. http://repo.bg.pw.edu.pl/index.php/en/r/info/bachelor/ WUTf8774ef527d74807a9d5c39e2a72dc5a
11. F. Sauli, GEM: a new concept for electron amplification in gas detectors. Nucl. Instrum. Methods Phys. Res. Sect. A Accel. Spectrom. Detect. Assoc. Equip. **386**(2–3), 531–534 (1997). https://doi.org/10.1016/S0168-9002(96)01172-2
12. A. Wojenski, K.T. Pozniak, G. Kasprowicz, P. Kolasinski, R. Krawczyk, W. Zabolotny, M. Chernyshova, T. Czarski, K. Malinowski, FPGA-based gem detector signal acquisition for SXR spectroscopy system. J. Instrum. **11**(11), C11035 (2016)

Journal of Fusion Energy (2019) 38:480–489
https://doi.org/10.1007/s10894-018-0181-2

ORIGINAL RESEARCH

FPGA and Embedded Systems Based Fast Data Acquisition and Processing for GEM Detectors

Wojciech M. Zabołotny[1] · Grzegorz Kasprowicz[1] · Krzysztof Poźniak[1] · Maryna Chernyshova[2] · Tomasz Czarski[2] · Michał Gąska[1] · Piotr Kolasiński[1] · Rafał Krawczyk[1] · Paweł Linczuk[2] · Andrzej Wojeński[1]

Published online: 13 August 2018

Abstract

Gas electron multiplier (GEM) detectors (Sauli in Nucl Instrum Methods Phys Res A 805:2–24, 2016. https://doi.org/10.1016/j.nima.2015.07.060 (special issue in memory of Glenn F. Knoll); Buzulutskov in Instrum Exp Tech 50(3):287–310, 2007. https://doi.org/10.1134/S0020441207030013) are widely used for detection of ionizing radiation. When used in the proportional mode, they provide information about time, location, and energy of a detected particle (Chernyshova et al. in Fusion Eng Design, 2017. https://doi.org/10.1016/j.fusengdes.2017.03.107; Altunbas et al. in Nucl Instrum Methods Phys Res A 490(1–2):177–203, 2002. https://doi.org/10.1016/S0168-9002(02)00910-5. http://linkinghub.elsevier.com/retrieve/pii/S0168900202009105). Modern technologies allow full utilization of detector properties, by acquiring the waveform of output current pulses and processing them using sophisticated digital signal processing (DSP) algorithms. The current pulses must be digitized at high speed (up to 125 MHz) with high resolution (up to 12-bits). Due to the high volume of the produced data, it is necessary to provide the high-performance data acquisition system (DAQ) to transmit the data to processing units. Efficient processing of the GEM data requires distributed parallel processing system to perform multiple tasks (Czarski et al. in Rev Sci Instrum 87(11), 11E336, 2016. https://doi.org/10.1063/1.4961559): (1) Filter out the background and transmit only hit related data. (2) Extract the parameters of a hit, describing the time and charge (related to energy). (3) Estimate the hit position by combining information from multiple anode pads. (4) In case of 2D GEM detectors, correlate pulses received from X and Y pads (pixels) or W, U and V pads (pixels). (5) Separate the hits overlapping in space or in time (if possible) to support detector operation at higher rates. The above functionalities may be achieved in different hardware architectures. The typical hardware platforms include FPGA chips, standard or embedded computer systems with different computation accelerators (Wojenski et al. in J Instrum 11(11):C11035, 2016. http://stacks.iop.org/1748-0221/11/i=11/a=C11035; Nowak et al. in J Phys Conf Ser 513(5):052–024, 2014. https://doi.org/10.1088/1742-6596/513/5/052024. http://stacks.iop.org/1742-6596/513/i=5/a=052024?key=crossref.c5912cfa72c30b309821e14c4384948f. The paper shows possible solutions with their feasibility for particular applications.

Keywords X-ray detectors · GEM detectors · Hot plasma diagnostics · FPGA

Introduction

Gas Electron Multiplier (GEM) detectors are widely used for detection of X-ray, gamma radiation or ionizing particles. Detection of fast neutrons is also possible after conversion via scattering on hydrogen (e.g., in a polyethylene foil). Detection of thermal neutrons may be achieved by using a borated aluminum cathode [1]. To fully utilize the potential of GEM detectors, including their energy resolution in proportional mode, it is necessary to provide a readout system able to extract the information about the

✉ Wojciech M. Zabołotny
wzab@ise.pw.edu.pl

1 Institute of Electronic Systems, Warsaw University of Technology, Nowowiejska 15/19, 00-665 Warszawa, Poland

2 Institute of Plasma Physics and Laser Microfusion, Hery 23, 01-497 Warszawa, Poland

location, time and energy of a detected particle. Additionally, for diagnostic and debugging activities, it is desirable that such a system should optionally provide access to raw GEM signals.

GEM Detectors

GEM detectors have been invented in 1997 by Fabio Sauli [2] in Gas Detector Development Group at CERN, and are under continuous development [3]. The structure and operating principle of a GEM detector are shown in Fig. 1. The GEM detectors consist of a few GEM foils placed between the cathode (located on the inner side of the detector's window) and the anode—created by readout strips, placed on the inner side of a readout board. The space inside the detector is divided into the conversion and drift area (between the anode and the first GEM foil), the transfer areas (between each two GEM foils) and the induction area (between the last GEM foil and the anode pads). Electrical voltages are applied, usually via a voltage divider, to the whole structure so that an electrical field is created in each area, and also between both sides of each GEM foil. The GEM foil is a thin (typically 50 μm) isolating kapton film, covered with conducting planes on both sides. The foil is perforated with regular circular holes. When the voltage is applied between both sides of the foil, the strong electrical field is created in the holes. Electrons produced by the ionizing particle in the conversion area, after reaching the first GEM foil, are accelerated in the strong electrical field in the holes and produce secondary electrons. The process repeats in each GEM foil. The charge gain in each foil may be between 10 and 100, giving the final gain for the triple GEM structure between 1000 and 10^5 [4]. The cloud of electrons finally reaches the readout board and produces an electrical current pulse in readout pads. The important feature of GEM detectors is that in properly chosen operating conditions, the charge delivered to the readout pads is proportional to the energy of the ionizing particle. The shape of the current pulse may be different [5, 6] depending on the gas used in the detector and on the operating conditions. The current pulses from each readout pad are received by the shaping charge amplifier with introduced low cut frequency, which forms the voltage pulses of standard shape. The properties of that amplifier are tuned according to the requirements of further signal processing stages. For example, lowering the upper cut-off frequency allows eliminating a high-frequency noise. The standard shape of the pulse allows separation of overlapping pulses described later.

Application of GEM Detectors

Depending on the application, the GEM detectors may be used either in one-dimensional (1-D) or in two-dimensional (2-D) configurations. The 1-D version is for example used for X-Ray spectroscopy, where a collimated X radiation is

Fig. 1 Structure and operating principle of the GEM detector

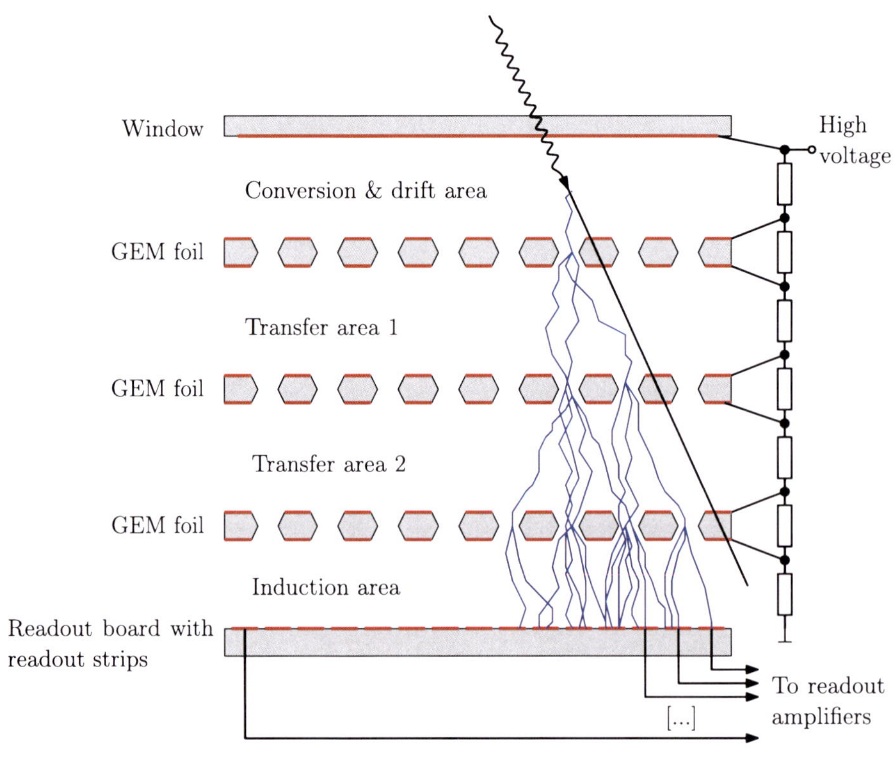

reflected from a Bragg's crystal. In that configuration, the readout pads have a form of strips. The position (the strip number) of the charge pulse informs about the reflection angle and hence about the energy of the photon. The total charge of the pulse provides redundant information about the energy of the photon and may be used to separate signals generated by different orders of reflection.

The 2-D version may be used for X-ray tomography. In that application, the radiation enters the detecting system via a pinhole and creates a 2-D image of the radiation source on the GEM detector. In that configuration, the readout pads should be arranged in a 2-D structure creating image pixels. Unfortunately, the number of pixels quickly increases with the increase of resolution. For rectangular pixels organized in a 16×16 matrix, the number of required readout channels is equal to 256. For higher resolutions, it is not possible to read each pad with a separate readout channel. Therefore, other readout systems must be used. The basic one uses two perpendicular systems of readout strips X and Y, or interlaced pads connected to X and Y lines (see Fig. 2). The more complex readout system uses pads organized in three sets of readout lines U, V, W (see Fig. 3) [7] and provides a better possibility to separate simultaneous pulses. Of course, finding the correlation between pulses in pads associated with different coordinates is a complex task at high rates, and poses significant requirements for a data processing part of the system.

Reception and Processing of GEM Pulses

The proper readout of GEM detectors depends on the operation mode of the detector. For triggered High Energy Physics (HEP) experiments, there are ASIC based solutions [8] that allow reading of the data collected from a GEM detector, after a global trigger was generated. However, the spectrography and tomography systems do not use a global trigger. They require the continuous acquisition of data during a longer period of time (several seconds or even a few tens of minutes, depending on the experiment). Therefore, another system for data acquisition was developed that allows continuous acquisition of digitized data from a GEM detector. First experiments and analyzes [9] have shown, that the shaped GEM signal may be digitized at the relatively low frequency, even 40 MHz and successfully processed to calculate the time and the charge of a single event. Therefore, the system was built with the direct simultaneous digitalization of all GEM channels with analog to digital (ADC) converters. The achievable sampling frequencies using reasonably priced components were between 77.7 and 125 MHz. The ADC converters are directly controlled by FPGA chips, which also preprocess data, detecting the events. Of course, the FPGAs must also perform the zero suppression to avoid transmission of a huge volume of empty data containing only noise. Therefore, the trigger that was avoided when reading analog data from the GEM detector is introduced here. It is based on digitized data, which allows introducing of flexible solutions like compensation of the low-frequency drift. The trigger used depends on the algorithms applied for further data processing. It may be a global trigger when sufficiently strong signal in any channel triggers transmission of data from all channels. The opposite is a local trigger when each channel individually detects the threshold crossing in its data. There is also an intermediate solution, where a trigger generated in one channel is also distributed to its neighbors, as each pulse is usually distributed between the adjacent channels.

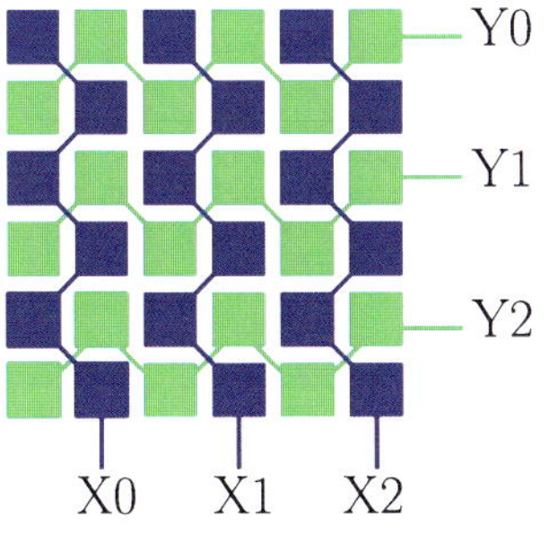

Fig. 2 Example of the pads layout for the XY readout

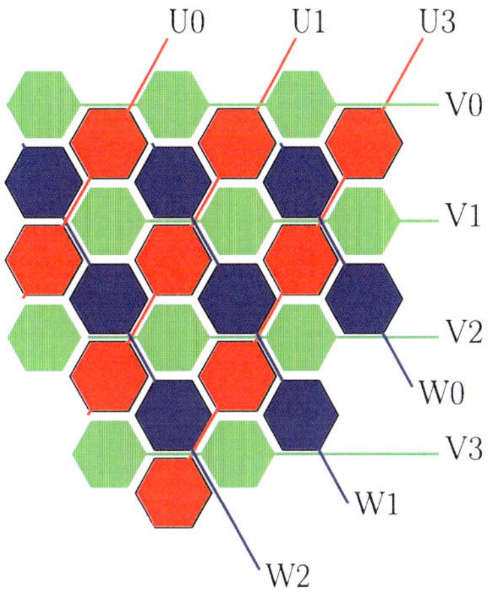

Fig. 3 Example of pads layout for the UVW readout

The acquired data are then processed, and the useful information for each event is extracted: the time (T), position (P) and charge (Q). If the pulse is distributed between multiple readout pads, the charge centroid may be calculated to achieve spatial resolution higher, than the distance between pads.

In both mentioned applications—the spectrography and the imaging, the extracted data are further used for histogramming. Histogram bins are defined by the energy range and by the position of an event.

The final data must be transmitted to an experiment data acquisition system (DAQ) for further processing and archiving. The communication with DAQ is performed by the embedded computer system (ES) because it usually involves complex communication protocols and handling of big data volumes.

As we can see, in the GEM data processing system we have four representations of data

- Raw GEM signal—a continuous stream of samples from all channels.
- Triggered GEM signal—the separated samples with detected events.
- Aggregated event data TPQ (time, position, charge)
- Calculated histograms

The boundary between the FPGA implemented processing, and the ES implemented processing depends on the particular application and requirements.

The general structure of the GEM data processing system is shown in Fig. 4.

Example Implementations of the FPGA Readout Systems

The described above concept of the FPGA and ES based DAQ for GEM detectors may be best explained by showing a few implementations elaborated by our team. Due to different requirements, different implementations have been used.

GEM System for KX1 Diagnostics at JET in Culham

The KX1 GEM system was built for the upgrade of KX1 diagnostics [10]. It is a high-resolution X-ray spectroscope able to detect and measure X-ray radiation emitted by different plasma impurities. The implemented system uses two detectors tuned for detection of the tungsten (W^{46+}) 2.4 keV radiation and nickel (Ni^{26+}) 7.8 keV radiation. High-resolution spectroscopy allows measuring the impurities concentration, the ion temperature—by the line broadening, and the plasma rotation velocity profile.

The system uses an additional ^{55}Fe source to perform calibration of the gain of the GEM detector. The system detects events and calculates soft X-ray (SXR) photons histograms using both—energy and position bins. All calculations are performed in FPGAs [11]. The histogramming algorithm detects and counts the signals that are corrupted (i.e., saturated or overlapped in time or in space).

The JET experiment uses relatively short plasma pulses (shots), and therefore all data could be stored in an FPGA-connected DDR RAM of the GEM system. Transmission of data to the ES is performed later, after the shot. The architecture of the system is shown in Fig. 5. Due to long pauses between plasma pulses at JET, it was possible to use a relatively slow, Python-based software system to control the detector and to transfer the data [13]. Object features of Python language allowed creating the reliable framework for configuration and monitoring of the hardware, which is also very important in measurement systems. Therefore, the data transfer from the GEM system DDR memory to the ES RAM could be done without using a DMA.

The GEM readout system built for KX1 was also able to detect and transfer extremely short sequences of raw pulses.

The important part of that algorithm was the detection of signals corrupted due to overlap or due to saturation. Such signals were not included in histograms but were counted as erroneous signals.

System Based on Serial Data Acquisition for ASDEX Upgrade

A similar hardware system as the one used for KX1 was also used for tests in the ASDEX Upgrade experiment in Germany [14]. The purpose of those tests was to study the detector's operation at high intensity of radiation. In that application, access to the histograms only was not sufficient. Investigating fast changes of the detector gain occurring at high radiation intensities, requires delivery of the information about each detected pulse (the T, P, Q data), to the ES. In this approach, events are detected in

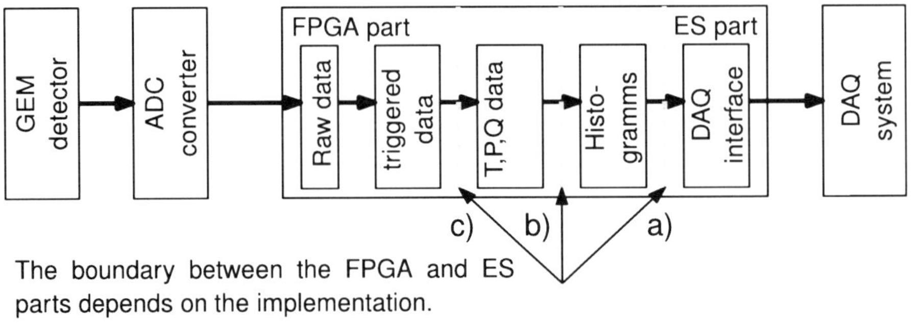

Fig. 4 General structure of the GEM readout system. **a** The system with the FPGA-based histogramming, see Sect. 5.1. **b** The system with event data extraction in FPGA, see Sect. 5.2. **c** The system with the ES-based processing of triggered data, see Sect. 5.3

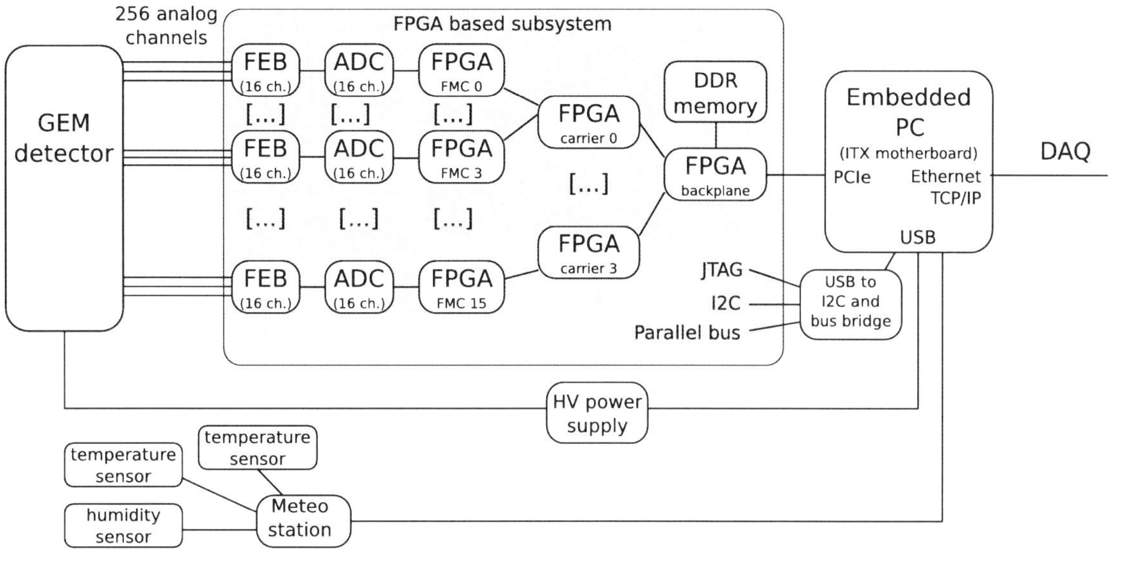

Fig. 5 Architecture od the GEM readout system for KX1 diagnostics [12]. FEB—Front-end boards, ADC—Analog to digital converters, DAQ—Data acquisition system

each channel individually with detection of erroneous (overlapping or saturated) pulses. Data of the correct pulses are transferred to the new "serial data acquisition" system [15, 16] consisting of a binary tree of registers. Each node accepts data from the two preceding nodes. In case if both preceding nodes provide data, the data with lower timestamp has priority. If the timestamps are equal, the data with the lower channel number has priority. Finally, data streams from different channels are concentrated into a single register. Therefore, if the overall event rate is too high, it may happen that it is not possible to transfer the new data from a certain channel because the associated register is still occupied with old data. In such a case the data is lost, and the appropriate flag is set in the data kept in the register. Such a system ensures the "best effort" delivery of data from all channels and warrants that each drop of data is noticed. This system also ensures that the output data are correctly sorted in time. The data are buffered in a DDR memory connected to the FPGA. The ES part of the system receives the data from that DDR memory via PCIe interface with the appropriate DMA engine.

One of the findings from the ASDEX Upgrade tests was the fact that at higher radiation intensity, the probability of overlapping pulses is not negligible. The further work was performed, aimed at a separation of overlapping pulses [17, 18] (see Fig. 6). Even though it is possible to split the overlapping pulses in FPGA, such a hardware implementation consumes too many resources. Therefore it should be done in the software in the ES part, which requires yet another change of the GEM readout architecture.

System Prepared for WEST Tokamak

The system designed for WEST tokamak is intended to provide plasma SXR tomography [19]. The system is supposed to provide plasma imaging with 1 ms time resolution. Raw GEM signals will be shaped using the second-order low-pass filtering with attenuation of 50 dB at 35 MHz. That forming of the signal should allow correct splitting of overlapped signals in the software. The sampling frequency may be selected between 80 and 125 MHz. The transmission of data used during the preliminary tests is described in [19]. The final transmission system will support the transmission of overlapped signals and is described below. The events are detected in each channel independently. When crossing the threshold level is detected, the frame of samples is recorded and transmitted to the ES part. Each frame is 32 samples long and contains up to 10 samples before the threshold trespassing. If the signal does not fall below the threshold level before the end of the frame, it is assumed that the pulse overlap occurred and the second frame with next 32 samples is transmitted.

Transmission of raw signal samples instead of extracted event's data results in significant increase of the data volume that must be transmitted from the FPGA part to the ES part. Therefore it is essential that the most efficient DMA engine is used to transfer the data from the FPGA part to the ES RAM [20]. The efficiency of data transfer and further data processing also depends on the implementation of the device driver in the ES. The data should be delivered directly to the memory buffer accessible for the data processing application, but the memory caching should remain active. Careful configuration of the operating system is needed to maximize the throughput and minimize the

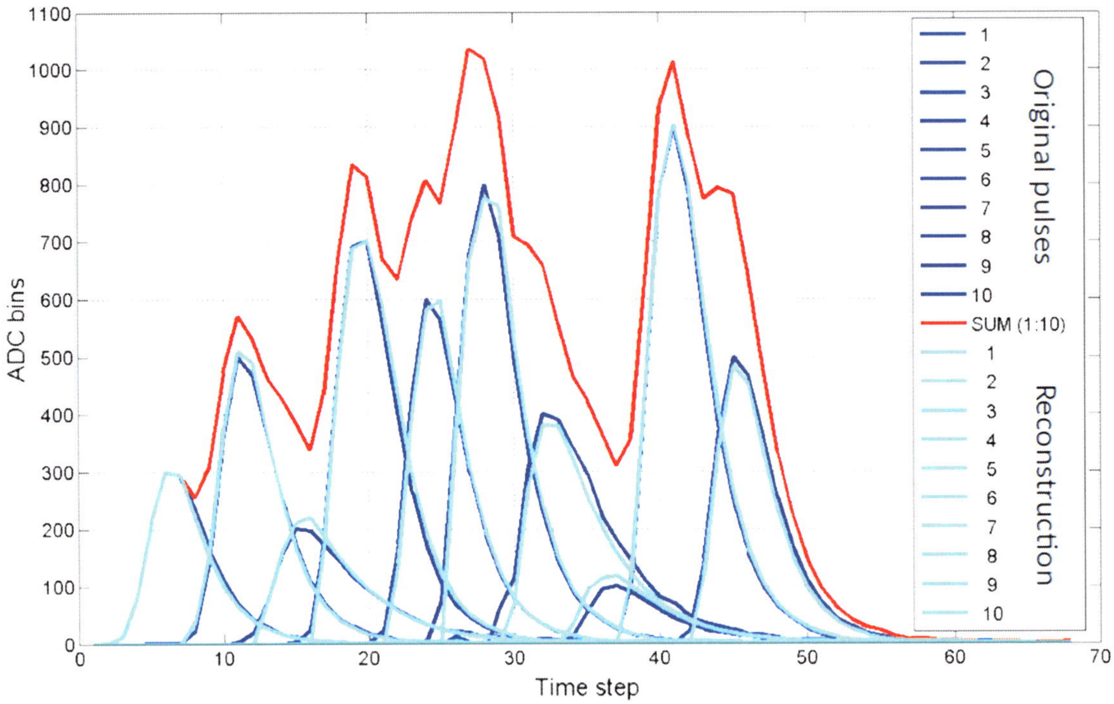

Fig. 6 Example of separation of multiple overlapped pulses [17]

latency [21]. It is especially important as the system is also supposed to deliver real-time plasma imaging for control of the experiment—the histograms with counting period of 1 ms delivered with the latency, not higher than 1 ms.

The GEM system for WEST tokamak delivers a huge stream of data. For 128 channels, 80 MHz sampling rate and 2 bytes per sample we get 20.48 GB/s. Transmission and processing of such stream are above the possibilities of the standard PC computers or embedded systems. Possible hardware solutions for acceleration of computations are discussed in the next section.

Possibilities to Accelerate Computations in the GEM Readout System

Processing of raw GEM data to produce histograms or a 2D image requires a huge number of computations, that must be divided between the FPGA part and the ES part of the system.

FPGA as a Computational Platform

The data processing algorithms for FPGA are usually written in hardware description languages (HDL) like VHDL or Verilog. The description itself is similar to a program written in a standard programming language, but after synthesis and implementation, the algorithms are executed in hardware, without the intermediate software layer. It is advantageous from the performance point of view, as it is possible to implement highly parallel, hard real-time and time-deterministic data processing. That is why FPGAs are perfect to interface ADC converters. Unfortunately, there are also certain drawbacks, because similarities between standard programming languages and HDLs are often misleading. A software engineer who starts to write code for FPGAs does not often understand how the truly parallel continuous assignment instructions work. Another error prone construct is a sequential assignment in the VHDL process that modifies the values of the signal only when the process suspends (usually at the end of the process). Those two are just the first pitfalls faced by the beginners, but there is much more. Therefore, porting of algorithms from computer languages like C or C++ to HDLs is rather difficult [22].

There are some attempts to allow software engineers to enter the world of FPGAs [23, 24]. The first of them was SystemC [25], introduced in 2002 and still being developed. Last years the most important FPGA vendors have provided a possibility to automatically translate C++ or C-described algorithms into HDL, using so-called High-Level Synthesis (HLS). Xilinx offers the Vivado HLS technology [26], while Altera (currently acquired by Intel) provides the Intel FPGA SDK for OpenCL [27] and Intel HLS Compiler [28]). Those approaches are worth investigating and may be used for further development. However, there is a risk of vendor lock-in, as the solutions of different vendors are not fully compatible.

Except for the problems related to the algorithm development, FPGAs also have certain hardware-related limitations. First of all the scalability of FPGA based designs is much worse, than in case of software solutions. If the algorithm fully utilizes resources provided by the FPGA chip, the only solution to add new functionality is to modify the hardware replacing the FPGA with a bigger (and more expensive) one or splitting the functionality between two FPGAs. Unfortunately, dividing the processing between a few FPGAs is also somewhat complicated, because the communication between different chips usually becomes a bottleneck. Therefore FPGAs are perfect for local processing of data, like filtering and triggering of data in individual GEM channels or concentrating the data from multiple channels into a single stream transmitted to the ES. Their performance is usually much worse when used for processing huge volumes of data (like correlating pulses in X and Y strips to reconstruct the 2D image).

Communication Between the FPGA and the ES Parts

Transmission of data from the FPGA part to the ES part of the system significantly affects the performance of the whole system. The communication channel should provide high throughput and (especially in case of real-time processing) low latency. The optimal technology may depend on the mechanical structure of the system. If the FPGA part is directly connected to the ES part, the optimal solution is the PCIe bus. However, if the FPGA part must be separated from the ES part, it may be reasonable to utilize the network interfaces.

PCIe as the FPGA-ES Communication Bus

Current FPGA chips offer high-speed PCIe interfaces. For example the chips from the newest Xilinx Ultrascale+ family [29] offer either x8 Gen4 or x16 Gen3 PCIe interfaces with throughput up to 16GB/s. Even the older chips, like the Artix 7, used in the current version of the GEM readout for WEST offer the x4 Gen2 PCIe with throughput up to 2 GB/s. Those interfaces are supported with dedicated IP cores equipped with the bus mastering DMA functionality. An excellent example of such IP core may be the DMA for PCI Express Subsystem [30].

However, current experiences show that even though PCIe may be the main interface for the high-performance data transmission, it should be supported with additional highly-reliable interfaces for FPGA configuration.

Network Interfaces for FPGA-ES Communication

The maximum length of the PCIe cable is 7 m, but for high-speed versions like Gen3 or Gen4, it is advisable to avoid cables longer than 1 m. Therefore, for longer connections, it is required to use either proprietary communication interfaces (which may be expensive) or standard network interfaces.[1] The network connection may be a perfect solution if due to high required computational power the ES part must be implemented as a computer farm. Such solution is rather not needed for currently built GEM detectors but is often used for big detectors in HEP experiments. The perfect solution for that purpose may be the Infiniband interface, which offers low latency transfer with RDMA. Unfortunately, there are no widely available, reasonably priced implementations of Infiniband IP core for FPGAs. The full implementation of the standard TCP/IP protocol consumes a lot of resources in the FPGA. Therefore other non-standard or simplified protocols are used [31, 32]. The network interface may also be used to control the FPGA part, for example via the IPbus [33] protocol.

Acceleration of Computations in the ES Part

The High-Performance Computing (HPC) is widely used in modern information technology. Therefore different accelerating technologies are available, with the price significantly reduced thanks to the mass production. Our team has performed a thorough review of possible solutions, and some of them have also been tested with the GEM processing algorithms. In [34, 35], the following approaches were investigated

- Using the additional Intel PCIe card with Xeon Phi MIC
- Using the graphical card with CUDA-compatible NVIDIA graphics card (GPU)
- Using the multicore Intel Xeon CPU

Extension of the ES part with additional cards (the first two points above) provides a certain amount of flexibility and scalability, because additional cards may be added as long as free extension slots (e.g., PCIe) are available. Unfortunately, this approach requires transmission of processed data to the accelerator card and reception of results. This may significantly load the communication bus. Therefore that approach is applicable only if the data may be split into blocks processed independently. The newest version of the Xeon Phi technology, formerly known as "Knights

[1] Use of proprietary communication interfaces usually requires FPGA-based boards at both sides of the link which increases the cost. In case of standard network interfaces, typical network interface cards may be used at the ES side of the link.

Landing" may be used as a standard CPU, avoiding the problems related to the limited PCIe bandwidth.

The tests [34–36] have shown, that to fully utilize the acceleration potential offered by the GPU or the vector instructions in the CPU, it is required that the algorithm processes regularly distributed data and that certain data are reused multiple times. All irregularities in the processing algorithm (e.g. branches) or in the data location impair the acceleration. In case of GPUs with SIMT (Single Instructions Multiple Threads), it is advisable that multiple (e.g., 32) threads are processing data in parallel. Therefore acceleration may be easily achieved in histogramming but is much more difficult in other parts of algorithms like separation of pulses. Additionally, implementation of the GEM signal processing with GPUs requires a special programming approach.

Summarizing, if the aim is not only the maximal performance but also the ease of development and maintenance, the multicore server processor like Intel Xeon may be a preferred choice.

Current Results for the WEST System

In the current version of GEM system for WEST two Artix-7 FPGA chips are connected to the PCIe switch. According to the information from Sect. 6.2.1, they should provide throughput of 4 GB/s. However, due to protocol overhead and internal bus inefficiencies, the measured throughput was 2.36 GB/s. This is 8.6 times less than the 20.48 GB/s data bandwidth produced by the detector (see Sect. 5.3). The Monte Carlo simulations have shown, that the optimized trigger system implemented in FPGA may reduce the data stream by a factor of 8.6 for photon rates below 220 thousand of photons per detector strip per second. That gives the limit of 28.16 million of photons per second for the whole detector. Further limitations are introduced by the ES part. Latest analyzes [37, 38] have shown that a system based on Xeon e5 S2600 v4 CPU with 10 cores is able to process up to 6.9 million photons per second. However, the system is still under development, and further improvement both regarding the PCIe bandwidth and processing throughput is expected.

Conclusion

Modern electronics provides technologies suitable for high-performance, digital readout of GEM detectors. Data may be digitized at a high sampling frequency and delivered to the FPGA-based preprocessing system. That allows both efficient data filtering in the standard mode of operation, but also a possibility to access raw detector signals for diagnostic purposes.

The system architecture may be selected depending on the requirements (e.g., pulsed or continuous operation, a possibility of off-line processing or the necessity of soft real-time or even hard real-time processing).

The processing tasks may be divided between the FPGA part (for time-deterministic tasks or simple and highly parallel tasks) and the processor-based ES part of the system (for more complex tasks using the huge volumes of data).

With the high-speed processors available today it is possible to increase the acceptable event-rate of the detector by splitting of overlapping events.

Usage of FPGA and ES technologies allows creating a highly flexible system, that may be tuned to the particular needs by modification of the software and firmware according to the specific requirements. [39–43]

Acknowledgements This scientific work was partly supported by Polish Ministry of Science and Higher Education within the framework of the scientific financial resources in the years 2016 and 2017 allocated for the realization of the international cofinanced Project—Agreement Nos. 332380/PnH/2016 and 3814/H2020-Euratom/2017/2.

References

1. G. Albani, G. Croci, C. Cazzaniga, M. Cavenago, G. Claps, A. Muraro, F. Murtas, R. Pasqualotto, E.P. Cippo, M. Rebai, M. Tardocchi, G. Gorini, Neutron beam imaging with GEM detectors. J. Instrum. **10**(04), C04040–C04040 (2015). https://doi.org/10.1088/1748-0221/10/04/C04040
2. F. Sauli, GEM: a new concept for electron amplification in gas detectors. Nucl. Instrum. Methods Phys. Res. A **386**(2–3), 531–534 (1997). https://doi.org/10.1016/S0168-9002(96)01172-2
3. F. Sauli, The gas electron multiplier (GEM): operating principles and applications. Nucl. Instrum. Methods Phys. Res. A **805**, 2–24 (2016). https://doi.org/10.1016/j.nima.2015.07.060. (special issue in memory of Glenn F. Knoll)
4. S. Bachmann, A. Bressan, M. Capeáns, M. Deutel, S. Kappler, B. Ketzer, A. Polouektov, L. Ropelewski, F. Sauli, E. Schulte, L. Shekhtman, A. Sokolov, Discharge studies and prevention in the gas electron multiplier (GEM). Nucl. Instrum. Methods Phys. Res. A **479**(2–3), 294–308 (2002). https://doi.org/10.1016/S0168-9002(01)00931-7
5. Y. Takeuchi, T. Tamagawa, F. Asami, A. Yoshikawa, T. Iwahashi, S. Konami, W. Iwakiri, Signal shape and charge sharing between electrodes of GEM in dimethyl ether. J. Instrum. **7**(03), C03042–C03042 (2012). https://doi.org/10.1088/1748-0221/7/03/C03042
6. G. Guedes, A. Breskin, R. Chechik, D. Mörmann, Effects of the induction-gap parameters on the signal in a double-GEM detector. Nucl. Instrum. Methods Phys. Res. A **497**(2–3), 305–313 (2003). https://doi.org/10.1016/S0168-9002(02)01924-1

7. F. Sauli, T. Meinschad, L. Musa, L. Ropelewski, Photon detection and localization with GEM. IEEE Nucl. Sci. Symp. **1**, 12–15 (2004). https://doi.org/10.1109/NSSMIC.2004.1462059
8. A. Wojenski, G. Kasprowicz, K.T. Poźniak, R. Romaniuk, Data acquisition methods for GEM detectors. Proc. SPIE **9290**, 9290–9290 (2014). https://doi.org/10.1117/12.2074970
9. W.M. Zabołotny, T. Czarski, M. Chernyshova, H. Czyrkowski, R. Dąbrowski, W. Dominik, K. Jakubowska, L. Karpiński, G. Kasprowicz, K. Kierzkowski, I.M. Kudła, K. Poźniak, J. Rzadkiewicz, Z. Sałapa, M. Scholz, Optimization of FPGA processing of GEM detector signal. Proc. SPIE **8008**, 8008–8008 (2011). https://doi.org/10.1117/12.905427
10. J. Rzadkiewicz, W. Dominik, M. Scholz, M. Chernyshova, T. Czarski, H. Czyrkowski, R. Dabrowski, K. Jakubowska, L. Karpinski, G. Kasprowicz, K. Kierzkowski, K. Pozniak, Z. Salapa, W. Zabolotny, P. Blanchard, S. Tyrrell, K.D. Zastrow, Design of T-GEM detectors for X-ray diagnostics on JET. Nucl. Instrum. Methods Phys. Res. A **720**, 36–38 (2013). https://doi.org/10.1016/j.nima.2012.12.041
11. K.T. Poźniak, A. Byszuk, M. Chernyshova, R. Cieszewski, T. Czarski, W. Dominik, K. Jakubowska, G. Kasprowicz, J. Rzadkiewicz, M. Scholz, W. Zabolotny, FPGA based charge fast histogramming for GEM detector. In: R.S. Romaniuk (ed.) Proceedings of SPIE, vol. 8903 (Wilga, Poland, 2013), p. 89032F. https://doi.org/10.1117/12.2037047
12. W.M. Zabołotny, A. Byszuk, M. Chernyshova, R. Cieszewski, T. Czarski, W. Dominik, K.L. Jakubowska, G. Kasprowicz, K. Poźniak, J. Rzadkiewicz, M. Scholz, Embedded controller for GEM detector readout system. Proc. SPIE **8903**, 8903–8903 (2013). https://doi.org/10.1117/12.2033281
13. W.M. Zabołotny, A. Byszuk, M. Chernyshova, R. Cieszewski, T. Czarski, S. Dalley, C. Hogben, K.L. Jakubowska, G. Kasprowicz, K. Poźniak, J. Rzadkiewicz, M. Scholz, A. Shumack, Python based integration of GEM detector electronics with JET data acquisition system. Proc. SPIE **9290**, 929024 (2014). https://doi.org/10.1117/12.2073379
14. M. Chernyshova, K. Malinowski, T. Czarski, A. Wojeński, D. Vezinet, K.T. Poźniak, G. Kasprowicz, D. Mazon, A. Jardin, A. Herrmann, E. Kowalska-Strzęciwilk, R. Krawczyk, P. Kolasiński, W. Zabołotny, P. Zienkiewicz, ASDEX Upgrade Team, Eurofusion MST1 Team: Gaseous electron multiplier-based soft X-ray plasma diagnostics development: preliminary tests at ASDEX Upgrade. Rev. Sci. Instrum. **87**(11), 11E325 (2016). https://doi.org/10.1063/1.4960305
15. P. Kolasinski, K.T. Pozniak, T. Czarski, M. Linczuk, A. Byszuk, M. Chernyshova, B. Juszczyk, G. Kasprowicz, A. Wojenski, W. Zabolotny, P. Zienkiewicz, D. Mazon, P. Malard, A. Herrmann, D. Vezinet, Serial data acquisition for GEM-2D detector. Proc. SPIE **9290**, 9290–9290 (2014). https://doi.org/10.1117/12.2075911
16. P. Kolasinski, K.T. Pozniak, T. Czarski, A. Byszuk, M. Chernyshova, G. Kasprowicz, R.D. Krawczyk, A. Wojenski, W. Zabolotny, Fast data transmission from serial data acquisition for the GEM detector system. In: R.S. Romaniuk (ed.) Proc. SPIE, vol. 9662 (Wilga, Poland, 2015), p. 96622J. https://doi.org/10.1117/12.2205264
17. T. Czarski, M. Chernyshova, K. Pozniak, G. Kasprowicz, W. Zabolotny, P. Kolasinski, R. Krawczyk, A. Wojenski, P. Zienkiewicz, Serial data acquisition for the X-ray plasma diagnostics with selected GEM detector structures. J. Instrum. **10**(10), P10013–P10013 (2015). https://doi.org/10.1088/1748-0221/10/10/P10013
18. T. Czarski, K.T. Pozniak, M. Chernyshova, K. Malinowski, G. Kasprowicz, P. Kolasinski, R. Krawczyk, A. Wojenski, W. Zabolotny, On line separation of overlapped signals from multi-time photons for the GEM-based detection system. In: R.S. Romaniuk (ed.) Proceedings of SPIE, vol. 9662 (Wilga, Poland, 2015), p. 96622W. https://doi.org/10.1117/12.2205804
19. M. Chernyshova, T. Czarski, K. Malinowski, E. Kowalska-Strzęciwilk, J. Król, K.T. Poźniak, G. Kasprowicz, W. Zabołotny, A. Wojeński, R.D. Krawczyk, P. Kolasiński, I.N. Demchenko, Y. Melikhov, Development of GEM detector for tokamak SXR tomography system: preliminary laboratory tests. Fusion Eng. Design (2017). https://doi.org/10.1016/j.fusengdes.2017.03.107
20. W.M. Zabołotny, DMA implementations for FPGA-based data acquisition systems. p. 1044548 (2017). https://doi.org/10.1117/12.2280937
21. P. Linczuk, W.M. Zabolotny, A. Wojenski, R.D. Krawczyk, K.T. Pozniak, M. Chernyshova, T. Czarski, M. Gaska, G. Kasprowicz, E. Kowalska-Strzeciwilk, K. Malinowski, Evaluation of FPGA to PC feedback loop. p. 104454B (2017). https://doi.org/10.1117/12.2280947
22. What every software programmer needs to understand about hardware design. https://www.nandland.com/articles/what-software-programmers-need-to-understand.html. Accessed 10 June 2018
23. D. Koch, F. Hannig, D. Ziener, (eds.): FPGAs for Software Programmers. Springer, Cham (2016). https://doi.org/10.1007/978-3-319-26408-0
24. International workshop on FPGAs for software programmers. http://www.fsp-workshop.org. Accessed 10 June 2018
25. SystemC. http://www.accellera.org/downloads/standards/systemc. Accessed 10 June 2018
26. Vivado High-Level Synthesis. https://www.xilinx.com/products/design-tools/vivado/integration/esl-design.html. Accessed 10 June 2018
27. Intel FPGA SDK for OpenCL - Overview. https://www.altera.com/products/design-software/embedded-software-developers/opencl/overview.html. Accessed 10 June 2018
28. Intel®HLS Compiler, Accelerate FPGA development with C++ using Intel HLS Compiler. https://www.altera.com/products/design-software/high-level-design/intel-hls-compiler/overview.html. Accessed 10 June 2018
29. PCI Express (PCIe). https://www.xilinx.com/products/technology/pci-express.html. Accessed 10 June 2018
30. DMA for PCI express (PCIe) subsystem. https://www.xilinx.com/products/intellectual-property/pcie-dma.html. Accessed 10 June 2018
31. G. Bauer, T. Bawej, U. Behrens, J. Branson, 10 Gbps TCP/IP streams from the FPGA for high energy physics. J. Phys. Conf. Ser. **513**(1), 012–042 (2014)
32. W. Zabolotny, Low latency protocol for transmission of measurement data from FPGA to Linux computer via 10 gbps Ethernet link. J. Instrum. **10**(07), T07005 (2015)
33. IPbus. https://ipbus.web.cern.ch/ipbus/. Accessed 10 June 2018
34. R. Krawczyk, P. Linczuk, P. Kolasinski, A. Wojenski, G. Kasprowicz, K. Pozniak, R. Romaniuk, W. Zabolotny, P. Zienkiewicz, T. Czarski, M. Chernyshova, The speedup analysis in GEM detector based acquisition system algorithms with CPU and PCIe Cards. Acta Phys. Pol. B **9**(2), 257 (2016). https://doi.org/10.5506/APhysPolBSupp.9.257
35. P. Linczuk, R.D. Krawczyk, K.T. Poźniak, G. Kasprowicz, A. Wojeński, M. Chernyshova, T. Czarski, Algorithm for fast event parameters estimation on GEM acquired data. p. 100313M (2016). https://doi.org/10.1117/12.2249058
36. P. Linczuk, R.D. Krawczyk, W. Zabolotny, A. Wojenski, P. Kolasinski, K.T. Pozniak, G. Kasprowicz, M. Chernyshova, T. Czarski, Feasibility of FPGA to HPC computation migration of plasma impurities diagnostic algorithms. Int. J. Electron. Telecommun. **63**(3), 323–328 (2017). https://doi.org/10.1515/eletel-2017-0043

37. R.D. Krawczyk, T. Czarski, P. Linczuk, A. Wojeński, P. Kolasiński, M. Gąska, M. Chernyshova, D. Mazon, A. Jardin, P. Malard, K. Poźniak, G. Kasprowicz, W. Zabołotny, E. Kowalska-Strzęciwilk, K. Malinowski, The software-defined fast post-processing for GEM soft X-ray diagnostics in the tungsten environment in steady-state tokamak thermal fusion reactor. Rev. Sci. Instrum. **89**(6), 063504 (2018). https://doi.org/10.1063/1.5022131
38. R. Krawczyk, P. Linczuk, A. Wojeński, K. Poźniak, G. Kasprowicz, W. Zabołotny, M. Gaska, D. Mazon, A. Jardin, T. Czarski, P. Kolasiński, M. Chernyshova, E. Kowalska-Strzęciwilk, K. Malinowski, Novel application of parallel computing techniques in soft x-rays plasma measurement systems for the WEST experimental thermal fusion reactor. In: 2018 26th Euromicro International Conference on Parallel, Distributed and Network-based Processing (PDP) (IEEE, 2018), pp. 118–125. https://doi.org/10.1109/PDP2018.2018.00024. https://ieeexplore.ieee.org/document/8374447/
39. A.F. Buzulutskov, Radiation detectors based on gas electron multipliers (review). Instrum. Exp. Tech. **50**(3), 287–310 (2007). https://doi.org/10.1134/S0020441207030013
40. C. Altunbas, M. Capéans, K. Dehmelt, J. Ehlers, J. Friedrich, I. Konorov, A. Gandi, S. Kappler, B. Ketzer, R. De Oliveira, S. Paul, A. Placci, L. Ropelewski, F. Sauli, F. Simon, M. van Stenis, Construction, test and commissioning of the triple-GEM tracking detector for COMPASS. Nucl. Instrum. Methods Phys. Res. A **490**(1–2), 177–203 (2002). https://doi.org/10.1016/S0168-9002(02)00910-5
41. T. Czarski, M. Chernyshova, K. Malinowski, K.T. Pozniak, G. Kasprowicz, P. Kolasinski, R. Krawczyk, A. Wojenski, W. Zabolotny, The cluster charge identification in the gem detector for fusion plasma imaging by soft X-ray diagnostics. Rev. Sci. Instrum. **87**(11), 11E336 (2016). https://doi.org/10.1063/1.4961559
42. A. Wojenski, K.T. Pozniak, G. Kasprowicz, P. Kolasinski, R. Krawczyk, W. Zabolotny, M. Chernyshova, T. Czarski, K. Malinowski, FPGA-based GEM detector signal acquisition for SXR spectroscopy system. J. Instrum. **11**(11), C11035 (2016)
43. A. Nowak, G. Bitzes, A. Dotti, A. Lazzaro, S. Jarp, P. Szostek, L. Valsan, M. Botezatu, J. Leduc, Does the Intel Xeon Phi processor fit HEP workloads? J. Phys. Conf. Ser. **513**(5), 052024 (2014). https://doi.org/10.1088/1742-6596/513/5/052024

Journal of Fusion Energy (2019) 38:490–498
https://doi.org/10.1007/s10894-018-0194-x

ORIGINAL RESEARCH

Evolution of the Pinched Column During Hard X-ray and Neutron Emission in a Dense Plasma Focus

P. Kubes[1] · M. Paduch[2] · M. J. Sadowski[2,3] · J. Cikhardt[1] · B. Cikhardtova[1] · D. Klir[1] · J. Kravarik[1] · V. Munzar[1] · K. Rezac[1] · E. Zielinska[2] · E. Skladnik-Sadowska[3] · A. Szymaszek[2] · K. Tomaszewski[4] · D. Zaloga[3]

Published online: 8 October 2018

Abstract
The paper summarizes important results of the recent experimental studies performed on the plasma-focus PF-1000 facility operated in Warsaw, Poland, mainly with the pure deuterium filling. Attention is focused on the evolution of toroidal and plasmoidal self-organized structures formed by internal closed currents inside the dense plasma column. The production of hard x-rays and neutrons corresponds with the formation and decay of plasmoids, in which charged particles can be accelerated effectively to high energies, during a release of the magnetic energy from current filaments of high energy density. It is noticed that the studies of laboratory fusion and cosmic plasmas deal with similar problems, e.g., the fast release of the magnetic energy in a form of high-energy charged particle beams.

Keywords Plasma focus · Neutron source · Magnetic reconnection · Self-generation

Introduction

Research on fusion plasmas produced in tokamaks or during interactions of powerful lasers with solid targets, as well as studies of z-pinch discharges and astrophysical jets—tries to solve a similar problem, namely the generation of high energy electrons and ions by mechanisms which have not been explained so far completely. Plasmas generated by plasma focus (PF) discharges have some advantages which can facilitate the solving of this problem, because they have parameters convenient for complementary diagnostics with good temporal-, spatial-, and energetic-resolution. When the deuterium filling is applied, the primary deuterons can reach energies sufficient for efficient fusion DD reactions, which can generate fast neutrons. These fusion-produced neutrons, together with hard x-rays (HXR) produced by fast electrons, provide an indirect information about the charged particles accelerated in plasma.

The comprehensive diagnostics applied in PF experiments have made it possible: (1) to determine instants of the generation and energies of the accelerated charged particles, (2) to observe organized structures of plasma and their spatial- and temporal-evolution, and (3) to describe their correlations with the HXR and fusion-neutron production.

In discharges of the Z-pinch type, as well as in PF-type experiments, the current of MA intensity, which flows through the plasma along the z-axis, forms its own magnetic field, which compresses plasma, usually with a relatively good axial symmetry. The plasma column of dimensions ranging several centimetres is usually compressed during hundreds nanosecond, reaching an electron temperature above (100–200) eV, and electron density above 10^{24} m^{-3}. The compression of a current sheath is accompanied by the conversion of a part of the kinetic energy into magnetic one associated with closed currents. This magnetic field is later released in a form of fast

✉ P. Kubes
kubes@fel.cvut.cz

1 Czech Technical University, 166-27 Prague, Czech Republic

2 Institute of Plasma Physics and Laser Microfusion, 01-497 Warsaw, Poland

3 National Centre for Nuclear Research, 05-400 Otwock-Świerk, Poland

4 ACS Ltd, 01-497 Warsaw, Poland

electrons and ions which can reach observed energy of hundreds keV [1–4].

Therefore, the main aim of this paper was to analyze evolution of organized structures, formed by closed currents with poloidal and toroidal components, and corresponding magnetic fields in the time of the HXRs and neutrons production. It should shed some light on specific features of the plasma transformation processes and mechanisms of the charged particle acceleration.

Experimental findings, concerning the release of the magnetic energy in the organized plasma structures in PF-discharges, can be an inspiration for researchers of tokamaks and laser-fusion experiments, as well as for the astrophysical community.

Experimental Set-Up and Diagnostics

The described studies have concerned plasma produced within the PF-1000 facility operated at the IPPLM in Warsaw, Poland [5–12]. The investigated discharges, as reported in this paper, were carried out at the deuterium filling. Plasmas had electron densities transparent for interferometric measurements in the visible wavelength region. The plasma transformation velocity, equal to $(1–2) \times 10^5$ m/s, made it possible to perform a detailed diagnostics with a temporal resolution of the order of 3 ns, and the spatial resolution of the order of 1 mm. A laser interferometric system enabled plasma internal structures to be observed, because it could record 15 frames from each shot, during the plasma evolution from the implosion phase to its decay [7].

Energy stored in a capacitor bank amounted to (250–350) kJ, and the peak current reached (1–2) MA. Time-resolved soft X-ray (SXR) signals were investigated with a filtered silicon PIN detector, which recorded photons of energy above 700 eV. Three scintillation detectors recorded the hard X-ray (HXR) and neutron signals lasting several tens ns, when measured side-on and in the both axial directions, at a distance of about 7 m from the pinch centre. Those detectors made it possible to determine instants of the emission of HXRs and fusion-produced neutrons, as well as to estimate their anisotropy and a mean neutron velocity, using a time-of-flight method [8]. The interferometric frames were recorded subsequently with a delay equal to 10 ns or 20 ns, during a period lasting 210 ns [9]. The applied laser interferometer used parallel mirrors in the Mach–Zehnder configuration without any wedge, and for shots without plasma it formed an image with widely distributed interferometric fringes. For shots with plasma, the closed dense fringes indicated regions of the density extremes. In some shots, absolute calibrated magnetic probes were placed in the anode front-plate at different radial distances from the z-axis, and recorded the azimuthal and axial components of the magnetic field [10, 11]. Electron temperature values were estimated from intensities of spectral lines which were emitted by H- and He-like ions of impurities C, N, and O, and recorded by means of an EUV spectrometer [12].

The presented results summarize results of 16 campaigns, which were performed during recent 8 years, when more than 1000 discharges were investigated. For the analysis of results we selected the shots of typical features and regularity.

Characteristics of the Recorded Fusion Neutrons

Particular attention was focused first on the studies of HRXs and fusion-neutrons. Typical HXRs and neutron signals are shown in Fig. 1.

The recorded waveforms showed usually two HXR peaks and two neutron signals. The beginnings of the these neutron signals (N1 and N2) corresponded to the beginnings of the corresponding HXR signals. The scintillation detectors, which were placed at the distance of 7 m from the pinch centre, recorded the fusion-produced neutrons (of an average energy equal to 2.45 MeV) with a delay of 300 ns. The first (HXR1) and neutron pulses (N1) were emitted at the stopping of the current sheath implosion, while the next pulses (HXR2 and N2) were produced during and after an interruption of the dense pinch column, which could occur several times.

The neutron signal, which was recorded in the downstream direction (i.e. along the z-axis), had the maximum delayed by less than 300 ns. It was caused by motion of fast deuterons (responsible for the neutrons production) in

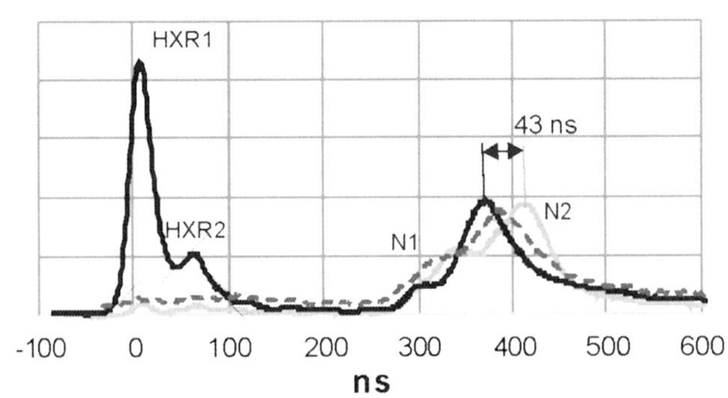

Fig. 1 Shot #8600. Two pulses of HXRs (HXR1, HXR2) and two neutron signals (N1, N2) recorded with the scintillation detectors placed at a distance of 7 m in front of the electrodes outlets in the downstream direction (black), behind the electrodes collector, i.e. in the upstream direction (shadow), and side-on (dashed). The shift marked by arrows shows the difference between the subsequent neutron peaks of the axial signals

the downstream direction mainly. The differences between the both neutron peaks, as measured in the downstream and upstream directions, were about (40–50) ns and corresponded to the mean deuteron velocity of (3–4.5) $\times$ 10^6 m/s and deuteron energy equal to about (100–200) keV.

Taking into consideration the observed velocity of deuterons and their anisotropy, one could deduce that the neutrons were produced by the beam-target mechanism. The HXRs pulses were produced by fast electrons, which bombarded the anode-surface and had a similar energy distribution as the deuterons, but were emitted mainly in the upstream direction. The start of the HXR pulse after the electrons acceleration was estimated to be about 2 ns (for 50 cm flight to the anode). The number of fast deuterons needed for the total neutron yield of the order of 10^{10}–10^{11} per shot was calculated on the basis of the known cross-section of DD reaction, the mean energy of deuterons (equal to about 200 keV), the mean ion density (about 3×10^{24} m^{-3}) and the assumed dimensions of the target. It amounted to about 1% of all deuterons contained inside the pinch column. Their energy was evaluated to be a few percentages of energy stored in the condenser bank [13].

Scattered neutrons, which were estimated to amount to about 20% of the total neutron yield [8], affected the tail of the neutron signals mainly. The total neutron yield changed from shot to shot within the range of one order of magnitude. This variation was caused by stochastic differences in the evolution of an individual shot from the breakdown to the pinch phase, probably due to some azimuthal asymmetry of the current sheath. The neutron signals and the total neutron yield could also be affected by the electrode configuration. For example, the neutron yield was considerably increased in shots performed after the fixing of an additional conical tip upon the anode end-plate. On the other hand, the neutron yield was decreased when an anti-anode was applied, despite of a higher energy density in the pinched column. The described changes in the electrode configuration, which were realized during (50–70) shots, did not change the main features of the neutron emission considerably.

It should be added that the fusion neutrons could also be produced in a rare and cold deuterium gas inside the discharge chamber. Their contribution was evaluated to amount to about 10% of the total neutron yield. The emission of such neutrons (at a distance of the order of 2 m from the plasma-focus centre) could be delayed by hundreds ns, and it (together with the scattered neutrons) could be responsible for long tails of time-resolved neutron signals.

Organized Structures in the Dense Plasma Column

The applied laser interferometry system made it possible to estimate distributions of the electron density in the observed quasi-symmetrical toroidal and plasmoidal plasma structures, and to correlate their evolution with instants of the HXR and neutron emission. As an example one can concern two interferometric images obtained from shot #11452, which are shown in Fig. 2.

The presented interferometric frames show typical toroidal (Fig. 2a) and plasmoidal (Fig. 2b) structures. The radial distribution of the electron density inside the toroidal structure, as calculated by means of the Abel transformation of the recorded interferometric fringes in the chosen cross-section, reached the maximal values of about 1.2×10^{24} m^{-3} at the axis of the toroidal structure, and 3.5×10^{24} m^{-3} at the centre of the plasmoidal structure, respectively.

The electron temperature was estimated from a ratio of H- and He-like Lyman α-lines of C-, N- and O-ions. It reached (50–150) eV [12]. The value above 100 eV was also obtained at the instance of the peak of SXRs recorded by means of a PIN detector. The known plasma density and electron temperature values made it possible to estimate that a relaxation time of the electron and ion temperatures [4] was about 20 ns. Hence, one can assume that (at slower transformations) the conditions inside the considered plasma structures were quasi-stationary during several tens nanoseconds.

The known radial distribution of the electron density n_e, which was determined in two cross-sections (marked by white dashed lines in Fig. 2a, b), and the evaluated electron temperatures T_e (on average about 70 eV) made it possible to estimate the plasma pressure in the toroidal and poloidal structures, as shown by grey and black curves in Fig. 2c, respectively.

The smooth shapes of the interferometric fringes and pressure gradient mentioned above could also indicate that the organized structure were created by closed internal currents. The formation of such organized structures was possible due to a magnetic dynamo effect in discharges with high Reynolds magnetic numbers, characterizing a strong magnetic field frozen in plasma, and ranging to 10^3. It showed the dominant role of an advection term in a comparison with a diffusion term for the magnetic field. For the whole pinch column the diffusion time was calculated to be of the order of 100 μs, i.e. much longer than the duration of the PF discharge. That value decreased, however, rapidly with the plasma dimensions and it reached about 10 ns at the dimension about 100 μm, i.e. much larger than the Debye length (equal to about 10^{-8} m).

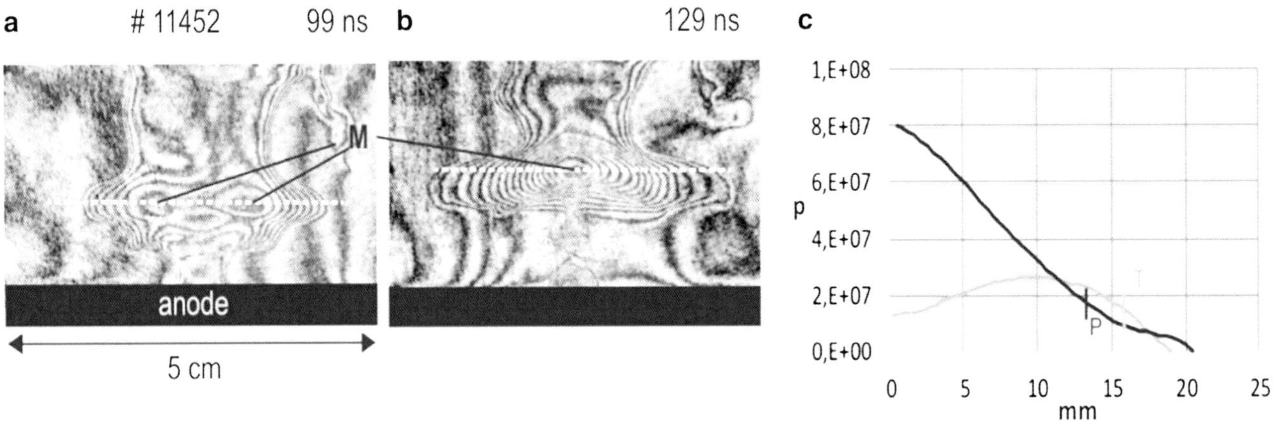

Fig. 2 Shot #11452; **a** Interferometric images of the toroidal structure recorded at 99 ns, **b** the plasmoidal structure recorded 30 ns later, and **c** a radial distribution of the plasma pressure p (in Pa units), as calculated for temperature of 70 eV along the cross-sections lines marked in the images **a**, **b**. The letter M and black lines indicate positions of the density maxima, and the letters P and T show the boundary radii of the plasmoidal and toroidal structures, respectively

The dimension of 100 μm corresponded as to the Larmor radius, as well as to a mean free path for electron–ion collisions, and to a deuteron skin layer. Consequently, it determined the spatial scale of microturbulences and collision-free plasma. It should be noted that the anisotropy of internal structures could also be influenced by high magnetization of electrons (ranging 20). In contrary, the magnetization of deuterons reached the value equal to about 1 only [4].

A high gradient of the plasma pressure in the considered organized structures was probably produced by the closed currents and their magnetic fields [7]. The dominant current inside the toroidal plasma structure could flow in the toroidal direction, having a small poloidal stabilizing component, similar as in the toroidal pinch or tokamak discharge. The current inside the plasmoidal structure should have the poloidal form, stabilized by a smaller toroidal component. At considering quasi-stationary conditions, the plasma pressure in the ordered structures was evidently compensated by the pinching pressure of the current flowing through the plasma column surface in an agreement with the virial theorem [14]. Hence, the pressure values (determined in points T and P shown in Fig. 2c) made it possible to estimate that the magnetic field B ranged up to 10 T, and the internal current I was of the order of hundred kA [15]. For those estimations the use was made of the known equations:

$$2n_e kT_e \approx B^2/2\mu \quad \text{and} \quad B \approx \mu I/2\pi r \tag{1}$$

where r means the radius and μ permeability. That value of the magnetic field was consistent with results of the experimental measurements performed by means of the absolute calibrated magnetic probes located within the anode end-plate [10, 11, 16]. Those measurements (carried out during 25 shots) showed the presence of a weak axial component of the magnetic field, which was generated spontaneously in front and in the imploding current sheath, as well as in the internal plasma structures.

Transformations of Organized Structures and Acceleration of Charged Particles

The interferometric images (shown in Fig. 2a, b) presented two basic plasma structures (toroidal and plasmoidal) which were observed during the same shot (with a time delay of 30 ns), and constituted two steps of the internal structures evolution. In this paragraph, we described the evolution of the considered internal structures, especially during the periods of the fast electrons and ions acceleration [17].

Many interferometric images, which were recorded during the plasma radial implosion, showed usually the formation of two toroidal structures, which had some radial lobules in front of the imploding current sheath, at different positions along the z-axis. Some examples of such interferometric frames, are presented in Fig. 3.

The toroidal plasma structures were produced by a concentration of an initially smoothly-distributed azimuthal current, which was flowing in the column surface layer, into toroidal-like shape due to a tearing instability. Its larger radial radius in form of the lobules was caused by a repulsive pressure of the internal toroidal current, which limited the radial implosion. The upper lobule (L1 in Fig. 3b, c) separated a region of the dense plasma column from the umbrella-shaped part of the imploding current sheath. The top of that lobule showed also some transfer of denser plasma from the pinch column to the external low-density region, what could be deduced from shapes of the interferometric fringes visible behind the lobule tops (L1

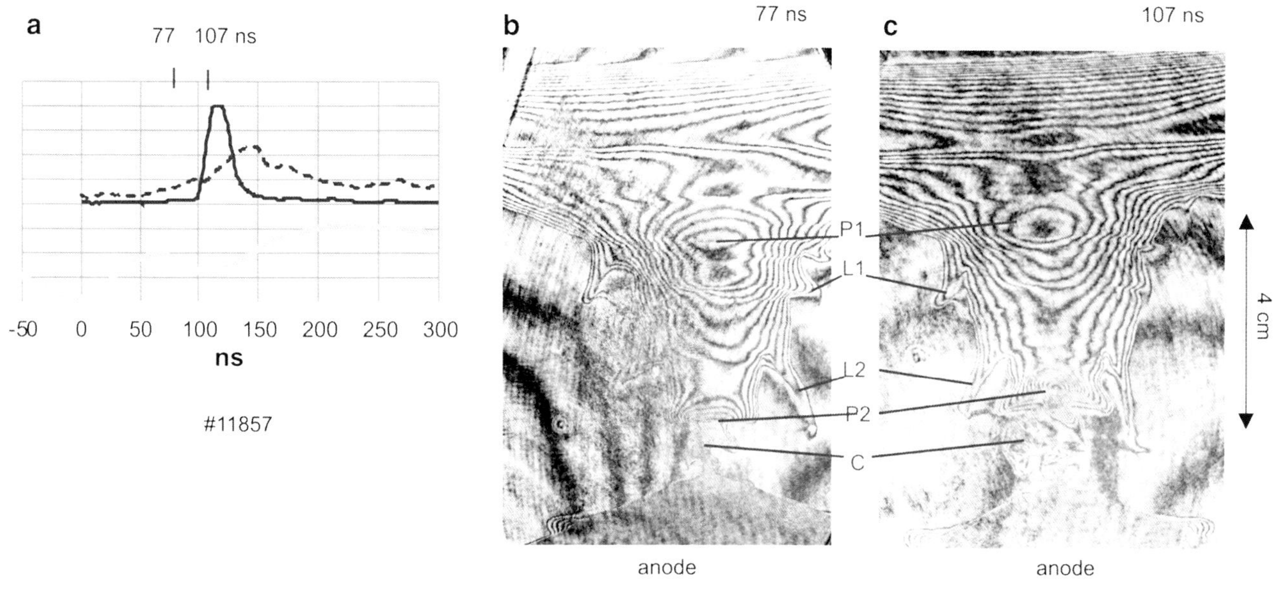

Fig. 3 Shot #11857; **a** Waveforms of the current derivative (thick grey), HXRs (thin black) and fusion-neutrons (dashed) as a function of time. Interferometric images were recorded at instants marked above the waveforms. The interferometric frame **b** was taken before the HXR and neutron pulses, and an interruption of the pinch constriction, while the frame **c** was recorded at the explosion of the constriction and production of HXRs and neutrons. The symbols P1 and P2 indicate the plasmoids, L1 and L2 the lobules, and C the constriction

and L2). The first plasmoid (P1 in Fig. 3b, c) was generated in the z-position of the L1 lobule. It confined plasma escaping in the axial direction from the bottom region, in the period of the implosion stopping.

The formation of the first plasmoid P1 coincided usually with the production of the first HXR and neutron pulses. Their production could be caused by energy released during the spontaneous transformation of internal magnetic fields at reconnections of the magnetic lines.

The formation of the first plasmoid P1 in the pinch column coincided usually with the production of the first HXR and neutron pulses. Their production could be caused by energy released during the spontaneous transformation of internal magnetic fields at the reconnection of magnetic lines. It should be noted that the formation of the first plasmoid must not necessarily be accompanied by the HXR and neutron emission. For example, it does not occur at the presence of a conical tip upon the anode end-plate. During such discharges the plasmoid is usually characterized by smaller density gradients, and internal energy is confined inside the pinch column of a higher density for a longer time, until an instability is developed (see Fig. 3) [31]. It should also be mentioned that the HXR and neutron pulses were not recorded when the dense plasma structures were absent.

After the formation of the first plasmoid, the stagnation phase was started, in which the internal part of the dense plasma column with sparse interferometric fringes (between the plasmoid and the anode) transported its mass to the plasmoid. This phase was characterized both, by the radial expansion and by absence of internal gradients of a plasma density. They manifested the presence of an axial component of the magnetic field inside the dense column, which was associated with the azimuthal current flowing in the column boundary layer. The stagnation phase was ended after a few tens of ns by the stratification, when the initially smooth surface of the pinch column was subject to an evolution of the Reyleigh-Taylor instability, mainly due to the tearing instability accompanied by an axial fragmentation of the azimuthal current. Circumstances of the cascading of plasma structures to smaller dimensions were enabled by the mentioned submillimeter value of Larmor radius and mean free path of electron ion collisions.

After the described phase a part of the column imploded towards the z-axis, and formed the constrictions C (marked in Fig. 3b, c). This evolution was accompanied by an axial transport of plasma from the constriction C, firstly into the toroid which was converted gradually in the second plasmoid (P2 in Fig. 3b, c) formed above the z-position of the bottom lobule (L2). This transfer increased the internal mass and probably its poloidal current component, similarly as in the first plasmoid. One could suspect that the poloidal current was flowing from the anode near the plasmoidal boundary, as the main discharge current. In the near-axial region it was flowing in the opposite direction, toward the anode. This orientation could be a consequence of the penetration of the main discharge current (flowing in the column boundary) into the internal layers.

Possible schemes of transformations of the kinetic energy of plasma (incoming across and along the magnetic field lines) into the magnetic energy, when that plasma is confined inside the organized structures, is presented in Fig. 4a. At this time, the motion of plasma along the magnetic lines could transform the axial line into toroidal one, as shown in Fig. 4b.

At the confinement of incoming plasma inside plasmoids, its kinetic energy could be transformed into a chaotic turbulence, and subsequently into the magnetic energy of the ordered internal currents.

In Fig. 4a, the kinetic energy was transformed into magnetic one at the moving of the plasma perpendicularly to the primary magnetic field lines. Both fluids in frozen state drive into a rotation both around the velocity direction and around the initial direction of the B-field line. This motion created a ball of the magnetic lines, which conserved the magnetic energy. That spontaneous relaxing ball was a building element of the plasma organization process.

This transformation can be described (for a high Reynolds magnetic number) by the equation of plasma frozen in the magnetic field [13]:

$$\partial \boldsymbol{B}/\partial t = \mathrm{rot}(v \times \boldsymbol{B}) \quad (2)$$

The motion of plasma along the magnetic line, as shown in Fig. 4b, describe the transformation of the toroidal structure into plasmoidal one.

During the final phase of the pinched constriction there was observed the second stratification of the constriction surface along its length, when the secondary plasmoids were formed and some secondary constrictions appeared between them. A similar cascade process of the instability evolution to the smaller plasma structures was also observed and discussed in X-pinch experiments, or in solar flares [18, 19].

A relatively stationary first plasmoid living about few hundred ns, the generation of which was accompanied by the emission of the first neutron pulse, could be characterized by some balanced internal poloidal- and toroidal-components of the local magnetic field. The secondary plasmoid, as formed during the constriction compression, could be developed at a higher velocity. Its central poloidal current component could overcome the stationary value. Its pinching was subject to the evolution of instabilities and finally led to a release of the magnetic energy in a form of the induced electric field $\boldsymbol{E}$ (in opposite direction to the central current). This field accelerated beams of charged particles during reconnections of magnetic field-lines in the organized plasma structures, according the Faraday law and energy density balance:

$$-\partial \boldsymbol{B}/\partial t = \mathrm{rot}\,\boldsymbol{E} \quad \text{and} \quad B^2/2\mu \approx (1/2)\rho v^2 \quad (3)$$

where ρ means the plasma mass density.

The plasma compression or expansion as well as the structures transformation inside the dense plasma column, which were observed in interferometric frames, occurred at the Alfvén velocity equal to (1–2) × 10^5 m/s. The mean energy density, as determined from Eq. (3), corresponded to deuteron energy reaching a few hundreds of eV. This energy was evidently insufficient for the production of fusion neutrons. Higher energies of the charged particles, ranging hundreds keV, might be reached at the considerably higher energy density at the non-uniform structure of the current.

It is well known that the current flow in PF discharges has often a filamentary form [2, 20–24]. It can be observed on VUV frames, particularly in discharges performed with an admixture of the a gas with a higher Z. Some examples are imaged in Fig. 5.

The observed filamentation was supported by high values of the Prandtl number (i.e. the ratio of the magnetic and viscose characteristics, ranging up to 10), what was important for the generation of tangled currents in sub-millimeter scales [4]. Detailed characteristics of various plasma-current filaments, which were recorded in the PF-1000 experiments, were described in earlier papers [23, 24]. The tiny filaments presented in Fig. 5 connected the external layers of the dense column with internal

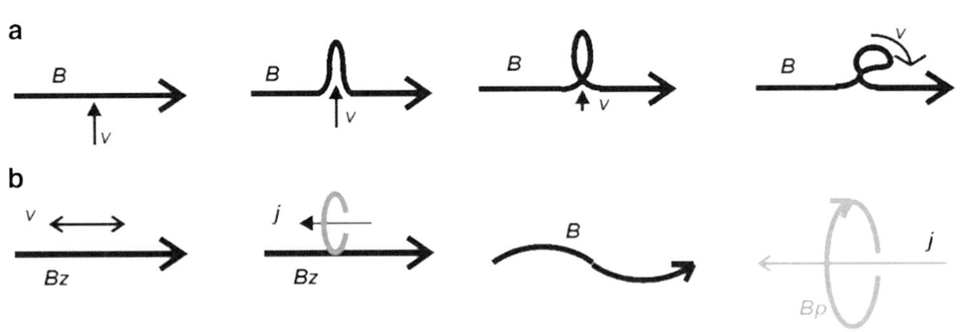

Fig. 4 **a** Sketch of transformations of the kinetic energy into the magnetic one in the case of: transversal motion, and **b** sketch of the transformation of the axial magnetic line in toroidal at the longitudinal motion of plasma in relation to the magnetic line. Symbols: $\boldsymbol{v}$ the velocity of the plasma flow, $\boldsymbol{B}$ the magnetic field line, and $\boldsymbol{j}$ a current density

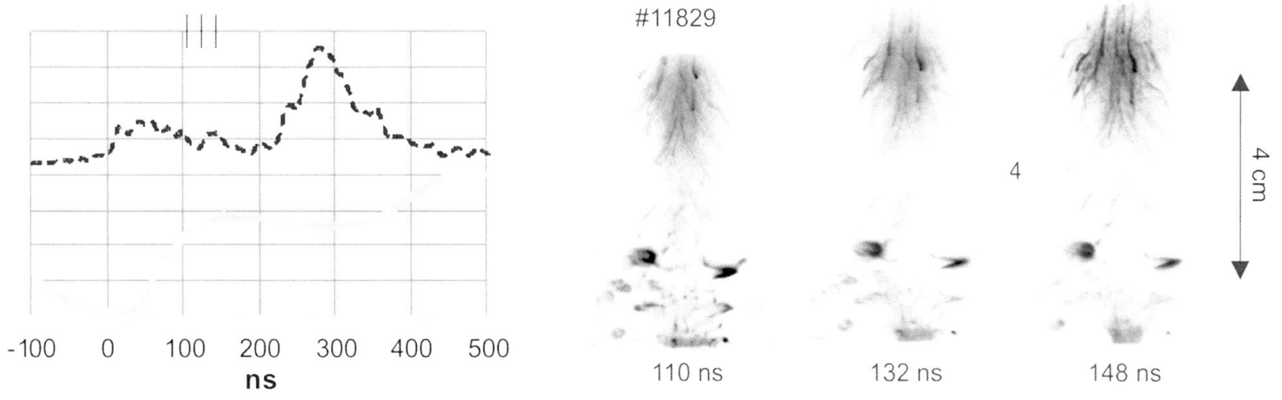

Fig. 5 Shot #11829: (left) Waveforms of the current derivative (thick grey) and fusion-neutrons (dashed) as a function of time, and (right) numerous plasma filaments visible on the VUV frames recorded at different instants during the neutron production

structures and they presented the transport of energy from the external current to the plasmoid, namely during the period of neutron production. The acceleration of deuterons during a decay of the plasmoids might be explained, as mentioned above, by the coalescence of the current filaments inside these plasmoids, which have high densities of the kinetic and magnetic energies, as can be estimated from Eq. (3). For the deuteron velocity of order of 4×10^6 m/s, their kinetic energies can reach about 200 keV. At the electron density equal to 3×10^{24} m^{-3} and the magnetic field $B \approx 500$ T, such deuteron energies would be sufficient for production of the recorded fusion neutrons. Such a magnetic field might be produced, e.g., by current filaments of the radius equal to about 40 μm, which carry the current of about 70 kA [25]. The electric field intensity $\boldsymbol{E}$, as can be estimated from a linearized Faraday law, can reach the value $E \approx v \times B \approx 2 \times 10^9$ V/m.

The deuterons are accelerated in the developing and decaying plasmoids, and the generation of fusion neutrons is occurring in dense plasma contained inside the same or other plasmoids. A delay of the neutron production after the deuteron acceleration can be calculated by taking into account the mean velocity of fast deuterons (equal to about 4×10^6 m/s) and the distance between neighbour plasmoids, which is from 0 to 3 cm. This delay can amount to 10 ns, and corresponds to a delay of the neutron peak appearing after the HXR peak.

It should be noted that the recorded interferometric images did not provide any experimental information about the acceleration of charged particles in the interrupted pinch region of low-density plasma. However, a possibility of such acceleration of deuterons to MeV energies has been discussed for z-pinch discharges, e.g. in a paper [32].

Examples of Organized Structures in Fusion Plasmas and Their Relations with Acceleration of Charged Particles

The questions of a fast release of the magnetic energy in the organized plasma structures, which might explain the non-thermal acceleration of energetic electron- and ion-beams, was also discussed in papers concerning other fusion and cosmic plasmas. For example, a dense plasma jet, which was produced by an intense laser beam focused upon a foil, was investigated by means of the laser interferometry and by the Faraday rotation method [26]. That jet was accompanied by a partially-closed poloidal current ranging hundreds kA and an azimuthal magnetic field of about 5 kT, which was observed in a sub-millimeter regions at the plasma density equal to about 10^{25} m^{-3}. Other laser experiment showed an attraction of two plasma jets (produced by two laser beams), whose coalescence and radiation could be interpreted by magnetic reconnections occurring in the sub-nanosecond scale [27]. The intense X-ray and neutron sources, which were observed in the NIF D-T fusion experiments [28], had the toroidal-like form oriented perpendicularly to the axis of the laser-irradiated pellet. In tokamak experiments the kinetic energy of charged particles, delivered by additional plasma heating systems, can be interpreted by conversion partially in the magnetic energy, which violates the smooth toroidal magnetic-flux surfaces. This energy is then periodically released in a form of the fast run-away electrons, which acceleration is interpreted also by magnetic reconnections. It enables to recover smooth surfaces of the magnetic fluxes repeatedly [29].

Therefore, the studies of organized plasma structures, their spontaneous evolution, and a release of the stored magnetic energy in the kinetic energy of fast charged particles, are important issues to be solved in the contemporary research on all fusion concepts. Also in cosmic space, e.g. the solar flares are generated in regions of interactions of different magnetic tubes. The evolution of such structures is powered by magnetic reconnections which transform the initial magnetic energy into heat and kinetic energy during the formation of so-called flux ropes, cascades of plasmoids and flaring loops. These transformations have been studied by computer simulations of magnetic reconnections in 1D and 2D approaches, and during the recent years—also in a 3D model [19, 30]. Unfortunately, such models have not yet been developed for PF discharges.

The observations and hypotheses mentioned above are similar to the results of the described PF-1000 experiments. Therefore, the studies of different plasma organized structures and their transformations are of importance not only for laboratory plasma research, but also for the astrophysics.

Summary and Conclusions

The most important results of this study can be summarized as follows: the convenient parameters of dense magnetized plasmas formed in PF-1000 device and the comprehensive diagnostics of a plasma density and a velocity of plasma column transformations, as well as the observed quasi-stationary states and short pulses of fast electrons and ions, made it possible to analyze the transformations of this fusion plasma, particularly during the periods of the emission of the fast electron- and ion-beams. The performed experiments proved the appearance of the toroidal and plasmoidal plasma structures, which were formed by closed internal currents with poloidal and toroidal components and their magnetic fields. Their spontaneous transformations have been explained by the magnetic dynamo and magnetic reconnections. Many experimental observations confirmed a filamentary structure of the current. The generation of fast particles was explained by the magnetic reconnections of the current filaments, in which a part of the magnetic energy was transformed into the accelerating electric field. It was noticed that the studies of laboratory fusion and cosmic plasmas solve similar problems, e.g. the fast release of the magnetic energy in a form of high-energy charged particles streams. A deeper collaboration of physicists engaged in the mentioned studies should give a new inspiration for reaching a desirable progress in the research on fusion plasma in a laboratory as well as astrophysical objects in space.

Acknowledgements This study was supported in part by the Research Program under Grants MSMT LTT17015, LTAUSA17084, CZ.02.1.01/0.0/0.0/16_019/0000778, GACR 16-07036S, IAEA CRP RC-19253, and SGS 16/223/OHK3/3T/13, as well by the Polish Ministry of Science and Higher Education within a framework of the financial resources allocated in 2018 for the realization of the international co-financed projects.

References

1. M.G. Haines, Plasma Phys. Control. Fusion **53**, 093001 (2011)
2. A. Bernard, H. Bruzzone, P. Choi, H. Chaqui, V. Gribkov, J. Herrera, K. Hirano, A. Krejci, S. Lee, C. Luo, F. Mezzetti, M. Sadowski, H. Schmidt, K. Ware, C.S. Wong, V. Zoita, J. Moscow Phys. Soc. **8**, 93 (1998)
3. D.D. Ryutov, M.S. Derzon, M.K. Matzen, Rev. Mod. Phys. **72**, 167 (2000)
4. D.D. Ryutov, IEEE Trans. Plasma Sci. **43**, 2363 (2015)
5. M. Scholz, L. Karpinski, M. Paduch et al., Nukleonika **46**, 35 (2001)
6. M.J. Sadowski, M. Scholz, Nukleonika **47**, 31 (2002)
7. P. Kubes, M. Paduch, D. Klir, J. Kravarik, K. Rezac, I. Ivanova-Stanik, L. Karpinski, M.J. Sadowski, K. Tomaszewski, E. Zielinska, IEEE Trans. Plasma Sci. **39**, 672 (2011)
8. K. Rezac, D. Klir, P. Kubes, J. Kravarik, Plasma Phys. Controlled Fusion **54**, 105011 (2012)
9. E. Zielinska, M. Paduch, M. Scholz, Contrib. Plasma Phys. **51**, 279 (2011)
10. V.I. Krauz, K.N. Mitrofanov, M. Scholz, M. Paduch, L. Karpinski, E. Zielinska, P. Kubes, Plasma Phys. Control. Fusion **54**, 025010 (2012)
11. V.I. Krauz, K.N. Mitrofanov, M. Scholz, M. Paduch, L. Karpinski, E. Zielinska, P. Kubes, Europhys. Lett. **98**, 45001 (2012)
12. P. Kubes, M. Paduch, J. Cikhardt, B. Cikhardtova, D. Klir, J. Kravarik, K. Rezac, E. Zielinska, M.J. Sadowski, A. Szymaszek, K. Tomaszewski, D. Zaloga, Phys. Plasmas **24**, 072706 (2017)
13. T.G. Cowling, *Magnetohydrodynamics* (Adam Hilger Ltd., Bristol, 1976)
14. V.D. Shafranov, *Theory of Plasma Physics* (Gosatomizdat, Moscow, 1963) in Russian
15. P. Kubes, M. Paduch, B. Cikhardtova, J. Cikhardt, D. Klir, J. Kravarik, K. Rezac, J. Kortanek, E. Zielinska, M.J. Sadowski, A. Szymaszek, K. Tomaszewski, Phys. Plasmas **25**, 032706 (2018)
16. P. Kubes, V.I. Krauz, K. Mitrofanov, M. Paduch, M. Scholz, T. Pisarczyk, T. Chodukowski, Z. Kalinowska, L. Karpinski, D. Klir, J. Kortanek, E. Zielinska, J. Kravarik, K. Rezac, Plasma Phys. Control. Fusion **54**, 105023 (2012)
17. P. Kubes, D. Klir, J. Kravarik, K. Rezac, J. Kortanek, V. Krauz, K. Mitrofanov, M. Paduch, M. Scholz, T. Pisarczyk, T. Chodukowski, Z. Kalinowska, L. Karpinski, E. Zielinska, Plasma Phys. Control. Fusion **55**, 035011 (2013)
18. S.A. Pikuz, T.A. Shelkovenko, D.A. Hammer, Plasma Phys. Rep. **41**, 291 (2015)
19. M. Janvier, J. Plasma Phys. **83**, 535830101 (2017)
20. M. Sadowski, H. Herold, H. Schmidt, M. Shakhatre, Phys. Lett. **105A**, 117 (1984)
21. H. Schmidt, M. Sadowski, L. Jakubowski, E. Skladnik-Sadowska, J. Stanislawski, Plasma Phys. Control. Fusion **36**, 13 (1994)
22. M.J. Sadowski, M. Scholz, Nukleonika **57**, 11 (2012)
23. P. Kubes, D.M. Paduch, J. Cikhardt, B. Cikhardtova, E. Zielinska, Phys. Plasmas **21**, 122706 (2014)
24. P. Kubes, M. Paduch, J. Cikhardt, B. Cikhardtova, D. Klir, J. Kravarik, K. Rezac, E. Zielinska, M.J. Sadowski, A. Szymaszek, K. Tomaszewski, D. Zaloga, Phys. Plasmas **24**, 032706 (2017)
25. P. Kubes, M. Paduch, B. Cikhardtova, J. Cikhardt, D. Klir, J. Kravarik, K. Rezac, J. Kortanek, E. Zielinska, M.J. Sadowski, K. Tomaszewski, Phys. Plasmas **23**, 082704 (2016)

26. T. Pisarczyk, SYu. Guskov, T. Chodukowski et al., Phys. Plasmas **24**, 102711 (2017)
27. P.M. Nilson, L. Willingale, M.C. Kaluza et al., Phys. Rev. Lett. **97**, 255001 (2006)
28. H.-S. Park, O.A. Hurricane, D.A. Callahan et al., Phys. Rev. Lett. **112**, 055001 (2014)
29. M. Yamada, R. Kulsrud, H. Ji, Rev. Mod. Phys. **82**, 603 (2010)
30. Y. Su, A.M. Veronig, G.D. Holman et al., Nat. Phys. **9**, 2675 (2013)
31. P. Kubes, M. Paduch, J. Cikhardt, B. Cikhardtova, D. Klir, J. Kravarik, K. Rezac, E. Zielinska, M.J. Sadowski, A. Szymaszek, K. Tomaszewski, D. Zaloga, Phys. Plasmas **24**, 092707 (2017)
32. D. Klir, J. Cikhardt, B. Cikhardtova, J. Kravarik, P. Kubes, V. Munzar, K. Rezac, O. Sila, T. Hyhlik, J. Stodulka et al., New J. Phys. **20**, 053064 (2018)

Journal of Fusion Energy (2019) 38:499–502
https://doi.org/10.1007/s10894-018-0196-8

ORIGINAL RESEARCH

Cryogenics in Fusion: Industrial Approach of Cryogenics Applied to Fusion Plants

D. D'Andrea[1]

Published online: 13 October 2018

Abstract

Cryogenics has a heavy impact on the overall cost of the experimental fusion plants, and the current approach to define cryogenics is not optimized in terms of energy efficiency. In order for future fusion reactors to produce energy for the grid, the energy consumption required by the *cryoplant* must be optimized to improve the final efficiency of the entire fusion power plant. We provide in this paper some background information useful in defining cryogenics for the future fusion power plants with a specific approach to capital and operational cost reduction.

Keywords Cryogenics · Cryoplant · Fusion power plant · Energy efficiency

Introduction

The purpose of cryogenics is to cool and maintain low temperatures of some fusion reactor elements such as magnets, thermal shields (see Fig. 1), vacuum pumps and feeders.

Superconducting magnets must be cooled to their critical temperature of around 5 K. Cryopumps operate at similar temperatures to achieve vacuum pumping of the cryostat, whereas thermal shields operate from 80 to 100 K.

All these elements are cooled by supercritical helium, although the liquid nitrogen temperature could be sufficient for cooling the thermal shields. However, liquid nitrogen cannot be used in the presence of neutron flux.

The fusion reactor *cryoplant* is the device that produces the cooling power required by all those parts, magnets, cryogenic pumps (*cryopumps*), thermal shields and feeders to correctly operate.

The *cryolines*, which are part of the entire *cryodistribution* system, distribute the cooler fluids at different temperatures produced by the cryoplant to the points of use of the fusion reactor. After the temperature exchange, the coolant is carried back to the cryoplant to be refrigerated again and restarts its cooling function. The energy needed to produce cold temperatures is high. We consider that to produce 1 kW at 4 K, 300 kW are needed.[1]

In order for future fusion reactors to produce energy for the grid, the energy consumption required by the *cryoplant* must be reduced to improve the efficiency of the entire fusion power plant. The high energy consumption of the cryoplant, due to the low energy efficiency, is the main reason why the cryoplant deserved to be considered as an important element of the fusion reactor.

Thus far, cryogenics has been considered as a utility on the fusion reactors and, consequently, as sub-equipment. The fusion reactor is designed by focusing on its own parameters and with limited considerations about the cryoplant.

Usually, at the end of the reactor's design phase, the cold power requirements are managed by the temperature range and flow rate, and the cryoplant is required to fit such parameters without other concerns. Such an approach is used to specify the cryoplant in the current experimental fusion reactors.

A significant change to this approach is required for the next generation of fusion power plants, and such a new approach must be guided by the end use: The purpose of the fusion experimental unit is to become a fusion

✉ D. D'Andrea
domenico.dandrea@airliquide.com

[1] Air Liquide Global Markets & Technologies, 2 Rue de Clémencières, 38360 Sassenage, France

[1] To be considered as a qualitative value only.

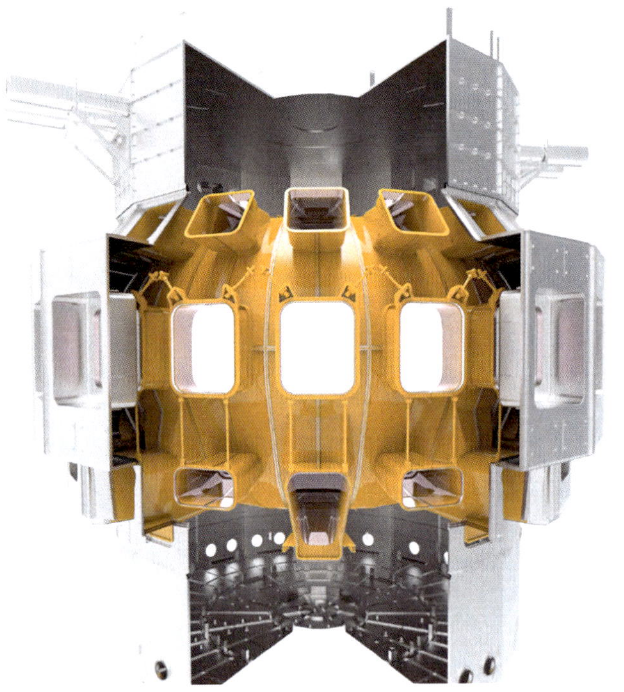

Fig. 1 The ITER thermal shield consists of stainless steel panels with a low emissivity surface (< 0.05) that are actively cooled by helium gas flowing inside of a cooling tube welded on the panel surface. During plasma operation, the temperature of the helium gas ranges from 80 to 100 K. One layer of the thermal shield will be installed between the vacuum vessel and the superconducting magnets; another will be installed between the magnets and the cryostat. The thermal shield covers a surface area of approximately 10,000 sqm. Once assembled, it will stand at 25 m at its highest point (Courtesy of the ITER Organization)

production unit responding to all the parameters required by the industry.

The modification of the ultimate destination is critical in terms of the final design for different factors. The main factors concern the energy efficiency, the reliability, the operation system and the frequency of maintenance.

Currently, the parameters affecting the reliability are similar between a cryoplant supplying an experimental unit and a cryoplant supplying an industrial unit. This is true because the industrial manufacturer of such plants utilizes similar designs for both. Cryogenic plant manufacturers are industrial companies for which standardization is a key element to improve the efficacy. Current cryoplants are conceived for mainly industrial uses.

The difference between a research- and an industrial-dedicated cryoplant exists because of special requests coming from the final user, which in this precise case means the owner of the fusion reactor. The purpose of the actual generation of fusion reactors is to validate the main parameters in order to later develop the next-generation reactor to produce energy. That finding actually means that several running modes are necessary, so the cryoplant could be specifically conceived to produce, for example, peaks of cooling power for a very short time, or other special running modes instead of a stable running mode 24/7.

In a research unit, the costs linked to the operation of the cryoplant (OPEX) are usually included in the budget of normal machine running. The final machine's user calculates the cost linked to the cryogenics in the same budget line as the costs linked to the other utilities such as HVAC, water cooling and others consumption. All those costs are classified as operation costs and estimated every year.

This factor becomes and acquires a completely different meaning when the unit is an industrial unit. In an industrial unit, the OPEX must be optimized because it directly impacts the electric production unit efficiency, and the procurement cost of the cryoplant (CAPEX) is spread over the lifetime of the unit.

Some topics affect the cryoplant design and must be considered to limit the cryoplant's cost and, consequently, increase the efficiency of the entire fusion energy plant.

The footprint and distance between the cryoplant and the reactor is another key parameter of the overall design of the fusion power plant, and the actual cryoplant's footprint could be large. To give an example, the ITER cryoplant occupies an area of 7800 sqm (see Fig. 2).

The fusion power plant's overall design must consider the cryoplant's footprint for the following reasons: Placing cryogenic unit inside the fusion reactor building is difficult because of the large footprint, moreover placing the cryogenic unit inside the restricted perimeter causes the cryoplant to be submitted to the nuclear regulation with a consequent huge impact on the costs.

At the same time, moving the cryoplant far from the final user creates extra costs because of the length of the cryogenic lines (*cryolines*).

The supercritical helium used as a coolant must be transferred from the cryogenic plant to the final points of use inside the Tokamak with limited heat loss and by maintaining under strict control the pressure drop. For this reason, the cryolines are components with highly technical solutions, so the cryolines' cost must be considered during the estimation because it could heavily affect the CAPEX.

In ITER, the length of the complex transfer of cryogenic lines means that the lines connecting the main stream coming out from the distribution cold boxes to each single point of use are around 1.6 km.

The energy efficiency of a cryogenic unit must be kept under strict control.

Two improvements may increase cryogenic plant efficiency: The first method involves improving cryoplant efficiency that results in the production of the same cold power but reduces the electric consumption.

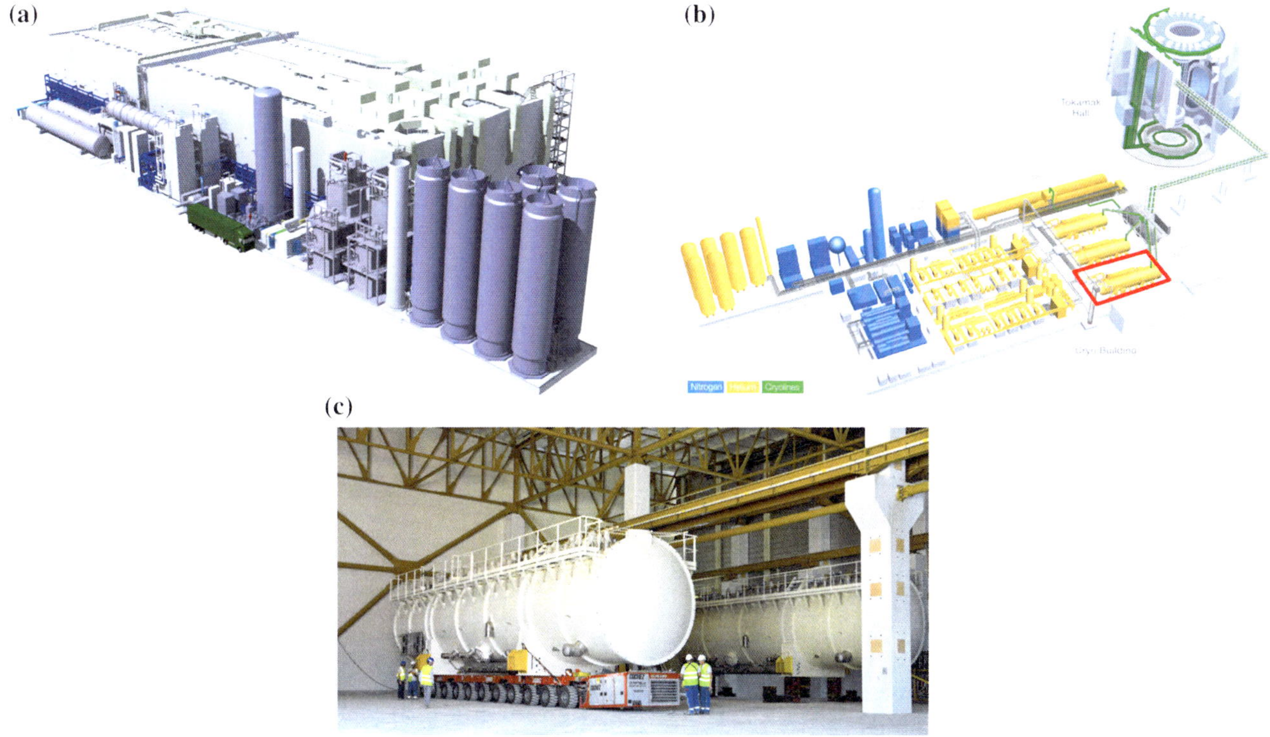

Fig. 2 **a** The cryogenic building at ITER, it consists of 5400 sqm of covered space and 2600 sqm of exterior storage (Courtesy of the ITER Organization). **b** Interior view of the cryogenic building at ITER (Air Liquide, courtesy of the ITER Organization/F4E). In the red frame, the coldbox detailed in Fig. 2c. **c** Photo of a cold box at ITER. The cryoplant is made up of three cold boxes. Each one is 24 m long and weighs over 130 tons (Air Liquide, courtesy of the ITER Organization)

The second procedure involves reducing the fusion reactor's cold power requirement without affecting the efficiency of the fusion reactor itself and without affecting its normal running.

Whereas the first way is the standard industrial path, followed by any industrial company as a normal upgrade of its unit and which can be completed independently without communicating with the final user, the second way is more difficult to achieve and requires a coordination between the fusion power plant designer and the cryoplant manufacturer.

In modern fusion reactors, the cryogenic plant consumes the most energy. ITER's cryogenic plant consumes around 40 MW of electric power.[2]

The only other utility with a comparable energy consumption is the water cooling system, which dissipates the power generated by the fusion reactor itself. In the next generation of fusion reactors, the energy dissipated by the water cooling system will be injected into the grid; therefore, the cryogenic plant will be the largest energy consumer of the entire fusion power plant.

It is important to note that improving the efficiency of the energy consumption of the cryogenic plant improves the efficiency of the entire fusion power plant.

The principle employed today to generate cold is the same one used since the beginning of cryogenics: The helium coolant is compressed and, as a result, the temperature increases. The excess temperature is extracted, and then the helium expands, lowering the temperature. This process involves several stages of compression and expansion until the operation temperature is reached; then, the coolant is transferred to the point of use. In this closed-loop process, after the helium cools down the point of use and becomes warmer, it is transferred back to the cryogenic plant and cooled down again.

The helium consumed in this process is minimal. The quantity of helium injected into the loop at the beginning of the cryogenic plant's operation remains stable and requires refilling only to compensate for leaks and maintenance when some sections require opening and reconditioning. The cryoplant does not require a significant amount of helium for operation, and the cost of helium is significantly lower than the energy cost.

The main power consumption of the cryogenic plant is used to supply the compressors. Large stacks of machines are used to compress a significant amount of gas. The

[2] Source: Air Liquide Advanced Technologies.

efficiency of the cycle is low, because helium is a light gas and, therefore, difficult to compress. Currently, the most common way to compress helium is by oil-injected screw compressors, but new technologies are under development to replace such compressors with other modern devices offering higher efficiency.

Although the design used for fusion reactor cryoplants is based on the old technology of the Brayton cycle, the continuous improvement of the efficiency required by the industrial market has optimized the overall efficiency of such cryoplants. Efficiency improvement work should be conducted now also on the fusion power energy plant devices using cryogenics. Magnets and thermal shields must be designed to minimize the pressure drops by considering the coolant flow.

In fact, the parameter that largely affects cryogenic efficiency is the pressure drop caused by the elements of the fusion reactor; the architecture of the magnets and the thermal shields shall be optimized not only regarding the function of the system itself but also by considering the coolant fluid flow.

A final element requiring cooldown through a high pressure drop requires cold circulators to transfer the coolant to the final destination at the correct flow rate.

The power absorbed by such cold compressors can reach up to 20% of the total cold power produced by the cryoplant.[3] Reducing the pressure drop means reducing the size of the cold circulators, and therefore the power required by the entire cryogenics.

Such improvements, as indicated previously, require cooperation among magnet designers, fusion power plant designers and cryogenic specialists in charge of the cryoplant manufacturing.

A common design since the beginning of the conceptual design process will result in an optimized system and in significant savings.

Furthermore, temperature is another key factor that significantly affects costs. Employing a coolant with a lower operating temperature results in a higher ratio between the cold power and electrical power required, which means a higher power consumption.

Producing the same cold power at 5.5 K instead of 4.5 K, results in a theoretical energy saving of 20%.[4]

This saving is even clearer by using High-Temperature Superconductors (HTS) instead of niobium-based superconductors. With HTS, working at 20 K, the energy saving is more consequent because also allow the utilization of smaller elements, such as compressors and turbines, reducing the capital expenditure.

The same principle should be applied to the thermal shields and other components requiring cryogenics and by keeping in mind that at a lower temperature such as 4 K, increase only 1 K could produce a significant saving, whereas at 80 K, the similar saving requires to increase the temperature by 10 K.[5] Increasing the temperature at the point of use has a double advantage in terms of CAPEX and OPEX. Consequently, when it is possible, HTS should be utilized.

Conclusion

Cryogenics is a utility for fusion power plants possessing available potential optimization. The next generation of fusion power plant must be designed in the way to reduce the capital expenditure and the operating expenses, and cryogenics is one of the key devices to be taken in consideration to reach the cost control.

To achieve this result, a closer collaboration between the cryogenic engineers and the fusion power plant designer should be considered since the conceptual design phase.

[3] This value is linked to many parameters and shall be considered as a qualitative value only.

[4] A general expression for the efficiency in Carnot Ideal Cycle can be written as $Efficiency = \frac{Th-Tc}{Tc}$. To cool down He from 300 to 4.5 K: $\frac{300\,K-4.5\,K}{4.5\,K} = 66$ W power needed to produce 1 W at 4.5 K. To cool down He from 300 to 5.5 K: $\frac{300\,K-5.5\,K}{5.5\,K} = 54$ W power needed to produce 1 W at 5.5 K. The saving power is ~ 20%. This value shall be considered as qualitative only. In real conditions, other parameters shall be taken into consideration.

[5] From 4 to 5 K and from 80 to 90 K = similar energy saving. To be considered as a qualitative value only.

MIX
Papier aus verantwortungsvollen Quellen
Paper from responsible sources
FSC® C105338

Printed by Books on Demand, Germany